Soil Ecology and Ecosystem Services

Soil Ecology and Ecosystem Services

EDITOR-IN-CHIEF

Diana H. Wall
Colorado State University, USA

SECTION EDITORS

Richard D. Bardgett
University of Manchester, UK

Valerie Behan-Pelletier
Agriculture and Agri-Food Canada, Canada

Jeffrey E. Herrick
USDA-ARS Jornada Experimental Range, USA

T. Hefin Jones
Cardiff University, UK

Karl Ritz
Cranfield University, UK

Johan Six
University of California, Davis, USA

Donald R. Strong
University of California, Davis, USA

Wim H. van der Putten
Netherlands Institute of Ecology, The Netherlands

Soil Ecology and Ecosystem Services. First Edition. Edited by Diana H. Wall *et al*.
© 2012 Oxford University Press. Published 2012 by Oxford University Press.

OXFORD
UNIVERSITY PRESS

Great Clarendon Street, Oxford, OX2 6DP,
United Kingdom

Oxford University Press is a department of the University of Oxford.
It furthers the University's objective of excellence in research, scholarship,
and education by publishing worldwide. Oxford is a registered trade mark of
Oxford University Press in the UK and in certain other countries

© Oxford University Press 2012

The moral rights of the authors have been asserted

First published 2012
First published in paperback 2013

All rights reserved. No part of this publication may be reproduced, stored in
a retrieval system, or transmitted, in any form or by any means, without the
prior permission in writing of Oxford University Press, or as expressly permitted
by law, by licence or under terms agreed with the appropriate reprographics
rights organization. Enquiries concerning reproduction outside the scope of the
above should be sent to the Rights Department, Oxford University Press, at the
address above

You must not circulate this work in any other form
and you must impose this same condition on any acquirer

Published in the United States of America by Oxford University Press
198 Madison Avenue, New York, NY 10016, United States of America

British Library Cataloguing in Publication Data
Data available

Library of Congress Cataloging in Publication Data
Data available

ISBN 978–0–19–968816–6

Links to third party websites are provided by Oxford in good faith and
for information only. Oxford disclaims any responsibility for the materials
contained in any third party website referenced in this work.

Contents

List of Contributors   xi

Introduction   1
Diana H. Wall

Section 1—The Living Soil and Ecosystem Services

Introduction   5
Karl Ritz and Wim H. van der Putten

1.1 Soil as a Habitat   7
Patrick Lavelle

1.1.1 Introduction   7
1.1.2 Conditions in soils   7
1.1.3 Adaptive strategies of soil organisms   12
1.1.4 Self-organization and the spatial organization of soils   15
1.1.5 Discrete scales in soil function   16
1.1.6 The challenge of an eco-efficient use of soils   18
1.1.7 Approaches to soil ecological research   21
1.1.8 Conclusions   21

1.2 Soil Biodiversity and Functions   28
Susanne Wurst, Gerlinde B. De Deyn, and Kate Orwin

1.2.1 Soil biodiversity   28
1.2.2 How to investigate soil communities   34
1.2.3 Diversity–function relationships   37
1.2.4 Taking a holistic view to soil diversity–ecosystem functioning   39
1.2.5 Conclusions   41

1.3 Ecosystem Services Provided by the Soil Biota   45
Lijbert Brussaard

1.3.1 Introduction   45
1.3.2 Understanding ecosystem functioning   46
1.3.3 Understanding ecosystem structure: revisiting the functional group concept   49
1.3.4 Understanding effects of environmental drivers and land management on ecosystem functioning and services   51
1.3.5 Working with nature   52

	1.3.6 Landscape context	54
	1.3.7 Conclusions	55

Synthesis   **59**
Karl Ritz and Wim H. van der Putten

Section 2—From Genes to Ecosystem Services

Introduction   **63**
Wim H. van der Putten and Karl Ritz

2.1 From Single Genes to Microbial Networks   **65**
Evelyn Hackl, Michael Schloter, Ute Szukics, Levente Bodrossy, and Angela Sessitsch

	2.1.1 Introduction	65
	2.1.2 Analyzing microbial genes to understand ecosystem functioning	66
	2.1.3 Methodological approaches to the gene-based study of microbial communities and networks	68
	2.1.4 Genes in microbial networks of organic matter decomposition and biodegradation of pollutants	69
	2.1.5 Microbial genes in nitrogen turnover cascades	71
	2.1.6 Genes underlying microbial communication	72
	2.1.7 Microbial genes for interacting in the plant environment	73
	2.1.8 From genes to microbial networks: future prospects	75

2.2 From Genes to Ecosystems: Plant Genetics as a Link between Above- and Belowground Processes   **82**
Jennifer A. Schweitzer, Michael D. Madritch, Emmi Felker-Quinn, and Joseph K. Bailey

	2.2.1 Introduction	82
	2.2.2 The role of plant functional traits in bridging species interactions with soil community dynamics	84
	2.2.3 The role of plant genetic variation on soil communities	85
	2.2.4 The role of plant genetic variation on ecosystem processes	87
	2.2.5 The evolutionary implications of plant–soil linkages	89
	2.2.6 Conclusions and future directions	92

2.3 Delivery of Soil Ecosystem Services: From Gaia to Genes   **98**
Katarina Hedlund and Jim Harris

	2.3.1 Introduction	98
	2.3.2 Ecosystem services delivery and Gaia theory	99
	2.3.3 At what biological levels are soil ecosystem services produced?	101
	2.3.4 At what spatial scales can we describe and quantify soil ecosystem services?	102
	2.3.5 Use of soil ecosystem services in a policy context	103
	2.3.6 Conclusions	105

Synthesis   **111**
Wim H. van der Putten and Karl Ritz

Section 3—Community Structure and Biotic Assemblages

Introduction — 115
Donald R. Strong and Valerie Behan-Pelletier

3.1 Succession, Resource Processing, and Diversity in Detrital Food Webs — 117
Justin Bastow
- 3.1.1 The surprising diversity of soil communities — 117
- 3.1.2 From litter and carrion to soil organic matter: detrital succession in soils — 118
- 3.1.3 Mechanisms and models for detrital succession — 121
- 3.1.4 Can successional specialization explain coexistence and the diversity in soils? — 126
- 3.1.5 Latitudinal gradients in soil diversity: detrital food webs thwart ecology's oldest pattern — 128
- 3.1.6 Future directions in understanding detrital succession — 130

3.2 Patterns of Biodiversity at Fine and Small Spatial Scales — 136
Matty P. Berg
- 3.2.1 The riddle of soil biodiversity — 136
- 3.2.2 It is all a matter of scale — 137
- 3.2.3 Spatial distribution of soil functions — 147
- 3.2.4 Spatial scales are nested — 149

3.3 Linking Soil Biodiversity and Human Health: Do Arbuscular Mycorrhizal Fungi Contribute to Food Nutrition? — 153
Pedro M. Antunes, Philipp Franken, Dietmar Schwarz, Matthias C. Rillig, Marco Cosme, Martha Scott, and Miranda M. Hart
- 3.3.1 Soil health is linked to human health and global food security — 153
- 3.3.2 Traditional ways of boosting crop nutrients — 154
- 3.3.3 A critical role for soil microbes — 155
- 3.3.4 Using rhizosphere microbes to create healthier food — 157
- 3.3.5 Negative effects of microbes on food quality — 162
- 3.3.6 The full potential of soil microbes to improve human health — 163
- 3.3.7 Conclusion — 164

3.4 Ecosystem Influences of Fungus-Growing Termites in the Dry Paleotropics — 173
Gregor W. Schuurman
- 3.4.1 Introduction — 173
- 3.4.2 Fungus-growers — 174
- 3.4.3 Fungus-grower influences on ecosystem processes — 177
- 3.4.4 Fungus-growers as ecosystem engineers — 179
- 3.4.5 Synthesis — 183
- 3.4.6 Take-home messages — 185
- 3.4.7 Future directions — 185

3.5 The Biogeography of Microbial Communities and Ecosystem Processes: Implications for Soil and Ecosystem Models — 189
Mark A. Bradford and Noah Fierer
- 3.5.1 Predicting environmental responses of soil processes — 189

	3.5.2 Misplaced physics envy in soil models	190
	3.5.3 Functional redundancy, similarity, equivalence, and biogeography	192
	3.5.4 Experimental tests of functional equivalence	195
	3.5.5 Putting ecology into soil models	197
	3.5.6 Revisiting the functional paradigm in soil ecology	198

3.6 Biogeography and Phylogenetic Community Structure of Soil Invertebrate Ecosystem Engineers: Global to Local Patterns, Implications for Ecosystem Functioning and Services and Global Environmental Change Impacts **201**
Lijbert Brussaard, Duur K. Aanen, Maria J.I. Briones, Thibaud Decaëns, Gerlinde B. De Deyn, Tom M. Fayle, Samuel W. James, and Tânia Nobre

	3.6.1 Introduction	201
	3.6.2 Macroecological patterns in soil invertebrate communities	203
	3.6.3 Termite biogeography and phylogenetic community structure	206
	3.6.4 Ant biogeography and phylogenetic community structure	211
	3.6.5 Earthworms	214
	3.6.6 Enchytraeids	218
	3.6.7 Trait-based ecology of soil invertebrate ecosystem engineers with a view to the possible effects on global environmental change and ecosystem functioning and services	222

Synthesis **233**
Donald R. Strong and Valerie Behan-Pelletier

Section 4—Global Changes

Introduction **239**
Richard D. Bardgett and T. Hefin Jones

4.1 Climate Change and Soil Biotic Carbon Cycling **241**
Nicholas J. Ostle and Susan E. Ward

	4.1.1 Introduction	241
	4.1.2 Climate change and plant–soil interactions	242
	4.1.3 Direct effects	243
	4.1.4 Indirect effects	245
	4.1.5 Making predictions	248
	4.1.6 Conclusions	249

4.2 The Impact of Nitrogen Enrichment on Ecosystems and Their Services **256**
Peter Manning

	4.2.1 Nitrogen—the Earth's most limiting resource?	256
	4.2.2 Direct impacts of nitrogen enrichment on soil chemistry and plant and microbial metabolism	258
	4.2.3 Effects of nitrogen enrichment on plants and the soil biota	259
	4.2.4 Net effects on ecosystem services	265
	4.2.5 Conclusion and future directions	267

4.3 Urbanization, Soils, and Ecosystem Services — 270
Mitchell A. Pavao-Zuckerman

    4.3.1 Introduction to urbanization and soils in cities — 270
    4.3.2 Urbanization effects on soils — 270
    4.3.3 Examples of ecosystem services in cities — 273
    4.3.4 Management for urban ecosystem services — 276
    4.3.5 Summary — 278

4.4 Management of Grassland Systems, Soil, and Ecosystem Services — 282
Phil Murray, Felicity Crotty, and Nick van Eekeren

    4.4.1 Introduction — 282
    4.4.2 Plant–soil interactions — 283
    4.4.3 Ecosystem services provided by the soil biota — 284
    4.4.4 Impact of management intensity of grassland systems — 288
    4.4.5 Trade-offs between ecosystem services — 288
    4.4.6 Conclusions — 290

Synthesis — 295
Richard D. Bardgett and T. Hefin Jones

Section 5—Sustainable Soils

Introduction — 299
Johan Six and Jeffrey E. Herrick

5.1 Soil Productivity and Erosion — 301
Kristof Van Oost and Martha M. Bakker

    5.1.1 Introduction — 301
    5.1.2 Soil gain versus soil loss, and accelerated versus natural erosion — 301
    5.1.3 Erosion's effect on agricultural productivity — 305
    5.1.4 The importance of erosion-induced productivity losses for agriculture — 309
    5.1.5 Summary — 312

5.2 Agroforestry and Soil Health: Linking Trees, Soil Biota, and Ecosystem Services — 315
Edmundo Barrios, Gudeta W. Sileshi, Keith Shepherd, and Fergus Sinclair

    5.2.1 Introduction — 315
    5.2.2 How trees influence soil properties and biota — 316
    5.2.3 Agroforestry systems increase abundance of soil biota — 318
    5.2.4 Soil biological processes and soil-based ecosystem services — 319
    5.2.5 Tree–soil biota interactions foster the provision of soil-based ecosystem services — 320
    5.2.6 Soil health monitoring systems — 324
    5.2.7 Conclusions and recommendations — 327

5.3 Soil Health: The Concept, Its Role, and Strategies for Monitoring — 331
Douglas L. Karlen

    5.3.1 The concept of soil health — 331

	5.3.2 The evolution of soil health	333
	5.3.3 Monitoring soil health	334
	5.3.4 Summary and conclusions	335

5.4 Managing Soil Biodiversity and Ecosystem Services — 337
Michel A. Cavigelli, Jude E. Maul, and Katalin Szlavecz

- 5.4.1 Introduction — 337
- 5.4.2 Edible crop diversity — 338
- 5.4.3 Plant selection impacts on ecosystem services — 338
- 5.4.4 Plant selection impacts on soil biodiversity — 340
- 5.4.5 Managing plant diversity — 341
- 5.4.6 Tillage impacts on ecosystem services — 342
- 5.4.7 Tillage impacts on soil biodiversity — 343
- 5.4.8 Chemical application impacts on ecosystem services — 344
- 5.4.9 Chemical application impacts on soil biodiversity — 345
- 5.4.10 Organic material application impacts on ecosystem services — 346
- 5.4.11 Organic material application impacts on soil biodiversity — 346
- 5.4.12 Organic cropping system impacts on ecosystem services — 347
- 5.4.13 Organic cropping system impacts on soil biodiversity — 348
- 5.4.14 Conclusions — 350

5.5 Soil Ecosystem Resilience and Recovery — 357
A. Stuart Grandy, Jennifer M. Fraterrigo, and Sharon A. Billings

- 5.5.1 Introduction — 357
- 5.5.2 Soil disturbance, resilience, and recovery — 358
- 5.5.3 Resilience and recovery: soil organic matter dynamics — 361
- 5.5.4 Resilience and recovery: soil nutrient cycling — 364
- 5.5.5 Future directions — 366

5.6 Applying Soil Ecological Knowledge to Restore Ecosystem Services — 377
Sara G. Baer, Liam Heneghan, and Valerie T. Eviner

- 5.6.1 Introduction — 377
- 5.6.2 Low to high legacy: lessons from restoration of mined land — 381
- 5.6.3 Moderate legacy: restoration of agricultural systems — 382
- 5.6.4 High legacy under dynamic change: preventing invasion and restoring invaded systems — 385
- 5.6.5 Novel legacy: no analog ecosystems and environmental conditions — 387
- 5.6.6 Conclusions — 389

Synthesis — 395
Jeffrey E. Herrick and Johan Six

Index — 397

List of Contributors

Duur K. Aanen Laboratory of Genetics, Wageningen University, P.O. Box 309, 6700 AH Wageningen, The Netherlands

Pedro M. Antunes Algoma University, Department of Biology, 1520 Queen Street East, Sault Ste. Marie, ON P6A 2G4, Canada

Sara G. Baer Department of Plant Biology & Center for Ecology, 420 Life Science II, 1125 Lincoln Dr., Southern Illinois University, Carbondale, IL 62901-6509, USA

Joseph K. Bailey Ecology and Evolutionary Biology, University of Tennessee, 569 Dabney Hall, Knoxville, TN 37996-1610, USA

Martha M. Bakker Wageningen University, P.O. Box 47, 6700 AA Wageningen, The Netherlands

Richard D. Bardgett Faculty of Life Sciences, University of Manchester, Oxford Road, Manchester, M13 9PL, UK

Edmundo Barrios World Agroforestry Centre, United Nations Avenue, Gigiri, PO Box 30677, Nairobi, 00100, Kenya

Justin Bastow Eastern Washington University, Department of Biology, 258 SCI, Cheney, WA 99004-2440, USA

Valerie Behan-Pelletier Invertebrate Biodiversity Program, Research Branch, Agriculture and Agri-Food Canada, Ottawa, ON, Canada K1A 0C6

Matty P. Berg VU University Amsterdam, Department of Ecological Science, De Boelelaan 1085, 1081 HV Amsterdam, The Netherlands

Sharon A. Billings Department of Ecology & Evolutionary Biology and Kansas Biological Survey, University of Kansas, Lawrence, KS 66047, USA

Levente Bodrossy CSIRO Marine and Atmospheric Research, GPO Box 1538, Hobart, Tasmania, Australia 7001

Mark A. Bradford School of Forestry and Environmental Studies, Yale University, New Haven, CT 06511, USA

Maria J.I. Briones Departamento de Ecología y Biología Animal, Universidad de Vigo, 36310 Vigo, Spain

Lijbert Brussaard Department of Soil Quality, Wageningen University, P.O. Box 47, 6700 AA Wageningen, The Netherlands

Michel A. Cavigelli USDA-ARS Sustainable Agricultural Systems Laboratory, Beltsville Agricultural Research Center, 10300 Baltimore Avenue, Beltsville, MD 20705, USA

Marco Cosme Freie Universität Berlin, Institute of Biology, Functional Biodiversity, Altensteinstr. 6, D-14195, Berlin, Germany

Felicity Crotty Rothamsted Research, North Wyke, Okehampton, Devon, EX20 2SB, UK

Thibaud Decaëns Laboratoire d'Ecologie, EA 1293 ECODIV, Fédération de Recherche SCALE, Bâtiment, IRESE A, UFR Sciences et Techniques, Université de Rouen, F-76821 Mont Saint Aignan Cedex, France

Gerlinde B. De Deyn Department of Soil Quality, Wageningen University, P.O. Box 47, 6700 AA Wageningen, The Netherlands

Valerie T. Eviner Department of Plant Sciences, University of California, Davis, 1210 PES, Mail Stop 1, One Shields Ave, Davis, CA 95616, USA

Tom M. Fayle Forest Ecology and Conservation Research Group, Imperial College London, Silwood Park Campus, Buckhurst Road, Ascot, Berkshire, SL5 7PY, UK

Emmi Felker-Quinn Ecology and Evolutionary Biology, University of Tennessee, 569 Dabney Hall, Knoxville, TN 37996-1610, USA

Noah Fierer Department of Ecology & Evolutionary Biology, Cooperative Institute for Research in Environmental Sciences, University of Colorado, 216 UCB, Boulder, CO 80309-0216, USA

Philipp Franken Leibniz-Institute of Vegetable and Ornamental Crops, Department for Plant

Nutrition, Theodor-Echtermeyer-Weg 1, D-14979 Grossbeeren, Germany

Jennifer Fraterrigo Department of Natural Resources and Environmental Sciences, University of Illinois, 1102 S. Goodwin, Urbana, IL 61801, USA

A. Stuart Grandy Department of Natural Resources and Environment, University of New Hampshire, Durham, NH 03824, USA

Evelyn Hackl AIT Austrian Institute of Technology, Bioresources Unit, Konrad-Lorenz-strasse 24, 3440 Tulln, Austria

Jim Harris Department of Environmental Sciences and Technology, School of Applied Sciences, Building 56b, Cranfield University, Cranfield, Bedfordshire MK43 0AL, UK

Miranda M. Hart Biology Unit, The University of British Columbia, Okanagan, SCI 311–3333 University Way, Kelowna, BC V1V 1V7, Canada

Katarina Hedlund Department of Biology, Lund University, S 22362 Lund, Sweden

Liam Heneghan Department of Environmental Science and Studies, DePaul University 1110 W Belden Chicago, IL 60614-3251, USA

Jeffrey E. Herrick USDA-ARS Jornada Experimental Range, P.O. Box 30003, MSC 3JER, NMSU Las Cruces, NM, USA

Samuel W. James Department of Biology, University of Iowa, Iowa City, IA 52242, USA

T. Hefin Jones Organisms and Environment Division, Cardiff School of Biosciences, Cardiff University, Cardiff CF10 3AX, UK

Douglas L. Karlen Agricultural Research Service, United States Department of Agriculture, 2110 University Blvd., Ames, IA 50011, USA

Patrick Lavelle UMR BIOEMCO, Université Pierre et Marie Curie (Paris 6) and Institut de Recherche pour le Développement, Centro Internacional de Agricultura Tropical, AA 6713, Cali, Colombia

Michael D. Madritch Department of Biology, Appalachian State University, 572 Rivers Street, Boone, NC 28608, USA

Peter Manning School of Agriculture, Food & Rural Development, Newcastle University, Newcastle upon Tyne, NE1 7RU, UK

Jude E. Maul USDA-ARS Sustainable Agricultural Systems Laboratory, Beltsville Agricultural Research Center, 10300 Baltimore Avenue, Beltsville, MD 20705, USA

Phil Murray Rothamsted Research, North Wyke, Okehampton, Devon EX20 2SB, UK

Tânia Nobre Laboratory of Genetics, Wageningen University, P.O. Box 309, 6700 AH Wageningen, The Netherlands

Kate Orwin Lancaster Environment Centre, Lancaster University, Lancaster, LA1 4YQ, UK

Nicholas J. Ostle Centre for Ecology and Hydrology, Lancaster Environment Centre, Library Avenue, Bailrigg, Lancaster LA1 4AP, UK

Mitchell A. Pavao-Zuckerman Biosphere 2, University of Arizona, P.O. Box 210088, Tucson, AZ 85721, USA

Matthias C. Rillig Freie Universität Berlin, Institut für Biologie, Plant Ecology, Altensteinstr. 6, D-14195, Berlin, Germany

Karl Ritz National Soil Resources Institute, School of Applied Sciences, Cranfield University, Cranfield, MK43 0AL, UK

Michael Schloter Research Unit Terrestrial Ecogenetics, Helmholtz Zentrum München, Ingolstädter Landstraβe 1, D-85764 Neuherberg, Germany

Gregor W. Schuurman Wisconsin Department of Natural Resources, Bureau of Endangered Resources, P.O. Box 7921, Madison, WI 53707, USA

Dietmar Schwarz Leibniz-Institute of Vegetable and Ornamental Crops, Department for Plant Nutrition, Theodor-Echtermeyer-Weg 1, D-14979 Grossbeeren, Germany

Jennifer A. Schweitzer Ecology and Evolutionary Biology, University of Tennessee, 569 Dabney Hall, Knoxville, TN 37996-1610, USA

Martha Scott Algoma University, Department of Biology, 1520 Queen Street East, Sault Ste. Marie, ON P6A 2G4, Canada

Angela Sessitsch AIT Austrian Institute of Technology, Bioresources Unit, Konrad-Lorenz-strasse 24, 3440 Tulln, Austria

Keith Shepherd World Agroforestry Centre, United Nations Avenue, Gigiri, PO Box 30677, Nairobi, 00100, Kenya

Gudeta W. Sileshi World Agroforestry Centre, United Nations Avenue, Gigiri, PO Box 30677, Nairobi, 00100, Kenya

Fergus Sinclair World Agroforestry Centre, United Nations Avenue, Gigiri, PO Box 30677, Nairobi, 00100, Kenya, and Bangor University, UK

Johan Six University of California, Davis, One Shields Avenue, Davis, CA 95616, USA

Donald R. Strong Department of Ecology and Evolution, University of California, Davis, CA 95616, USA

Katalin Szlavecz Department of Earth and Planetary Sciences, Johns Hopkins University, Baltimore, MD 21218, USA

Ute Szukics Institute of Ecology, University of Innsbruck, Sternwartestrasse 15, A-6020 Innsbruck, Austria

Wim H. van der Putten Netherlands Institute of Ecology, Department of Terrestrial Ecology, Droevendaalsesteeg 10, 6708 PB Wageningen, The Netherlands

Nick van Eekeren Agrobiodiversity and Sustainable Animal Production Systems, Louis Bolk Institute, Hoofdstraat 24, 3972 LA Driebergen, The Netherlands

Kristof Van Oost Earth & Life Institute, TECLIM, Université catholique de Louvain, Place Louis Pasteur, 3B-1348 Louvain-la-Neuve, Belgium

Diana H. Wall Natural Resource Ecology Laboratory and School of Global Environmental Sustainability, Colorado State University, Fort Collins, Colorado 80523-1499, USA

Susan E. Ward Lancaster Environment Centre, Lancaster University, Lancaster LA1 4YQ, UK

Susanne Wurst Freie Universität Berlin, Institute of Biology, Functional Biodiversity, Altensteinstr. 6, D-14195, Berlin, Germany

Introduction

Diana H. Wall

Desertification, lack of fertile land to feed a rapidly growing population, insufficient water quality and availability, atmospheric nitrogen deposition, and climate change are together impacting ecosystems and their ability to deliver goods and services for humans. The last of these, climate change, has emerged as one the most damaging and menacing. It directly affects the flow of benefits that people receive and are dependent upon from ecosystems and their biodiversity, at local and global scales. Moreover, these global challenges occur simultaneously and interact, and require an infusion of new approaches, out-of-the-box thinking, and the best scientific information, if we are to rapidly arrive at solutions for sustaining people and our environment. Addressing environmental challenges gains a greater urgency when we consider that the benefits provided by ecosystems and their biodiversity are the basis of human well-being and substantial economic activity—the environmental capital of all nations.

It is striking that all of these environmental issues intersect around one common resource, soils and their biodiversity. There is no doubt that the global challenge of sustaining soils and their biodiversity, while often ignored in scientific discussions of global environmental issues, is gaining new attention as being key to sustaining ecosystem functioning and human well-being. While scientists have long recognized soils as living and of central importance to food production, there is now wide appreciation that they are a foundation of human and ecosystem survival. The ecosystem services that flow from soils and their biodiversity include soil formation and renewal of its fertility, maintenance of the composition of the atmosphere through carbon storage and greenhouse gas flux, erosion prevention, the regulation of the distribution and populations of pathogens and pests of humans, animals, and crops, the decontamination and bioremediation of wastes and toxic chemicals, and habitat and food for a variety of wildlife. Additionally living soil is a global receptacle of genetic diversity that is yet to be fully explored by humans.

Despite this, soils are being degraded at a rapid pace due to human activities. This degradation includes increasing desertification, erosion, and depletion of fertile land for food production. Policy makers involved in ongoing international agreements, such as the United Nations Convention to Combat Desertification, the Convention on Biodiversity, and the Intergovernmental Panel on Climate Change, are seeking multiple solutions and need reliable scientific information on soils, their biodiversity and the many services they provide, as well as their resilience under the interacting environmental challenges.

This book offers a unique synthesis of state-of-the-art information on soil ecology and ecosystem services. Research in soil ecology has matured and continues to accelerate, while the study of living soils as a provider of ecosystem services is still emerging. This advancement is reflected in this book by the authors' expertise, analyses, and perspectives of how soils and their biodiversity, scaling from single genes up to communities of hundreds of interacting species, function to provide ecosystem services across local to global scales, and how soils are affected by global change. The discussions of new technologies, molecular tools, models, and lab and field studies show a progression of increasing understanding of the functions of many species, both animals and microbes, which comprise the biodiversity in our soils. The information in this book also reveals the relevance of soil biodiversity to humankind, and contributes data, examples, and perspectives that will be useful in scientific and policy discussions on sustaining soils and people. My hope is that these discussions will indeed stimulate the recognition that the living organisms in soil are central to resolving many of the environmental challenges we face.

A book such as this only comes into being when many believe that accumulating evidence on an emerging topic needs to be brought to the forefront of scientific discussions. The intent is to catalyze further examination and discourse, and to generate new hypotheses and concepts, as well as to provide a baseline for options useful in the real world. Hopefully, this will occur. I have many people to thank for assuring these new ideas made it to this book. I want to acknowledge particularly my colleagues, the authors, and the section editors, Wim van der Putten, Karl Ritz, Richard Bardgett, Hefin Jones, Val Behan-Pelletier, Don Strong, Jeff Herrick, and Johan Six, for contributing their cutting-edge knowledge, hard work, encouragement, criticisms, discussion, and vision: they were terrific to work with! I can only express to Ian Sherman my profound and sincere thanks for his continued guidance and support throughout the preparation of the book. His ideas and experience added greatly to my vision. Helen Eaton has been superb in every aspect of the evolution of the book and all of us owe her a great deal of thanks. From day one until we finished, Helen assured us through her patience, wise counsel, and careful reviews that we were making progress, even when we sometimes felt otherwise. I am deeply indebted to Dr. Susan Melzer whose expertise in soils and her careful and intelligent editing of the chapters assured a higher quality book. Kerri Minatre meticulously provided her knowledge and technical editing, which was a great assistance for all of us, and I appreciate this greatly. The full citation reference for this book is as follows: Wall, D.H., *et al.* (eds.) (2012) *Soil Ecology and Ecosystem Services*, Oxford University Press, Oxford.

SECTION 1

The Living Soil and Ecosystem Services

SECTION EDITORS: **Karl Ritz and Wim H. van der Putten**

Introduction

Karl Ritz and Wim H. van der Putten

Soils have played significant roles in the development of the Earth system, life has evolved in the context of soil systems, and civilizations have risen and fallen by virtue of their exploitation and management of the earth they have inhabited. Soils continue to support the needs of contemporary societies, and this requirement will unquestionably prevail. From the human perspective, soil is literally fundamental to human civilization. This fact is underappreciated in many contemporary societies, which are increasingly becoming city inhabitants and disconnected from the rural context. The transition of more than half of the global population from rural to urban environments has now occurred, and the gross awareness of the importance of soils has become impaired due to a lack of contact and familiarity with the land. As a result, this rural:urban turning point may be a significant point in our current epoch.

Soils are remarkable materials, constituted of an extraordinarily diverse range of mineral and organic components, a tiny fraction of which are alive, but organized and interactive in particular ways that result in the delivery of the range of ecosystem services upon which sustained functioning of the contemporary Earth system depends. Soils form the outermost and extremely thin layer of the terrestrial system. They are the interface between the atmospheric and subsurface zones, connecting them to the hydrosphere, and provide the land surfaces which physically support all terrestrial biomes. Soils are highly heterogeneous in space and time with many different types, and concomitant properties, distributed across the planet. Nonetheless, all soils underpin to a greater (but rarely lesser) extent the basic suite of supporting, provisioning, regulating, and cultural services which are the subject of this book.

Soils are remarkable systems. They are extremely complex in both constitution and spatial organization. A significant property of soils that affects the way they function is that they are porous. Soil pore networks comprise hugely complex, ultimately always connected, three-dimensional labyrinths that span size ranges over many orders of magnitude. Such networks regulate the way that gases, liquids, solutes, particles, and organisms are held and can move through the matrix; they thus regulate many aspects of soil function. The pore networks also provide a huge range of niches which result in the extreme levels of soil biodiversity, far exceeding that which occurs aboveground. Soils are expressly *living* systems and the key roles that their belowground biota play in ecosystem service delivery is a theme which is prevalent in many chapters of this volume. Furthermore, soils are also *adaptive* systems, apparently able to respond to changes in environmental conditions in a manner that their functions are generally maintained. Such resistance and resilience to perturbation and stress may, however, have limits, and it is important to understand the mechanistic basis of such phenomena, the circumstances where such limits are exceeded, and how to manage or restore them.

The role that soils and their biodiversity play in supporting human societies has been conceptualized by naming and describing the ecosystem services that they provide. The Millennium Ecosystem Assessment plays a key role in this process. This concept has been developed in order to enhance communication of ecologists and conservation biologists with users and stakeholders on the consequences of rapid and ongoing loss of biodiversity for human society. It is now becoming increasingly applied by a large community varying from scientists investigating the relationship between species, their traits, and the functioning of ecosystems, to socioeconomists and policy-makers operating at local, regional, and global levels, and beyond to a wide range of professionals tasked with the challenge of managing land sustainably.

This first section of the Handbook provides a primer on soil systems from three complementary perspectives which provide a framework for understanding the other themes which are considered in more detail in the volume. Firstly, Patrick Lavelle considers the soil as a habitat, which is a crucial perspective and an appropriate one to commence with, since it emphasizes the fact that functional soils are founded on the biota that live within them. He describes the soil system in terms of the conditions it provides for life therein, and how organisms and their associated activity are governed by the structure of their habitat, across a range of size scales. The various adaptive strategies of soil organisms are explained, including those related to how soil biota effect ("engineer") soil structure and thence the physical arrangement of their habitat. He then explores the concept that soils may be self-organizing systems, arising from interactions between organisms that structure the soil habitat, and the resultant functions that ensue which appear to optimize ecosystem service delivery.

In Chapter 1.2, Susanne Wurst, Gerlinde De Deyn, and Kate Orwin review the remarkable and beguiling levels of biodiversity which occur belowground. They consider this from the perspective of body-size classes, which has a strong and well-rooted basis but is also logical in terms of functional pertinence, not least for reasons of spatial-constraints imparted by the soil pore network described in the preceding chapter. They also explain how we can characterize and study soil biota at the individual and community level, which is challenged by the huge biodiversity which prevails, the opacity of the soil matrix, and the intriguing fact that only a minute fraction of the bacterial and archaeal soil flora will apparently grow in culture media. The recent advances in genetics and the various guises of "-omics" are revolutionizing soil biology as much as other branches of the life sciences. They then explain some of the key basic functions which the biota carry out, and from the perspective of primary productivity, decomposition, and nutrient cycling consider what is arguably one of the principal ecological questions of our time, which is that of the relationship between biodiversity and function. They show that diversity–function relationships within the breadth of soil biota follow similar trends, in that the community composition, related then to the traits of key species or groups and their relative abundance and complementarity, appear to be the most significant drivers of soil processes and functions.

In Chapter 1.3, Lijbert Brussaard then moves up and across a number of scales to the primary theme of the volume. He reviews how the soil biota actually deliver ecosystem services. He explores the contrasting "soil biogeochemistry" and "soil biology" perspectives of how ecosystems function, and reviews the functional group concept. This leads him to consider the importance of traits rather than species *per se* as being the most appropriate way of conceptualizing such relationships, and that the trait concept can also be usefully applied to a higher-order of the soil community, particularly with respect to trophic levels. The practical implications of such concepts are then discussed with consideration of how we might more effectively harness natural processes to deliver a range of goods *and* services in the form of what could be designated "smarter" agro-ecosystems.

CHAPTER 1.1

Soil as a Habitat

Patrick Lavelle

1.1.1 Introduction

Soils provide an immense array of habitats that contain a vast and still largely unknown biodiversity, arranged in a highly-organized combination of a solid mineral phase, a network of water and air filled pores, and dead decomposing organic matter. Since the early days of the study of soil science, much has been done to describe and understand the progressive transformation of bedrocks into soils, including the distribution, nature and specific characteristics of mineral and organic components, and the opportunities and constraints that they represent for organisms (Lavelle & Spain 2001; Buol et al. 2003; Coleman et al. 2004; Bardgett et al. 2005). Nevertheless, analyzing how soil functions as a habitat is a more recent emphasis, including the way organisms create their habitats and live in them.

Soil scientists have described the vertical and sometimes horizontal gradation of soil horizons via a wide variety of soil classification systems (Baize & Girard 1995; Soil Survey Staff 1999; Brady & Weil 2007). Soil physicists have provided detailed descriptions of the solid and porous phases in their attempts to model soil hydric processes (Hillel 1980). Agronomists and ecologists have detailed how soils can provide nutrients and a physical substrate for plant growth. There is an extensive body of literature that details the biological, physical, and chemical transformations that occur during soil organic matter decomposition (mineralization and humification) (Swift et al. 1979; Lavelle & Spain 2001).

Adaptation to living in soils has generated a wide range of life-forms and specific biological traits. Understanding these adaptations requires a holistic view of the nature of soils, linking physical, chemical, and biological processes and trying to understand what being a bacterium or a collembolan in this environment actually entails. The understanding of these adaptations is a mandatory step before conceptual models of soil as a habitat are developed within different environmental contexts, or before intuitive theory is described (Pachepsky et al. 2001; Lavelle 2009). For example, there is considerable recent interest in stoichiometry and the ecological consequences of differences between supplies of soil nutrients and the needs of living organisms for these nutrients (Gignoux et al. 2001; Guillaume et al. 2001; Sterner & Elser 2002, Osler & Sommerkorn 2007; Hättenschwiler et al. 2008; Marichal et al. 2011). Soil is a refuge for a large number of very old (in evolutionary terms) taxa and an environment with a large, enigmatic, biodiversity (Anderson 1975). This issue has received much interest. Subsequently, biodiversity in soils and its relationship to soil function has also become a highly debated topic (e.g. Lavelle et al. 1995; Brussaard et al. 1997; de Ruiter et al. 1998; Andrén & Balandreau 1999; Wolters 2001; André et al. 2002; Wall 2004; Coleman & Whitman 2005; Crawford et al. 2005; Fitter et al. 2005).

This chapter first describes the main constraints that organisms face when living in soil. The way soil organisms have overcome these constraints, through self-organization across scales is discussed. Finally, the consequences for soils and their management of their auto-organized nature are explained.

1.1.2 Conditions in soils

Soils are compact environments where life is concentrated in a porous space that typically comprises

Soil Ecology and Ecosystem Services. First Edition. Edited by Diana H. Wall *et al.*
© 2012 Oxford University Press. Published 2012 by Oxford University Press.

30% to more than 60% of the soil volume in the upper horizons. Living and moving in this space is the first important constraint for soil organisms. Pore size is highly variable as is its connectivity, which may serve both as an opportunity or an obstacle to having an extended home. The available space theoretically accessible to a given organism may be further reduced by the amphibious nature of the system, partly filled with water and partly with air. Soil organisms should be adapted first to this very important double constraint.

The second important constraint in soil is feeding on mainly low-quality food. The most abundant resources are leaf and woody litter deposited at the soil surface, root material, and soil organic matter, a very diverse set of materials with different particle sizes and chemical composition. Though relatively abundant, these elements are mostly of a low nutritive quality and potentially difficult to digest as they often combine stoichiometric imbalances and high concentrations of compounds that are very resistant to digestion. Leaves and also roots can have low resource quality owing to the secondary metabolites that serve, amongst others, as defenses against herbivory (Harborne 1990; Hol et al. 2004). These chemical constraints are often exacerbated by the discrete distribution of the resource when combined into the soil matrix.

1.1.2.1 Dwelling and moving in soils

Total soil porosity gradually decreases from the upper (often litter-containing) strata made of accumulated decomposing organic material pieces of different sizes, to the deeper mineral soil. While the structural porosity, that is, the porosity constructed by physical or biological processes, tends to decline sharply, porosity that naturally separates particles of different sizes and shapes becomes progressively dependent on soil texture (i.e. grain size distribution). Structural pores are generally large in size mostly ranging from micrometers to decimeters for some root channels or earthworm and termite galleries, whilst textural pore sizes vary from fractions of micrometers in pure clay soils to a maximum of tens of micrometers in sandy soils. Surprisingly, the overall distribution of pores among size classes and complete measurements of their connectivities have yet to be achieved. Different methodologies are used for different size classes, and only parts of the porosity spectrum are considered, while connectivity based on three-dimensional assessments from next-generation imaging technologies such as computed tomography is still under development (e.g. Grimaldi 1993; Vogel & Roth 2001; Pierret et al. 2002; Capowiez et al. 2003, 2011; Young & Crawford 2004; Braudeau et al. 2005; Chan & Govindaraju 2011).

Soil porosity does not seem to exhibit a continuous distribution pattern across the whole range of size classes; distribution is rather discontinuous and seemingly follows a fractal pattern, with concentration of porosities at discrete scales (Fig. 1.1.1; Rieu & Sposito 1991; Grimaldi et al. 1993). Three pore classes may be defined according to their sizes, the associated matric potential, and the subsequent force to be applied to extract water molecules held by surface tension forces: micro-, meso-, and macropores. Micropores are textural pores of a size below 0.15 µm. They correspond to the spaces accumulated among soil particles and retain water at potentials of less than -1.5 MPa (i.e. beyond the limit where plants can extract water). These pores may comprise a large proportion of porosity in fine textured soils, for example 1/3 of total porosity of a clay soil, and none for a sandy soil (Yong & Warkentin 1975; Lavelle & Spain 2001). Mesopores are the smallest fraction of the structural porosity, ranging from 0.15 to ca. 30 µm. They retain capillary water, held at tensions of -1.5 MPa to -0.05 MPa, and are available to plants. These pores comprise textural voids left by assemblages of silt and sand particles and are built by a number of small soil organisms (like channels made by the finest roots or fungal mycelia (Cabidoche & Guillaume 1998) or intervals between microbial constructs as observed by Young and Crawford (2004) and by physical processes (cracks and spaces left among physical aggregates). Macropores comprise the large-sized soil porosity from which water drains almost freely for as long as the connectedness of pores allows. They are rather diverse in origin, having been created by physical or biological processes. The formation of cracks by the physical retraction of drying or thawing soils is a rather frequent process that creates physical macropores. Most macropores are due to soil bioturbation, which leaves macropores between the compact aggregates created by earthworm casts or termite compacted runways, and burrowing activities of soil ecosystem engineers (as defined by Jones *et al.* 1994), predominantly roots, earthworms, termites and ants (Capowiez *et al.* 2003, 2011; Nahmani *et al.* 2005; Cerda & Jurgensen 2008; Colloff *et al.* 2010; Jarvis *et al.* 2010).

Connectivity of pores across different sizes is a very important prerequisite for life in soils, since it determines the physical domain within which a given organism may possibly move without having to dig its way out. Although little is known about the discontinuous distribution of pore networks in soils, it is likely a major limitation to movement and dispersal of organisms. At the same time, natural isolations of populations may protect them from their possible predators or competitors and favor the huge biodiversity observed among small soil organisms (Postma & van Veen 1990; Young *et al.* 2008).

1.1.2.2 Respiration in soils

After a rainfall or irrigation event, soil pores may be completely filled with water. Drainage will first empty the largest pores where water is retained at very low tensions, provided that there are connections allowing drainage. Evaporation and plant evapotranspiration will then progressively remove water from the pores, first the largest and then the smallest ones (Papendick & Campbell 1981). This is a result of water retention forces that are directly linked to the distance between the water molecule and the nearest solid particle. Soil water status may be extremely variable and environmental conditions may shift from aerobic to aquatic in a few hours while the inverse dynamics is usually slower. At field capacity (pF 2.5 with a water tension of −0.035 Mpa) all pores of a size >10 μm are filled with water. At pF4.2 (permanent wilting point of plants), water tension is ca. −1.5 MPa and the size of pores still completely filled with water is 0.15 μm. Organisms living in water with aquatic respiration have higher chances to survive if they are small and thus live in the smallest pores that remain filled with water for longest time periods. Consequently, in a soil at field capacity (ca. pF 2.5) in which gravimetric water has been transferred to lower parts of the soil profile or topography, the water film is 10 μm thick thus providing aquatic environments to microbes and a number of microfauna elements. Larger invertebrates will find continuous moisture and meso- and macropores will contain air that allows respiration. At pF 4.2, the water film is only 0.15 μm and tension of 2000 MPa is beyond the capacity of roots to extract water. This explains why at pF 4.2 plants start to wilt, microorganisms have

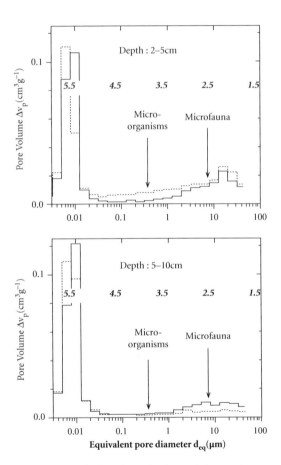

Figure 1.1.1 Distributions of pore sizes in an oxisol supporting primary forest (continuous line) and a pasture (dotted line) near Manaus (Amazonia, Brazil) (after Grimaldi *et al.* 1993). pF values corresponding to saturation of pore sizes are indicated in bold italics. Arrows indicate minimum size limit of microorganisms and microfauna. Note that soil compaction in pastures significantly reduces habitats for microfauna at 5–10 cm depth while habitat for micro-organisms increases at 2–5 cm.

decreased activities and invertebrates suffer water stress. Water may, however, be retained in pores of a larger size than expected when they have a bottle-neck shaped opening since the diameter of the opening determines the tension that should be exerted to extract water.

Amphibiosis in soils imposes specific constraints on respiration physiology. Respiration may be either of an aquatic type, based on the use of dissolved oxygen with gill or direct membrane exchanges (e.g. in bacteria and earthworms) or of an aerial type involving air circulation into trachea systems or lung-like structures (e.g. arthropods or Gasteropoda). A number of physiological and behavioral characteristics allow species to make best use of the soil environment in spite of rather unstable conditions considering the relative proportions of air and water in pores.

1.1.2.3 Feeding in soils

On average, 90% of all primary production returns to the soil after the death of plants or plant organs. Dead soil microbes and fauna also follow the same pattern. Soils and their litter layers are therefore environments with potential high resource availability. This, however, does not mean that feeding is easy. Soil resources may be divided into three groups according to the kind of difficulties opposed to their digestion: leaf, root, and woody litters; soil organic matter and soluble resources; the exudates and mucus produced by soil ecosystem engineers or leachates from the canopy and leaf litter environments.

Litter comprises dead plant organs deposited at the surface or inside the soil. It is the major source of carbon that may represent inputs of several tens of Mg per ha and per year. This resource, however, is not easy to use as it frequently presents severe stoichiometric limitations, and C components are present as polymers that request specific digestive abilities (although with large variations among plant species and environmental conditions; Lavelle and Spain (2001)). Plants generally translocate a large proportion of the nutrients contained in leaves before abscission, and similar processes probably happen in branches too, which explains why litter generally has low nutrients concentrations. Polyphenols and lignin are sometimes abundant components of plant materials and they are generally difficult to digest, if not toxic (Lavelle *et al.* 1993; Tian *et al.* 1995; Hättenschwiler & Vitousek 2000; Schweitzer *et al.* 2004). Furthermore, following the death of leaves, polyphenols combine with proteins creating highly resistant polyphenol protein compounds (Handley 1961; Toutain 1987; Palm & Sanchez 1991). Polyphenols present in vacuoles of living cells are mixed with the cell content when leaves die. They combine with cytoplasmic proteins and complex up to 72% of the nitrogen of freshly dead leaves and 85% in roots (François *et al.* 1986). Efficient release of nitrogen from these compounds is done by only very few organisms, principally the Basidiomycete fungi known as white rot fungi. Bacterial activity in earthworm guts may also degrade these substrates (Toutain 1987). This decomposition is carried out under physical limitations due to the occasional toughness of tissues and structures.

Soil organic matter (SOM) is another abundant resource that comprises 20 to >100 Mg ha^{-1} and is distributed across the whole soil profile at concentrations decreasing with depth. Detailed analyses reveal that this resource, though abundant, presents a number of severe drawbacks when used as a food resource. Apart from having similar stoichiometric limitations as leaf and woody litter, SOM is enriched in polymerized humic compounds that may be highly resistant to decomposition and also form highly stable complexes with clay minerals. SOM is also diluted in large volumes of mineral material that naturally represent over 90% of the soil weight in most cases. Digestion of SOM by soil feeding invertebrates thus requires specific enzymatic activities plus the ability to ingest high volumes of soil enriched in OM by selective foraging as polyhumic endogeic earthworms do (Lavelle 1983).

Although a number of models consider a continuous rather than discrete evolution of SOM (Gignoux *et al.* 2001), observation and modeling exercises tend to rather support a discontinuous classification (Smith *et al.* 1997, 2002). For example, the CENTURY model proposes a discontinuous classification of organic matter into three different pools (Parton *et al.* 1983; Jenkinson *et al.* 1987; Smith *et al.* 2002). The first is a metabolic pool mainly of microbial biomass, which has a relatively fast turnover time (2–4 years in conditions of temperate climates) and only comprises 3–5% of the soil C (Lavelle and Spain 2001). Microbial biomass, a high-quality resource in theory, would only represent 0.1–0.3% of the total soil volume in most soils, probably not enough to sustain a geophagous invertebrate that would exclusively use this resource in the absence of highly adapted foraging strategies.

The second is a physically-protected pool that comprises organic residues included in compact soil aggregates and is thus isolated from sites where microbial activities would easily decompose them. This pool, which generally includes around half of the non-microbial soil C and has a longer turnover (20–50 years in temperate climate conditions), may vary significantly according to the state of soil physical structure. In vertisols from the Caribbean Martinique island for example, intensive market gardening with limited soil cover may decrease soil organic matter content from 36 g C kg^{-1} in semi permanent pastures to 16 g C kg^{-1} after 15 years of intensive market gardening, following the destruction of aggregates and the mineralization of OM that had been physically protected (Blanchart et al. 2004). After 4 years of pasture restoration, C content had increased up to 24.8 g kg^{-1} while in a treatment with no plants or earthworms, C content further decreased down to 14.1g kg^{-1}. The difference between pasture value and the minimum value in market gardening corresponds to physically protected OM. The physically protected fraction is only accessible to soil organisms if aggregates are broken down; furthermore, a rather large energetic investment is required to use organic matter that only comprises a limited proportion (ca. 5% in favorable cases) of the aggregates.

The third is a chemically-protected pool generally equivalent to the amount of the physically protected pool comprising recalcitrant humic compounds with a very long turnover time (800–1200 years). This OM stays when soils have been highly degraded by, for example, intensive agriculture practices with intensive tillage, limited return of organic residues and soil maintained bare for several months each year. In the example of the Martinique vertisol, this pool would essentially comprise C left after 15 years of market gardening plus 4 years with no plant cover, 39% of total SOM in this case. This pool may be decomposed if fresh OM allows a priming effect on microbial activity. The scarcity of these compounds, especially in deep soils horizons would at least partly explain the large residence time of this OM (Fontaine et al. 2007). Mucus in earthworm guts (Martin et al. 1987) and exudates in plant rhizospheres (Brown et al. 2000; Kuzyakov 2002) are, however, known to trigger such priming effects.

Soluble resources, or easily assimilated resources, exist in soils in sizeable amounts. Directly issued from organism metabolism, they comprise root or mycorrhizal exudates, earthworm mucus, and a few other minor sources. Root exudates and other solid elements of rhizodeposition are a rather abundant source that comprises on average 17% of the net C fixed by photosynthesis (Nguyen 2003). They are released in soils mainly as polysaccharides with relatively low molecular weights. This represents a total amount of several Mg dry mass ha^{-1}, a rather considerable mass of easily assimilable C. Although reported N contents in exudates are generally low, Wichern et al. (2008) indicate that 4–71% of total assimilated plant N may be found in root deposits, with respective medians of 16% in legumes and 14% in cereals.

Although many studies have focused on the elements present at very low concentrations in exudates (hormones and micronutrient elements), they actually represent a very large and easily available C source that has profound effects on microbial and faunal activities in the rhizosphere (Gregory 2006). Using rhizodeposition as a resource for soil organisms causes a number of constraints to rhizosphere organisms. Rhizodeposition-based resources generally have low contents in nutrients, which force organisms living on them to find complementary sources outside the rhizosphere and/or adapt their C metabolism to this condition (Kuzyakov et al. 2000; Muhammad et al. 2007). They are also a pulsed resource produced at specific times, and at root tips with a highly discrete pattern of distribution in the soil volume. Their uneven distribution in the soil volume and unpredictable temporal pattern of production requires highly specific strategies for their use by heterotrophic organisms. Mycorrhizal fungi that directly benefit from plant C allocation to roots at taking C from inside the root probably use a large part of resources translocated from plants to soils (Högberg & Read 2006).

Earthworm intestinal and cutaneous mucus are similar in function to root exudates although with different compositions. Mucus are glycoproteins of relatively low molecular weights produced at the earthworm body surface as a lubricant, or added to the gut content in the digestion process (Cortez & Bouché 1987; Martin *et al*. 1987). Although the composition and amount of cutaneous mucus lost to the soil have been very seldom estimated (Jegou *et al*. 2001), their effects on microbial activities in earthworm burrow linings are well known. Intestinal mucus is perhaps a more significant component in ecosystems where earthworms are abundant than cutaneous mucus. Intestinal mucus is a glycoprotein 40–60 thousands Da in size that is added at rates of 5–37% in equivalent dry weight to the ingested food in the anterior part of the gut (Martin *et al*. 1987; Trigo *et al*. 1999). Although largely reabsorbed in the posterior gut, this mucus represents a rich resource that is used by ingested microflora during the 20 min to 4 h transit time of soil through this part of earthworm guts (Lavelle *et al*. 1995).

1.1.3 Adaptive strategies of soil organisms

According to their respective sizes, life forms, and physiologies, soil organisms have evolved different kinds of adaptive strategies in order to deal with the three major constraints of the soil habitat on an individual basis or via mutualistic associations. Some are actually better adapted to face constraints of the soil environment than others. This situation has led to the development of associations in which complementary adaptive strengths are shared. An interesting consequence is that self-organization based mainly on intense mutualistic and/or non-trophic interactions among organisms seems to be a prevailing process in soils that in turn has promoted a mode of functioning that is unique in the biosphere. None of the groups of soil organisms is optimally adapted to each of the three main soil constraints. Size largely reflects the ability to face each kind of constraint (Lavelle *et al*. 1997; Fig. 1.1.2; Table 1.1.1). A remarkable feature of soils development has been the coevolution among organisms with vastly different adaptive strategies that has allowed them to adapt to a severely constraining environment. Since none of the organisms individually could fully adapt to the three major constraints, interactions have made the best use of major abilities of each one.

1.1.3.1 Digestion and cooperation

The outstanding ability of the microbial community to digest the most recalcitrant substrates (such as cellulose, chitin, lignin, or phenol protein compounds), and produce a wide range of hormone-like products, is used by all other organisms. They, in turn, provide more or less sophisticated mechanisms for selection and stimulation of microbial population activities (predation included) and their transport to new substrates to decompose. Generally, invertebrates have relatively limited abilities to digest organic compounds. Some earthworms, for example, do not have proper cellulolytic enzymes (Lattaud *et al*. 1999). Cellulases found in the gut are at least partly released by the ingested soil microflora in the medium part of the gut. These microorganisms are first selectively stimulated in the anterior part of the gut where one volume of water and 5–35% of the weight of ingested soil as intestinal mucus are added and energetic mixing occurs (Barois & Lavelle 1986; Trigo *et al*. 1999). Once their full activity is restored, they start to digest organic substrates and release substrates in the gut that can be assimilated by themselves and by the worms. Other examples of digestion and cooperation are transfer of enzymes from bacteria to nematodes, or symbiotic mutualists. An example of horizontal transfer between bacteria and phytoparasitic nematodes suggests that cooperation among invertebrates and microorganisms in digesting low-quality resources may be much broader and diverse than currently imagined so far (Smant *et al*. 1998). Mycorrhizal fungi play special roles in soils by providing plants with nutrients that they forage well outside the root zone, in association with free microorganisms living in their mycorhizosphere (Högberg & Read, 2006). More information on these types of interactions is provided in the Chapter 1.2 by Wurst *et al*. (this volume).

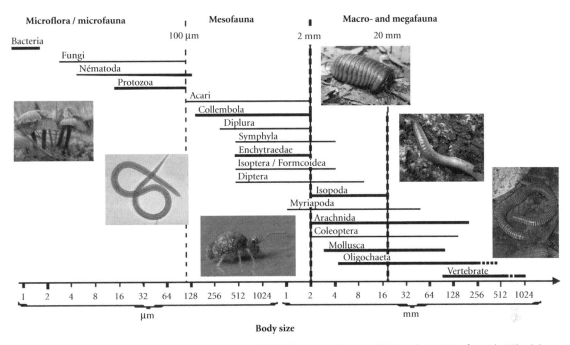

Figure 1.1.2 Size distribution of soil organisms by Swift *et al.* (1979) illustrated by Decaëns (2010), with permission from John Wiley & Sons.

1.1.3.2 Movements, habitat building, and bioturbation

The ability of soil ecosystem engineers to dig and burrow and constantly mix the soil is essential to the expansion and maintenance of the structural pore space and to the selection and redistribution of microorganisms. Roots do considerable "biological drilling" (Cresswell & Kirkegaard 1995; Angers & Caron 1998), a well-recognized but still poorly quantified process. Earthworms are the most efficient bioturbators in soil, able to ingest and egest as solid macro aggregates up to 1200 Mg dry soil per ha (Lavelle & Spain 2001) and drilling up to 900 m of galleries per m^2 (Kretzschmar 1982). They produce macroaggregates with rather complex properties as regards organic matter distribution and sequestration (Martin 1991; Fonte *et al.* 2007). Termites and ants are other powerful ecosystem engineers that complete or replace earthworm activities in most ecosystems where they are present (Lobry de Bruyn & Conacher 1990). In specific conditions, isopods have been claimed to be responsible for important transfers of soil (Yair 1995; Jones *et al.* 2006; Davidson *et al.* 2010). Besides the major activities of these important engineers, some degree of engineering activities is developed by most soil organisms though at smaller scales. Microbial physical engineering, for example, may be important when very small-scale changes in particle distribution and drilling of microscopic channels by fungal mycelia are accurately measured (Cabidoche & Guillaume 1998; Young & Crawford 2004).

1.1.3.3 Amphibiotic conditions

Depending on the dependence on free water in soil, organisms have developed different degrees of resistance, from cysts and spores to resting stages or non-specific stage involving different degrees of desiccation and/or the construction of specific structures (Jimenez *et al.* 2000; Fig. 1.1.3). Resistance to desiccation generally increases as size decreases (Lavelle & Spain 2001) although some relatively large invertebrates, like ants, may survive in extremely dry environments.

Table 1.1.1 Main parameters of the adaptive strategies of organisms in soils (Lavelle & Spain 2001)

Functional group	Microflora	Microfauna	Mesofauna	Macrofauna
Body width	0.3–20 μm	<0.2 mm	0.2–10mm	>10mm
Taxa	Bacteria	Protozoa	Microarthropods	Termites
	Fungi	Nematodes	Enchytraeidae	Earthworms
				Myriapoda,
				Ants, etc.
Water relationships	Hydrobiont	Hydrobiont	Hygrobiont	Hygrobiont
Interactions with micro-organisms	Antibiosis	Predation	Predation	Mutualism (external rumen, facultative or obligate internal mutualism)
Ability to change the physical environment	Very limited	None	Limited (faecal pellets)	High (galleries, burrows, macroaggregates)
Resistance to environmental stresses	High (cysts, spores…)	High (cysts, spores…)	Intermediate	Low (with possibility of behavioral compensation)
Intrinsic digestive capabilities	High	Intermediate	Low	Low

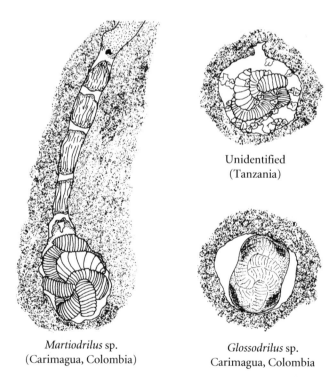

Martiodrilus sp. (Carimagua, Colombia)

Unidentified (Tanzania)

Glossodrilus sp. Carimagua, Colombia

Figure 1.1.3 Resting stage in three different earthworm species (a) *Martiodrilus* sp. (Carimagua, Colombia) rests in a chamber at the end of a gallery closed with several separations; (b) *Glossodrilus* sp. rests in a circular chamber coated with fine textured casts; (c) an unidentified species from Tanzania rests on small gravel stuck to the walls that prevents contact with the wall.

1.1.4 Self-organization and the spatial organization of soils

Self-organization has led soil organisms to develop interactions and adapt to soils constraints (Lavelle *et al.* 2004a; Lavelle *et al.* 2006). The energy mobilized through microbial activities and photosynthesis (for plants and their roots) is used by soil ecosystem engineers to build habitats in the compact soil matrix and live there in mutuality with the organisms associated with them. The theoretical question of whether by doing so they actually extend their phenotype and increase their fitness or simply have an accidental effect on other organisms is still debated (Jouquet *et al.* 2006).

1.1.4.1 Soils as self-organized systems

Soils may be considered as self-organized systems according to criteria defined as follows (Perry 1995; Lavelle *et al.* 2006):

1. They are characterized by order where disorder would have been predicted. The organization of soil horizons, the distribution of pores among size classes and their spatial arrangement, the structure of invertebrate and microbial communities and food webs are among the many examples of "organization" and "order" in soils.

2. Structures and processes mutually reinforce one another. An obvious example is the building of epigeic nests by social insects that protect them from hazardous climate conditions and from predators. Another example is the maintenance of structural soil porosity by invertebrates and roots that, in turn, enhances their own activities and other associated biological activities by enhancing infiltration and water storage capacity.

3. The system maintains order within boundaries through internal interactions. Specific observations indeed tend to show that functional domains of soil ecosystem engineers, that is the volume of soil that is shaped by their activities (Lavelle 2002), have recognizable limits. These can be drawn for example by looking at the near infrared spectral signatures (NIRS) of macroaggregates that comprise their domain (Hedde *et al.* 2005). At a large scale, earthworm, termite, or root populations are often distributed in patches inside which soils have notably specific characteristics and functions (Elridge & Greene 1994; Rossi *et al.* 1996; Jouquet *et al.* 2006). At the smallest scales, accurate observation of microbial environments also reveals some degree of habitat construction (Cabidoche & Guillaume 1998; Young and Crawford 2004). Despite the difficulty to rigorously classify the type of interactions developed in these systems, it seems to be largely based on mutualism and/or non-trophic relationships akin to ecosystem engineering (Jones *et al.* 1994).

4. Far from equilibrium, these systems are effectively in a state of dynamic equilibrium. Experiments show that soil physical function can be profoundly modified when disturbances affect the activity of the invertebrates (Blanchart *et al.* 1997; Mando 1997; Barros *et al.* 2001). Invasive species, for example, may disproportionally enhance one function (e.g. producing large compact structures or mineralizing organic matter accumulated in humus layers) in a way that the system no longer sustains its dynamic equilibrium (Chauvel *et al.* 1999; Hendrix 2006). When eliminated by aggressive land management practices, the environmental conditions that they maintained in their sphere of influence may drastically change; an example may be the disappearance of control exerted by plant parasitic nematode communities on the most aggressive species of their own community when nematicides are applied and ultimately leave the most aggressive species with no competitors (Lavelle *et al.* 2004b).

5. Natural systems can be seen as comprising a hierarchy of embedded self-organizing systems, stabilized by cooperative relationships and focused at spatial and temporal boundaries. Soil function is thus envisaged as a combination of processes that links small-scale, fast developing processes to progressively larger-scale and slower processes (Fig. 1.1.4). The analogy to mechanical devices is supported by the observation of discrete scales for interactions in soils, and different speeds in processes. This view is clearly opposite to that supposed by models that present soil function as complex webs of interactions with no specific spatial organization or specific spatial constraints.

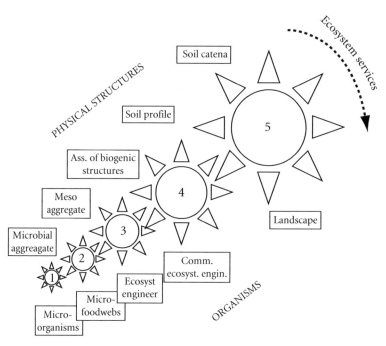

Figure 1.1.4 Self-organizing systems in soils at different scales from microbial biofilms and aggregates (1), to intermediate aggregates (2), individual ecosystem engineer functional domain (3), mosaics of functional domains in an ecosystem (4) and the landscape (5) where ecosystem services are delivered (Lavelle *et al.* 2006). Each scale is defined by a community of interacting organisms, a set of physical structures that they inhabit and, in some cases, have created themselves, and processes that operate at this scale of time and space.

1.1.5 Discrete scales in soil function

Five relevant scales have been identified in soil function (Fig. 1.1.4). At each scale, interactions among organisms of one or several groups develop within boundaries of such structures as biofilms, mesoaggregates, or functional domains of invertebrate ecosystem engineers (Lavelle *et al.* 2004a). Although these scales are implicitly acknowledged, based on observations, the evidence of such a discrete organization of the soil environment is still rather weak. While discrete patterns are observed in both porosity and aggregate fractionation distribution (Menendez *et al.* 2005; Globus 2006; Fedotov *et al.* 2007) data are still rather scarce and fragmentary. Studies considering application of fractal model to, e.g. soil porosity, have been successfully applied to part of the range of soil porosity, but surprisingly, no study has ever considered the entire range of porosities.

Scale 1: microbial biofilms. The smallest habitat in soils is represented by assemblages of mineral and organic particles approximately 20 μm in size, called microaggregates (Fig. 1.1.4, Scale 1). Most chemical transformations that result in organic matter cycling and soil chemical fertility are operated by microorganisms in microsites and biofilms. With a few exceptions (ex. surface crusts made by cyanobacteria; Leys & Elridge 1998), these structures are embedded into larger-scale structures made and/or inhabited by larger organisms that organize space at larger scales.

Scale 2: micro food webs. Microorganisms may live inside (e.g. in micropores filled with water) or outside soil mesoaggregates (ca. 100–500 μm; Hattori & Hattori 1976), which in turn determines their access to resources, exposure to predators and inclusion as prey in micro-food web systems (Scale 2). At this scale microaggregates form assemblages that leave spaces among them where micropredators use to live, such as nematodes and protists that live on microbial biomass and control their populations and activities. Specific assemblages of microorganisms in the rhizosphere also belong to this scale. Mycorrhizae and their mycorhizosphere operate at the same scale representing an autotrophic option for interactions at this particular scale (Högberg & Read 2006).

Scale 3: functional domains of ecosystem engineers (Lavelle 2002). At the scale of decimeters to decameters, ecosystem engineers and abiotic factors determine the architecture of soils through the accumulation of aggregates and pores of different sizes, microaggregates of Scale 1, mesoaggregates of Scale 2 and macroaggregates that they produce. These spheres of influence (= functional domains) (Scale 3) extend horizontally over areas ranging from decimeters (e.g. the rhizosphere of a grass tussock) to 20–30 m (drilosphere of a given earthworm population) or more, and from a few centimeters up to a few meters in depth, depending on the organism (Decaëns & Rossi 2001; Jimenez et al. 2001).

Scale 4: mosaics of functional domains at plot scale. Functional domains are distributed in patches that may have discrete or nested distributions and form together a mosaic of patches (Scale 4). Such a mosaic has been described, for example, by Rossi (2003) who observed the distribution of two groups of earthworms with opposite effects on soils. One group called "compacting" would stimulate soil macroaggregation through the accumulation of large (ca. 1 cm) compact casts and reduce soil macroporosity with resulting high values of the soil bulk density (Blanchart et al. 1997). Another group referred to as "decompacting," would have an opposite effect breaking large aggregates into smaller pieces with a resulting decrease in bulk density and increased density of rootlets that find a more suitable environment in these patches. More complex designs mixing termites, ants, earthworms, and plant roots spatial domains probably exist, although their structure and the relationships between their different constituents have very seldom been addressed. Still, however, little is known on the spatial organization of functional domains of the entire ecosystem engineer communities. For example, there is a need to investigate whether root rhizospheres and functional domains of invertebrate ecosystem engineers are separate entities. Near infrared spectrometric signals of soil macroaggregates show clear different signals for biogenic structures made by different species of macroinvertebrates. They also show that within the soil matrix, some macroaggregates may have been built in common by roots and earthworms while others exhibit single specific signals (Zhang et al. 2009; Hedde et al. 2005; Zangerlé et al. 2011).

Scale 5: landscape/watershed. At the landscape level, different ecosystems coexist in a mosaic with clearly defined patterns (Scale 5). The occurrence and distribution of land cover types in landscapes may result from natural variations in the environment and/or human land management. Soil formation processes, for example, are very sensitive to topography, which generates the formation of catenas of soils from upper to lower-lying areas. Significant differences in soil type at this scale often determine different vegetation types and the formation of a mosaic of ecosystems (Sabatier et al. 1997). In savannah regions of Western Africa, plateaus that have a thick soil and a gravel horizon are often covered with open woodland. Slopes that have more shallow soils have fewer trees; low-lying areas have fine textured soils resulting from the transport and accumulation of fine elements from the upper lying areas. They are also moister environments where vegetation is comprised of grasses and forbs. In riparian zones of river catchments, gallery forests may utilize constant water availability from a water table located close to the surface (Brabant 1991). On the other hand, there is growing evidence that the composition and structure of artificial mosaics created with land management have effects on biodiversity and the distribution of soil habitats (Lavelle et al. 2010).

Soil formation at regional scales is one of the ecosystem services that integrates processes at all scales; it is a slow process that extends over long periods of time and is largely determined by climatic conditions and the nature of the parent material. In temperate areas, for example, it takes 20,000 years to transform alumino-silicate parent material into a 1m thick soil, but it takes half that time to develop carbonate rich material (Chesworth 1992). Most soils in Northern Europe and America that have been formed after the retreat of glaciers 20,000 years ago still have properties of relatively young soils, as compared to soils from Australia or some parts of Africa that began forming millions of years ago (Fyfe et al. 1983).

1.1.6 The challenge of an eco-efficient use of soils

The close interdependence at all integration levels from organisms, communities, their habitats in soils, and ecosystem processes developed in these environments, has strong scientific challenges and practical implications. Scientists should develop theories and tools to characterize, classify, and evaluate soils while considering all these integration levels involved. Then, these tools can be used by scientists and practitioners to define conditions for an eco-efficient use of soils aiming at a combination of economic and social development while conserving or enhancing the natural capital. Ultimately, practice may create agro-ecosystems with all the necessary functional units assembled to optimize the production of ecosystem goods and services.

1.1.6.1 Developing the theory

The self-organization model for soil systems is still an intuitive view, although fed by a large amount of accurate observations and proven facts. A key element that should be tested is the reality of the discrete distribution of functional units across all scales by testing the discontinuous organization of the porous space across spatial scales. Although evidence has been given for some scales from observation (Grimaldi et al. 1993) or modeling exercises (Menendez et al. 2005), assessments of all porosities from the tiniest textural pores to the largest macro-pores made by ecosystem engineers and physical processes are required. Significant covariations among biodiversity measurements and diversity of the pore sizes would indicate that the habitat composition and structure matters in determining biodiversity in soils. As a matter of fact, theoretical developments on biodiversity in soils tend to largely overlook physical constraints on biological processes.

Theory should then describe how evolution driven by soil constraints has led communities to form self-organized systems with specific, although largely ignored interaction mechanisms. The production of exudates by roots is a simple example of a trophic mechanism that organizes decomposer communities and food webs in the rhizosphere (Lambers et al. 2009). Exudate composition and fluxes are plant traits that certainly influence microbial activities in soils and their interactions with other organisms. More complex interactions, involving horizontal transfers of genes or effects on gene expression has appeared as research accumulates on the topic (Smant et al. 1998; Ashelford et al. 2001). The creation of physical structures in soil acting as extensions of phenotypes is another mechanism that may have increased the fitness at species and community levels (Jouquet et al. 2006) by influencing processes at their scales.

Scientists should thus investigate communities, the habitat, possible creation of physical structures and participation to processes operated for each functional unit illustrated as elements of the conceptual model of Fig. 1.1.4. This would allow testing of the existence of general patterns in the constitution of functional units at each scale and evaluate auto-organized systems, their biodiversity and participation in the formation of soil physical structures and in the provision of ecosystem services at their scales.

Finally, an important issue is the adaptation of functional units observed at one scale to the adjacent ones. Where plants, micro-organisms, and fauna had coevolved for very long periods of time, the introduction of alien elements, like crop plants and introduced micro-organisms in managed systems, and disappearance of local fauna may change and/or impair functions based on interactions among these organisms (e.g. adaptation of local mycorrhizae to introduced plants or response of an introduced cultivar to local earthworm-induced changes in gene expression). Mismatch among elements artificially introduced may impair some ecosystem functions. Soil compaction by invasive earthworms in tropical pastures (Chauvel et al. 1999), or the low ability of some recent soybean cultivars to select for the most efficient rhizobium strands (Kiers et al. 2007), are example of such mismatches.

1.1.6.2 Composition of self-organized communities at a given scale

It is expected that long periods of coevolution have selected for the best combination of organisms, based on the general principle proposed by Allee *et al.* (1949) that interactions among organisms progressively shift from adverse to neutral and mutualist. The highest grade of interaction is represented by symbiogenesis, that is, the merging of several organisms into a single one that benefits from the abilities of each component (Mereschkowsky 1926; Margulis & Sagan 2002). Following this logic, plants must have coevolved, e.g. with mycorrhizal fungi that have become best adapted to a given plant population in a given environment. Bezemer *et al.* (2010) show that nematode communities in rhizospheres of eight different plant species are plant specific (self-organization at Scale 3) and also depend to a certain extent on composition of the plant community where these plant species grow (Scale 4). These examples show the interest of testing the existence of specific interactions among components of communities inside each scale and identifying the traits of organisms that make species (of plants, invertebrates, or micro-organisms) more or less compatible and mutually beneficial. Physiological and genetic traits are probably deeply involved in the quality of interspecific interactions. For example, positive effects of earthworms on plant growth may be mediated by specific bacterial colonies that release plant growth promoters in earthworm casts, with species-specific or generalist effects on plants (Tomati *et al.* 1988; Blouin *et al.* 2005). On the other hand, differences have been observed in the response of five rice cultivars to inoculation of the earthworm *Pontoscolex corethrurus* and biochar application (Noguera *et al.* 2011). This result leads to the question of whether the adaptation of plant cultivars to conditions of soils under conventional intensive practices has not led to the loss of specific genes involved in plant earthworm interactions. In similar conditions of intensive plant selection, Kiers *et al.* (2007) show that legumes tend to lose their defense against ineffective rhizobia. The quality of interactions inside each scale should be evaluated with specific indicators. They would be based on such attributes as the diversity of the different components (microbial and faunal communities in functional units at Scales 1 to 3) and their compatibility assessed by standard laboratory or field tests (e.g. response of plants to microbial or faunal communities present in the soil, length and complexity of food webs at Scale 2).

1.1.6.3 Physical structures as habitats for organisms

A second important attribute of the functional units represented in Fig. 1.1.4 is the creation of a set of structures that form habitats for the organisms (Ritz & Young 2011). These structures may be described with physical and chemical parameters such as size, shape, texture, nutrient and organic contents, and resistance to different types of physical aggressions, enzymatic activities, or near infrared spectrometric signals (Decaëns *et al.* 2001; Hedde *et al.* 2005).

Their localization and spatial arrangement, the importance, size distribution, and connectivity of the porous space that they produce are also important attributes to analyze, at the different scales from 1 to 4 at least. The mere existence of structures, embedded at nested spatial scales that form the fundament of the conceptual view, should be verified and patterns compared to search for general laws and indicators of their organization in different ecosystems. Composite indicators should characterize the habitats created at the different scales. The distribution and diversity of pore sizes and shapes and the average volume of interconnected pores of a given minimal size would evaluate the ability of soils to host organisms of different sizes and the actual volume of interconnected pores that they may physically access. Large volumes of interconnected pores would allow larger populations to live with greater possibilities for exchanging genetic material. Stability in time of these structures that determines the number of generations that would live in this space probably depends on their physical resistance and the rate at which new structures are produced by ecosystem engineers. Resilience of soil structures and patterns of biodiversity at the different scales would probably be related to these aspects of the composition and dynamics of their habitats.

1.1.6.4 Processes in functional units at different scales

An interesting consequence of self-organization in soils is the creation of discrete units where functions may operate with different and even opposite dynamics due to the diversity of habitats that they represent and of the resources that they offer. This characteristic allows, for example, keeping anaerobic microsites in a soil with an overall aerobic situation (Sextone *et al.* 1985), and get mineralization of P and N in microsites of soils with high average C:nutrient ratios. Earthworm casts, the unit structures of the drilosphere functional domain, present such conditions for the specific composition and organization of particles that they encompass. A practical output of the self-organized model is that assessment of microbial communities and activities done on bulk sieved soil may not provide a realistic assessment of activities. In this approach, all microbes are provided with optimal and homogeneous conditions for their activity where active and inactive microsites and also sites with possibly opposite patterns were organized in space and time. For example, microbial composition and activities in the rhizosphere varies according to plant species and in time, from the moment a root tip reaches the microsite, to different phases of a successional process (Garrett *et al.* 2001; Bonkowski 2004; Innes *et al.* 2004; Lambers *et al.* 2009).

1.1.6.5 Connection of functional units within and across scales

The image of connected units (numbered from 1 to 5 in Fig. 1.1.4) representing the soil functional units and their dynamics at different scales, suggests that connection among functional units may depend on their shape and the presence of lubricant to ease the transmission of the movement. Similar effects probably exist in soils, illustrated, for example, by the concept of synchrony in decomposition processes (Swift *et al.* 1979) that postulates that, in nutrient efficient systems, nutrient release from decomposition in natural systems is adjusted to the needs of plants at the same moment, thus preventing nutrient losses (Fig. 1.1.5). In this case, nutrients are released in microbial sites (Scale 1), possibly regulated by food web processes (Scale 2) inside the functional domain of a tree (leaf litter system, Scale 3) or an invertebrate ecosystem engineer (drilosphere of an earthworm population) and transferred to another functional domain, the rhizosphere of a plant. Synchrony in time and synlocalization in

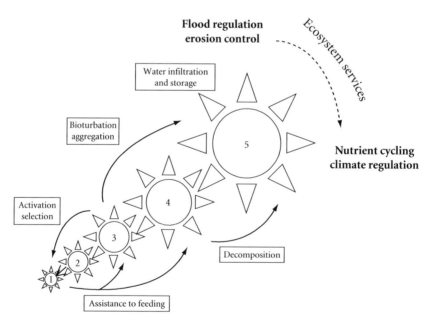

Figure 1.1.5 Intervention at Scale 3 (ecosystem engineers) determines habitats, communities and processes at Scales 1 (microbial biofilms and colonies) and 2 (micro-food webs); activities triggered at these scales enhance activities of ecosystem engineers at Scales 3 and 4 resulting in the provision of ecosystem services at Scale 5 through the control and operation of chemical and physical processes.

space (van Noordwijk *et al.* 1993) may be promoted by synchronous responses to major climatic drivers and the spatial organization that determines localization of organisms and the circulation of nutrients in water flows. Other such interfaces must be numerous and many of them are still largely unknown or ignored.

1.1.7 Approaches to soil ecological research

Life in soil only occurs in the porous space. Organisms that can dig create their proper space while smaller or weaker ones have to use porosity made by other living or physical agents or restrain their activity to the litter layer where circulation is easier. Our current ignorance of the spatial extent and structure of the porous space is a major limitation to the real understanding of organism interactions in soils. Not taking into account spatial limitations in soils has led research to underestimate isolation among organisms living in non-connected porous domains and the expected protection from predators or competitors that this isolation may provide (Postma & van Veen 1990; Crawford *et al.* 2005). This also ignores the effect of isolation from food sources and any mutual partnership that assists in solving this problem. Although it is intuitively perceived that general models of predator/prey relationships cannot fully apply to this situation of potential isolation, we have no idea of the extent to which this assumption is true and how important it is not to take it into consideration in general models of soil function. Research specially focused on this point would provide fast answers and open new areas in which to develop specific soil ecological theory.

Much soil ecology experimentation has actually been based on laboratory or small-scale field experiments (micro/mesocosms) since natural field observations and experimental designs, generally time consuming and costly, hardly provided conditions to test a single effect in isolation by an experimentation. These approaches provided a great deal of interesting results. However, they did not consider the soil physical organization and multiple biological interactions across space and time scales that likely make soils much different from other terrestrial and aquatic environments (Swift *et al.* 2004). Conditions of our experiments, with highly disturbed soils and modified communities, may have reproduced the ones that mostly occur at the very beginning of ecosystem adaptive cycles (as defined by Oldeman (1990) or Gunderson & Holling (2002)); in this phase of ecosystem reorganization that occurs after a strong disturbance or, for example, after the natural death of trees in forests, adverse relationships among organisms are expected to be more prevalent since mutualisms take longer periods to get organized than adverse conditions and involve more specialized organisms (Allee *et al.* 1949; Lavelle 1986).We may have thus only explored the part of soil function that follows general laws otherwise well described, e.g. in freshwater ecosystems; another possibly more important part, hidden in the always large part of the non-explained variance may follow patterns that are unique to the soil.

A new challenge is now to integrate the habitat constraints and their changes over time in our approaches, by testing hypotheses that had been verified in previous works in the real soil conditions, at the different scales proposed in the model shown in Fig. 1.1.4. Much remains to be done to adequately understand the soil environment, its opportunities, and constraints. This requires a renewed conceptual framework such as the one proposed by the self-organization theory and collaborations with molecular biologists, soil chemists, physicists, and hydrologists who also can apply their expertise to ecology driven questions.

1.1.8 Conclusions

The soil habitat has very specific characteristics that make it different from any other one found in the biosphere, with a possible exception of fresh water and marine sediments that have some commonalities (Wall *et al.* 2004). Low qualities of feeding resources and spatial constraints have stimulated the ecosystem engineering as a central process. While the physical effects of ecosystem engineers on smaller organisms have been reasonably explored, the effect of chemical mediators (e.g. root exudates, earthworm mucus, termite saliva), on other communities need more comprehensive description and interpretation. Accumulating evidence of genetic adjustment induced by organisms on others, whether through modifications in the expression of genes (Blouin *et al.* 2005; Jana *et al.* 2009) or by horizontal gene transfer (Smant *et al.* 1998; Wang *et al.* 2007; Hart *et al.* 2009; Sundaramurthy 2010; Pritchard 2011) opens wide perspectives for testing new soil ecological theory and improving agricultural practices. This biological engineering may turn out to be more important than the mere physical engineering. This integrated view of the soil functioning not only aims at bridging scales and disciplines inside soil science; the growing evidence that auto-organization is a powerful force in the development of all social, economic, and environmental domains sets new objectives and opens wide avenues for soil ecology, at considering external societal drivers of sustainability in addition to the biophysical factors.

References

Allee, W.C., Emerson, A.E., Park, O., Park, T., & Schmidt, K.P. (1949) *Principles of Animal Ecology*. W.B. Saunders, Philadelphia, PA.

Anderson, J.M. (1975) The enigma of soil animal species diversity. In: J. Vanek (eds.) *Progress in Soil Zoology*, pp. 51–8. Akademia, Praha.

André, H.M., Ducarme, X., & Lebrun, P. (2002) Soil biodiversity: myth, reality or conning? *Oikos* 96: 3–24.

Andrén, O., & Balandreau, J. (1999) Biodiversity and soil functioning—from black box to can of worms? *Applied Soil Ecology* 13: 105–8.

Angers, D.A., & Caron, J. (1998) Plant-induced changes in soil structure: Processes and feedbacks. *Biogeochemistry* 42: 55–72.

Ashelford, K.E., Learner, M.A., & Fry, J.C. (2001) Gene transfer and plasmid instability within pilot-scale sewage filter beds and the invertebrates that live in them. *FEMS Microbiology Ecology* 35: 197–205.

Baize, D., & Girard, M.C. (1995) *Référentiel pédologique*. INRA, Paris.

Bardgett, R., Usher, M., & Hopkins, D. (2005) *Biological diversity and function in soils*. Cambridge University Press, Cambridge.

Barois, I., & Lavelle, P. (1986) Changes in respiration rate and some physicochemical properties of a tropical soil during transit through *Pontoscolex corethrurus* (Glossoscolecidæ, Oligochæta). *Soil Biology & Biochemistry* 18 (5): 539–41.

Barros, E., Curmi, P., Hallaire, V., Chauvel, A., & Lavelle, P. (2001) The role of macrofauna in the transformation and reversibility of soil structure of an oxisol in the process of forest to pasture conversion. *Geoderma* 100: 193–213.

Bezemer, T.M., Fountain M.T., Barea, J.M., *et al.* (2010) Divergent composition but similar function of soil food webs of individual plants: plant species and community effects. *Ecology* 91: 3027–36.

Blanchart, E., Albrecht, A., Brown, G., *et al.* (2004) Effects of tropical endogeic earthworms on soil erosion. *Agriculture, Ecosystems and Environment* 104: 303–15.

Blanchart, E., Lavelle, P., Braudeau, E., Le Bissonais, Y., & Valentin, C. (1997) Regulation of soil structure by geophagous earthworm activities in humid savannas of Cote d'Ivoire. *Soil Biology & Biochemistry* 29: 431–9.

Blouin, M., Zuily-Fodil, Y., Pham-Thi, A.T., *et al.* (2005) Belowground organism activities affect plant aboveground phenotype, inducing plant tolerance to parasites. *Ecology Letters* 8: 202–8.

Bonkowski, M. (2004) Protozoa and plant growth: the microbial loop in soil revisited. *New Phytologist* 162: 617–31.

Brabant, P. (1991) *Le sol des forêts claires du Cameroun*. Orstom-Mesires. Paris.

Brady, N.C., & Weil, R.R. (2007) *The nature and properties of soils*. Pearson Prentice Hall, New York.

Braudeau, E., Sene, M., & Mohtar, R.H. (2005) Hydrostructural characteristics of two African tropical soils. *European Journal of Soil Science* 56: 375–88.

Brown, G.G., Barois, I., & Lavelle, P. (2000) Regulation of soil organic matter dynamics and microbial activity in the drilosphere and the role of interactions with other edaphic functional domains. *European Journal of Soil Biology* 36: 177–98.

Brussaard, L., Behan-Pelletier, V.M., Bignell, D.E., *et al.* (1997) Biodiversity and ecosystem functioning in soil. *Ambio* 26: 563–70.

Buol, S.W., Southard, R.J., Graham, R.C., & McDaniel, P.A. (2003) *Soil Genesis and Classification, 5th Edition*. Iowa State Press–Blackwell, Ames, IA.

Cabidoche, Y.M., & Guillaume, P. (1998) A casting method for the three-dimensional analysis of the intraprism structural pores in vertisols. *European Journal of Soil Science* 49: 187–96.

Capowiez, Y., Pierret, A., & Moran, C.J. (2003) Characterisation of the three-dimensional structure of earthworm burrow systems using image analysis and mathematical morphology. *Biology and Fertility of Soils* 38: 301–10.

Capowiez, Y., Sammartino, S., & Michell, E. (2011) Using X-ray tomography to quantify earthworm bioturbation non-destructively in repacked soil cores. *Geoderma* 162: 124–31.

Cerda, A., & Jurgensen, M.F. (2008) The influence of ants on soil and water losses from an orange orchard in eastern Spain. *Journal of Applied Entomology* **132**: 306–14.

Chan, T.P., & Govindaraju, R.S. (2011) Pore-morphology-based simulations of drainage and wetting processes in porous media. *Hydrology Research* **42**: 128–49.

Chauvel, A., Grimaldi, M., Barros, E., *et al.* (1999) Pasture damage by an Amazonian earthworm. *Nature* **398**: 32–3.

Chesworth, W. (1992) Weathering systems. In: Chesworth, & I. Martini. (eds.) *Weathering, Soils and Palaeosols*, pp. 19–40. Elsevier, Amsterdam.

Coleman, D.C., Crossley, D.A., & Hendrix, P.F. (2004) *Fundamentals of Soil Ecology*. Elsevier Academic Press, San Diego, CA.

Coleman, D.C., & Whitman, W.B. (2005) Linking species richness, biodiversity and ecosystem function in soil systems. *Pedobiologia* **49**: 479–97.

Colloff, M.J., Pullen, K.R., & Cunningham, S.A. (2010) Restoration of an ecosystem function to revegetation communities: The role of invertebrate macropores in enhancing soil water infiltration. *Restoration Ecology* **18**: 65–72.

Cortez, J., & Bouché, M.B. (1987) Composition chimique du mucus cutané de *Allolobophora chaetophora chaetophora* (Oligochaeta: Lumbricidae). *Comptes Rendus de l'Académie des Sciences, Paris* **305**: 207–10.

Crawford, J.W., Harris, J.A., Ritz, K., & Young, I.M. (2005) Towards and evolutionary ecology of life in soil. *Trends in Ecology & Evolution* **20**: 281–7.

Cresswell, H.P., & Kirkegaard, J.A. (1995) Subsoil Amelioration by Plant Roots—The Process and the Evidence. *Australian Journal of Soil Research* **33**: 221–39.

Davidson, T.M., Shanks, A.L., & Rumrill, S.S. (2010) The composition and density of fauna utilizing burrow microhabitats created by a non-native burrowing crustacean (*Sphaeroma quoianum*). *Biological Invasions* **12**: 1403–13.

De Ruiter, P.C., Neutel, A.M., & Moore, J.C. (1998) Biodiversity in soil ecosystems: the role of energy flow and community stability. *Applied Soil Ecology* **10**: 217–28.

Decaëns, T. (2010) Macroecological patterns in soil communities. *Global Ecology and Biogeography* **19**: 287–302.

Decaëns, T., Galvis, J.H., & Amezquita, E. (2001) Properties of the structures created by ecological engineers at the soil surface of a Colombian savanna. *Comptes Rendus De L Academie Des Sciences Serie III Sciences De La Vie Life Sciences* **324**: 465–78.

Decaëns, T., & Rossi, J.P. (2001) Spatio-temporal structure of earthworm community and soil heterogeneity in a tropical pasture. *Ecography* **24**: 671–82.

Elridge, D.J., & Greene R.S.B. (1994) Microbiotic soil crusts: a review of their roles in soil and ecological processes in the rangelands of Australia. *Australian Journal of Soil Research* **32**: 389–415.

Fedotov, G.N., Tretyakov, Y.D., Putliyaev, V.I., Pakhomov, E.I., Kuklin, A.I., & Islamov, A.K. (2007) Origin of fractal organization in soil colloids. *Doklady Chemistry* **412**: 55–8.

Fitter, A.H., Gilligan, C.A., Hollingworth, K., Kleckzowski, A., Twyman, R.M., & Pitchford, J.W. (2005) Biodiversity and ecosystem function in soil. *Functional Ecology* **19**: 369–77.

Fontaine, S., Barot, S., Barré, P., Bdioui, N., Mary, B., & Rumpel, C. (2007) Stability of organic carbon in deep soil layers controlled by fresh carbon supply. *Nature* **450**: 277–80.

Fonte, S.J., Kong, A.Y.Y., van Kessek, C., Hendrix, P.F., & Six, J. (2007) Influence of earthworm activity on aggregate-associated carbon and nitrogen dynamics differs with agroecosystem management. *Soil Biology and Biochemistry* **39**: 1014–22.

François, C., Rafidison, Z., Villemin, G., & Toutain, F. (1986) The accumulation and fate of brown pigments in leaves from a litter of *Fagus sylvatica* L.: a morphological and chemical study. In: G. Giovannozzi-Sermanni & P. Nannipieri (eds.) *Current Perspectives in Environmental Biogeochemistry*, pp. 317–27. CNR-IPRA, Rome.

Fyfe, W.S., Kronberg, B.I., Leonardos, O.H., & Olorunfemi, N. (1983) Global tectonics and agriculture: a geochemical perspective. *Agriculture, Ecosystems and Environment* **9**: 383–99.

Garrett, C.J., Crossley, D.A., Coleman, D.C., Hendrix, P.F., Kisselle, K.W., & Potter, R.L. (2001) Impact of the rhizosphere on soil microarthropods in agroecosystems on the Georgia piedmont. *Applied Soil Ecology* **16**: 141–8.

Gignoux, J., House, J., Hall, D., Masse, D., Nacro, H.B., & Abbadie, L. (2001) Design and test of a generic cohort model of soil organic matter decomposition: the SOMKO model. *Global Ecology and Biogeography* **10**: 639–60.

Globus, A.M. (2006) Fractal character of some physical parameters of soils. *Eurasian Soil Science* **39**: 1116–26.

Gregory, P.J. (2006) Roots, rhizosphere and soil: the route to a better understanding of soil science? *European Journal of Soil Science* **57**: 2–12.

Grimaldi, M., Sarrazin, M., Chauvel, A., *et al.* (1993) Effets de la déforestation et des cultures sur la structure des sols argileux d'Amazonie brésilienne. *Cahiers Agriculture* **2**: 36–47.

Guillaume, K., Huard, M., Gignoux, J., Mariotti, A., & Abbadie, L. (2001) Does the timing of litter inputs determine natural abundance of C-13 in soil organic matter? Insights from an African tiger bush ecosystem. *Oecologia* **127**: 295–304.

Gunderson, L.H., & Holling, C.S. (2002) *Panarchy. Understanding transformations in human and natural systems.* Island Press, New York.

Handley, W.R.C. (1961) Further evidence for the importance of residual protein complexes on litter decomposition and the supply of N for plant gowth. *Plant and Soil* **15**: 37–73.

Harborne, J.B. (1990) Role of secondary metabolites in chemical defense mechanisms in plants. *Ciba Foundation Symposia* **154**: 126–39.

Hart, M.M., Powell, J.R., Gulden, R.H., et al. (2009) Detection of transgenic cp4 epsps genes in the soil food web. *Agronomy for Sustainable Development* **29**: 497–501.

Hättenschwiler, S., Aeschlimann, B., Coûteaux, M.M., Roy, J., & Bonal, D. (2008) High variation in foliage and leaf litter chemistry among 45 tree species of a neotropical rainforest community. *New Phytologist* **179**: 165–75.

Hättenschwiler, S., & Vitousek, P.M. (2000) The role of polyphenols in terrestrial ecosystem nutrient cycling. *Trends in Ecology & Evolution* **15**: 238–43.

Hattori, T., & Hattori, R. (1976) The physical environment in soil microbiology. An attempt to extend principles of microbiology to soil micro-organisms. *Critical Reviews in Microbiology* **26**: 423–61.

Hedde, M., Lavelle, P., Joffre, R., Jimenez, J.J., & Decaëns, T. (2005) Specific functional signature in soil macroinvertebrate biostructures. *Functional Ecology* **19**: 785–93.

Hendrix, P. F. (2006) Biological invasions belowground—earthworms as invasive species. *Biological Invasions* **8**: 1201–4.

Hillel, D. (1980) Additional factors affecting aggregation. In: *Fundamentals of Soil Physics*, pp. 101–2. Academic Press, New York.

Högberg, P., & Read, D.J. (2006) Towards a more plant physiological perspective on soil ecology. *Trends in Ecology & Evolution* **21**: 548–54.

Hol, W.H.G., Macel, M., Van Veen, J.A., & Van Der Meijden, E. (2004) Root damage and aboveground herbivory change concentration and composition of pyrrolizidine alkaloids of Senecio jacobaea. *Basic and Applied Ecology* **5**: 253–60.

Innes, L., Hobbs, P.J., & Bardgett, R.D. (2004) The impacts of individual plant species on rhizosphere microbial communities in soils of different fertility. *Biology and Fertility of Soils* **40**: 7–13.

Jana, U., Barot, S., Blouin, M., Lavelle, P., Laffray, D., & Reppellin, A. (2009) Earthworms influence the production of above- and belowground biomass and the expression of genes involved in cell proliferation and stress responses in Arabidopsis thaliana. *Soil Biology and Biochemistry* **42**: 244–52.

Jarvis, N.J., Taylor, A., Larsbo, M., Etana, A., & Rosén, K. (2010) Modelling the effects of bioturbation on the redistribution of ^{137}Cs in an undisturbed grassland soil. *European Journal of Soil Science* **61**: 24–34.

Jegou, D., Schräder, S., Diestel, H., & Cluzeau, D. (2001) Morphological, physical and biochemical characteristics of burrow walls formed by earthworms. *Applied Soil Ecology* **17**: 165–74.

Jenkinson, D.S., Hart, P.B.S., Rayner, J.H., & Parry, L.C. (1987) Modelling the turnover of organic matter in a long-term experiment of Rothamsted. *Intecol Bulletin* **15**: 1–8.

Jimenez, J.J., Brown, G.G., Decaëns, T., Feijoo, A., & Lavelle, P. (2000) Differences in the timing of diapause and patterns of aestivation in tropical earthworms. *Pedobiologia* **44**: 677–94.

Jimenez, J.J., Rossi, J.P., & Lavelle, P. (2001) Spatial distribution of earthworms in acid-soil savannas of the eastern plains of Colombia. *Applied Soil Ecology* **17**: 267–78.

Jones, C.G., Gutierrez, J.L., Groffman, P.M., & Shachak, M. (2006) Linking ecosystem engineers to soil processes: a framework using the Jenny State Factor Equation. *European Journal of Soil Biology* **42**: S39–53.

Jones, C.G., Lawton, J.H., & Shachak, M. (1994) Organisms as ecosystem engineers. *Oikos* **69**: 373–86.

Jouquet, P., Dauber, J., Lagerlöf, J., Lavelle, P., & Lepage, M. (2006) Soil invertebrates as ecosystem engineers: Intended and accidental effects on soil and feedback loops. *Applied Soil Ecology* **32**: 153–64.

Kiers, E.T., Hutton, M.G., & Denison, R.F. (2007) Human selection and the relaxation of legume defences against ineffective rhizobia. *Proceedings of the Royal Society B: Biological Sciences* **274**: 3119–26.

Krestzschmar, A. (1982) Description des galeries de vers de terre et variations saisonnières des réseaux (observations en conditions naturelles). *Revue d'Ecologie et de Biologie du Sol* **19**: 579–91.

Kuzyakov, Y. (2002) Review: Factors affecting rhizosphere priming effects. *Journal of Plant Nutrition and Soil Science-Zeitschrift Fur Pflanzenernahrung Und Bodenkunde* **165**: 382–96.

Kuzyakov, Y., Friedel, J.K., & Stahr, K. (2000) Review of mechanisms and quantification of priming effects. *Soil Biology and Biochemistry* **32**: 1485–98.

Lambers, H., Mougel, C., Jaillard, B., & Hinsinger, P. (2009) Plant-microbe-soil interactions in the rhizosphere: an evolutionary perspective. *Plant and Soil* **321**: 83–115.

Lattaud, C., Mora, P., Garvin, M., LocatiI, S., & Rouland, C. (1999) Enzymatic digestive capabilities in geophagous earthworms—origin and activities of cellulolytic enzymes. *Pedobiologia* **43**: 842–50.

Lavelle, P. (1983) The structure of earthworm communities. In: J.E. Satchell (ed.) *Earthworm Ecology: from Darwin to Vermiculture* pp. 449–66. Chapman and Hall, London.

Lavelle, P. (1986) Mutualism with soil microflora and species richness in the tropics—the 1st link hypothesis. *Comptes Rendus de l'Académie Des Sciences Serie III* **302**: 11–14.

Lavelle, P. (2002) Functional domains in soils. *Ecological Research* **17**: 441–50.

Lavelle, P. (2009) Ecology and the challenge of a multifunctional use of soil. *Pesquisa Agropecuaria Brasileira* **44**: 803–10.

Lavelle, P., Blanchart, E., Martin, A., *et al.* (1993) A hierarchical model for decomposition in terrestrial ecosystems: application to soils of the humid tropics. *Biotropica* **25**: 130–50.

Lavelle, P., Bignell, D., Lepage, M., *et al.* (1997) Soil function in a changing world: the role of invertebrate ecosystem engineers. *European Journal of Soil Biology* **33**: 159–93.

Lavelle, P., Bignell, D.E., Austen, M.C., *et al.* (2004a) Connecting soil and sediment biodiversity: The role of scale and implications for management. *Sustaining Biodiversity and Ecosystem Services in Soils and Sediments* **64**: 193–224.

Lavelle, P., Blouin, M., Boyer, J., *et al.* (2004b) Plant parasite control and soil fauna diversity. *Comptes Rendus Biologies* **327**: 629–38.

Lavelle, P., Decaëns, T., Aubert, M., *et al.* (2006) Soil invertebrates and ecosystem services. *European Journal of Soil Biology* **42**: 3–15.

Lavelle, P., Lattaud, C., Trigo, D., & Barois, I. (1995) Mutualism and biodiversity in soils. *Plant and Soil* **170**: 23–33.

Lavelle, P., & Spain, A.V. (2001) *Soil Ecology.* Amsterdam, Kluwer Scientific Publications.

Lavelle, P., Veiga, I., Ramirez, B., *et al.* (2010) Socioeconomic determinants of landscapes and consequences for biodiversity and the provision of ecosystem goods and services in the Arch of deforestation of Amazonia. (online) http://ciat-library.ciat.cgiar.org/Articulos_CIAT/Rapport_finalANR_AMAZ_BD.pdf.

Leys, J.F., & Elridge, D.J. (1998) Influence of cryptogamic crusts disturbance to wind erosion on sand and loam rangeland soils. *Earth Surface Processes and Landforms* **23**: 963–74.

Lobry De Bruyn, L.A., & Conacher, A.J. (1990) The role of termites and ants in soil modification: A review. *Australian Journal of Soil Research* **28**: 55–93.

Mando, A. (1997) Effect of termites and mulch on the physical rehabilitation of structurally crusted soils in the Sahel. *Land Degradation and Development* **8**: 269–78.

Margulis, L., & Sagan, D. (2002) *Acquiring Genomes: A Theory of the Origins of Species.* Perseus Books Group, Amherst, MA.

Marichal, R., Mathieu, J., Couteaux, M.M., Mora, P., Roy, J., & Lavelle, P. (2011) Earthworm and microbe response to litter and soils of tropical forest plantations with contrasting C:N:P stoichiometric ratios. *Soil Biology and Biochemistry* **43**: 1528–35.

Martin, A. (1991) Short- and long-term effects of the endogeic earthworm *Millsonia anomala* (Omodeo) (Megascolescidae, Oligochaeta) of tropical savannas, on soil organic matter. *Biology and Fertility of Soils* **11**: 134–8.

Martin A, Cortez, J., Barois, I., & Lavelle, P. (1987) Les mucus intestinaux de Ver de Terre, moteur de leurs interactions avec la microflore. *Revue d'Ecologie et de Biologie du Sol* **24**: 549–58.

Menendez, I., Caniego, J., Gallardo, J.F., & Olechko, K. (2005) Use of fractal scaling to discriminate between and macro- and meso-pore sizes in forest soils. *Ecological Modelling* **182**: 323–35.

Mereschkowsky, K. (1926) *Symbiogenesis and the Origin of Species.* T.H. Morgan.

Muhammad, S., Jöergensen, R.G., Mueller, T., & Muhammad, T.S. (2007) Priming mechanism: Soil amended with crop residue. *Pakistan Journal of Botany* **39**: 1155–60.

Nahmani, J., Capowiez, Y., & Lavelle, P. (2005) Effects of metal pollution on soil macroinvertebrate burrow systems. *Biology and Fertility of Soils* **42**: 31–9.

Nguyen, C. (2003) Rhizodeposition of organic C by plants: mechanisms and controls. *Agronomie* **23**: 375–96.

Noguera, D., Laossi, K.R., Lavelle, P., *et al.* (2011) Amplifying the benefits of agroecology by using the right cultivars. *Ecological Applications* **21**: 2349–56.

Oldemann, R.A.A. (1990) *Forests: elements of silvology.* Springer Verlag, Berlin.

Osler, G.H.R., & Sommerkorn, M. (2007) Toward a complete soil C and N cycle: Incorporating the soil fauna. *Ecology* **88**: 1611–21.

Pachepsky, E., Crawford, J.W., Bown, J.L. and Squire, G. (2001) Towards a general theory of biodiversity. *Nature* **410**: 923–6.

Palm, C.A., & Sanchez, P.A. (1991) Nitrogen relaase from the leaves of some tropical legumes as affected by their lignin and polyphenolic contents. *Soil Biology & Biochemistry* **23**: 83–8.

Papendick, R.I., & Campbell, G.S. (1981) Theory and measurement of water potential. In: J.F. Parr, W.R. Gardner and L.F. Elliot (eds.) *Water Potential Relations in Soil Microbiology,* pp. 1–22. Soil Science of America, Madison, WI.

Parton, W.J., Anderson, D.W., Cole, C.V., & Stewart, J.W.B. (1983) Simulation of soil organic matter formation and mineralization in semiarid agroecosystems. In: R.R. Lowrance, R.L. Todd, L.E. Asmussen & R.A. Leonard (eds.) *Nutrient Cycling in Agricultural Ecosystems.* Special Publication n° 23 of the Georgia Experimental Station. Athens, GA.

Perry, D.A. (1995) Self-organizing systems across scales. *TREE* **10**: 241–5.

Pierret, A., Capowiez, Y., Belzunces, L., & Moran, C.J. (2002) 3D reconstruction and quantification of macropores using X-ray computed tomography and image analysis. *Geoderma* **106**: 247–71.

Postma, J., & van Veen, J.A. (1990) Habitable pore space and survival of Rhizobium leguminosarum biovar trifolii introduced into soil. *Microbial Ecology* **19**: 149–61.

Pritchard, S.G. (2011) Soil organisms and global climate change. *Plant Pathology* **60**: 82–99.

Rieu, M., & Sposito, G. (1991) Fractal fragmentation, soil porosity, and soil-water properties. 1. Theory. *Soil Science Society of America Journal* **55**: 1231–8.

Ritz, K., & Young, I. (eds.) (2011) *Architecture and Biology of Soils. Life in Inner Space.* CAB International, Wallingford.

Rossi, J.P. (2003) The spatiotemporal pattern of a tropical earthworm species assemblage and its relationship with soil structure. *Pedobiologia* **47**: 497–503.

Rossi, J.P., Albrecht, A., & Lavelle, P. (1996) Relationships between spatial pattern of the endogeic earthworm Polypheretima elongata and soil heterogeneity in a tropical pasture of Martinique (French West Indies). *Soil Biology and Biochemistry* **29**: 485–8.

Sabatier, D., Grimaldi, M., Prévost, M.F., *et al.* (1997) The influence of soil cover organization on the floristic and structural heterogeneity of a Guianan rain forest. *Plant Ecology* **131**: 81–108.

Schweitzer, J.A., Bailey, J.K., Rehill, B.J., *et al.* (2004) Genetically based trait in a dominant tree affects ecosystem processes. *Ecology Letters* **7**: 127–34.

Sextone, A.J., Revsbeich, N.P., Parkin, T.B., & Tiedje, J.M. (1985) Direct measurement of oxygen profiles and denitrification rates in soil aggregates. *Soil Science Society of America Journal* **49**: 645–51.

Smant, G., Stokkermans, J., Yan, Y.T., *et al* (1998) Endogenous cellulases in animals: Isolation of beta-1,4-endoglucanase genes from two species of plant-parasitic cyst nematodes. *Proceedings of the National Academy of Sciences of the United States of America* **95**: 4906–11.

Smith, J.U., Smith, P., Monaghan, R., & MacDonald, J. (2002) When is a measured soil organic matter fraction equivalent to a model pool? *European Journal of Soil Science* **53**: 405–16.

Smith, P., Smith, J.U., Powlson, D.S., *et al.* (1997) A comparison of the performance of nine soil organic matter models using datasets from seven long-term experiments. *Geoderma* **81**: 153–225.

Soil Survey Staff (1999) *Soil Taxonomy: A Basic system of soil Classification for making and interpreting Soil Survey.* (2nd Ed.) U.S. Department of Agriculture, Soil Conservation Service, Washington DC.

Sterner, R.W., & Elser, J.J. (2002) *Ecological stoichiometry. The Biology of elements from molecules to the biosphere.* Princeton University Press, Princeton, NJ.

Sundaramurthy, V.T. (2010) The impacts of the transgenes on the modified crops, non-target soil and terrestrial organisms. *African Journal of Biotechnology* **9**: 9163–76.

Swift, M.J., Heal, O.W., & Anderson, J.M. (1979) *Decomposition in terrestrial ecosystems*, Blackwell Scientific, Oxford.

Swift, M.J., Izac, A.M.N., & Van Noordjwik, M. (2004) Biodiversity and ecosystem services in agricultural landscapes—are we asking the right questions? *Agriculture Ecosystems & Environment* **104**: 113–34.

Tian, G., Brussaard, L., & Kang, B.T. (1995) Breakdown of plant residues with contrasting chemical compositions; effects of earthworms and millipedes. *Soil Biology and Biochemistry* **27**: 731–7.

Tomati, U., Grappeli, A., & Galli, E. (1988) The hormone-like effect of earth worm casts on plant growth. *Biology and Fertility of Soils* **5**: 288–94.

Toutain, F. (1987) Les litières: sièges de systèmes interactifs et moteurs de ces interactions. *Revue d'Ecologie et de Biologie du Sol* **24**: 231–42.

Trigo, D., Barois, I., Garvin, M.H., Huerta, E., Irisson, S., & Lavelle, P. (1999) Mutualism between earthworms and soil microflora. *Pedobiologia* **43**: 866–73.

Van Noordwijk, M., De Ruiter, P. C., Zwart, K. B., *et al.* (1993) Synlocation of biological activity, roots, cracks and recent organic inputs in a sugar beet field. *Geoderma,* **56:** 265–76.

Vogel, H.J., & Roth, K. (2001) Quantitative morphology and network representation of soil pore structure. *Advances in Water Resources* **24**: 233–42.

Wall, D.H. (2004) *Sustaining biodiversity and ecosystem services in soils and sediments.* Island Press, Washington, DC.

Wall, D.H., Bardgett, R.D., Covich, A.P., & Snelgrove, P.V.R. (2004) Understanding the functions of biodiversity in soils and sediments will enhance global ecosystem sustainability and societal well-being. *Sustaining Biodiversity and Ecosystem Services in Soils and Sediments* **64**: 249–54.

Wang, Y.J., Xiao, M., Geng, X.L., Liu, J.Y., & Chen, J. (2007) Horizontal transfer of genetic determinants for degradation of phenol between the bacteria living in plant and its rhizosphere. *Applied Microbiology and Biotechnology* **77**: 733–9.

Wichern, F., Eberhardt, E., Mayer, J., Jöergensen, R.G., & Muller, T. (2008) Nitrogen rhizodeposition in agricultural crops: Methods, estimates and future prospects. *Soil Biology and Biochemistry* **40**: 30–48.

Wolters, V. (2001) Biodiversity of soil animals and its function. *European Journal of Soil Biology* **37**: 221–7.

Yair, A. (1995) Short and long-term effects of bioturbation on soil-erosion, water-resources and soil development in an arid environment. *Geomorphology* **13**: 87–99.

Yong, R.N., & Warkentin, B.P. (1975) *Soil Properties and Behaviour*. Amsterdam, Elsevier.

Young, I.M., & Crawford, J.W. (2004) Interactions and self-organization in the soil-microbe complex. *Science* **304**: 1634–7.

Young, I.M., Crawford, J.W., Nunan, N., Otten, W., & Spiers, A. (2008) Microbial Distribution in Soils: Physics and Scaling. *Advances in Agronomy* **100**: 81–121.

Zangerlé, A., Pando, A., Lavelle, P. (2011) So earthworms and roots cooperate to build soil macroaggregates? A microcosm experiment. *Geoderma* **167–68**: 303–309.

Zhang, C., Langlest, R., Velasquez, E., Pando, A., Brunet, D., Dai, J., & Lavelle, P. (2009) Cast production and NIR spectral signatures of *Aporrectodea caliginosa* fed soil with different amounts of half-decomposed *Populus nigra* litter. *Biology and Fertility of Soils* **45**: 839–44.

CHAPTER 1.2

Soil Biodiversity and Functions

Susanne Wurst, Gerlinde B. De Deyn, and Kate Orwin

1.2.1 Soil biodiversity

1.2.1.1 Introduction

Soil harbors a huge variety of organisms, many of them still unknown. A major constraint of studying soil organisms, in addition to limited taxonomic knowledge, is their physical inaccessibility. This is due to the complex spatial structure of the soil. The sizes of soil organisms range from <100 μm body width for the microbes (e.g. bacteria, archaea, and fungi) and microfauna (e.g. protozoa, nematodes), to mesofauna with body width between 100 μm and 2 mm (e.g. Collembola, mites), macrofauna with body widths >2 mm (e.g. earthworms, myriapods, and insects) and up to megafauna with body width >2 cm (e.g. moles, voles). As discussed in Chapter 1.1 (Lavelle), soil structure is a major driver of adaptations of individual soil organisms, but soil organisms also influence soil structure. The chemical, physical, and spatial heterogeneity of soil as a habitat and the multiple adaptations of soil organisms to this complex environment may be the primary cause of their great diversity.

In this chapter, we will discuss diversity and functions of soil organisms within and across size classes in more detail. We follow the classical approach of grouping soil organisms according to size, since this is often linked to how they experience and change the chemical and physical properties of their habitat, as well as how they are typically studied. We introduce the main functions of the soil organisms, and present their known and estimated species richness. We also highlight that ecosystem processes and functions, as well as the activity and survival of many soil organisms depend on the presence of other soil biota, which can act at different spatial scales (Fig. 1.2.1).

1.2.1.2 Microbes

Soil microbes have body widths of <100 μm and are amongst the most abundant and diverse groups of soil organisms (Table 1.2.1), with a single gram of soil estimated to contain tens of thousands of species (Fierer *et al.* 2007a; Roesch *et al.* 2007b). The most studied microbes are bacteria and fungi. Bacteria are single-celled prokaryotes that require soil water films to live and move within the soil matrix, whereas filamentous fungi are less constrained in this manner and can cross air-filled pore spaces (Ritz 2007). At least 25 different bacterial phyla have been found in soil to date, but most soils appear to be strongly dominated by bacteria from the Acidobacteria, Proteobacteria, Actinobacteria, Bacteriodes, and Firmicutes phyla (Lauber *et al.* 2009). Fungal species in soil derive from a range of phyla, including the Basidiomycota, Ascomycota, and Glomeromycota. Soil also contains a high abundance and diversity of archaea and viruses (Fierer *et al.* 2007b; Table 1.2.1). Archaea are single-celled prokaryotes but are distinct from bacteria in their evolutionary history. Viruses are typically extremely small, and consist of nucleic acids surrounded by a protein coat.

Soil microbes contribute to many essential ecosystem functions, including decomposition, carbon and nutrient cycling, disease suppression, and regulation of plant growth and primary productivity (Fig. 1.2.1). Bacteria and fungi play a major role in soil as the primary degraders of organic matter,

Soil Ecology and Ecosystem Services. First Edition. Edited by Diana H. Wall *et al.*
© 2012 Oxford University Press. Published 2012 by Oxford University Press.

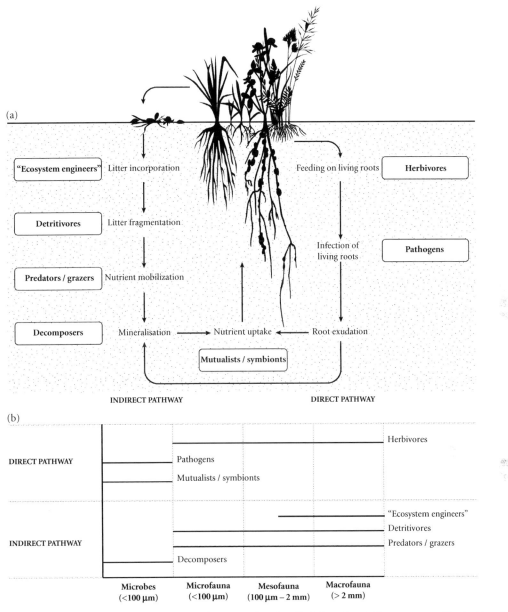

Figure 1.2.1 a) Soil biota belong to different functional groups (i.e. groups of species with similar traits and effects on processes) involved in carbon and nutrient mobilization from litter, i.e. dead plant residues ("indirect pathway") and from living plant roots ("direct pathway"). Soil biota have complementary functions and their interactions often increase process rates. b) Some functional groups are restricted to one size class (the microbes), while other functional groups such as detritivores, herbivores, and predators/grazers involve organisms of the micro-, meso-, and macrofauna.

which determines both the rate at which nutrients become available to plants and the amount of carbon stored in soils. Microbes are particularly abundant and active in the rhizosphere, where they influence plant growth through multiple mechanisms. Over 80% of plant species form root associations with mycorrhizal fungi, which provide the plant with essential nutrients for growth in exchange for plant photosynthates (Smith & Read 2008). Some plant species host nitrogen-fixing bacteria in their

Table 1.2.1 Soil biota groups based on body size, their global diversity, and main food sources

Size class	Organism group	Global no. of described species[a]	% known of expected	Main food source	Refs
Microbes	*Viruses*	5,000	5		1
	Archaea	180			2
	Bacteria	10,000	1	Detritus, roots	2, 3, 4
	Fungi	72,000	5	Detritus, roots	2, 5
Microfauna	*Nematodes*	25,000	6	Roots, bacteria, fungi, soil fauna	2, 6, 7, 8
	Protozoa	40,000	20	Bacteria, fungi	1, 5
	Rotifera	2,000	Low	Detritus, algae, bacteria	2
Mesofauna	*Acari (mites)*	45,000	4	Plants, bacteria, fungi, soil fauna	7, 9, 10
	Collembola	8,000	15	Fungi, roots, algae	9, 11, 12
	Tardigrada	930		Plants, bacteria, fungi, soil fauna	13, 14
	Protura	500		Detritus, fungi	13
	Diplura	800		Soil fauna, plants	13
	Enchytraeidae	700	10	Detritus, bacteria, fungi	15, 16
Macrofauna	*Isoptera (termites)*	2,700	70	Wood	2, 17, 18
	Formicidae (ants)	14,000	50	Plants, fungi, soil fauna	19
	Isopoda	5,000[b]		Detritus	13
	Chilopda	3,200	46	Soil fauna	20
	Diplopoda	11,000	15	Detritus	20
	Coleoptera	350,000	30	Soil fauna, roots, detritus	21, 22
	Earthworms	3,500	50	Detritus	23, 24
	Araneae (spiders)	38,000		Soil fauna	25

Empty cells indicate that the information is either unknown or not in the public domain.

[a] To date based upon reference sources cited, and rounded to two significant figures where >1000.
[b] not including marine species. Global number of described species and expected numbers from ref. 2 include terrestrial, fresh water and marine species.

[1] Cox, C.B. & Moore, P.D. (2005) Patterns of biodiversity. In: C.B. Cox & P.D. Moore (eds.) Biogeography, An Ecological and Evolutionary Approach (7th edn), pp. 45–71. Blackwell, Oxford.
[2] Groombridge, B. & Jenkins, M.B. (2002) *World Atlas of Biodiversity: Earth's Living Resources in the 21st Century*. University of California Press, Berkeley, CA.
[3] Finlay, B.J. & Clarke, K.J. (1999) Ubiquitous dispersal of microbial species. *Nature* **400**: 828.
[4] Horner-Devine, M.C., Carney, K.M., & Bohannan, B.J.M. (2004) An ecological perspective on bacterial biodiversity. *Proceedings of the Royal Society of London B* **271**: 113–22.

5 Finlay, B.J. (2002) Global dispersal of free-living microbial eukaryote species. *Science* **296**: 1061–3.
6 Procter, D.L.C. (1984) Towards a biogeography of free-living soil nematodes. I. Changing species richness, diversity and densities with changing latitude. *Journal of Biogeography* **11**: 103–17.
7 Hillebrand, H. (2004) On the generality of the latitudinal diversity gradient. *American Naturalist* **163**: 192–211.
8 Boag, B. & Yeates, G.W. (1998) Soil nematode biodiversity in terrestrial ecosystems. *Biodiversity and Conservation* **7**: 617–30.
9 Courtright, E.M., Wall, D.H., & Virginia, R.A. (2001) Determining habitat suitability for soil invertebrates in an extreme environment: the McMurdo Dry Valleys, Antarctica. *Antarctic Science* **13**: 9–17.
10 Walter, D.E. & Procter, H.C. (eds.) (1999) *Mites: Ecology, Evolution and Behaviour.* CABI Publishing, Wallingford.
11 Hopkin, S.P. (1998) Collembola: the most abundant insects on earth. *Antenna* **22**: 117–21.
12 http://www.collembola.org
13 http://tolweb.org (Tree of life web project)
14 Hebert, P. & Bledzki, L.A. (2008. Tardigrada. In: C.J. Cleveland (ed.) *Encyclopedia of Earth.* Environmental Information Coalition, National Council for Science and the Environment, Washington, DC. http://www.eoearth.org/article/Tardigrada
15 Briones, MJI (2006) Enchytraeidae. In: R. Lal (ed.) *Encyclopedia of Soil Science,* pp. 514–18. Taylor and Francis, New York.
16 M.J.I Briones, personal communication.
17 Hawkins, B.A., Field, R., Cornell, H.V., *et al.* (2003) Energy, water, and broad-scale geographic patterns of species richness. *Ecology* **84**: 3105–17.
18 Kambhampati, S. & Eggleton, P. (2000) Taxonomy and phylogeny of termites. In: T. Abe, D.E. Bignell, & M. Higashi (eds.) *Termites: Evolution, Sociality, Symbiosis, Ecology,* pp. 1–25. Kluwer Academic Publishers, Dordrecht.
19 http://www.antweb.org
20 http://www.mnhn.fr/assoc/myriapoda/KEYCLE.HTM
21 Gaston, K.J. (1991) The magnitude of global insect species richness. *Conservation Biology* **5**: 283–96.
22 http://www.coleoptera.org
23 Lavelle, P., Lattaud, C., Trigo, D., and Barois, I. (1995) Mutualism and biodiversity in soils. *Plant and Soil* **170**: 23–33.
24 P. Lavelle, personal communication.
25 Platnick, N.I. (2003) *The world spider catalog, version* 3.5. American Museum of Natural History. http://research.amnh.org/entomology/spiders/catalog81-87/index.html

roots, which provide an important source of soil nitrogen (an element that frequently limits plant growth) in many systems (van der Heijden et al. 2008). Other rhizosphere bacteria stimulate plant growth through, for example, the production and regulation of plant hormones, and by acting as biocontrol agents of plant diseases (Lugtenberg & Kamilova 2009). Soil microbes can also suppress plant growth through pathogenic interactions; indeed, microbial pathogens have recently been implicated in driving reduced plant community productivity at low plant species diversity (Maron et al. 2011). The role of archaea and viruses in ecosystem functioning is less understood, but it is known that archaea produce methane and contribute to nitrification in soil (Philippot et al. 2009), and that viruses can influence the population sizes of plant pathogens and beneficial bacteria, and may play a role in the biogeochemical cycle through causing microbial cell lysis (Kimura et al. 2008).

1.2.1.3 Microfauna

Soil microfauna comprise soil animals with body widths <100 μm, and the most abundant groups are Nematoda, Protozoa, and Rotifera (Swift et al. 1979). Amongst these three groups, many thousands of species are known globally, but it is expected that these are still only a fraction of the number of species actually present on the globe (Table 1.2.1). Nematodes are small roundworms that live in soil water films or plant roots. Nematodes differ widely in their feeding strategy, and they are present at all levels in the food web. Nematodes have specialized mouthparts which are indicative of their feeding habits, hence whether they are plant feeders, bacterivores, fungivores, carnivores, or omnivores (Yeates et al. 1993).

Protozoa are single-celled eukaryotes and are reliant upon water-filled soil pores in order to move. They comprise four main morphologically distinct groups: flagellates, ciliates, naked amoeba, and testate amoeba; only the latter have the ability to form an outer (protective) shell made of silica, soil particles, or calcium carbonate (Foissner 1999). Most protozoa feed on bacteria, but fungal feeders, predators, and saprophytic species of protozoa also exist.

Rotifera are multicelled animals, often globular in shape. Many species possess cilia that help with the acquisition of food and movement through soil in water films. Most rotifer species feed on bacteria and algae, but some species are predators.

Soil microfauna support several ecosystem functions. Their best known function is the promotion of nutrient cycling by feeding on various food sources and the release of nutrients via their excrements. The majority of microfauna feed on bacteria, fungi, and (dead) plant material (Table 1.2.1; Fig. 1.2.1). In doing so, they can regulate the population size and activity of soil microbes and can promote the competitive ability and dispersal of beneficial rhizosphere microbiota by selective grazing on detrimental soil microorganisms (Scheu et al. 2005; Bonkowski et al. 2009). Microfauna that live in the rhizosphere of plants can also affect plant productivity and composition by feeding on plant roots (De Deyn et al. 2003) and by altering the production of plant hormones and defences (Bonkowski et al. 2009). Soil microfauna are themselves a food source for other soil organisms, such as soil mesofauna, and once dead, saprophytic soil microbes (Piskiewicz et al. 2008). The population size and diversity of microfauna can thus be controlled top-down by other soil biota, bottom-up by the host plant and their mutualists, or through competitive interactions (van der Putten et al. 2006). Finally, soil microfauna, such as entomopathogenic nematodes, can contribute to insect pest suppression. These nematodes inject their endosymbiotic bacteria into insect larvae, where the bacteria kill and pre-digest the insect host which then serves as food for the nematodes.

1.2.1.4 Mesofauna

Soil mesofauna are soil animals with body widths between 100 μm and 2 mm (Swift et al. 1979). The main groups are Acari, Collembola, Tardigrada, Protura, Diplura, and Enchytraeidae. Acari, or mites, are often the most abundant and the most species-rich group of the soil mesofauna (Table 1.2.1). Mites inhabit air-filled soil pores and litter layers, and like spiders, have a hard body and eight legs. Species of acari are found in different trophic levels (Fig. 1.2.2): herbivores (feeding on plants or algae from the first trophic level), bacterivores and fungivores (feeding on bacteria or fungi which belong to the second trophic level), and predators (feeding on small soil-dwelling animals of the second and higher trophic levels). Collembola, or springtails, are primitive insects, with six legs and antennae, but no wings. The number of known collembolan species is much lower than that of the acari, but they may reach the same abundances (Petersen & Luxton 1982). Collembola live in air-filled soil pores and in litter layers. Compared to acari, collembola are more dependent on high humidity and are more restricted in their diet, as most species feed on fungi and algae; but a few feed on plants or are predators (Petersen & Luxton 1982).

Tardigrada, or waterbears, have a soft, plump body and eight poorly articulated limbs with claws. Tardigrades need a waterfilm to be active and are often found in mosses and humid detritus. Tardigrades feed by piercing plant cells, mosses, or animals. Protura are primitive insects without eyes or antennae. They live in airspaces and litter and mainly feed on detritus and fungi. Diplura resemble members of the Protura, but have antennae and often feed on soil fauna or plants rather than on detritus. Enchytraeidae, or potworms, resemble small earthworms. They are most abundant in wet, organic soils and generally feed on detritus, soil, and microorganisms.

Soil mesofauna contribute to nutrient cycling (Fig. 1.2.1), suppress pests and diseases by selective feeding on pathogenic microorganisms, serve as a food source for other organisms, such as small vertebrates, and distribute smaller soil biota throughout the soil (Scheu *et al.* 2005; Coleman 2008; Bonkowski *et al.* 2009). Different species of mesofauna can contribute differentially to nutrient cycling because of preferential feeding on specific organisms at particular times and locations in the soil (Petersen & Luxton 1982).

1.2.1.5 Macrofauna

Soil macrofauna have body widths >2 mm (Swift *et al.* 1979) and consist of decomposers, predators, herbivores, and so called "ecosystem engineers." Insects, spiders, isopods, myriapods, and others belong to the "macroarthropods," a taxonomically diverse group of larger arthropods. Other important macrofauna include soft-bodied, legless soil biota such as annelids and gastropods. Diplopoda (millipedes), Isopoda, and Gastropoda are important decomposers of dead plant material both in temperate and tropical ecosystems. Soil- and litter-inhabiting insects such as scarab beetles and dipteran larvae are involved in decomposing dung and animal carcasses. Chilopoda (centipedes) are predators in the soil and the litter layer.

Another group of predators in soil are larger Arachnida (e.g. scorpions, spiders, and opiliones) and insects. Beetles such as ground beetles (Carabidae) and rove beetles (Staphylinidae) are predators at the soil surface and in the litter layer. Wireworms, the larvae of click beetles (Elateridae), or grubs, the larvae of scarab beetles, can cause considerable damage to roots in forest and agro-ecosystems. Larvae of true flies (Diptera) can also be phytophagous on roots, such as the leatherjackets (larvae of *Tipula paludosa*), whereas the majority of dipteran larvae in soil are saprophagous.

Ecosystem engineers are organisms that have profound impacts on their habitat and often change its chemical, physical, and structural properties with impacts on other biota and ecosystem functions (Jones *et al.* 1994). Earthworms, ants, and termites are the major ecosystem engineers in soil. In temperate zones earthworm biomass often exceeds the biomass of all other soil biota; by active burrowing and mixing of soil and organic matter earthworms have major impacts on other soil biota and soil conditions. Ants and termites represent high proportions of the animal biomass in ecosystems, particularly in the tropics, where they mix organic litter into the soil and have profound effects on soil structure. Soil macrofauna contribute to diverse ecosystem functions such as decomposition and nutrient cycling, soil structure, water infiltration, suppression of pests and diseases, regulation of other biota by predation or serving as a food source, and can have both negative and positive effects on plant growth and primary productivity (Fig. 1.2.1).

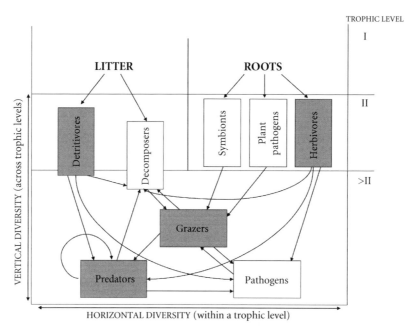

Figure 1.2.2 Studies on diversity–function relationships in soil have almost exclusively focused on diversity within individual functional groups (i.e. groups of species with similar traits and effects on processes). Diversity across functional groups within a trophic level ("horizontal diversity") and across trophic levels ("vertical diversity") has been rarely studied, but is probably important for ecosystem functions. Note that some functional groups are restricted to the second trophic level, while others cross trophic levels. For example, decomposers may degrade both dead organic matter from plants and other soil fauna, while larger detritivores might ingest other soil biota while degrading plant material. In the higher trophic levels (>II) no clear levels are defined, since soil organisms such as generalist predators often feed on several levels of the soil food web.

1.2.2 How to investigate soil communities

1.2.2.1 Introduction

Methods for quantifying the taxonomy of soil biota depend to a large extent on which size class is being assessed. There are also a range of techniques, many of which are still being developed, that allow links between soil biota and the ecosystem functions they mediate to be quantified. Next, we give a brief summary of common methods and highlight approaches that are providing new insights into the role of soil biota in ecosystem functioning.

1.2.2.2 Taxonomic approaches

1.2.2.2.1 Soil microbes
Quantifying and describing the taxonomy of the soil microbial community is a major challenge due to their huge diversity, variability in space and time, and the difficulty of studying individual species in isolation. It is even difficult to define exactly what a bacterial species is, as the traditional definitions used for larger organisms are based on reproduction and morphological characteristics that cannot be easily applied to many microbes (Kirk *et al.* 2004). Nevertheless, there are a wide range of techniques available that allow various aspects of community structure to be assessed. Traditional approaches are based on culturing bacteria and fungi on selective media. Although the ecology of individual species can be studied in detail, it is limited to the 1% of the total microbial community that is culturable. More recently, techniques using phospholipid fatty acids (PLFA), a constituent of microbial membranes, have been developed. The main advantage of the PLFA technique is that it is quantitative, and particular PLFAs can nominally be assigned to broad-scale fungal and bacterial groups (Kirk *et al.* 2004). Quantifying community structure at a higher resolution

requires the use of molecular techniques. These techniques use variation in DNA sequences to determine broad community structure of the dominant species in soil (e.g. denaturing gradient gel electrophoresis (DGGE), and terminal restriction fragment length polymorphism (T-RFLP)), the abundance of particular groups (e.g. quantitative polymerase chain reaction (qPCR)), and detailed information on microbial composition and community structure (e.g. microarrays, sequencing; Hackl et al. Chapter 2.1, this volume).

1.2.2.2.2 Micro-, meso-, and macrofauna
In contrast to soil microbes, the larger soil organisms can be extracted from soil and identified to species level by eye. Extraction methods for these organisms can be active, where movement of the soil organism is involved, or passive, where soil organisms are extracted mechanically or chemically. The active method used depends on the behavior of the group in question. For example, active methods to extract microfauna generally use water in combination with funnels and sieves (Petersen & Luxton 1982; Esteban et al. 2006). In contrast, most mesofauna (e.g. acari and collembolans) and some macrofauna (e.g. insect larvae) can be extracted by gradual soil drying, for example, in Tullgren funnels, with or without a temperature gradient (Petersen & Luxton 1982). Litter-inhabiting macrofauna such as carabid beetles or myriapods can be quantified by placing pitfall traps in soil, i.e. small open vessels containing water or formalin to catch and kill animals as they forage. Earthworms can be collected from the field by pouring irritants (e.g. mustard-based suspensions) on the ground or by providing an electric current (electrical octet method). Passive extraction methods include centrifugation techniques using a density gradient of aqueous solutions. Such methods are frequently used for microfauna but to a lesser extent for mesofauna. Passive methods for macrofauna entail collection and excavation of soil and hand-sorting (Petersen & Luxton 1982).

Once extracted, organisms are typically classified based on their morphology. This often requires a high level of operator skill, and morphological characteristics may need to be supplemented with biochemical analysis of enzyme and body lipid profiles to enable identification for some soil organisms (Briones 2006). More recently, many of the molecular techniques that have been applied to soil microbes (e.g. DGGE, T-RFLP, and sequencing) are now being investigated for use with micro- and mesofauna communities (van Megen et al. 2009; Wu et al. 2009; Porazinska et al. 2010). Although these techniques allow assessment of overall community profiles, the accuracy of translating T-RFLP and sequencing data into species composition and abundance is currently constrained by difficulties in separating species based on their DNA, and the availability of databases with species-specific T-RFLP lengths and DNA sequences. The quantification of species relative abundance is also still difficult (Gibb et al. 2008; Hamilton et al. 2009). Nevertheless, given the rapid developments in this research area, these molecular techniques may eventually provide a faster way of identifying these soil organisms that is less reliant on specialist taxonomists.

1.2.2.3 Functional approaches

Understanding which species are present in a system may not translate easily into quantifying their role in ecosystem functions. To address this issue various other techniques have been developed. These range from altering how communities are described, to focusing more on distinguishing what different organisms or groups of organisms are doing in a given system.

1.2.2.3.1 Describing soil biota based on their functional roles
One way to link soil biota more explicitly to ecosystem functions is to classify them based on their functionality (i.e. their effect on ecosystem processes) rather than solely on their taxonomy. For soil microbes, this has largely taken the form of quantifying community structure based on short-term responses to the addition of a range of carbon substrates (e.g. community level physiological profiling [CLPP] such as Biolog® (Garland & Mills 1991) or multiple substrate-induced respiration [MSIR] by GC (Degens & Harris 1997) or Multiresp® (Campbell et al. 2003)), and more recently, using

molecular techniques to quantify differences among communities in the relative abundance of genes linked to specific functions. Attempts are now also being made to classify soil microbes based on their functional traits, e.g. it has been suggested that variation in a set of morphological and physiological traits of arbuscular mycorrhizal fungi could be used to predict how their community structure impacts on ecosystem functioning (van der Heijden & Scheublin 2007). Functional approaches for larger soil organisms have also been commonly used, and new ones are in development. These tend to use life history traits such as feeding mode, dispersal and reproductive strategies to describe functional groups of organisms (i.e. groups of species with similar traits and effects on processes) (Yeates *et al.* 1993, Siepel 1994). Such approaches have helped to elucidate, for example, the role of microfauna in nutrient cycling by enabling the assembly of communities with low and high levels of similarity in feeding strategy and testing the associated response of nutrient mineralization.

1.2.2.3.2 Experimental manipulation of soil communities

These techniques have to date aimed to determine what role soil communities as a whole, or different groups within the soil community, have on ecosystem functions. These include "removal studies" where particular groups are excluded from soil (e.g. via the addition of chemicals such as fungicides (e.g. Maron *et al.* 2011), or via altering mesh sizes to limit access of different sized organisms to litter bags (Kampichler & Bruckner 2009)), and "addition studies" where soil communities with different complexity or characteristics are created by inoculating soil with selected biota (e.g. Bradford *et al.* 2002; Wertz *et al.* 2006). Other approaches such as reciprocal transplantation, where soil cores are shifted among environments, have been used to determine the relative role of soil community structure and abiotic factors in driving ecosystem functions (e.g. Balser & Firestone 2005). Finally, plant–soil feedback approaches, where soil is conditioned by one plant species/community and then re-planted with the same or a different plant species/community, have been used to determine how changes in soil biota affect plant-related functions (Kulmatiski *et al.* 2008).

1.2.2.3.3 Tracing and observational approaches

Tracing studies use isotopes of key elements to ascertain the relative role of different groups in a given function. Some techniques take advantage of the fact that the trophic position of organisms and their feeding preferences can be related to their stable isotope ratios. For example, natural stable isotope ratios have been used to determine the trophic position of many species in soil and revealed that taxa such as Acari have much broader feeding spectra than previously assumed (Schneider *et al.* 2004). Actively introducing labeled substrates into the soil (e.g. stable isotopes such as ^{15}N and ^{13}C or radioactive isotopes such as ^{14}C and ^{32}P) can allow the cycling of these elements to be traced from living plants and/or litter through the soil food web by investigating the incorporation of the enriched elements in organism tissues. This can provide insights into trophic links within soil food webs, process rates, and residence times of particular substrates. Because ^{13}C, ^{15}N, and ^{18}O are often incorporated into the DNA and/or cell membranes of microbes that use the labelled substrate, tracer studies combined with molecular and PLFA techniques can also be used to distinguish which microbes are actively involved in a given process (stable isotope probing; Neufeld *et al.* 2007).

There are also techniques available that allow the spatial distribution of soil biota to be ascertained. One such technique, fluorescence *in situ* hybridization (FISH) involves adding a fluorescently-labeled probe to soil, which binds to highly similar microbial nucleic acid sequences. Fluorescence is then detected using microscopy, allowing the position of individual cells within the environment to be determined. The technique can be used to target specific strains, phylogenetic groups, or functional genes. Similar techniques are also being developed that allow the position of cells labeled with stable or radioactive isotopes to be detected *in situ* (e.g. nanoSIMS) (Neufeld *et al.* 2007). There are also techniques such as nano sensors (Coleman 2008) and X-ray computed tomography (Young & Crawford 2004; Peth 2010) that allow the direct observation of interactions between soil biota, plants and soil biota without destruction of the soil's physical three-dimensional structure.

1.2.2.3.4 Modeling approaches
Most mathematical models of ecosystem processes treat soil as a "black box," and do not explicitly incorporate the traits of soil biota or the links between the different trophic levels. However, some models have been developed that include more detail of the soil food web. These have been used to elucidate, for example, the impact of different feeding guilds of soil biota on processes such as nitrogen cycling (Hunt *et al.* 1987; Hunt & Wall 2002) and how variation in the traits of mycorrhizal fungi (Orwin *et al.* 2011) affect carbon and nutrient cycling. Because these kind of models often allow us to manipulate individual components of the soil food web independently of abiotic factors, which is very difficult to do in real-life systems, they have the potential to provide significant insight into the relationships of soil biota to ecosystem functioning.

1.2.3 Diversity–function relationships

1.2.3.1 Introduction

A very important, but still largely unanswered question concerns the relationships between soil biodiversity and ecosystem functioning. Basic functions of terrestrial ecosystems such as decomposition, nutrient and matter cycling, and primary productivity rely on the activity of soil organisms (Fig. 1.2.1). Although many organisms spend part or their whole life in soil, the majority of theories and empirical knowledge on biodiversity focus on aboveground organisms. The two theories most commonly applied to soil biota contradict each other, with some researchers suggesting that functional redundancy is common, and others that complementarity between soil organisms promotes ecosystem processes.

In this section we focus on what is known about the effects of soil biodiversity on ecosystem functions such as primary productivity, decomposition, and nutrient cycling. The impact of soil biodiversity on plant community composition and diversity is covered elsewhere in the volume (Schweitzer *et al.* Chapter 2.2). We summarize the current knowledge on diversity–function relationships of soil organisms within the microbes, microfauna, mesofauna, and macrofauna. We show that diversity–function relationships within the different size classes show a similar trend, i.e. that the composition of the community, the traits of key species or groups and their relative abundance and complementarities rather than species richness per se appear to be significant drivers of soil processes and functions.

1.2.3.2 Microbes

It has been argued that there must be a high level of functional redundancy in soil microbial communities, given their high diversity, ability to adapt quickly, and ubiquitous presence in soil (Wertz *et al.* 2006; Strickland *et al.* 2009). Although intuitively appealing, evidence for this argument is varied. For example, studies that have manipulated microbial diversity using removal approaches have shown that soils with up to a 99% reduction in microbial diversity had similar carbon mineralization, nitrification, and denitrification rates as compared to soils with the full range of species present (Wertz *et al.* 2006). However, other studies using similar methods found that reduced species richness did affect plant productivity and nutrient content (albeit in the opposite direction to that expected; Hol *et al.* 2010), and nitrification rates (Griffiths *et al.* 2001). Similarly, studies that have manipulated the species richness of bacteria, fungi, or mycorrhizal fungi by building model communities tend to show that increases in species richness at the low end of the scale can increase ecosystem functions such as decomposition and plant growth (e.g. van der Heijden *et al.* 1998; Setälä & McLean 2004; Bell *et al.* 2005). However, as these studies also show that this relationship often weakens considerably at higher, more realistic species richness levels, it is unclear what role microbial species richness actually plays in real systems.

What is becoming increasingly obvious is that microbial community composition can be a significant driver of ecosystem functioning. Observational studies suggest that changes in the relative abundance of three of the five dominant bacterial phyla are correlated to carbon mineralization rates, indicating that each phylum differs in its traits (Fierer *et al.* 2007a) and therefore its likely impact on functioning. Experimental studies that have inoculated litter with microbial communities from different soils suggest sufficient microbial specialization to cause differences in ecosystem functions such as decomposition (e.g. Strickland *et al.* 2009), and reciprocal transplant studies show that microbial community composition can be important for functions such as nitrogen mineralization and denitrification rates (e.g. Balser & Firestone 2005). Similarly, plant-soil feedback studies have shown that the composition of fungal communities may alter plant productivity (Kardol *et al.* 2007). A recent modeling study showed that variation in the traits of mycorrhizal fungi, and in particular their ability to access organic nutrients, may significantly alter soil carbon storage (Orwin *et al.* 2011). Thus, although the role of species richness of soil microbes in determining ecosystem functions remains unclear, it is evident from a wide range of approaches that the composition of the microbial community can play an important role.

1.2.3.3 Microfauna

The role of soil microfauna diversity for plant growth and ecosystem process rates has been tested at the level of species richness across and within feeding guilds (Postma-Blaauw *et al.* 2005; Huhta 2007). At low levels of diversity, diversity across feeding guilds stimulates ecosystem process rates, while at higher diversity levels functional redundancy for ecosystem processes likely increases because many species tend to be food generalists rather than specialists (Wardle 2002; Scheu *et al.* 2005). To date, only a few studies have investigated the effects of species richness within feeding guilds of soil microfauna and have covered only small gradients of species richness. However, the results indicate that within feeding guilds species composition is more important than species richness per se, as found for bacterial-feeding nematodes and the rate of N mineralization (Postma-Blaauw *et al.* 2005). Moreover soil communities of different composition due to plant species specific influences can still be remarkably similar in their impact on soil C and N cycling (Bezemer *et al.* 2010). This latter study demonstrates functional similarity of the soil communities for these general ecosystem processes, which may require more drastic differences in soil communities before the functional impact becomes apparent.

Bacterial-feeding protozoan communities show the same pattern, with the life history strategy of the component species being the main driver of effects on nutrient mineralization (Scheu *et al.* 2005). Within plant-feeding nematodes, more species do not imply stronger effects on plant growth. A field study by Brinkman *et al.* (2005) found that the presence of less damaging plant-feeding nematode species reduced the impact of the most damaging plant-feeding nematode species on *Ammophila arenaria*; the effects of nematode communities were dependent on nematode species composition and not on their species richness.

In soil, local species richness of entomopathogenic nematodes is low, and individuals of insects are usually infected and killed by one single nematode species and its bacterial symbiont (Gaugler *et al.* 1997). However, across entomopathogenic nematode groups, different host-finding strategies can be distinguished: "ambushers" forage near the soil surface and "cruisers" forage within the soil profile. Thus, ambushers are more effective against surface dwelling insects and cruisers against sedentary root-feeding insects (Gaugler *et al.* 1997). The level of ecosystem functioning in terms of pest suppression is hence dependent on the match between the behavior of the insect pest and the entomopathogenic nematode species, rather than being directly related to their species richness. Overall, the results indicate that, within the microfauna, species richness across feeding guilds has stronger effects on ecosystem functions than species richness within feeding guilds. However, the relationship between soil microfauna diversity and ecosystem functioning is not easy to predict and is dependent on the characteristics of the species and their abundance in the community (Wardle 2002; Scheu *et al.* 2005; Coleman 2008).

1.2.3.4 Mesofauna

Across ecosystems the Acari and Collembola, which represent the vast majority of the micro-arthropods, are the most species rich and abundant groups within the mesofauna (Petersen & Luxton 1982). Larger species richness of microarthropods can promote litter decomposition and plant growth, but effects appear to be limited to the lower range of the diversity gradient (Laakso & Setälä 1999; Liiri et al. 2002). There is also evidence for the importance of dominant species traits within mesofauna species assemblages of the same feeding guild. In microcosms containing fungal-feeding collembolan assemblages with different levels of species richness, rate of litter decomposition, microbial activity, and availability of mineral N depend on community composition and the presence of specific collembolan species rather than on species richness (Cragg & Bardgett 2001). Amongst the mesofauna, enchytraeids have been proven to act as a keystone group in boreal forests and peat soils, as they have effects disproportional to their biomass regarding decomposition, nutrient mineralization, and plant growth in organic soils (Laakso & Setälä 1999; Briones 2006). Within the Enchytraeidae different functional groups may exist according to their life history strategy, but such distinctions need further study (Briones 2006). The suppression of soil pathogens by soil mesofauna is accomplished by selective grazing on pathogenic soil fungi, and the indirect promotion of beneficial mycorrhizal fungi (Scheu et al. 2005). The level of pathogen suppression is thus expected to be driven by the traits of the mesofauna species in the community, i.e. their selectivity and grazing pressure, rather than on the species richness, but empirical evidence is still needed. Based on the studies available there does not seem to be a general relation between the species diversity of mesofauna and ecosystem functioning. The combinations of traits of the component mesofauna species in the community, rather than their species richness, appear to determine the level of functioning (Hättenschwiler et al. 2005; Coleman 2008).

1.2.3.5 Macrofauna

Studies which have investigated the diversity–function relationship within the macrofauna are scarce. Functions such as litter incorporation, decomposition, water infiltration, and primary production are affected by macrofaunal activity. However, whether the diversity of macrofauna has an impact on these functions is largely unexplored. The study by Heemsbergen et al. (2004) is an exception where three species of earthworms, two millipede species, and three isopod species were assembled using different species combinations of macro-detritivores (zero, one, two, four, eight species per microcosm) and the effects were measured on soil ecosystem process variables such as leaf litter matter loss, leaf litter fragmentation, soil respiration, and nitrification. A high functional dissimilarity between the detritivorous species and not the species numbers enhanced leaf litter mass loss and soil respiration. Also, one species, the epigeic earthworm (*Lumbricus rubellus*) disproportionally affected the measured soil processes. In all species combinations where *L. rubellus* was present, the decomposition was enhanced due to its efficiency in incorporating the litter into the soil. Thus, in this simplified experimental community, there was a key species which was not redundant in its function. Other studies (Zimmer et al. 2005; de Oliveira et al. 2010) reported complementary effects of contrasting species of detritivorous macrofauna on litter decomposition depending on litter quality and diversity; however, the interactions of only two species of detritivores were studied. Schuurman (2005) showed that community composition of termites can affect decomposition rates more than climatic parameters. A taxonomic subfamily of the termites, the fungal-cultivating termites (Macrotermitinae), were the drivers of decomposition of wood litter indicating that certain taxonomic and functional groups might be more important than others for certain soil functions. Litter bag studies have been conducted where the exclusion of macrofauna affected litter decomposition (e.g. Joergensen 1991); however, it remains largely unstudied whether functional or taxonomic diversity within the macrofauna is the best predictor of soil functions.

1.2.4 Taking a holistic view to soil diversity–ecosystem functioning

We have so far described soil biota and their relationship to functioning within each size class. However, soil organisms belonging to different size classes and functional groups interact with each other (Fig. 1.2.1). How these complex interactions affect ecosystem processes and functions is still widely unknown, as most studies on diversity–function relationships in soil have been conducted under controlled conditions and have restricted manipulations to soil organisms of the same size class and functional group. In the field, however, the diversity is much higher and ecosystem processes are performed by multiple interactions between soil biota of different sizes and functions under non-constant abiotic conditions (Fig. 1.2.1). Despite recognition that ecosystem functioning in soil is a product of multitrophic, multispecies interactions, there remain remarkably few studies that have examined how these interactions may affect functioning.

There are, however a few studies that give an indication of how diversity across size classes, functional groups, and trophic levels might influence ecosystem functioning. Bradford *et al.* (2002) manipulated the composition of a complex soil community and measured its impact on the productivity of a model grassland ecosystem within mesocosms. The authors assigned the soil biota according to size classes and constructed three communities: 1) microfauna, 2) microfauna and mesofauna, 3) micro, meso-, and macrofauna. This study investigated soil communities with increasing size and functional complexity, but did not address whether species diversity per se is crucial for ecosystem functions. Since key ecosystem processes such as net primary and ecosystem productivity were resistant to the changes in soil community composition, the authors proposed that positive and negative faunal-mediated effects may cancel each other out, resulting in no net ecosystem effects. Another study (Wurst *et al.* 2008) used abundant representatives of the soil biota belonging to different size classes (microbes, nematodes, and earthworms) and investigated their individual and combined effects on plant community composition and productivity. The negative effects of nematodes and microbes on grass growth and diversity of the plant community, respectively, were outweighed by the mainly positive effects of earthworms. Ladygina *et al.* (2010) investigated the individual and combined effects of functional groups of soil biota that interact differently with plants (symbionts, detritivores, and root herbivores) on a grassland plant community. These functional groups had mainly independent and therefore additive effects on nutrient uptake and productivity of the plant community, probably because they occupy distinct niches. Since the additive effects of the functional groups on productivity worked in opposing directions, no net effect was observed in the combination treatment. However, at higher resolutions, interactive, non-additive effects of the functional groups were detected on microbial community structure and growth of individual plant species. The results of the mentioned studies indicate that the net effects on ecosystem functions will depend on the direction and strength of the effects of the individual components and their degree of interaction. However, more studies focusing on interactions among multiple size classes, functional groups and trophic levels are needed to see how impacts of individual components of the soil community change in the presence of other components.

Although we are beginning to appreciate the extent to which soil biota affect ecosystem functioning, there are still large gaps in our understanding. For example it is largely unknown whether diversity effects within one trophic level depend on the presence or diversity of other trophic levels (Duffy *et al.* 2007). The majority of experimental studies on the impacts of soil diversity on ecosystem processes have focused on soil biota belonging to the same trophic level. However, both diversity within a trophic level ("horizontal diversity") and across trophic levels ("vertical diversity") is likely to be crucial for ecosystem functioning (Duffy *et al.* 2007) in a wide range of ecosystems including soil (Fig. 1.2.2). Srivastava and Bell (2009) showed that reducing horizontal and vertical diversity in an aquatic food web affects ecosystem functions and triggers extinctions. However, studies that assess both the impact of horizontal and vertical diversity of soil communities on ecosystem processes simultaneously are still scarce (but see Laakso & Setälä 1999). Another key challenge for studies on soil diversity-ecosystem functioning will be to quantify and describe functional diversity. It has been argued that a set of (functional) traits should be ascribed to species rather than classifying species into functional groups defined a priori

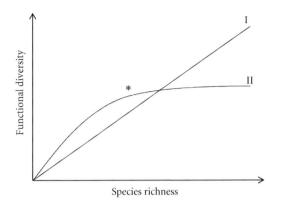

Figure 1.2.3 Theoretical relationship between species richness and functional diversity (= number of functional traits). In Line I the number of considered functional traits or trait combinations ≥ number of species. In Line II the numbers of considered functional traits < number of species, e.g. this might be the case when the traits are weighted (based on the importance of a given trait for a given function) or functional groups are considered. The point * where Line II reaches a plateau marks a high functional diversity with low species redundancy. Note that the lines strongly depend on the number of traits considered which may be related to the ecological process under investigation (modified after Naeem 2002).

(Petchey & Gaston 2002). However, when the traits are not weighted based on their importance for a given function, and their number or combinations are high, an increase in species numbers will inevitably result in a linear increase in functional diversity, since each species added to a community will likely bring a new trait or trait combination (Fig. 1.2.3). Therefore, it is crucial to determine appropriate traits with regard to the ecosystem process under investigation, and also to weigh the impact of individual traits (Petchey & Gaston 2006) to achieve an expedient measure of functional diversity.

1.2.5 Conclusions

The most important reason to consider the relationship of diversity and function is the alarming decline in species numbers worldwide and its potential impact on the sustainability of ecosystems. Soil is one of the most diverse habitats on earth. However, the extent of soil diversity is still largely unknown, as are the underlying mechanisms resulting in so many different species and whether this diversity is actually important for ecosystem processes.

The present chapter highlighted that ecosystem processes, which result in ecosystem services such as nutrient and matter (re)cycling, soil fertility, and pest control, rely on multitrophic interactions between functionally dissimilar soil organisms across all size classes and trophic levels. So far most of the studies on biodiversity–function relationships have focused on organisms belonging to the same feeding guild, functional group, size class, and trophic level. Future studies are recommended which involve manipulation of a wider range of soil organisms with different functional traits, from different size classes and trophic levels, to increase their relevance to real systems and to give more realistic answers to the diversity–function question.

Experimental investigations of the effects of non-random loss of soil species as a consequence of external factors such as global change may also provide more realistic indications of the impact of species loss from soil on ecosystem functions. Our understanding from such studies could be enhanced by using a variety of methods, including, for example, molecular tools and observational *in situ* techniques and/or modeling. This calls for much more interdisciplinary cooperation of researchers that often focus on their "favorite" group of organisms without taking interactions with other organisms into account. Combining these approaches is likely to significantly enhance our understanding of the role of soil biota in ecosystem functioning, and the provision of ecosystem services in current and changing environments.

References

Balser, T.C., & Firestone, M.K. (2005) Linking microbial community composition and soil processes in a California annual grassland and mixed-conifer forest. *Biogeochemistry* **73**: 395–415.

Bell, T., Newman, J.A., Silverman, B.W., et al. (2005) The contribution of species richness and composition to bacterial services. *Nature* **436**: 1157–60.

Bezemer, T.M., Fountain, M.T., Barea, J.M., et al. (2010) Divergent composition but similar function of soil food webs of individual plants: plant species and community effects. *Ecology* **91**: 3027–36.

Bonkowski, M., Villenave, C., & Griffiths, B. (2009) Rhizosphere fauna: the functional and structural diversity of intimate interactions of soil fauna with plant roots. *Plant and Soil* **321**: 213–33.

Bradford, M.A., Jones, T.H., Bardgett, R.D., et al. (2002) Impacts of soil fauna community composition on model grassland ecosystems. *Science* **298**: 615–18.

Brinkman, E.P., Duyts, H., & van der Putten, W.H. (2005) Consequences of variation in species diversity in a community of root-feeding herbivores for nematode dynamics and host plant biomass. *Oikos* **110**: 417–27.

Briones, M.J.I. (2006) Enchytraeidae. In R. Lal (ed.) *Encyclopedia of Soil Science*, pp. 514–18. Taylor and Francis, New York.

Campbell, C.D., Chapman, S.J., Cameron, C.M., et al. (2003) A rapid microtiter plate method to measure carbon dioxide evolved from carbon substrate amendments so as to determine the physiological profiles of soil microbial communities by using whole soil. *Applied and Environmental Microbiology* **69**: 3593–9.

Coleman, D.C. (2008) From peds to paradoxes: Linkages between soil biota and their influences on ecological processes. *Soil Biology and Biochemistry* **40**: 271–89.

Cragg, R.G., & Bardgett, R.D. (2001) How changes in soil faunal diversity and composition within a trophic group influence decomposition processes. *Soil Biology and Biochemistry* **33**: 2073–81.

De Deyn, G.B., Raaijmakers, C.E., Zoomer, H.R., et al. (2003) Soil invertebrate fauna enhances grassland succession and diversity. *Nature* **422**: 711–13.

De Oliveira, T., Hättenschwiler, S., & Handa, I.T. (2010) Snail and millipede complementarity in decomposing Mediterranean forest leaf litter mixtures. *Functional Ecology* **24**: 937–46.

Degens, B.P., & Harris, J.A. (1997) Development of a physiological approach to measuring the catabolic diversity of soil microbial communities. *Soil Biology and Biochemistry* **29**: 1309–20.

Duffy, J.E., Cardinale, B.J., France, K.E., McIntyre, P.B., Thébault, E., & Loreau, M. (2007) The functional role of biodiversity in ecosystems: incorporating trophic complexity. *Ecology Letters* **10**: 522–38.

Esteban, G.F., Clarke, K.J., Olmoc, J.L., & Finlay, B.J. (2006) Soil protozoa—An intensive study of population dynamics and community structure in an upland grassland. *Applied Soil Ecology* **33**: 137–51.

Fierer, N., Bradford, M.A., & Jackson, R.B. (2007a) Toward an ecological classification of soil bacteria. *Ecology* **88**: 1354–64.

Fierer, N., Breitbart, M., Nulton, J., et al. (2007b) Metagenomic and small-subunit rRNA analyses reveal the genetic diversity of bacteria, archaea, fungi, and viruses in soil. *Applied and Environmental Microbiology* **73**: 7059–66.

Foissner, W. (1999) Soil protozoa as bioindicators: pros and cons, methods, diversity, representative examples. *Agriculture, Ecosystems and Environment* **74**: 95–112.

Garland, J.L., & Mills, A.L. (1991) Classification and characterisation of heterotrophic microbial communities on the basis of patterns of community-level sole carbon-source utilisation. *Applied and Environmental Microbiology* **57**: 2351–9.

Gaugler, R., Lewis, E., & Stuart, R.J. (1997) Ecology in the service of biological control: The case of entomopathogenic nematodes. *Oecologia* **109**: 483–9.

Gibb, K., Beard, J., O'Reagain, P., Christian, K., Torok, V., & Ophel-Keller, K. (2008) Assessing the relationship between patch type and soil mites: A molecular approach. *Pedobiologia* **51**: 445–61.

Griffiths, B.S., Ritz, K., Wheatley, R., et al. (2001) An examination of the biodiversity-ecosystem function relationship in arable soil microbial communities. *Soil Biology and Biochemistry* **33**: 1713–22.

Hamilton, H.C., Strickland, M.S., Wickings, K., Bradford, M.A., & Fierer, N. (2009) Surveying soil faunal communities using a direct molecular approach. *Soil Biology and Biochemistry* **41**: 1311–14.

Hättenschwiler, S., Tiunov, A.V., & Scheu, S. (2005) Biodiversity and litter decomposition in terrestrial ecosystems. *Annual Review of Ecology Evolution and Systematics* **36**: 191–218.

Heemsbergen, D.A., Berg, M.P., Loreau, M., van Haj, J.R., Faber, J.H., & Verhoef, H.A. (2004) Biodiversity effects on soil processes explained by interspecific functional dissimilarity. *Science* **306**: 1019–20.

Hol, W.H.G., de Boer, W., Termorshuizen, A.J., et al. (2010) Reduction of rare soil microbes modifies plant-herbivore interactions. *Ecology Letters* **13**: 292–301.

Huhta, V. (2007) The role of soil fauna in ecosystems: A historical review. *Pedobiologia* **50**: 489–95.

Hunt, H.W., Coleman, D.C., Ingham, E.R., et al. (1987). The detrital food web in a shortgrass prairie. *Biology and Fertility of Soils* **3**: 57–68.

Hunt, H.W., & Wall, D.H. (2002) Modelling the effects of loss of soil biodiversity on ecosystem function. *Global Change Biology* **8**: 33–50.

Joergensen, R.G. (1991) Organic matter and nutrient dynamics of the litter layer on a forest Rendzina under beech. *Biology and Fertility of Soils* **11**: 163–9.

Jones, C.G., Lawton, J.H., & Shachak, M. (1994) Organisms as ecosystem engineers. *Oikos* **69**: 373–86.

Kampichler, C., & Bruckner, A. (2009) The role of microarthropods in terrestrial decomposition: a meta-analysis of 40 years of litterbag studies. *Biological Reviews* **84**: 375–89.

Kardol, P., Cornips, N.J., van Kempen, M.M.L., et al. (2007) Microbe-mediated plant-soil feedback causes historical contingency effects in plant community assembly. *Ecological Monographs* **77**: 147–62.

Kimura, M., Jia, Z.J., Nakayama, N., & Asakawa, S. (2008) Ecology of viruses in soils: Past, present and future perspectives. *Soil Science and Plant Nutrition* **54**: 1–32.

Kirk, J.L., Beaudette, L.A., Hart, M., et al. (2004) Methods of studying soil microbial diversity. *Journal of Microbiological Methods* **58**: 169–88.

Kulmatiski, A., Beard, K.H., Stevens, J.R., & Cobbold, S.M. (2008) Plant-soil feedbacks: a meta-analytical review. *Ecology Letters* **11**: 980–92.

Laakso, J., & Setälä, H. (1999) Sensitivity of primary production to changes in the architecture of belowground food webs. *Oikos* **87**: 57–64.

Ladygina, N., Henry, F., Kant, M.R., et al. (2010) Additive and interactive effects of functionally dissimilar soil organisms on a grassland plant community. *Soil Biology and Biochemistry* **42**: 2266–75.

Lauber, C.L., Hamady, M., Knight, R., & Fierer, N. (2009) Pyrosequencing-based assessment of soil pH as a predictor of soil bacterial community structure at the continental scale. *Applied and Environmental Microbiology* **75**: 5111–20.

Liiri, M., Setälä, H., Haimi, J., Pennanen, T., & Fritze, H. (2002) Relationship between soil microarthropod species diversity and plant growth does not change when the system is disturbed. *Oikos* **96**: 137–49.

Lugtenberg, B., & Kamilova, F. (2009) Plant-growth-promoting Rhizobacteria. *Annual Review of Microbiology* **63**: 541–56.

Maron, J.L., Marler, M., Klironomos, J.N., Cleveland, C.C. (2011) Soil fungal pathogens and the relationship between plant diversity and productivity. *Ecology Letters* **14**: 36–41.

Naeem, S. (2002) Ecosystem consequences of biodiversity loss: the evolution of a paradigm. *Ecology* **83**: 1537–52.

Neufeld, J.D., Wagner, M., & Murrell, J.C. (2007) Who eats what, where and when? Isotope-labelling experiments are coming of age. *ISME Journal* **1**: 103–10.

Orwin, K.H., Kirschbaum, M.U.F., St John, M.G., Dickie, I.A. (2011) Organic nutrient uptake by mycorrhizal fungi enhances ecosystem carbon storage: a model-based assessment. *Ecology Letters* **14**: 493–502.

Petchey, O.L., & Gaston, K.J. (2002) Functional diversity (FD), species richness and community composition. *Ecology Letters* **5**: 402–11.

Petchey, O.L., & Gaston, K.J. (2006) Functional diversity: back to basics and looking forward. *Ecology Letters* **9**: 741–58.

Petersen, H., & Luxton, M. (1982) A comparative-analysis of soil fauna populations and their role in decomposition processes. *Oikos* **39**: 287–388.

Peth, S. (2010) Applications of microtomography in soils and sediments. In: B. Singh, M. Gräfe (eds.) *Developments in Soil Science Volume 34*, pp. 73–101. Elsevier, Amsterdam.

Philippot, L., Hallin, S., Borjesson, G., & Baggs, E.M. (2009) Biochemical cycling in the rhizosphere having an impact on global change. *Plant and Soil* **321**: 61–81.

Piskiewicz, A.M., Duyts, H., & van der Putten, W.H. (2008) Multiple species-specific controls of root-feeding nematodes in natural soils. *Soil Biology and Biochemistry* **40**: 2729–35.

Porazinska, D.L., Giblin-Davis, R.M., Esquivel, A., Powers, T.O., Sung, W., & Thomas, W.K. (2010) Ecometagenetics confirm high tropical rainforest nematode diversity. *Molecular Ecology* **19**: 5521–30.

Postma-Blaauw, M.B., De Vries, F.T., de Goede, R.G.M., Bloem, J., Faber, J.H., & Brussaard, L. (2005) Within-trophic group interactions of bacterivorous nematode species and their effects on the bacterial community and nitrogen mineralization. *Oecologia* **142**: 428–39.

Ritz, K. (2007) Spatial organisation of fungi in soils. In: Franklin, R.B. and Mills, A.L. (eds.) *Spatial Distribution of Microorganisms in the Environment*, pp. 179–202. Kluwer, Dordecht.

Roesch, L. F., Fulthorpe, R. R., Riva, A., et al. (2007) Pyrosequencing enumerates and contrasts soil microbial diversity. *ISME Journal* **1**: 283–90.

Scheu, S., Ruess, L., & Bonkowski, M. (2005) Interactions between microorganisms and soil micro- and mesofauna. In: F. Buscot and A. Varma (eds.) *Soil Biology, Microorganisms in Soils: Roles in Genesis and Functions*, pp. 253–75. Springer-Verlag, Berlin.

Schneider, K., Migge, S., Norton, R.A., et al. (2004) Trophic niche differentiation in soil microarthropods (Oribatida, Acari): evidence from stable isotope ratios (N-15/N-14). *Soil Biology and Biochemistry* **36**: 1769–74.

Schuurman, G. (2005) Decomposition rates and termite assemblage composition in semiarid *African Ecology* **86**: 1236–49.

Setälä, H., & McLean, M.A. (2004) Decomposition rate of organic substrates in relation to the species diversity of soil saprophytic fungi. *Oecologia* **139**: 98–107.

Siepel, H. (1994) Life-history tactics of microarthropods. *Biology and Fertility of Soils* **18**: 263–78.

Smith, S.E., & Read, D.J. (2008) *Mycorrhizal symbiosis*, 3rd edition. Elsevier, London.

Srivastava, D.S., & Bell, T. (2009) Reducing horizontal and vertical diversity in a foodweb triggers extinctions and impacts functions. *Ecology Letters* **12**: 1016–28.

Strickland, M.S., Lauber, C., Fierer, N., & Bradford, M.A. (2009) Testing the functional significance of microbial community composition. *Ecology* **90**: 441–51.

Swift, M.J., Heal, O.W., & Anderson, J.M. (1979) *Decomposition in Terrestrial Ecosystems*. University of California Press, Berkeley, CA.

van der Heijden, M.G.A., Klironomos, J.N., Ursic, M., *et al.* (1998) Mycorrhizal fungal diversity determines plant biodiversity, ecosystem variability and productivity. *Nature* **396**: 69–72.

van der Heijden, M.G.A., & Scheublin, T.R. (2007) Functional traits in mycorrhizal ecology: their use for predicting the impact of arbuscular mycorrhizal fungal communities on plant growth and ecosystem functioning. *New Phytologist* **174**: 244–50.

van der Heijden, M.G.A., Bardgett, R.D., & van Straalen, N.M. (2008) The unseen majority: soil microbes as drivers of plant diversity and productivity in terrestrial ecosystems. *Ecology Letters* **11**: 296–310.

van der Putten, W.H., Cook, R., Costa, S., *et al.* (2006) Nematode Interactions in nature: models for sustainable control of nematode pests of crop plants? *Advances in Agronomy* **89**: 227–60.

van Megen, H., van den Elsen, S., Holterman, M., *et al.* (2009) A phylogenetic tree of nematodes based on about 1200 full-length small subunit ribosomal DNA sequences. *Nematology* **11**: 927–50.

Wardle, D.A. (2002) *Communities and Ecosystems: Linking the Aboveground and Belowground Components.* Princeton University Press, Princeton, NJ.

Wertz, S., Degrange, V., Prosser, J.I., *et al.* (2006) Maintenance of soil functioning following erosion of microbial diversity. *Environmental Microbiology* **8**: 2162–9.

Wu, T.H., Ayres, E., Li, G., Bardgett, R.D., Wall, D.H., & Garey, J.R. (2009) Molecular profiling of soil animal diversity in natural ecosystems: Incongruence of molecular and morphological results. *Soil Biology and Biochemistry* **41**: 849–57.

Wurst, S., Allema, B., Duyts, H., & van der Putten, W.H. (2008) Earthworms counterbalance the negative effect of microorganisms on plant diversity and enhance the tolerance of grasses to nematodes. *Oikos* **117**: 711–18.

Yeates, G.W., Bongers, T., de Goede, R.G.M., *et al.* (1993) Feeding-habits in soil nematode families and genera-an outline for soil ecologists. *Journal of Nematology* **25**: 315–31.

Young, I.M., & Crawford, J.W. (2004) Interactions and self-organization in the soil-microbe complex. *Science* **304**: 1634–7.

Zimmer, M., Kautz, G., & Topp, W. (2005) Do woodlice and earthworms interact synergistically in leaf litter decomposition? *Functional Ecology* **19**: 7–16.

CHAPTER 1.3

Ecosystem Services Provided by the Soil Biota

Lijbert Brussaard

1.3.1 Introduction

Ecosystem services are the benefits that people derive from ecosystems, manifest as supporting, provisioning (i.e. producing ecosystem goods), regulating, and cultural services; they are closely connected to human well-being and sustainability (MEA 2005). These goods and services are the utilitarian outcomes of ecosystem functioning (i.e. the natural processes which occur in ecosystems), which in turn depend on ecosystem structure (i.e. the biotic and abiotic constituents of ecosystems). The drivers of ecosystem structure are partly natural and partly of human origin, including threats to soil quality (e.g. CEC 2006) and land management practices (e.g. Wickings et al. 2010). Sustainability and human well-being depend on these ecosystem characteristics as conceptually represented in Fig. 1.3.1. The living soil is represented by soil biota that interacts with aboveground biota and with the abiotic constructs of soil, represented as soil structure, organic matter, and nutrients. Many ecosystem goods and services can be intuitively linked to the functioning of the soil. Extensive lists of such services have been published recently (Swift et al. 2004; Wall et al. 2004; Haygarth & Ritz 2009; Dominati et al. 2010).

Ecosystem goods and services are delivered by different ecosystem functions, which in aggregate have been associated with groups of the soil biota (Fig. 1.3.2). Because of their perceived association with ecosystem functions, these groups are often called "functional groups" or "functional assemblages" (Lavelle et al. 1995; Brussaard 1998; Kibblewhite et al. 2008). An alternative way of categorizing the soil biota is size (cf. Chapter 1.2), but such categorization cannot be unequivocally connected with function. Functional groupings, like body-size groupings, result from the need for pragmatic ways to categorize the soil biota in the face of the potentially overwhelming biodiversity in soil, which is still largely unknown, especially for the smaller-sized organisms. Functional group-based concepts are not without problems either. Firstly, some functional groups are, like body-size groups, associated with more than one aggregate function (e.g. the micro-food web and the litter transformers; Fig. 1.3.2). Secondly, different life-stages of the same species may be associated with different functions. Thirdly, within one group, species may occur with different effects on soil functions. For example, both harmful and beneficial organisms may occur in terms of plant performance within the root/rhizosphere biota. Splitting this group into harmful and beneficial soil biota does not always help; for example, root herbivores are both detrimental to plant growth and at the same time elicit systemic defense responses against aboveground herbivores or parasitoids (Bezemer & Van Dam 2005). The subdivision of groups based on feeding habits and microhabitat (e.g. Faber 1991; Siepel & de Ruiter Dijkman 1993) would seem appropriate, but is hampered by lack of knowledge. Therefore, functional groups could more aptly be described as "broad habitat groups," where "habitat" is synonymous with "spheres" *sensu* Beare et al. (1995) or "domains" *sensu* Lavelle (2002). These should be considered functional in that the aggregate functions to which they are connected

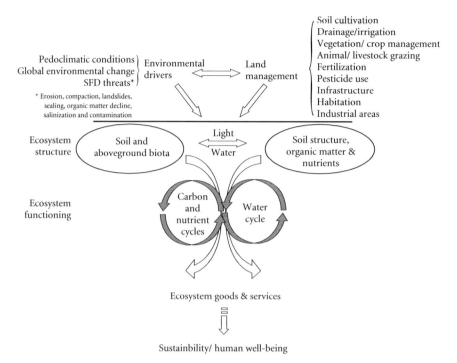

Figure 1.3.1 Conceptual diagram of drivers, state variables (ecosystem structure), processes (ecosystem functioning), and ecosystem goods and services, determining ecosystem sustainability/human well-being. SFD = Soil Framework Directive as proposed in CEC (2006). (Modified after Brussaard et al. 2007a, with permission from Elsevier Science Ltd.)

predominantly occur in these soil habitats. The exception is biological population regulation, which is a unifying aggregate function. The utility of these groups will be reviewed in the following section, and an alternative principle for grouping will be presented after considering the ecosystem functions underlying ecosystem goods and services.

1.3.2 Understanding ecosystem functioning

The effects of changes in environmental drivers and land management decisions on the sustainability of a system are predominantly mediated by the soil biota (Fig. 1.3.1). With very few exceptions, all soil organisms are ultimately driven by energy which is derived from reduced forms of carbon (C); the C transfer with associated energy flows is the main integrating factor in ecosystem functioning (Kibblewhite et al. 2008). This implies that manipulations of the soil biota, induced by the living plant, plant litter, and soil organic matter, would affect ecosystem functioning with possible feedback to aboveground biota. Is there indeed such a role for the soil biota? Broadly speaking, there are two perspectives which can be taken, viz. a "soil biogeochemistry" perspective on ecosystem functioning, which plays down the importance of functional group detail, versus a "soil biology" view, which considers such detail as a necessary perspective.

The soil biogeochemistry perspective emphasizes the apparent consistency, or limited importance of differences, in microbial communities and soil organic matter dynamics across widely different ecosystems. These variations in local organic matter dynamics are mainly driven by constraints over the physiology and metabolic activity of soil biotic communities, resulting in the energetics of metabolism and the nutrient stoichiometry[1] of both plants and microbes and the whole food web, plus the effects of physical-chemical processes in soils on organic

---

[1] Nutrient stoichiometry is the theory of elemental ratios, the balance of elements of organisms in interactions with the environment. This ratio differs between species.

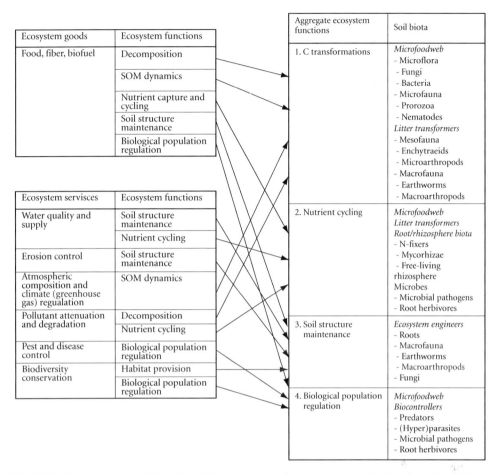

Figure 1.3.2 Relationships between the activities of the soil biota and a range of ecosystem goods and services that society might expect from soils. OM = organic matter; SOM = soil organic matter. (Modified after fig. 1 on p. 687 in Kibblewhite et al. 2008.)

matter stabilization (Fierer et al. 2009, and literature cited therein). As a result, organic matter passing through the soil food web undergoes predictable transformations that reduce variations in the chemical structure (Fierer et al. 2009). Hence, to understand C and nutrient transfers, there does not seem to be a need to know the identity of the organisms and what they do.

The soil biology perspective has a different basis. Community ecologists often use population dynamic models in explaining community structure. In contrast, ecosystem ecologists (largely representing the soil biogeochemistry view) have not routinely used community-based models, which in turn are based on individual-level behavior, in understanding population phenomena. This separation between ecosystem and population ecologists has long hampered our understanding of linkages between biodiversity—which is a function of population and community ecology—and ecosystem processes, and the effects of changes in species or functional groups of soil organisms on ecosystem processes and associated ecosystem services (Fitter 2005; Loreau 2010). Population and community ecologists (largely representing the soil biology view) provide accumulating evidence that soil biodiversity (in terms of functional groups and functional composition) can matter for *all* aggregate ecosystem functions in Fig. 1.3.2. Examples of functional groups with reference to these aggregate ecosystem functions include:

- the litter-transformers and the micro-food web, affecting C transformations and nutrient cycling (Heemsbergen *et al.* 2004; De Vries *et al.* 2006; Postma-Blaauw *et al.* 2006; van Eekeren *et al.* 2008);
- the root/rhizosphere biota, such as nitrogen fixing bacteria (Giller 2001) and mycorrhizal fungi (Cardoso & Kuyper 2006) and root herbivores, affecting nutrient cycling and plant performance (Bezemer *et al.* 2005; Piśkiewicz *et al.* 2008);
- the ecosystem engineers, such as earthworms and plant roots, affecting soil structure maintenance, the soil as a habitat for other soil organisms and plants (Brussaard *et al.* 2007a; van Eekeren *et al.* 2008) and the greenhouse balance of the soil (Giannopoulos *et al.* 2010);
- the antagonists of soil-borne diseases, (Garbeva *et al.* 2004; Raaijmakers *et al.* 2009), affecting biological population regulation.

Yet, the relationships between the soil functional groups and the aggregate ecosystem functions and services are not straightforward. For example, using data on seven ecosystem functions from the BIODEPTH project (Spehn *et al.* 2005), Hector and Bagchi (2007) found appreciable differences in the sets of species influencing different ecosystem functions, resulting in a positive saturating relationship between the number of ecosystem functions and the number of species influencing overall functioning. Moreover, the various spatial and temporal scales over which soil (biological) processes take place and interact, result in momentary observations at particular scales which are difficult to interpret (Ettema & Wardle 2002; Brussaard *et al.* 2007a).

Modeling has proved useful to better understand the relations between the soil biota and ecosystem functioning and to identify gaps in our knowledge (de Ruiter *et al.* 1993, 1995). The ensuing insight holds promise for answering fundamental questions related to the partitioning of C and nutrients during decomposition among the various functional groups of the soil biota and associated ecosystem processes and services in soil. Such fundamental questions on the aggregate ecosystem function "C transformations" (questions 1–3; Kibblewhite *et al.* 2008) and "nutrient (N) cycling" (questions 4–6; Schimel & Bennett 2004) include:

1) How might the allocation of soil carbon among the various functional groups regulate functional outputs?
2) What quantities and qualities of organic matter are needed to support soil system performance?
3) How do the forms and flows of soil carbon to and between different functional groups of soil organisms exert control over the physical condition of the soil habitat?
4) How are biotic processes, such as depolymerization, mineralization, microbial uptake, and root uptake linked?
5) How important are the physical and spatial processes that are occurring at the microsite scale in regulating macroscale characteristics of ecosystem N cycling?
6) How important are roots and mycorrhizas in creating high-N or low-N microsites and in mediating the biochemical/biological processes and their linkages?

These questions are relevant for *all* soil biota performing the aggregate ecosystem functions of Fig. 1.3.2, but questions 3 and 5 are particularly relevant for the soil ecosystem engineers, such as earthworms, ants, termites, and enchytraeids. The input of carbon in the soil provides them with the energy to create macro- and microaggregates. However, the effects of the engineering soil fauna on carbon and nutrient cycling are rarely quantitatively integrated with their effects on soil physical properties and transport processes, although plant growth and productivity are mediated by the biogenic structures they create (Brussaard *et al.* 2007a; Fig. 1.3.3). The lack of integration may lead to serious flaws in the estimation of soil organic matter under environmental change (Smith *et al.* 1998; Lavelle *et al.* 2004). Earthworms play a decisive role where they occur (Six *et al.* 2004). Different species of soil ecosystem engineers may create contrasting structures in terms of aggregation (aggregate size, stability and quality, and content of organic matter), porosity (pore

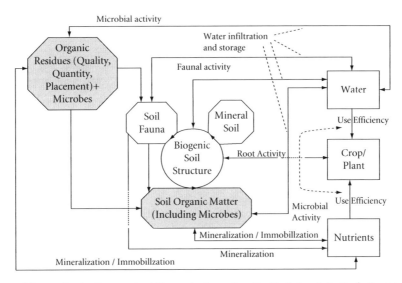

Figure 1.3.3 Conceptual diagram showing that organic additions and soil organic matter (shaded) are indicative for the state of the soil, affecting (interactions between) soil organisms with impact on the carbon, water and nutrient cycles and soil structure, which in turn affect plant water and nutrient use efficiencies and, hence, plant performance. (Modified after Brussaard et al. 2007a, with permission from Elsevier Science Ltd.)

volume and pore size distribution), and soil biota which inhabit these structures. Yet, differences between structures produced by broad taxonomic groups (earthworms, ants, termites, and enchytraeids) appear to be larger than those between structures produced within such groups (Lavelle 2002). Hence, in terms of the dynamics of biogenic structures and the ecosystem functions associated with them, earthworms, termites, ants, and enchytraeids should be considered as different functional groups within the soil ecosystem engineers. Further functional group division will be important as some earthworm species have been described as compacting and others as de-compacting (Blanchart et al. 2004).

The questions listed earlier require model-assisted research with a scope beyond the available environment-driven soil biogeochemistry or trophic interactions-driven soil biology models. Required models should not be merely related to (the persistence of) organic matter transformations that result in certain levels of carbon and nutrient cycling, but also to the partitioning of energy to soil structure maintenance and biological population regulation (Fig. 1.3.2). They should also define thresholds beyond which the described soil ecosystem moves to a new (and usually undesirable) stability domain (de Ruiter et al. 2005; Solé & Bascompte 2006). This is the domain of the upcoming field of interaction web modeling of the whole suite of ecological functions (e.g. Thébault & Fontaine 2010).

1.3.3 Understanding ecosystem structure: revisiting the functional group concept

The ecosystem functions depicted in Fig. 1.3.2 are ultimately driven by two biological processes of overriding importance, viz. photosynthesis (i.e. composition/C fixation, largely occurring aboveground, associated with plant growth) and respiration (i.e. decomposition/C dissipation, largely occurring belowground, inasmuch as associated with plant death). Recognizing C as the common denominator and main factor that integrates ecosystem functions suggests that our concept of soil functional groups conducting the ecosystem processes that result in ecosystem services cannot be discussed without accounting for a link to the vegetation. In nature, sets of plant species commonly occur together (as communities) and they are predictably characterized by common sets of plant traits, such as leaf size, specific leaf area, canopy height, life span, seed shape, etc. (Grime 2006; Díaz et al. 2007). A trait is a well-defined property of organisms, usually measured at the individual level and used comparatively across species. In trait-based ecology (Webb et al. 2010), organisms are characterized in terms of their multiple biological attributes such as physiological, morphological, behavioral, or life-history traits. The conceptual foundation consists of trait distributions (initially derived from the pool of possible traits of individual organisms) and performance filters (i.e. environmental filters eliminating traits with inadequate local fitness), resulting in associated community composition and ecosystem functioning (e.g. Grime 2006). Likewise, soil communities are then physically manifested via environmental filters that determine the proving-ground for the expression of genes, ultimately associated with ecosystem services (Fig. 1.3.4). It is important to note that, among the filtered plant traits, several have a direct bearing on the decomposition process (i.e. C dissipation) and the associated nutrient cycling. Whereas differences in decomposition rates of litter between biomes can largely be explained by differences in climate (i.e. temperature and precipitation) regimes, differences *within* biomes can largely be ascribed to the legacy of plant functional traits as litter quality (Cornwell et al. 2008; Fortunel et al. 2009), which in turn filter the traits represented in the soil microbial (e.g. Eskelinen et al. 2009) and soil faunal (e.g. Hättenschwiler & Gasser 2005) community. Taking dead wood as a quantitatively important example of plant litter globally, Cornwell et al. (2008) reviewed the importance of microbial, faunal, and physical degradation and fire as decomposition pathways. Whilst absolute figures are difficult to obtain (and hence, in need of collection), Cornwell et al. (2008) argue that plant species with different litter traits have afterlife effects selecting for soil microbial and insect traits associated with decomposition, with quantitatively important effects on global carbon fluxes. This relationship is affected by historical biogeography. For example, termites, which may be responsible for the decomposition of half of the wood in tropical forests (Cornwell et al. 2008), are represented by fungus-growing species in Africa, Madagascar, and Asia, but not in America and Australia. Biogeographic effects on the relationships between soil biota and ecosystem services are further dealt with in this volume (cf. Chapter 3.6). *Vice versa*, the filtered traits of the soil biota will have a direct bearing on the vegetation, influencing net photosynthesis, i.e. carbon fixation (e.g. Bezemer et al. 2005; Soler et al. 2008). Hence, it is not just abiotic, but also biotic environmental filters that determine the morphological, physiological, behavioral, and life-history traits present in both plant and soil communities. These traits determine the fitness (growth, reproduction, survival), performance (interactions with other species and the environment), and information (genome size and content, mutation rate) of the constituent species which are at the base of ecosystem functions.

In human-dominated systems, the environmental filters are to a large extent imposed by land management, which is essentially soil management (Fig. 1.3.1). The challenge to soil ecology is to better understand the sorting of traits in *natural* communities (comprising both plants, animals, and microbes) and apply those in *managed* ecosystems. Rather than starting from a preconception of functional groups associated with a number of ecosystem functions, trait-based ecology delineates functional groupings for every ecosystem function under consideration, focusing on the relevant traits and using a number of explicit steps (Petchey & Gaston 2006; Fig. 1.3.5). Hence, the concept of functional diversity has developed into functional *trait* diversity, i.e. the value and range of those organismal and species traits that influence ecosystem functions (Petchey & Gaston 2006). Most functional trait diversity measures are related to resource use complementarity,

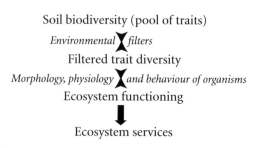

Figure 1.3.4 Environmental filters determine trait diversity which, thorough properties and activities of organisms, results in ecosystem functioning and services.

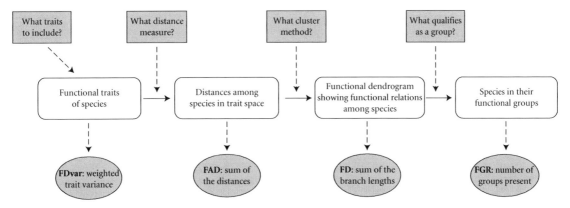

Figure 1.3.5 The process of ascribing a functional grouping (unshaded objects) and estimating different measures of functional diversity (shaded ellipses). The shaded rectangular boxes represent decisions in the process of making a classification, so that the number of decisions required for each measure increases from left to right. (From Petchey and Gaston 2006, with permission from John Wiley & Sons.)

which would imply that greater trait diversity means a more complete and/or efficient use of resources and that changes in functional diversity are an indicator of changes in ecosystem functioning, inasmuch as process rates are not determined by abiotic conditions (Petchey & Gaston 2006). The choice of traits is critical and obviously restricted by lack of knowledge. The quantitative methods for selecting traits and the appropriate number of traits associated with a certain ecosystem function are in development (e.g. Naeem & Bunker 2009).

There is evidence that the sorting of traits in natural communities occurs according to assembly rules, rather than by chance (e.g. Cavender-Bares et al. 2009). As traits arise along the Tree of Life, often reflecting their biogeographical origins, and tending to be shared by species that have common ancestry (phylogenetic history), it can be argued that assemblages based on phylogenetic diversity are better correlated with ecosystem functions than species richness or other measures of diversity (Maherali & Klironomos 2007; Cavender-Bares et al. 2009; Donoghue et al. 2009). The very existence of numerous different natural communities suggests that, over evolutionary time scales, a variety of trade-off possibilities have been operational for both plants and microorganisms and, indeed, the whole food web, resulting in the persistence of the aggregate functions and favoring sustainability of the ecosystem over ecological time scales. One driving force behind the partitioning of energy is modulation of plant mineral nutrient availability by the plant itself, in addition to modulation by the soil food web, ultimately matching the nutrient stoichiometry of the plant species in the community. Another driving force is enhanced plant systemic defense induced by beneficial microbes interacting with plant roots, and there is evidence that the two reinforce each other (Phelan 2009). Considering the aggregate soil functions of Fig. 1.3.2, it seems likely that yet another driving force is soil porosity and associated water-holding capacity as important determinants of plant growth, for which part of the organic matter is allocated to feed ecosystem engineers (Lavelle et al. 2001), creating biogenic structures in soil (Fig. 1.3.3).

1.3.4 Understanding effects of environmental drivers and land management on ecosystem functioning and services

1.3.4.1 Environmental drivers

Most of the functional groups in Fig. 1.3.2 are restricted to one trophic level. However, there is accumulating evidence that the type, range, and especially relative abundance of functional traits in biotic communities exert control over different ecosystem functions, meaning that there are multiple associations between traits and ecosystem services across different trophic levels (de Bello et al. 2010).

For example, associated traits for plants and soil organisms show close associations with C and nutrient cycling (through effects on C sequestration and decomposition), herbivory, and productivity on the one hand, and water flow, and soil and sediment formation on the other (de Bello et al. 2010).

An important question arises: how do trait–function relationships perform under the influence of (global) environmental change drivers? The answer is not straightforward for two reasons. Firstly, changes in the environment can affect ecosystem functions directly through effects on abiotic controls and indirectly through effects on the physiology, morphology, and behavior of individual organisms, the structure of populations, and the composition of communities, meaning that scale is important (for a more elaborate treatment of the scale issue, see Swift et al. 2004; Young & Crawford 2004; Brussaard et al. 2007a; and Suding et al. 2008). Secondly, the traits affected by these drivers (response traits) are not predictably related to the traits affecting ecosystem functions and services (effect traits).

The degree to which individuals with response traits favored by the changed environment differ in their effect traits compared with the initial assemblage will determine the extent to which community change influences ecosystem functioning (Suding et al. 2008).

1.3.4.2 Land management

Land management constitutes another important category of change drivers. Agriculture is a case in point because the six *fundamental* questions listed earlier in Section 1.3.2, have also surfaced in recent years as *practical* questions, mostly framed as concern over the possible decrease of organic matter contents in agricultural soils worldwide (Dawson & Smith, 2007; Hanegraaf et al. 2009; Reijneveld et al. 2009). Farmers' options to bypass the natural functioning of the soil by external inputs, such as fertilizers, pesticides, and water have decreased, at least in industrialized countries, due to environmental regulations to restore ecosystem services that do not have a market value such as most of the ecosystem services in Fig. 1.3.2. All these inputs are energy-intensive and becoming increasingly expensive.

Concomitantly, the direct costs of fossil fuel to operate farm machinery have risen sharply, which is one of the reasons for the current interest in conservation tillage. Hence, the need to significantly reduce levels of external inputs to the soil is widespread. The relevance of this quest for the post-carbon economy needs no further explanation.

The questions referred to previously are pertinent from a high-input as well as a low-input starting point, the latter of which is prevalent in countries where external inputs are not available and/or affordable to the farmers (i.e. in many developing countries). Under such conditions, agriculture has resulted in soil fertility decline, especially in sub-Saharan Africa, where soil fertility is inherently low. External inputs should not be excluded, wherever the natural conditions are too poor to allow acceptable yield levels (Vanlauwe & Giller 2006; Vanlauwe et al. 2010). However, such inputs should be instrumental to the reinforcement of natural processes in the (re)-design and optimization of the management of agricultural and non-productive landscape components, so as to avoid any disadvantages of external inputs the developed world is just trying to repair.

1.3.5 Working with nature

What knowledge does trait-based ecology offer to respond to the global quest for measures and means to work with nature? An obvious first step is to analyze existing agro-ecosystems from a trait-based ecology perspective. In an illuminating study by Glover et al. (2010) it was found that unfertilized native North American prairie grasslands, annually harvested for roughly 75 years, yielded similar amounts of nitrogen per hectare in biomass as adjacent high-input wheat fields yielded in grain. The grasslands, despite being harvested each year, supported a range of ecosystem functions, among which were greater C storage and N retention in soil than the annual croplands. Culman et al. (2010) found that these harvested perennial grasslands also supported higher levels of soil fertility and soil structure and more complex biological communities than the annual crops produced for similar periods of time on adjacent farm fields. In a grassland conversion study, which allowed for direct comparison between native grasslands and grasslands converted to no-till (i.e. never-tilled) or conventional-tillage arable agriculture, DuPont et al. (2010) found that even the conversion of the harvested grasslands to no-till (never-tilled) annual crop production resulted in reduced root biomass and decreased active soil C stocks, and negatively impacted the soil biota and food webs important in nutrient cycling after just three years. However, the level of plant species diversity found in natural grasslands was not necessary to maintain high productivity. Over the 11-year period of the study, average yields of perennial legume and perennial warm-season grass bi-cultures were similar to those of 16-species plots, leading DeHaan et al. (2010) to conclude that by selecting plant species or cultivars with "key characteristics", farmers might be able to achieve many of the advantages and services of higher-diversity natural ecosystems using lower diversity agricultural production systems. It would appear that a trait-based ecology approach offers promise in providing a sound scientific base to the choice of species or cultivars, not only in view of maintaining aboveground productivity, but also belowground ecosystem services. For example, De Deyn et al. (2009) found that vegetation C and N and soil microbial biomass increased significantly with a greater number of plant species and functional groups, but soil C and N pools were affected only by functional group composition. From this set of studies it can be inferred that a judicious choice of crops/cultivars, combining the traits for production aboveground and ecosystem services belowground is feasible and this clearly warrants further study.

Likewise, analysis of grazing systems, inter/mixed/relay cropping systems and perennial *versus* annual arable systems should be illuminating in terms of trait associations below- and aboveground. Grazing can affect plant composition and consequent C storage. For example, Bagchi and Ritchie (2010) found that, despite comparable grazing intensities, the displacement of native herbivores in Trans-Himalayan watersheds by livestock reduced soil C by almost 50%, driven by livestock diet selection, which leads to vegetation shifts that lower plant production and likely reduce C inputs in soil. Many cash crops are produced in annual cropping systems where soil disturbance by tillage and other field operations are detrimental to the larger soil fauna and perennial plants. However, the more agriculture moves into the direction of conservation agriculture (i.e. agriculture characterized by minimal soil disturbance), keeping the soil continuously covered by mulching/cropping and applying crop rotations and diversification (Hobbs et al. 2008, but see Giller et al. 2009), the more similar the ecological processes in agriculture will be to those in "natural" systems and the better the prospect will be of applying ecological knowledge to natural ecosystems in agriculture.

In conservation agriculture systems, mulches (Kruidhof et al. 2009), cover crops (Kruidhof et al. 2008), and intercrops (Baumann et al. 2000) that are increasingly applied as components in an integrated weed management strategy, can also serve the production of additional marketable crops and the delivery of various ecosystem services (e.g. Schepers et al. 2005). The scientific challenge here is to expand the study of genotype–management–environment (GxMxE) trait-based interactions from the individual crop level to the relevant multi-species trait associations (Ong et al. 2004; Bastiaans et al. 2007). So, while perennial cropping systems are generally considered preferable in terms of productivity and delivery of other ecosystem services (van Eekeren et al. 2008; Glover 2010), a judicious choice of crops in arable agriculture, animals in livestock husbandry and grazing systems, and crops and animals in mixed farming, holds promise to optimize the delivery of ecosystem goods and services, and to make agricultural production systems robust (i.e. adaptable to changing climate variability and environmental risks). In this context, the judiciousness hinges on the outcome of research for options to choose crops/varieties and livestock/breeds and to choose the amounts and qualities of organic matter entering the soil that optimize the match between the provision of agricultural goods and ecosystem services at the desired productivity level in a way for generations of farmers to thrive. The evidence discussed here, albeit limited, suggests that a trait-based approach, addressing multiple associations between traits and ecosystem services across different trophic levels, is promising for the design of agro-ecosystems.

1.3.6 Landscape context

The prospects are even brighter when we put these developments and opportunities into a landscape perspective. A landscape is characterized by both agricultural crop and livestock diversity and "wild" (be it planned or unplanned) biodiversity in a certain spatial configuration. Landscape composition and configuration determine to what extent agriculture benefits from biodiversity and the associated ecosystem services (e.g. by providing habitat for natural enemies of pests; Tscharntke *et al.* 2005) and, *vice versa*, to what extent agriculture can be improved to do less damage and contribute to conservation of the biodiversity that can be associated with such land use (e.g. by providing habitat for farm birds) and ecosystem services (e.g. water storage and nutrient retention).

For measures to be effective, entry points for biological management need to be identified (Fig. 1.3.6). The framework of Fig. 1.3.4 can be used to analyze the dependence of the functioning of ecosystems on the existence of trait filters and resulting trait distributions, using a procedure developed by Díaz *et al.* (2007). This procedure and other ones (e.g. van Eekeren *et al.* 2008), using the principle of parsimony, are aimed at identifying those cases in which trait diversity matters for ecosystem services. As trait-based ecology theory develops towards projection of performance filters across environmental gradients to make predictions about response and effect traits of plants and the soil biota, so will our ability to target management for ecosystem services on the environmental filters selecting for the species assemblages carrying the traits and trait values needed to sustain the associated ecosystem functions. Indeed, a further prospect is to apply and further develop trait-based ecology to design (agro)-ecosystems at the

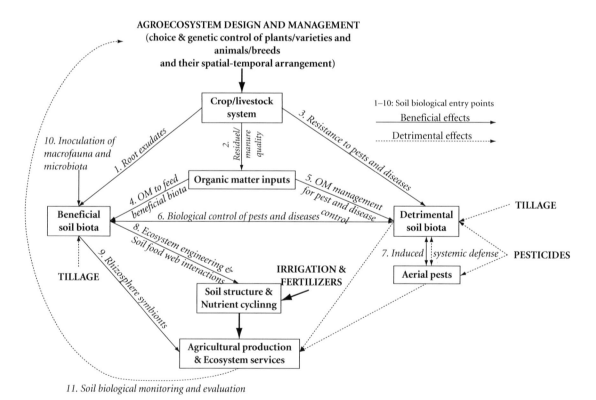

Figure 1.3.6 The potential entry points (1–10) for biological management of crop/livestock systems, organic matter inputs and soil organisms, aimed at sustainable agricultural production and ecosystem services, and feedback to agro-ecosystem design and management using monitoring and evaluation (11). OM = organic matter. (Modified after Brussaard *et al.* 2007b, with permission from Elsevier Science Ltd.)

landscape scale in ways that are conducive to wild biodiversity and to the use of currently un-utilized or under-utilized crops and cultivars and livestock breeds which enhance food security, improve environmental health, and foster social well-being (Brussaard *et al.* 2010).

1.3.7 Conclusions

The delivery of ecosystem goods and services depends on the structure and functioning of ecosystems, which are affected by (global) environmental change and land management effects on the soil biota. To understand and ultimately be able to manipulate the relationships between the soil biota and the provision of ecosystem goods and services, the utility of the concept of "functional groups" of the soil biota is limited. Functional groups are usually defined at one trophic level only (e.g. fungal feeders) and the relationship between such groups and ecosystem functioning is not straightforward. There is evidence that the type, range, and relative abundance of functional organismal and species traits in biotic communities exert control over different ecosystem functions, meaning that there are multiple associations between traits and ecosystem services across different trophic levels. From a functional trait-based ecology perspective, existing agro-ecosystems can be analyzed in terms of the sustained delivery of ecosystem goods and services. Such analyses will be helpful in the design of resource use-efficient and biodiverse agro-ecosystems and agricultural landscapes. A range of entry points for biological management of such systems is available for sustainable production of goods and ecosystem services.

Acknowledgments

I thank Thomas Kuyper for stimulating discussions and constructive comments on an earlier version of the chapter. The comments of two anonymous reviewers and the section editors Karl Ritz and Wim van der Putten are gratefully acknowledged. Lammert Bastiaans and Felix Bianchi also made some useful suggestions. Caroline Wiltink edited the references. This work was supported by the European Commission within the project EcoFINDERS - Ecological Function and Biodiversity Indicators in European Soils (FP7-264465).

References

Bagchi, S., & Ritchie, M.E. (2010) Introduced grazers can restrict potential soil carbon sequestration through impacts on plant community composition. *Ecology Letters* **13**: 959–68.

Bastiaans, L., Zhao, D.L., den Hollander, N.G., Baumann, D.T., Kruidhof, H.M., & Kropff, M.J. (2007) Exploiting diversity to manage weeds in agro-ecosystems. In: J.H.J. Spiertz, P.C. Struik, & H.H. van Van Laar (eds.) *Scale and Complexity in Plant Systems Research: Gene-Plant-Crop Relations*, pp. 267–84. Springer, Dordrecht.

Baumann, D.T., Kropff, M.J., & Bastiaans, L. (2000) Intercropping leeks to suppress weeds. *Weed Research* **40**: 359–74.

Beare, M.H., Coleman, D.C., Crossley, D.A., Hendrix, P.F., & Odum, E.P. (1995) A hierarchical approach to evaluating the significance of soil biodiversity to biogeochemical cycling. *Plant and Soil* **170**: 5–22.

Bezemer, T.M., de Deyn, G.B., Bossinga, T.M., van Dam, N.M., Harvey, J.A., & van der Putten, W.H. (2005) Soil community composition drives aboveground plant-herbivore-parasitoid interactions. *Ecology Letters* **8**: 652–61.

Bezemer, T.M., & van Dam, N.M. (2005) Linking aboveground and belowground interactions via induced plant defenses. *Trends in Ecology & Evolution* **20**: 617–24.

Blanchart, E., Albrecht, A., Brown, G., et al. (2004) Effects of tropical endogeic earthworms on soil erosion. *Agriculture Ecosystems and Environment* **104**: 303–15.

Brussaard, L. (1998) Soil fauna, guilds, functional groups and ecosystem processes. *Applied Soil Ecology* **9**: 123–35.

Brussaard, L., Caron, P., Campbell, B., et al. (2010) Reconciling biodiversity conservation and food security: scientific challenges for a new agriculture. *Current Opinion in Environmental Sustainability* **2**: 34–42.

Brussaard, L., de Ruiter, P.C., & Brown, G.G. (2007b) Soil biodiversity for agricultural sustainability. *Agriculture, Ecosystems and Environment* **121**: 233–44.

Brussaard, L., Pulleman, M.M., Ouédraogo, E., Mando, A., & Six, J. (2007a) Soil fauna and soil function in the fabric of the food web. *Pedobiologia* **50**: 447–62.

Cardoso, I.M., & Kuyper, T.W. (2006) Mycorrhizas and tropical soil fertility. *Agriculture, Ecosystems and Environment* **116**: 72–84.

Cavender-Bares, J., Kozak, K.H., Fine, P.V.A., & Kembel, S.W. (2009) The merging of community ecology and phylogenetic biology. *Ecology Letters* **12**: 693–715.

CEC (2006) *Proposal for a Directive of the European Parliament and of the Council Establishing a Framework for the Protection of Soil and Amending Directive 2004/35/EC.*

COM(2006) 232 final, 2006/0086 (COD), pp. 1–30. Commission of the European Communities, Brussels.

Cornwell, W.K., Cornelissen, J.H.C., Amatangelo, K., et al. (2008) Plant species traits are the predominant control on litter decomposition rates within biomes worldwide. *Ecology Letters* **11**: 1065–71.

Culman, S.W., Young-Mathews, A., Hollander, A.D., et al. (2010) Biodiversity is associated with indicators of soil ecosystem functions over a landscape gradient of agricultural intensification. *Landscape Ecology* **25**: 1333–48.

Dawson, J.J.C., & Smith, P. (2007) Carbon losses from soil and its consequences for land-use management. *Science of the Total Environment* **382**: 165–90.

De Bello, F., Lavorel, S., Díaz, S., et al. (2010) Towards an assessment of multiple ecosystem processes and services via functional traits. *Biodiversity and Conservation* **19**: 2873–93.

De Deyn, G.B., Quirk, H., Yi, Z., Oakley, S., Ostle, N.J., & Bardgett, R.D. (2009) Vegetation composition promotes carbon and nitrogen storage in model grassland communities of contrasting soil fertility. *Journal of Ecology* **97**: 864–75.

De Ruiter, P.C., Neutel, A.M., & Moore, J.C. (1995) Energetics, patterns of interaction strengths and stability in real ecosystems. *Science* **269**: 1256–60.

De Ruiter, P.C., van Veen, J.A., Moore, J.C., Brussaard, L., & Hunt, H.W. (1993) Calculation of nitrogen mineralization in soil food webs. *Plant and Soil* **157**: 263–73.

De Ruiter, P.C., Wolters, V., Moore, J.C., & Winemiller, K.O. (2005) Food web ecology: playing jenga and beyond. *Science* **309**: 68–71.

De Vries, F.T., Hoffland, E., van Eekeren, N., Brussaard, L., & Bloem, J. (2006) Fungal/bacterial ratios in grasslands with contrasting nitrogen management. *Soil Biology and Biochemistry* **38**: 2092–103.

DeHaan, L.R., Weisberg, S., Tilman, D., & Fornara, D. (2010) Agricultural and biofuel implications of a species diversity experiment with native perennial grassland plants. *Agriculture, Ecosystems and Environment* **137**: 33–8.

Díaz, S., Lavorel, S., De Bello, F., Quétier, F., Grigulis, K., & Robson, T.M. (2007) Incorporating plant functional diversity effects in ecosystem service assessments. *Proceedings of the National Academy of Sciences of the United States of America* **104**: 20684–9.

Dominati, E., Patterson, M., & Mackay, A. (2010) A framework for classifying and quantifying the natural capital and ecosystem services of soils. *Ecological Economics* **69**: 1858–68.

Donoghue, M.J., Yahara, T., Conti, E., et al. (2009) *Providing an Evolutionary Framework for Biodiversity Science—bioGenesis Science Plan and Implementation Strategy*. Report No. 6, pp. 52. Diversitas, Paris.

DuPont, S.T., Culman, S.W., Ferris, H., Buckley, D.H., & Glover, J.D. (2010) No-tillage conversion of harvested perennial grassland to annual cropland reduces root biomass, decreases active carbon stocks, and impacts soil biota. *Agriculture, Ecosystems and Environment* **137**: 25–32.

Eskelinen, A., Stark, S., & Männistö, M. (2009) Links between plant community composition, soil organic matter quality and microbial communities in contrasting tundra habitats. *Oecologia* **161**: 113–23.

Ettema, C.H., & Wardle, D.A. (2002) Spatial soil ecology. *Trends in Ecology & Evolution* **17**: 177–83.

Faber, J.H. (1991) Functional classification of soil fauna: a new approach. *Oikos* **62**: 110–17.

Fierer, N., Grandy, A.S., Six, J., & Paul, E.A. (2009) Searching for unifying principles in soil ecology. *Soil Biology and Biochemistry* **41**: 2249–56.

Fitter, A.H. (2005) Darkness visible: reflections on underground ecology. *Journal of Ecology* **93**: 231–43.

Fortunel, C., Garnier, E., Joffre, R., et al. (2009) Leaf traits capture the effects of land use changes and climate on litter decomposability of grasslands across Europe. *Ecology* **90**: 598–611.

Garbeva, P., van Veen, J.A., & van Elsas, J.D. (2004) Microbial diversity in soil: selection of microbial populations by plant and soil type and implications for disease suppressiveness. *Annual Review of Phytopathology* **42**: 243–70.

Giannopoulos, G., Pulleman, M.M., & van Groenigen, J.W. (2010) Interactions between residue placement and earthworm ecological strategy affect aggregate turnover and N_2O dynamics in agricultural soil. *Soil Biology and Biochemistry* **42**: 618–25.

Giller, K.E. (2001) *Nitrogen Fixation in Tropical Cropping Systems*. CABI Publishing, Wallingford.

Giller, K.E., Witter, E., Corbeels, M., & Tittonell, P. (2009) Conservation agriculture and smallholder farming in Africa: the heretics' view. *Field Crops Research* **114**: 23–34.

Glover, J.D. (2010) Harvested perennial grasslands: ecological models for farming's perennial future. *Agriculture, Ecosystems and Environment* **137**: 1–2.

Glover, J.D., Culman, S.W., DuPont, S.T., et al. (2010) Harvested perennial grasslands provide ecological benchmarks for agricultural sustainability. *Agriculture, Ecosystems and Environment* **137**: 3–12.

Grime, J.P. (2006) Trait convergence and trait divergence in herbaceous plant communities: mechanisms and consequences. *Journal of Vegetation Science* **17**: 255–60.

Hanegraaf, M.C., Hoffland, E., Kuikman, P.J., & Brussaard, L. (2009) Trends in soil organic matter contents in Dutch grasslands and maize fields on sandy soils. *European Journal of Soil Science* **60**: 213–22.

Hättenschwiler, S., & Gasser, P. (2005) Soil animals alter plant litter diversity effects on decomposition. *Proceedings of the National Academy of Sciences of the USA* **102**: 1519–24.

Haygarth, P.M., & Ritz, K. (2009) The future of soils and land use in the UK: soil systems for the provision of land-based ecosystem services. *Land Use Policy* **26**: S187–97.

Hector, A., & Bagchi, R. (2007) Biodiversity and ecosystem multifunctionality. *Nature* **448**: 188–90.

Heemsbergen, D.A., Berg, M.P., Loreau, M., van Hal, J.R., Faber, J.H., & Verhoef, H.A. (2004) Biodiversity effects on soil processes explained by interspecific functional dissimilarity. *Science* **306**: 1019–20.

Hobbs, P.R., Sayre, K., & Gupta, R. (2008) The role of conservation agriculture in sustainable agriculture. *Philosophical Transactions of the Royal Society B* **363**: 543–55.

Kibblewhite, M.G., Ritz, K., & Swift, M.J. (2008) Soil health in agricultural systems. *Philosophical Transactions of the Royal Society B* **363**: 685–701.

Kruidhof, H.M., Bastiaans, L., & Kropff, M.J. (2008) Ecological weed management by cover cropping: effects on weed growth in autumn and weed establishment in spring. *Weed Research* **48**: 492–502.

Kruidhof, H.M., Bastiaans, L., & Kropff, M.J. (2009) Cover crop residue management for optimizing weed control. *Plant and Soil* **318**: 169–84.

Lavelle, P. (2002) Functional domains in soils. *Ecological Research* **17**: 441–50.

Lavelle, P., Barros, E., Blanchart, E., *et al.* (2001) SOM management in the tropics: why feeding the soil macrofauna? *Nutrient Cycling in Agroecosystems* **61**: 53–61.

Lavelle, P., Charpentier, F., Villenave, C., *et al.* (2004) Effects of earthworms on soil organic matter and nutrient dynamics at a landscape scale over decades. In: C.A. Edwards (eds.) *Earthworm ecology*, pp. 145–60. CRC Press, Boca Raton, FL.

Lavelle, P., Lattaud, C., Trigo, D., & Barois, I. (1995) Mutualism and biodiversity in soils. *Plant and Soil* **170**: 23–33.

Loreau, M. (2010) Linking biodiversity and ecosystems: Towards a unifying ecological theory. *Philosophical Transactions of the Royal Society B: Biological Sciences* **365**: 49–60.

Maherali, H., & Klironomos, J.N. (2007) Influence of phylogeny on fungal community assembly and ecosystem functioning. *Science* **316**: 1746–8.

MEA (2005) *Millenium Ecosystem Assessment. Ecosystems and Human Well-being: Current Status and Trends: Findings of the Condition and Trends Working Group*, pp. 831. Island Press, Washington, DC.

Naeem, S., & Bunker, D.E. (2009) TraitNet: furthering biodiversity research through the curation, discovery, and sharing of species trait data. In: S. Naeem, D.E. Bunker, A. Hector, M. Loreau & C. Perrings (eds.) *Biodiversity, ecosystem functioning and human wellbeing: an ecological and economic perspective*, pp. 281–9. Oxford University Press, New York.

Ong, C.K., Kho, R.M., & Radersma, S. (2004) Ecological interactions in multispecies agroecosystems: concepts and rules. In: M. van Noordwijk, G. Cadisch & C.K. Ong, (eds.) *Below-ground interactions in tropical agroecosystems*, pp. 1–15. CABI Publishing, Wallingford.

Petchey, O.L., & Gaston, K.J. (2006) Functional diversity: back to basics and looking forward. *Ecology Letters* **9**: 741–58.

Phelan, P.L. (2009) Ecology-based agriculture and the next green revolution—is modern agriculture exempt from the laws of ecology? In: P.J. Bohlen & G. House (eds.) *Sustainable agroecosystem management—integrating ecology, economics and society*, pp. 97–135. CRC Press, Boca Raton, FL.

Piskiewicz, A.M., Duyts, H., & van der Putten, W.H. (2008) Multiple species-specific controls of root-feeding nematodes in natural soils. *Soil Biology and Biochemistry* **40**: 2729–35.

Postma-Blaauw, M.B., Bloem, J., Faber, J.H., van Groenigen, J.W., de Goede, R.G.M., & Brussaard, L. (2006) Earthworm species composition affects the soil bacterial community and net nitrogen mineralization. *Pedobiologa* **50**: 243–56.

Raaijmakers, J.M., Paulitz, T.C., Steinberg, C., Alabouvette, C., & Moënne-Loccoz, Y. (2009) The rhizosphere: a playground and battlefield for soil-borne pathogens and beneficial microorganisms. *Plant and Soil* **321**: 341–61.

Reijneveld, A., van Wensem, J., & Oenema, O. (2009) Soil organic carbon contents of agricultural land in the Netherlands between 1984 and 2004. *Geoderma* **152**: 231–8.

Schepers, J.S., Francis, D.D., & Shanahan, J.F. (2005) Relay cropping for improved air and water quality *Journal of Biosciences* **60**: 186–9.

Schimel, J.P., & Bennett, J. (2004) Nitrogen mineralization: challenges of a changing paradigm. *Ecology* **85**: 591–602.

Siepel, H., & de Ruiter Dijkman, E.M. (1993) Feeding guilds of oribatid mites based on their carbohydrase activities. *Soil Biology and Biochemistry* **25**: 1491–7.

Six, J., Bossuyt, H., Degryze, S., & Denef, K. (2004) A history of research on the link between (micro)aggregates, soil biota, and soil organic matter dynamics. *Soil and Tillage Research* **79**: 7–31.

Smith, P., Andrén, O., Brussaard, L., *et al.* (1998) Soil biota and global change at the ecosystem level: describing soil biota in mathematical models. *Global Change Biology* **4**: 773–84.

Solé, R.V., & Bascompte, J. (2006) *Self-organization in Complex Ecosystems*. Princeton University Press, Princeton, NJ.

Soler, R., Harvey, J.A., Bezemer, T.M., & Stuefer, J.F. (2008) Plants as green phones: novel insights into plant-mediated communication between below- and above-ground insects. *Plant Signaling and Behavior* **3**: 519–20.

Spehn, E.M., Hector, A., Joshi, J., *et al.* (2005) Ecosystem effects of biodiversity manipulations in European grasslands. *Ecological Monographs* **75**: 37–63.

Suding, K.N., Lavorel, S., Chapin, F.S., *et al.* (2008) Scaling environmental change through the community-level: a trait-based response-and-effect framework for plants. *Global Change Biology* **14**: 1125–40.

Swift, M.J., Izac, A.M.N., & van Noordwijk, M. (2004) Biodiversity and ecosystem services in agricultural landscapes—are we asking the right questions? *Agriculture, Ecosystems and Environment* **104**: 113–34.

Thébault, E., & Fontaine, C. (2010) Stability of ecological communities and the architecture of mutualistic and trophic networks. *Science* **329**: 853–6.

Tscharntke, T., Klein, A.M., Kruess, A., Steffan-Dewenter, I., & Thies, C. (2005) Landscape perspectives on agricultural intensification and biodiversity—ecosystem service management. *Ecology Letters* **8**: 857–74.

van Eekeren, N., Bommelé, L., Bloem, J., *et al.* (2008) Soil biological quality after 36 years of ley-arable cropping, permanent grassland and permanent arable cropping. *Applied Soil Ecology* **40**: 432–46.

Vanlauwe, B., Bationo, A., Chianu, J., *et al.* (2010) Integrated soil fertility management: operational definition and consequences for implementation and dissemination. *Outlook on Agriculture* **39**: 17–24.

Vanlauwe, B., & Giller, K.E. (2006) Popular myths around soil fertility management in sub-Saharan Africa. *Agriculture, Ecosystems and Environment* **116**: 34–46.

Wall, D.H., Bardgett, R.D., Covich, A.P., & Snelgrove, V.R. (2004) The need for understanding how biodiversity and ecosystem functioning affect ecosystem services in soils and sediments. In: D.H. Wall (ed.) *Sustaining biodiversity and ecosystem services in soils and sediments*, pp. 1–12. Island Press, Washington, DC.

Webb, C.T., Hoeting, J.A., Ames, G.M., Pyne, M.I., & Poff, N.L. (2010) A structured and dynamic framework to advance traits-based theory and prediction in ecology. *Ecology Letters* **13**: 267–83.

Wickings, K., Stuart Grandy, A., Reed, S., & Cleveland, C. (2010) Management intensity alters decomposition via biological pathways. *Biogeochemistry* **104**: 365–79.

Young, I.M., & Crawford, J.W. (2004) Interactions and self-organization in the soil-microbe complex. *Science* **304**: 1634–7.

Synthesis

Karl Ritz and Wim H. van der Putten

The chapters in this section amply demonstrate that in terms of the delivery of ecosystem services, it is the living constituent of soils which are of paramount importance. This is not to downplay the roles of the abiotic constructs and circumstances in soils as they provide the basic framework and "ground rules" in which the biota operate. However, the particular feature of the soil biota is that it is adaptive, both in the short term and up to evolutionary timescales, in a manner which enhances the resilience of soil systems to perturbation. Thus, evolutionary adaptation of the soil biota is promoting soil ecosystem functioning and the delivery of ecosystem services. Part of this adaptation mechanism is underpinned by the manner in which soil biota shape the physical arrangement of their habitat (i.e. the pore network or architecture), which has an impact on both the architects and the other residents in the system, across scales of the individual, the population, and the community. The concept of self-organization of soil systems, as discussed by Lavelle, is potentially profound and important. If correct, it has major implications for how we should manage soils and ultimately the ecosystems they support. Crucially, it shows how management strategies need to consider the system-level context, the soil as a habitat, and that monotonously targeted single-point interventions, mapped by Brussaard (but noting that he also espouses the system-level approach) are unlikely to be effective in the round. Point-interventions may fix the immediate issue, but in complex systems the interconnectedness between components tends to result in a dissipation and compensation of such effects, and the resultant trade-offs may not be optimal at an overall level. This is notwithstanding that "optimal" may be perceived as different between different end-users.

In terms of studying and characterizing the huge diversity of soil biota, and particularly in the microbial domains, it is almost becoming a cliché to state that there is so much yet to discover in these realms. However, the rate of development in techniques in molecular genetics, and their ensuing application into the soil environmental arena, will undoubtedly result in a more sophisticated understanding of the biological basis of soil function. Wurst *et al.* highlight that the roles of soil archaea and viruses in soil functioning are barely considered, and both groups may play significant roles. Notably, molecular analysis is a critical tool in their study. Multispectrum (i.e. broad-scale to specific) molecular probes of increasing fidelity are required to underpin other key methodological advances which are needed in terms of visualizing the spatial location of organisms and associated potential or actual activity, within the soil matrix. Stable isotope probing is also a significant approach which will enable the links between the genetics of organisms and communities to their function.

A prevalent theme across the three chapters is that of the relative roles of biodiversity in underpinning ecosystem service delivery with respect to basic taxonomic richness versus the functional properties that the organisms impart. Evidence is increasing that species richness *per se* (i.e. a simple quantification of the number of taxonomically distinct organisms) is not that important; rather, it is much more subtle in that it is the functional repertoire of the organisms that is the critical issue. As Wurst *et al.* explain, species richness across feeding guilds has stronger effects on ecosystem functions at the microfaunal scale than species richness within feeding guilds. The relationship between soil microfaunal diversity and ecosystem functioning is not easy to predict and is dependent on the characteristics of the species and their abundance in the community. Such functional capabilities are manifest as the traits associated with individual organisms. The trait-based concept can be applied at both the individual organismal scale, as Wurst *et al.* show, but also at a community scale, as reviewed by Brussaard. At this community scale, trait-related issues of the relationships both within and between trophic levels also become important. As Brussaard explains, the community scale collection of traits, which undoubtedly is the key link to resultant function, is underwritten by the (community) genotype. When this genome is run through the local environmental sieve, the traits are then manifest as the phenotypes of the constituent organisms. However, there is also a community-scale phenotype expressed, which if considered as an "extended phenotype," starts to link to self-organization concepts, and system-level controls and feedbacks which could optimize overall delivery of services. The concept of trait-based approaches to characterize and formalize the functional consequences of soil biodiversity is of clear utility. It is in many ways affiliated to the feeding-guild approach which soil ecologists have previously espoused, but arguably considers such groupings at a finer scale. The issue of the level of resolution of description and analysis required to adequately understand biodiversity–function relationships is essentially another scaling issue (as indeed so many matters are when considering soils), and this is particularly so in modelling approaches. Brussaard argues the case that the soil biology perspective is likely to be superior to the soil biogeochemistry perspective. There remains however the need for empirical evidence—and its manifestation in models—to be realized.

These complexities and subtleties are of great consequence since they demonstrate a likely high degree of context-dependency for the relationships between biodiversity and ecosystem function, and a simple "more biodiversity is better" policy is, whilst doubtless convenient, unlikely to be effective. Thus in many instances case-specific solutions or strategies will be required, contingent on the local circumstances. It also explains why the considerations for further study in all three chapters tend to involve increasing the range, scale, and scope of research. Soils are complex systems. This means that system-level approaches have to be considered both in their study and in any consideration of their management. Research approaches that move back-and-forth from systems to community and individual approaches at the species and at the trait level must be chosen. These approaches, as will be shown in the next section, ultimately need to be connected to policy-relevant decision loop mechanisms, thereby ensuring continuous feedback interactions between science, policy, and society.

SECTION 2

From Genes to Ecosystem Services

SECTION EDITORS: **Wim H. van der Putten and Karl Ritz**

Introduction

Wim H. van der Putten and Karl Ritz

Ecosystem services arise as a result of the processes that determine the functioning of ecosystems. During the past two decades, there has been an exponential increase in the number of studies, from many perspectives, on the relationships between ecosystem functioning and biodiversity. Ultimately ecosystem functioning results from gene activities; hence, in this section the relationships between genes and ecosystem services are explored. Soil biota are at the basis of many ecosystem services. For example, soil biodiversity may relate to primary productivity, whereas (food) production that results from primary productivity is considered an ecosystem service. Ecosystem services can be subdivided into a number of categories, including provisioning services (e.g. food production), regulatory services (e.g. pest and climate control), and cultural services. The latter are perhaps more difficult to envision as a result of soil ecology. For example, orchids have a special meaning to humans, and are entirely dependent upon symbioses with mycorrhizal fungi, thus these soil biota are unequivocally important for maintaining a cultural service.

In general, there are six major groups of ecosystem services to which soil biota contribute, *viz.* soil organic matter and fertility, regulation of carbon flux and climate control, regulation of the water cycle, decontamination and bioremediation, pest control, and human health. The role of soil biodiversity, or belowground genetic diversity, in providing these services can vary depending on underlying processes. For example, the decomposition of organic matter is a process carried out by a large variety of belowground species, whereas pest control may be due to a relatively smaller number of species that can do this particular job. In all cases, the functions ultimately relate to the genes that underlie the processes. Nitrogen-fixing bacteria have specific genes that enable them to convert atmospheric nitrogen into mineral forms of this element. Information on genes involved in such specific processes can be detected by molecular markers. This genomic information can be used to qualify and quantify soil microbial communities based on the genetic information present at an individual and community scale. Therefore, the genes are not only driving the processes, they may also be used to characterize the microbial community. The genetic code and evolution of soil biota ultimately relate to the ecosystem services provided by these organisms. The processes that underlie these services can be related to the functioning of the entire planet. These issues are explored in this section across a range of scales from genes to the entire planet.

At the scale of microbial characterization, enormous progress has been made by microbial ecologists in using molecular tools to unravel the identity and functioning of microbial species and microbial communities. In Chapter 2.1, Evelyn Hackl and colleagues explain how microbial genes relate to functioning and how genetic information can be used to detect the functional and structural composition of microbial communities. They explain community profiling techniques up to 454-based pyrosequencing, and high-throughput methods leading to the quantification of both the structure and functionality of microbial communities in soil at accelerating rates and hitherto unprecedented detail. This contemporary overview is followed by a discussion on the applications, with respect to microbial degradation, of pollutants, nitrogen transformations, and genes involved in microbial signaling. Finally, there is a focus on microbial genes involved in microbe–plant relationships, especially on specific effects of symbioses and on pathogen control in the rhizosphere, since these arenas are making rapid progress, for example, in identifying how the pathogen-suppressive capacity of some soils may be explained via knowledge of soil community composition.

In Chapter 2.2 Jennifer Schweitzer and colleagues provide new views on the order of scale from genes to plant–soil–aboveground interactions. Plant community dynamics can be strongly influenced by soil biota and plant–soil interactions also influence the ecology of species that live predominantly aboveground. However, addressing the question of how evolutionary processes are involved in these complex interactions is still in its infancy. The concept of community genetics offers an opportunity to develop an understanding of how evolution may take place in a "multi-trophic selection arena." Plants may benefit from accumulating symbiotic mutualists in their root zone, as well as from developing a soil community of decomposer organisms that can convert their own litter most effectively into mineral nutrients. This is a so-called "home-field advantage." However, when plants are not able to do proper "house-keeping" they accumulate soil-borne pests and pathogens that can develop negative legacy effects in the soil, which is known for many wild plant species and particularly for early successional forms. In this chapter, the view has been developed on how plant genetic, species, and trait variation can influence and be influenced by the interactions with soil biota. Ultimately, it is demonstrated how soil biota are part of the community genetics involving plants and associated aboveground biota.

Finally, in Chapter 2.3 Katarina Hedlund and Jim Harris scale-up the relationship between genes and ecosystem services to a global scale. They show that there is a relationship between gene characteristics and the functioning of processes at the scale of the entire planet. Two global approaches are linked, that of the Millennium Ecosystem Assessment (MEA) and that of considering the world as as if it were a super-organism (i.e. Gaia). In both MEA and Gaian approaches, the importance of soils has been acknowledged; however, the contribution of soil ecology and soil biodiversity has not been elucidated in any real detail. They show how four ecosystem services (carbon sequestration, nutrient retention, resistance to diseases and pests, and the formation of soil structure) depend on soil biota at field, regional, and global scales. When focusing on plant–pathogen relations, it is shown how these interactions develop at the scale of individuals, but also how they depend on community composition and how these interactions may translate to a global scale. For example, the sensitivity of crop species for soil-borne pathogens is a major driver of the need for crop rotation. Therefore, plant–soil pathogen interactions have a strong influence on the composition of the landscape. However, this is not the only driver, as international trade and the increasing need for more bio-based economies may influence prices on the world market. These prices make some crops more profitable than others, which can result in a narrowing of crop rotations and, consequently, an increase in pesticide use to control the increased incidence of soil-borne diseases. The result is that surface and ground water will become more polluted with consequences for the role of soils in providing ecosystem services.

CHAPTER 2.1

From Single Genes to Microbial Networks

Evelyn Hackl, Michael Schloter, Ute Szukics,
Levente Bodrossy, and Angela Sessitsch

2.1.1 Introduction

The functioning of terrestrial ecosystems depends heavily on the activities of micro-organisms residing in the soil. Soil micro-organisms are basal components of the soil food web. Moreover, they hold key positions in the terrestrial biogeochemical cycles, where they are heavily involved in processes such as nutrient acquisition, nutrient cycling, and soil formation. Via these activities, soil micro-organisms mediate the provision of vital ecosystem services, including the balancing of greenhouse gases, the regulation of soil carbon sequestration, and functions that secure plant health and productivity.

Soil micro-organisms form interactive and interdependent networks when accomplishing wide-ranged and multilayered ecosystem processes. Collectively, micro-organisms exert an enormous metabolic force, which arises from their great genetic diversity and wide enzymatic versatility. For example, metabolically diverse soil micro-organisms build functional guilds when they perform biogeochemical cycling; and multiple microbial groups act as an apparently coordinated metabolic workforce in the degradation of complex organic substrates. As well as this involvement in enabling inter-microbial networking, micro-organisms use their metabolic repository to interact with members of other organismal groups. In the soil environment, they commonly build beneficial (i.e. symbiotic), neutral or deleterious (pathogenic/antagonistic) relationships with plants. The vast metabolic capabilities of soil microbial communities are important for ecosystem functioning, and they grant additional benefits to human society via their potential use in biotechnological, agricultural, and pharmaceutical applications.

Notwithstanding the numerical dominance of micro-organisms and their acknowledged importance, we are only gradually gaining knowledge on the mechanisms and processes underlying microbial controls on ecosystem functioning. With the recent advent of molecular, gene-based methods we have started to unravel the phylogenetic diversity of soil micro-organisms beyond the boundaries of laboratory cultivation, together with a growing understanding of their functional roles on the community and organism level. Members of microbial assemblages have been identified according to their phylogenetic or functional genomic traits through the use of an ever-growing armament of genomic and metabolomic tools together with the accompanying bioinformatics frame; and the ecological inter-relations of microbial assemblages have been described at various levels of organization.

The present chapter portrays our current understanding of the functional networks that are formed by soil micro-organisms both from intra- and inter-species perspectives, and across microbial and other organismal groups, which is strongly founded on gene-based analysis (Fig. 2.1.1). First, we illustrate the various roles of soil microbial communities for ecosystem functioning; and we consider how the analysis of microbial genes has added to our understanding of the soil ecosystem. Then, we present an overview of the methodological toolbox

Soil Ecology and Ecosystem Services. First Edition. Edited by Diana H. Wall *et al*.
© 2012 Oxford University Press. Published 2012 by Oxford University Press.

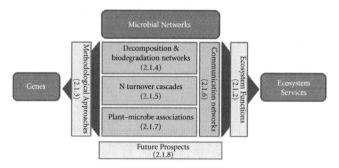

Figure 2.1.1 Functional networks that are formed by soil micro-organisms both from intra- and interspecies perspectives, and across microbial and other organismal groups.

that is currently available for studying microbial communities in the soil environment. We then exemplify how microbial networks and their functions in the soil system are increasingly becoming tangible via the application of the molecular toolbox. We elucidate the genomic basis underlying microbial interactions in organic matter degradation and biogeochemical cycling, with a focus on nitrogen cycling processes. We further show how functional gene-based analysis enables the exploration of microbial communication networks, and how it allows insights into the complex interrelations between micro-organisms and plants. In conclusion, we discuss future prospects, and argue that the current concept of microbial controls on ecosystem functioning is continuously challenged by the discovery of novel processes and organisms. As new insights evolve into how soil microbial communities contribute to soil functioning, this will enhance our abilities to sustain and promote soil ecosystem services.

2.1.2 Analyzing microbial genes to understand ecosystem functioning

Since the vast majority of micro-organisms resist cultivation on standard laboratory media, access to the soil microbial communities and their functions nowadays is gained primarily via the analysis of microbial genes. The advent of molecular techniques and their application to ecology have enabled the *in situ* exploration of soil microbial communities by using nucleic acid-based methods, without the need for cultivation. Analyses based on the highly conserved 16S rRNA gene serving as a phylogenetic marker have unraveled a previously unknown diversity within the phylum of Bacteria (Woese 1987), with complex habitats harboring an estimated 10^4–10^6 prokaryotic species in a single gram (e.g. Torsvik *et al.* 1998; Gans *et al.* 2005). Thus, molecular techniques in most cases do not allow an exhaustive sampling of complex microbial assemblages but target the most dominant community members. However, the introduction of next-generation high-throughput sequencing has improved the detection of rare population members down to 0.1% (Zagordi *et al.* 2010).

How the enormous biodiversity displayed by soil micro-organisms is linked to ecosystem functioning, however, is not yet fully resolved. According to the "insurance hypothesis" a large biodiversity is considered beneficial for ecosystem functioning, because the presence of species that are functionally redundant enhances the probability that some will maintain functioning even if others fail (Yachi & Loreau 1999). At short timescales, however, species complementarity has been considered the most important aspect of biodiversity (Loreau *et al.* 2001). Studies of the abundance and diversity of microbial communities at various scales have revealed that they are heterogeneously distributed (e.g. Horner-Devine *et al.* 2004), and that they display biogeographic patterns that extend to community functions (Fierer 2008). Thus, it seems that the taxa in many microbial communities are not in fact functionally redundant, and that different communities are not functionally similar (Allison & Martiny 2008; Strickland *et al.* 2009).

Certainly, the great diversity and abundance of soil micro-organisms bestow on them a powerful metabolic machinery for performing essential ecosystem processes. Soil micro-organisms are the key performers in decomposition, and they catalyze important transformations in the carbon, nitrogen, sulfur, and phosphorus cycles. Thus, soil micro-organisms are responsible for providing "supporting" ecosystem services, which are necessary for the production of all other ecosystem services, including the provision of food and fuel and non-material, cultural benefits (Millenium Ecosystem Assessment 2005).

The microbial groups mediating individual reaction steps in soil nutrient cycling can be specifically tracked via the analysis of functional marker genes, which encode enzymes catalyzing the processes of interest (McDonald & Murrell 1997; Bothe et al. 2000). For instance, studies investigating C degrading or CH_4 oxidizing micro-organisms make use of the *mmoX* and *pmoA* functional marker genes, encoding soluble methane mono-oxygenase and particulate methane mono-oxygenase, respectively (McDonald et al. 2008). Denitrifiers performing individual steps in the reduction of NO_3^- to NO_2^- and further to NO, N_2O, and N_2 are commonly targeted via the *narG* or *napA*, *nirK* or *nirS*, *norB*, and *nosZ* genes, coding for reductases catalyzing the respective sequential reduction steps (Philippot 2002); and the *nifH* gene encoding the nitrogenase enzyme is used in the analysis of bacteria performing N_2 fixation (Zehr et al. 2003).

The structure, diversity, and abundance of functional microbial communities can be assessed through the analysis of these functional marker genes. The community composition and abundance of functional microbial groups as well as the corresponding nutrient turnover rates are influenced by environmental factors such as soil pH, temperature, water content, and nutrient availability. Accordingly, soil denitrifying communities were found to change in their structural composition and abundance depending on season (Wolsing & Priemé 2004) and following fertilizer amendments (Avrahami et al. 2002). Notably, changes in denitrifier community composition and abundance effected alterations in N turnover rates, as has been evidenced via N-process measurements performed alongside *narG*, *nosZ*, and *nirK* gene analyses (Enwall et al. 2005; Stres et al. 2008; Szukics et al. 2010). By targeting the respective gene transcripts, the active subset of specific functional groups can be characterized regarding their abundance and diversity. Thus, for instance, plant species-specific effects on active *nirK* denitrifiers were shown in the rhizosphere of grain legumes (Sharma et al. 2004).

Microbial activities respond rapidly to changing environmental conditions, whereas the respective shifts in the community composition are usually delayed over time (Mendum et al. 1999; Avrahami et al. 2002). This is because microbial communities may adjust more rapidly to a changing environment by modifying gene expression patterns than by undergoing shifts in the population composition. Hence, a slow adaptation to changing environmental conditions has been attested to soil nitrifying communities, which in a two-step enzymatic process (encoded by *amoA* and *hoaA* genes) oxidize NH_3 via NH_2OH and NO_2^- to NO_3^-. Homologues of the *amoA* gene, coding for the small subunit of the ammonia-mono-oxygenase enzyme, are carried both by members of the Bacteria (Rotthauwe et al. 1997) and Archaea domains (Treusch et al. 2005). However, archaeal- versus bacterial-derived *amoA* genes and transcripts were differentially affected by changes in soil pH (Nicol et al. 2008), water content, and temperature (Szukics et al. 2010), as well as by N fertilization (Shen et al. 2008; Schauss et al. 2009), indicating that the two nitrifying microbial groups have different environmental and nutritional requirements.

Changes in soil microbial communities and their activities in response to human interventions such as land use change or N enrichment may have far-reaching consequences for multiple ecosystem services that are often interrelated. For instance, increased rates of industrial N fixation and atmospheric N deposition, resulting primarily from fossil-fuel combustion, have an impact on microbial N cycling processes and cause changes in soil N balances. As a consequence, nitrate is increasingly leached into the groundwater, representing a potential human health hazard, and the biogenic greenhouse gases NO and N_2O are increasingly emitted from soils into the atmosphere (Mosier & Parkin 2007). N enrichment may additionally affect plant primary production by suppressing the growth and activity of decomposer and mycorrhizal fungi (Wall et al. 2010).

Significant advances have been made in understanding how microbial involvement in key ecological processes responds to disturbance through the analysis of the abundance and composition of functional genes and their transcripts. Increasingly, shifts in microbial gene expression are being recognized as key drivers of ecosystem processes. Still, a strong predictive framework is missing to interpret the consequences of changing microbial communities. In particular, we urgently need to better understand impacts of climate change on soil micro-organisms and potential microbe feedbacks to C exchange and greenhouse gas production (Bardgett et al. 2008).

2.1.3 Methodological approaches to the gene-based study of microbial communities and networks

The information on the composition of soil microbial communities, which is provided by phylogenetic and functional gene diversity, can be accessed via application of numerous methods. The ones that are most broadly used are summarized in this section. Most of these methods are based on a consensus polymerase chain reaction (PCR) amplification of the gene of interest.

Cloning and sequencing via the traditional Sanger dideoxy-method (Sanger et al. 1977) has been extensively used in the last two decades. The information generated by hundreds of such studies has established the basis for our current understanding of microbial functional diversity. Cost and labor requirements have set the practical size limits of clone libraries to a few hundred clones in most cases. Clone libraries provide the ultimate resolution in terms of sequence differentiation. In most cases, however, their sampling breadth (a few hundred clones) provides only a limited coverage of the diversity studied. As a consequence, only the dominant taxa are detected with a reasonable certainty; any other sequences are detected at random. Constraints in the sequence output set a limit to the time- and cost-efficiency of the method.

Among the methods used for molecular fingerprinting of soil microbial communities, both denaturing gradient gel electrophoresis (DGGE; Muyzer et al. 1995) and thermal gradient gel electrophoresis (TGGE; Muyzer & Smalla 1998) are based on the differential melting properties of similar nucleotide sequences. They provide a higher throughput than cloning and sequencing, and do resolve the dominant members of the community (within the limitations of the PCR). DGGE and TGGE are however labour intensive, require a skilled operator, and miss the rare members of the community (Casamayor et al. 2000). DGGE and TGGE enable the comparison of similar samples, resolving temporal or spatial changes within a single environment. These methods are also very powerful in combination with stable isotope probing (SIP; Neufeld et al. 2008). It is difficult, however, to compare a large number of samples that are run on different gels, as no internal standards are applied, and the large variation in sample motility between gels is hard to correct. The identities of the bands on a DGGE/TGGE gel cannot be predicted from sequence information, thus each individual band needs to be extracted, cloned, and sequenced in order to confirm the identity of the related organisms. Due to its high resolution, DGGE/TGGE is well suited to studies involving genes with a low evolutionary rate, such as *dsrA* (Dar et al. 2005) or *amoA* (Avrahami et al. 2002). The high resolution of DGGE/TGGE may also be a drawback in some cases with a single nucleotide difference resulting in two separate bands in the same way as a much larger difference, for example a 10% sequence divergence. Single-strand-conformation polymorphism (SSCP) utilizes the conformation polymorphism of single stranded sequences, and the resulting differences in their migration speed in a polyacrylamide gel (Schwieger & Tebbe 1998; Junca & Pieper 2004). SSCP enables the detection of about 90% of single nucleotide differences (Sunnucks et al. 2000).

Terminal restriction fragment length polymorphism (T-RFLP) analysis is another high-throughput method, which is based on variations in the restriction endonuclease recognition sites within the gene of interest (Liu et al. 1997). Other than the DGGE/TGGE gels, the T-RFLP fingerprints provide community data that can be directly related to sequence information. This is possible because fragment lengths (thus peak positions) are predictable *in silico*, even though artifacts in the form of pseudo-TRFs may occur due to single-stranded DNA (Egert & Friedrich 2003) or the failure of some predicted restriction sites. It is recommended to start a T-RFLP analysis with the construction of a clone library representative of the environment, and to identify the best combination of primers and restriction endonuclease(s) to yield the highest resolution within the given environment. The internal standards utilized in T-RFLP improve the reproducibility over other techniques, making it suitable for studies involving a large number of samples.

Intergenic spacer analysis is based on a consensus PCR amplification of a region between two genes of a given gene cluster (Norton *et al.* 2002; Tavormina *et al.* 2010). Depending on the variability of the intergenic region, the method is capable of a reasonably high resolution with a high throughput. The drawback of the approach is its difficulty to relate the new formation to the existing knowledge on the diversity of the coding regions, even if intergenic clone libraries are sequenced.

Microarrays utilize short or long oligonucleotide probes or gene fragments and hybridization to detect the diversity of microbial phylogenetic or functional genes (Zhou 2003; DeSantis *et al.* 2007; Bodrossy & Sessitsch 2004). Short oligonucleotide microarrays rely on PCR amplification of the gene(s) of interest and can achieve specificity at the level of single nucleotide differentiation (Kostic *et al.* 2007). Long oligonucleotide and gene fragment based microarrays, on the other hand, provide enough sensitivity to detect genes from a whole metagenome background. The drawback of this approach is the specificity being in the range of 10–15% sequence divergence (Kane *et al.* 2000). Microbial diagnostic microarrays provide reasonably high throughput; current methodologies enable 40-plus samples to be analyzed per week per operator from samples to final results (McDonald *et al.* 2008).

The various next-generation sequencing technologies have been developed as a cost-effective, fast alternative to classical dideoxy sequencing. These technologies circumvent the need to build clonal templates in plasmid libraries and process them individually. Instead, molecular clones are created *in vitro* by physically separating single template molecules, followed by (usually) replication and automated sequence analysis. New sequencing chemistries enable that the template sequence is read as a series of discrete events in real time. Pyrosequencing, for example, uses the detection of light production that is induced during DNA synthesis by released pyrophosphate, which is hence proportional to the number of incorporated nucleotides. Next-generation platforms that are now in widespread contemporary use include the technology from 454 Life Sciences and the IG instrumentation developed by Solexa, and several other platforms are in the development stage.

Among the methods currently available for microbial community analysis, next-generation sequencing provides by far the most detail for a particular selected sample. The capabilities of 454 pyrosequencing were quickly adopted for microbial whole-genome sequencing (e.g. Hiller *et al.* 2007); and improvements in sequence quality and read length have enabled application to the high-resolution analysis of complex microbial populations (Edwards *et al.* 2006; Warnecke *et al.* 2007). Barcoding of pyrosequencing templates allows that multiple samples are sequenced in parallel, further increasing throughput. In spite of the large sequence output and dropping costs, substantial sampling efforts may still be required to ensure sequence representativeness. While relatively low sequencing effort (as few as 100 sequences) is needed to compare very different environments according to their microbial diversity, thousands of reads might be necessary for comparing closely related communities (Lemos *et al.* 2011).

2.1.4 Genes in microbial networks of organic matter decomposition and biodegradation of pollutants

Networks of interacting micro-organisms accomplish the transfer of soil organic matter into its inorganic constituents and living biomass during the decomposition of organic compounds, thus recycling energy and chemicals; and groups of interdependent microbial degraders accomplish the bioremediation of polluted environments.

Micro-organisms that subsist on simple, easily degradable organic compounds cross-feed on the decomposition products of other microbial groups, which feed on more recalcitrant compounds. While a broad phylogenetic range of micro-organisms is equipped with extracellular enzymes mediating the degradation of labile compounds such as soluble sugars in early decomposition stages, only members of certain taxonomic groups (mainly fungi, a few bacteria, and plant parasitic nematodes) are endowed with a more specialized enzymatic repository, empowering them to degrade cellulose and hemicelluloses as well as recalcitrant substrates like lignin in mid and late stages of decomposition (Berg & McClaugherty 2003; Danchin et al. 2010). The genetic information underlying these enzymatic capabilities has been utilized to track and characterize the various microbial groups performing specific degradation steps in their *in situ* environments. Oligonucleotide probes were designed based on genes encoding known degradation pathways and metabolites for the construction of biochips. The GeoChip, as the most comprehensive of these microarrays, covers 410,000 genes in 4,150 functional groups involved in biochemical cycling as well as metal reduction and resistance, and organic contaminant degradation (He et al. 2007; He et al. 2010).

Microarray-based analyses have proven useful in the high-throughput monitoring of C and N cycling microbial communities in soils across geographical locations and land use types, revealing that different vegetation types were associated with distinct microbial communities involved in organic carbon decomposition (Zhang et al. 2007; Yergeau et al. 2007). Insight into the driving forces of nutrient cycling processes may be gained by combining the microarray approach with real-time PCR, and with enzymatic and process rate measurements. For instance, a higher prevalence of fungal decomposition genes in lower versus higher latitude Antarctic plots substantiated the role of fungi as the dominant decomposers of Antarctica; and abundance data of microbial C fixation and cellulase genes together with cellulolytic assays reflected the dependence of organic matter accumulation on temperature and vegetation cover (Yergeau et al. 2007). Microarray applications rely on sequence information from a limited number of gene families that are already known; and hence, unexplored but ecologically important biodegradation functions are left disregarded. Novel sequence types retrieved from soil habitats by using a shotgun-sequencing approach included many cellobiose phosphorylase orthologs (Tringe et al. 2005), providing evidence that additional microbial functions in biogeochemical cycling are still awaiting discovery.

As in the decomposition of plant- or animal-derived organic residues, microbial communities transfer substrates and products between each other when they degrade organic pollutants. This process has been referred to as *metabolic cooperation* (Pelz et al. 1999). Individual strains can metabolize only a limited range of hydrocarbons that make up pollutants such as crude oil or petroleum, whereas mixed microbial assemblages encompassing manifold enzymatic capabilities may utilize the collective compounds including their degradation products. Genes encoding enzymes along various degradation pathways of xenobiotic chemicals have been cloned and characterized, subsequently to the isolation of degrading strains and the purification of the respective enzymes. PCR-based detection and quantification of these genes allow the monitoring of the biodegradation potential and activity of soils. This has been demonstrated for polyaromatic hydrocarbon (PAH) and benzene, toluene, ethylbenzene, and xylene (BTEX) biodegradation (Andreoni & Gianfreda 2007).

Gaining an in-depth understanding of the catabolic functions of innate microbial communities in dependence of environmental influences may pave the way towards the development and successful application of bioremediation strategies. The biodegradation gene composition of various soils contaminated with PAHs, BTEX, and heavy metals was characterized by the use of a microarray based on 2,402 known genes and pathways involved in biodegradation, which was hence suggested as a tool for the high-throughput monitoring of bioremediation processes *in situ* (Rhee et al. 2004). In addition, profiling techniques such as DGGE or T-RFLP have proven useful in the monitoring of microbial community members responsible for particular processes during ongoing bioremediation. For instance, the diversity of sulfate-reducing bacteria at a metal-precipitation site was surveyed via DGGE analysis of *dsrB* (dissimilatory sulphite reductase β-subunit)-genes (Geets *et al.* 2006). 16S rRNA gene based T-RFLP analysis revealed that the bacterial community changed significantly when nutrients were added to a diesel oil-contaminated Antarctic soil, when at the same time catechol degrading and denitrification genes were increasingly detected. Hence, this study suggested that biostimulation with nutrients may be a viable means to accelerate the recovery of contaminated soils (Vázquez *et al.* 2009).

Furthermore, the recent advances in genomics techniques are conducive to characterizing known and identifying novel biodegradation pathways. Genomic sequence information has been obtained from numerous microbial strains significant to biodegradation processes and, via metagenomic approaches, also from whole communities, elucidating the genomic basis for catabolic and resistance functions and providing access to novel biocatalysts and other gene products involved in biodegradation (Desai *et al.* 2010).

2.1.5 Microbial genes in nitrogen turnover cascades

The best studied nutrient cycle in soil is the turnover of nitrogen. This is because all biota require nitrogen for growth, and in particular in natural ecosystems the availability of ammonia and nitrate is still considered a bottleneck for growth and for the activity of most of the organisms present (Jackson *et al.* 2008). However, nitrogen has often been perceived as highly problematic in many agricultural systems of industrialized countries. According to the *UN Millennium report* (UN Millennium Project 2005) the nitrogen input to the environment has doubled in the last 50 years, accompanied by a significant reduction of biodiversity. Nitrate that is not taken up by plants or microbes can leach to the groundwater, leading to the contamination of drinking water resources, or may be reduced by microorganisms, resulting inter alia in the formation of NO and N_2O, which are both gases contributing to global warming (Boyer *et al.* 2006).

Five major microbial processes contribute to the biogeochemical cycling of nitrogen in aerobic terrestrial soils: nitrogen fixation ($N_2 \rightarrow NH_4^+$), mineralization ($N_{org} \rightarrow NH_4^+$), nitrate ammonification ($NO_3^- \rightarrow NH_4^+$), nitrification ($NH_4^+ \rightarrow NO_3^-$) and denitrification ($NO_3^- \rightarrow N_2$) (Hallin *et al.* 2009). Other processes, like the recently discovered anaerobic oxidation of ammonium (ANAMMOX) are of minor importance for terrestrial ecosystems (Humbert *et al.* 2010). All steps involved in nitrogen turnover are closely interlinked. The rates at which these metabolic pathways occur determine the availability of nitrogen in the soil and influence therefore the activity of soil micro-organisms and plant growth in agricultural ecosystems. Besides bacteria, fungi and archaea are key players in nitrogen turnover. The interplay of the different processes, which together form the microbial nitrogen cycle, are highly complex: chronosequence studies in developing ecosystems like glacier forefields indicate that the completion of a complete nitrogen cycle may take more than 200 years (Kandeler *et al.* 2006.).

The genes encoding enzymes that catalyze the different chemical reactions in nitrogen cycling processes have in many cases been well described, and the regulation pathways involved are well understood in pure cultures. However, it is still unclear whether the data published so far, which are based on single isolates, can be transferred to complex microbial communities in soil, where a large part of the microbes is still not known or only described via phylogenetic markers. For example, the discovery of ammonium oxidizing archaea has completely changed the previously well-established picture of nitrification. At the turn of the century, it was believed that only some autotrophic proteobacteria were able to catalyze the transformation of ammonia to nitrite (Rotthauwe *et al.* 1997); today we are faced with a hitherto unknown group of organisms (ammonia oxidizing archaea) that perform the same transformative step, obviously under different environmental conditions and different regulatory pathways. This raises questions of general ecological relevance, concerning issues of "functional redundancy," "resilience," or "niche differentation" of these two groups (Schauss *et al.* 2009).

Similar questions have been raised regarding denitrification. For nitrite reductases, two groups have been described so far, a copper containing nitrite reductase (NirK) and a *hem* type nitrite reductase (NirS). Although the distribution of both types among different phyla is not well understood, there is convincing evidence that microbes harboring the NirK type mainly colonize habitats that are rich in easily available carbon like the rhizosphere, whereas microbes harboring the NirS type mainly occur in bulk soil (Babić et al. 2008; Hai et al. 2009). However, the question is left unanswered whether diversity or abundance of a functional gene is more important for process stability under changing environmental conditions.

The induction of expression of selected genes coding for different enzymes involved in nitrogen cycling has been studied mainly in laboratory experiments in the last few years, because transcription rates are highly variable and change rapidly depending on the given environmental conditions. In many cases a correlation was found between the abundance of transcript rates and turnover rates. Sharma et al. (2006), for example, showed a close correlation between increased transcription levels of the nitrite reductase *nirK* after freezing and thawing of a soil and the formation of N_2O; in the same time period the amount of DNA remained unchanged, indicating increased activity pattern but no growth of the corresponding community. The factors driving the induction of gene expression have been elucidated in case of the nitrous oxide reductase (NosZ), which catalyzes the transformation of N_2O to N_2. Soils that were high in nitrate were low in transcription rates of *nosZ* even under anaerobic conditions; whereas high expression levels of NosZ were found in soils with reduced nitrate content under O_2 limiting conditions (Morales et al. 2010). This has been related to the fact that the reduction of nitrate to nitrite as an alternative electron acceptor is more energy efficient as compared to the reduction of N_2O to N_2.

The overall turnover of microbial nitrogen is not only linked to the availability of different nitrogen forms but also to the carbon amount and quality. The best studied example in this field is the symbiosis between rhizobia and legumes (Prell & Poole 2006), where high amounts of energy are needed to accomplish the transformation of the highly inert N_2 into NH_4^+. Specifically, 1 kg of glucose equivalents has to be provided by the plant to enable that 20 g of N_2 are fixed by the bacteria. As the nitrogenase (the enzyme which converts N_2 into NH_4^+) is highly sensitive to oxygen, a highly complex regulation system is needed, which ensures an efficient nutrient utilization on the one hand, and a best possible protection of the nitrogenase complex on the other hand.

Today more than 50 genes expressed either by the bacterial or plant partner have been identified that are needed to form this symbiosis, including genes of the *nod* (nodulation) family, the *nif* (nitrogenase) operon, and several regulatory elements. Newer studies have also proven communication pathways between rhizobia, the plant, and associated mycorrhizal fungi (De Hoff et al. 2009) to regulate nutrient uptake and distribution. A close connection between carbon and nitrogen cycles has not only been described for nitrogen fixation. Microbes involved in the mineralization of dead biomass require huge amounts of energy to transform proteins into amino acids or ammonium; and denitrification as a redox process is mainly regulated by the activity status of the microbes (Russell 2007). Only nitrification, being an autotrophic process, is not directly linked to the availability of carbon. Latest studies have proven that processes which so far have been linked only to carbon availability, like the anaerobic oxidation of methane in soil, are closely linked also to soil nitrate availability (Zhu et al. 2010).

2.1.6 Genes underlying microbial communication

The interactive and cooperative behavior displayed by soil microbial communities at various levels of taxonomic and functional organization, which results from the coordination of the activities of individual cells, is made possible through complex communication systems. In addition to the unidirectional, plasmid-coded response mechanism that facilitates plasmid transfer in bacterial conjugation (Clewell & Dunny 2002), bacteria employ a range of cell-to-cell communication systems, which involve signal sensing also of the producing organism. Recent discoveries have unveiled the molecular underpinnings of so-called quorum-sensing systems, which involve small diffusible signal molecules (sometimes termed pheromones or autoinducers) which are instrumental in the regulation of gene expression. Oligopeptides, acyl homoserine lactones (AHLs), and autoinducer-2 (AI-2) are the three classes of molecules currently associated with quorum sensing in bacteria, which bind to receptors on, or in, the bacterial cell, and upon reaching some threshold concentration ("quorum") lead to changes in gene expression (Keller & Surette 2006). In the most widely studied, AHL-mediated communication system, a *luxI*-homologue gene mediates the synthesis of AHL molecules, which bind *luxR* homologue encoded LuxR proteins to regulate a variety of downstream processes.

Among the many bacterial functions that are regulated via a quorum sensing circuitry are the expression of virulence factors by pathogenic bacteria, behavioral responses to the environment via swarming or swimming motility, symbiotic interactions, and the aggregation of bacterial cells to biofilms. The coordination of population behavior thus enhances access to nutrients or favorable environmental niches, and enables collective defense against competitors or community escape. Strategies counteracting quorum sensing include quorum quenching through degradation of AHLs (Dong *et al.* 2000; Huang *et al.* 2003), while interference with peptide signaling and AI-2-mediated communication have also been reported (Bassler & Losick 2006). Although the production, sensing, and function of quorum sensing signals have been more intensely studied within cultures of single bacterial strains, cross-conversation among species belonging to different genera has been evidenced as well. Intriguingly, various bacterial signal molecules can modify the behavior also of fungal, plant, and animal cells, which in return may secrete molecules that either activate or inhibit bacterial quorum sensing (Dudler & Eberl 2006).

Quorum-sensing signal molecules are chemically diverse; and bacteria may use multiple signal molecules in regulatory hierarchies. AHL-based quorum sensing systems have been experimentally identified in more than 40 different bacterial species, and putative *luxI/luxR* homologues have additionally been revealed in genome sequencing projects. Previously unknown quorum sensing synthase genes as well as novel quorum sensing quenchers and inhibitors have been uncovered in metagenomic screenings (Williamson *et al.* 2005; Riaz *et al.* 2008; Schipper *et al.* 2009; Hao *et al.* 2010). Notably, most low molecular mass organic microbial compounds including antibiotics have been implicated a role as quorum sensing signaling molecules (e.g. Goh *et al.* 2002), and the discovery of an even wider range of small organic molecules with intercellular signaling activity is anticipated (Bassler & Losick 2006).

Understanding the mechanisms underlying the quorum sensing circuitry forms a basis for future practical applications. Evidence shows that quorum sensing inhibitors limited bacterial infection in mice by disrupting virulence gene expression (Lesic *et al.* 2007). This may give rise to the design of selective anti-infectives that attenuate virulence of clinically relevant pathogens. Quorum sensing quenching genes retrieved from soil metagenomic libraries actually inhibited biofilm formation when expressed in *Pseudomonas aeruginosa*, suggesting possible applications for the prevention of microbial biofilm formation on surfaces (Schipper *et al.* 2009). Even though quorum sensing inhibitors/antagonists alone are not expected to have a bactericidal effect, their ability to attenuate virulence and biofilm formation and to render pathogens more susceptible to conventional treatment may complement the mode of action of currently available antibiotics, and may thus represent a strategy for fighting antimicrobial resistance.

2.1.7 Microbial genes for interacting in the plant environment

Plants live in association with specific assemblages of micro-organisms, which may confer deleterious (i.e. pathogenic/antagonitic), neutral or beneficial effects on plant growth or health. Plant-associated microbial communities typically comprise micro-organisms with diverse functions: some community members possess specific genes to support plant growth or provide plant nutrients or hormones, whereas the majority of micro-organisms are important for out-competing phytopathogens without directly affecting plant growth (e.g. Francis *et al.* 2010). Plant–microbe interactions have been studied most often in relation to plant pathogens. Among the beneficial plant–microbe interactions most prominently studied are the mycorrhizal symbioses, where a fungal partner provides better access to nutrients for the plant, and symbioses between legumes and rhizobia, which are fixing atmospheric nitrogen.

In the nitrogen (N_2)-fixing rhizobia *nod* (nodulation) genes are activated in the presence of specific plant-produced signaling molecules such as flavonoids. This triggers the production of signal molecules by the bacteria, which stimulate the onset of root hair curling and nodule formation in the plant. Within the nodules rhizobial nitrogen fixation (*nif*) genes are active, encoding the nitrogenase reductase enzyme. The *nif* genes are thus responsible for the nitrogen fixation process, where rhizobia reduce atmospheric N_2 to ammonia, which is then used by the plant as a N source. The *nifH* gene within this gene cluster has been widely used as a phylogenetic and functional marker for plant-associated rhizobial communities, and in addition serves as a marker for free-living nitrogen-fixing communities. Mycorrhizal associations, among which the endomycorrhiza (arbuscular mycorrhiza, AM) is largely dominant and the most ancient type of mycorrhizal symbiosis, are established in almost all land plants. In this symbiosis, the fungal partner facilitates plant uptake of mineral nutrients from the soil, and in return receives photosynthetically fixed carbon from the plant. As with the root-nodule symbiosis of rhizobia, chemical signal exchange between the symbionts initiates the colonization of plant roots by AM hyphae. Fungal hyphae that have started to branch in response to plant root exudates and strigolactone derivatives emit a signal of unknown nature, which activates the promoter of the symbiosis-related *ENOD11* gene in the plant and causes upregulation of a number of fungal genes associated with increased cell activity. Notably, the successive establishment of a functional interaction in the AM symbiosis involves a signaling pathway that is shared with the rhizobia symbiosis (Reinhardt 2007).

In addition to undergoing symbiotic and other relationships with micro-organisms in the rhizosphere, plants are internally colonized by bacteria and fungi. Endophytic micro-organisms by definition reside within living plant tissues without causing harm to their plant hosts, and in fact often exhibit plant growth promoting or other beneficial qualities. The structural composition of the plant-associated communities in the rhizosphere as well as in the plant interior is shaped by root exudates and other plant metabolites, which serve as nutrients for the micro-organisms and may activate certain microbial activities. The associated micro-organisms eventually, vice versa, induce specific pathways in the plant, and may then evoke systemic responses. However, (non-pathogenic) microbe-plant interactions are still poorly understood at the gene level. Studies performed with non-pathogenic, beneficial bacteria inoculated onto plants showed that the bacteria frequently activate or deactivate plant processes, and, for instance, induce defense-related pathways (Cartieaux *et al.* 2003) or increase photosynthesis efficiency (Zhang *et al.* 2008). Among the few bacterial metabolites that have been identified as modulators of plant gene expression are homoserine lactones (Mathesius *et al.* 2003; von Rad *et al.* 2008), bacterial volatiles (Ryu *et al.* 2004), and siderophores (Audenaert *et al.* 2002). The plant physiology (e.g. expressed as the amount or composition of root exudates) is a determining factor for structuring the plant associated microbial community (Badri *et al.* 2009); and different plant species as well as different plant growth stages typically show differentially structured associated microbial populations (Gyamfi *et al.* 2002).

The plant environment provides important substrates, nutrients and signaling compounds that sustain bacterial growth and determine bacterial community composition (Garbeva *et al.* 2004). In addition, plant-associated communities are shaped by the biotic interactions in the rhizosphere, phyllosphere, and endosphere. Features like pathogenicity, symbiotic functions, and the production of antibiotics and siderophores, which all play an important role in the plant environment, are typically regulated by quorum sensing circuits, and can be substantially influenced by micro-organisms that are able to degrade quorum sensing signals and thereby modify microbial activities. Several soil and plant-associated micro-organisms produce antibiotics, which may render the producing strain more persistent and more competitive. In the plant environment the production of antibiotics means an important biocontrol function, because it enables out-competing and inactivating invading plant pathogens. Two types of antibiotics have frequently been found in plant-associated bacteria: 2,4-diacetylphloroglucinol (2,4-DAPG) and phenazines. 2,4-DAPG has been particularly detected in pseudomonads that are involved in the natural suppression of various diseases such as take-all decline in wheat (Raaijmakers & Weller 1998). The key gene in 2,4-DAPG synthesis, *phlD*, has been used to study the genetic diversity of *phlD* +-pseudomonads (De la Fuente *et al.* 2006) as well as their abundance (Mavrodi *et al.* 2007). Among the phenazines, which have been detected mainly in pseudomonads but also in other bacteria (Mavrodi *et al.* 2010), phenazine-1-carboxylic acid (PCA) is well known for its biocontrol activity against various fungal phytopathogens such as *Gaeumannomyces graminis* var. *tritici* (Thomashow & Weller 1988). The *phzF* gene in the phenazine (*phz*) biosynthesis cluster has been successfully applied as a marker to study the diversity of PCA producers in the plant environment (Mavrodi *et al.* 2010).

Most genes involved in plant growth promotion and in various interactions with the plant or other microbial community members are either yet unknown or known only from individual strains. An exception is the 1-aminocyclopropane-1-carboxylate (ACC) deaminase gene (*acdS*), which has been studied in more detail in various plant-associated bacteria (Hontzeas *et al.* 2005). ACC deaminase is produced in particular by plant growth-promoting bacteria, and can reduce the level of ethylene production in developing or stressed plants. This is accomplished through cleaving ACC, which is the precursor in plant ethylene production (Glick 2005). While in the last few years the genomes of an increasing number of plant-associated bacteria have been fully sequenced, still only few gene markers are available for studying the functions attributed to plant-associated microbial communities. Furthermore, interactions within plant-associated microbial communities are still poorly understood, particularly at the gene level. The advance of next generation sequencing technologies together with comparative genomics and metagenomics approaches will help to identify new marker genes for the study of plant associated microbes. This may improve our understanding of plant–microbe interactions and their significance for plant health and open up new strategies of pest control and biofertilization.

2.1.8 From genes to microbial networks: future prospects

The emergence of molecular, sequence-based technologies during the last few decades has enabled soil micro-organisms to be studied in their *in situ* habitats for the first time. Thus, studying microbial genes in their environmental context has opened up a new perspective on the functional networks that micro-organisms form within and among microbial groups, and in associations with other inhabitants of the soil environment. Nevertheless, in view of the fundamental roles of the soil microbial communities in terrestrial processes such as nutrient cycling or carbon sequestration, remarkable gaps in knowledge remain, particularly in relation to their contribution to ecosystem functioning. Our knowledge is incomplete, for example, about the importance of micro-organisms for the resistance and resilience of soils to perturbations and effects of climate change; and we need a better understanding of the ecology of micro-organisms that are beneficial to, or threaten, plant health and crop quality. Determining the key functions of soil microbial communities and understanding the importance of diversity and abundance within functional networks is critical for elaborating strategies that sustain or promote beneficial microbial activities. It seems, for instance, that the apparently rare soil microbial species can play a particular role in crop protection by enhancing plant defense against herbivores (Hol *et al.* 2010). Other open questions in this context relate to the interchangeability of microbial species, and to the effects of invaders into functional networks.

There has been a notable development of new research possibilities and novel databases that hold promise for further advancing functional knowledge on the soil microbial communities. The use of high-density DNA microarrays (DNA chips), which screen genes and gene products linked to organisms and known microbial processes, for instance, provides great promise in analyzing the composition and function of microbial communities in their native habitat in a highly parallel and high-throughput manner. Extending this approach to the analysis of mRNA molecules allows surveying the *in situ* metabolic activities of soil microbial communities. Critical issues associated with the application and further development of this technique include the design of appropriate probes that are highly specific against the intended target genes, and achieving a high sensitivity of analysis and effective visualization of data.

The latest, most powerful advances in the sequence-based analysis of complex communities of micro-organisms have been driven by the invention of next-generation sequencing technology, to the effect that obtaining genetic sequence information of a whole ecosystem may be deemed reachable by 2020 (Tautz *et al.* 2010). However, major challenges will be to manage the current exponential growth in sequence data, and to understand how the many genomes interact with each other. Thus, the anticipated accumulation of large quantities of data demands that sequencing efforts be directed towards the systematic and comprehensive exploration by national and international collaborations, as has been realized within the International Soil Metagenome Sequencing Project, or "Terragenome" initiative (Kyrpides 2009; Yilmaz *et al.* 2011). Sequence data from communities and genomic information obtained from sequencing the DNA from single cells complement each other and are of added value for gaining a more holistic understanding of the community and its members (Kyrpides 2009). Furthermore, great insight is anticipated by analyzing the meta-transcriptome in soil. Studying mRNA and rRNA molecules simultaneously from the same sample allows relating taxonomic groups to their ecological function, and may finally enable us to monitor structural and functional community shifts caused by environmental changes (Urich *et al.* 2008).

The analytical power of sequence-based analysis of the various functional networks formed by soil micro-organisms will continuously improve as sequence lengths obtained by next-generation sequencing technologies increase and rRNA reference and genome databases continue to grow. *In situ* exploration of the genomic and metabolic potentials of whole microbial consortia has revolutionized microbial ecology, and continued research is bound to lead to exciting and non-foreseen discoveries. As previously unknown organisms and novel processes continue to be discovered, new implications for microbial ecosystem regulation will continuously evolve. Finally, incorporating this information into models will enhance our abilities to manage specific soil functions and to predict impacts of environmental change. Thus, the gene-based exploration of microbial communities may result in the implementation of management practices that sustain and promote soil ecosystem services.

References

Allison, S.D., & Martiny, J.B.H. (2008) Resistance, resilience, and redundancy in microbial communities. *Proceedings of the National Academy of Sciences of the United States of America* **105**: 11512–19.

Andreoni, V., & Gianfreda, L. (2007) Bioremediation and monitoring of aromatic-polluted habitats. *Applied Microbiology and Biotechnology* **76**: 287–308.

Audenaert, K., Pattery, T., Cornelis, P., Höfte, M. (2002) Induction of systemic resistance to Botrytis cinerea in tomato by Pseudomonas aeruginosa 7NSK2: role of salicylic acid, pyochelin, and pyocyanin. *Molecular Plant Microbe Interactions* **15**: 1147–56.

Avrahami, S., Conrad, R., & Braker, G. (2002) Effect of soil ammonium concentration on N_2O release and on the community structure of ammonia oxidizers and denitrifiers. *Applied Environmental Microbiology* **68**: 5685–92.

Babić, K.H, Schauss, K., Hai, B., *et al.* (2008) Influence of different Sinorhizobium meliloti inocula on abundance of genes involved in nitrogen transformations in the rhizosphere of alfalfa (*Medicago sativa* L.). *Environmental Microbiology* **10**: 2922–30.

Badri, D.V., Weir, T.L., van der Lelie, D., Vivanco, J.M. (2009) Rhizosphere chemical dialogues: plant-microbe interactions. *Current Opinion in Biotechnology* **20**: 642–50.

Bardgett, R.D., Freeman, C., & Ostle, N.J. (2008) Microbial contributions to climate change through carbon cycle feedbacks. *The ISME Journal* **2**: 805–14.

Bassler, B.L., & Losick, R. (2006) Bacterially speaking. *Cell* **125**: 237–46.

Berg, B., & McClaugherty, C. (2003) *Plant Litter—Decomposition, Humus Formation, Carbon Sequestration*. Springer, New York.

Bodrossy, L., & Sessitsch, A. (2004) Oligonucleotide microarrays in microbial diagnostics. *Current Opinion in Microbiology* **7**: 245–54.

Bothe, H., Jost, G., Schloter, M., Ward, B.B., & Witzel, K.P. (2000) Molecular analysis of ammonia oxidation and denitrification in natural environments. *FEMS Microbiology Reviews* **24**: 673–90.

Boyer, E.W., Alexander, R.B., Parton, W.J., et al. (2006) Modeling denitrification in terrestrial and aquatic ecosystems at regional scales. *Ecological Applications* **16**: 2123–42.

Cartieaux, F., Thibaud, M.C., Zimmerli, L., et al. (2003) Transcriptome analysis of Arabidopsis colonized by a plant-growth promoting rhizobacterium reveals a general effect on disease resistance. *The Plant Journal* **36**: 177–88.

Casamayor, E.O., Schafer, H., Baneras, L., Pedros-Alio, C., & Muyzer, G. (2000) Identification of and spatio-temporal differences between microbial assemblages from two neighboring sulfurous lakes: comparison by microscopy and denaturing gradient gel electrophoresis. *Applied and Environmental Microbiology* **66**: 499–508.

Clewell, D.B., & Dunny, G.M. (2002) Conjugation and genetic exchange in enterococci. In: M.S. Gilmore, D.B. Clewell, P. Courvalin, G.M. Dunny, B.E. Murray, and L.B. Rice (eds.) *The enterococci: pathogenesis, molecular biology, and antibiotic resistance* pp. 265–300. American Society for Microbiology, Washington, DC.

Danchin, E.G.J., Rosso, M.N., Vieira, P., et al. (2010) Multiple lateral gene transfers and duplications have promoted plant parasitism ability in nematodes. *Proceedings of the National Academy of Sciences of the United States of America* **107**: 17651–56.

Dar, S.A., Kuenen, J.G., & Muyzer, G. (2005) Nested PCR-denaturing gradient gel electrophoresis approach to determine the diversity of sulfate-reducing bacteria in complex microbial communities. *Applied and Environmental Microbiology* **71**: 2325–30.

De Hoff, P.L., Brill, L.M., & Hirsch, A.M. (2009) Plant lectins: the ties that bind in root symbiosis and plant defense. *Molecular Genetics and Genomics* **282**: 1–15.

De la Fuente, L., Mavrodi, D.V., Landa, B.B., Thomashow, L.S., Weller, D.M. (2006) phlD-based genetic diversity and detection of genotypes of 2,4-dacetylphloroglucinol-producing Pseudomonas fluorescens. *FEMS Microbiology Ecology* **56**: 64–78.

Desai, C., Pathak, H., & Madamwar, D. (2010) Advances in molecular and "-omics" technologies to gauge microbial communities and bioremediation at xenobiotic/anthropogen contaminated sites. *Bioresource Technology* **101**: 1558–69.

DeSantis, T.Z., Brodie, E.L., Moberg, J.P., Zubieta I.X., Piceno, Y.M., & Andersen, G.L. (2007) High-density universal 16S rRNA microarray analysis reveals broader diversity than typical clone library when sampling the environment. *Microbial Ecology* **53**: 371–83.

Dong, Y.H., Xu, J.L., Li, Z., & Zhang, L.H. (2000) AiiA, an enzyme that inactivates the acylhomoserine lactone quorum-sensing signal and attenuates the virulence of Erwinia carotovora. *Proceedings of the National Academy of Sciences of the United States of America* **97**: 3526–31.

Dudler, R., & Eberl, E. (2006) Interactions between bacteria and eukaryotes via small molecules. *Current Opinion in Biotechnology* **17**: 268–73.

Edwards, R.A., Rodriguez-Brito, B., Wegley, L., et al. (2006) Using pyrosequencing to shed light on deep mine microbial ecology. *BMC Genomics* **7**: 57.

Egert, M., & Friedrich, M.W. (2003) Formation of pseudo-terminal restriction fragments, a PCR-related bias affecting terminal restriction fragment length polymorphism analysis of microbial community structure. *Applied and Environmental Microbiology* **69**: 2555–62.

Enwall, K., Philippot, L., & Hallin, S. (2005) Activity and composition of the denitrifying bacterial Community respond differently to long-term fertilization. *Applied and Environmental Microbiology* **71**: 8335–43.

Fierer, N. (2008) Microbial biogeography: patterns in microbial diversity across space and time. In: K. Zengler (ed.) *Accessing Uncultivated Microorganisms: from the Environment to Organisms and Genomes and Back*, pp. 95–115. ASM Press, Washington DC.

Francis, I., Holsters, M., Vereecke, D. (2010) The Gram-positive side of plant–microbe interactions. *Environmental Microbiology* **1**: 1–12.

Gans, J., Wolinsky, M., Dunbar, J. (2005) Computational improvements reveal great bacterial diversity and high metal toxicity in soil. *Science* **309**: 1387–90.

Garbeva, P., van Veen, J.A., & van Elsas, J.D. (2004) Microbial diversity in soil: selection of microbial populations by plant and soil type and implications for disease suppression, *Annual Reviews of Phytopathology* **42**: 243–70.

Geets, J., Borremans, B., Diels, L., et al. (2006) DsrB gene-based DGGE for community and diversity surveys of sulfate-reducing bacteria. *Journal of Microbiological Methods* **66**: 94–205.

Glick, B.R. (2005) Modulation of plant ethylene levels by the bacterial enzyme ACC deaminase. *FEMS Microbiology Letters* **252**: 1–7.

Goh, E.B., Yim, G., Tsui, W., McClure, J., Surette, M.G., & Davies, J. (2002) Transcriptional modulation of bacterial gene expression by subinhibitory concentrations of antibiotics. *Proceedings of the National Academy of Sciences of the United States of America* **99**: 17025–30.

Gyamfi, S., Pfeifer, U., Stierschneider, M., Sessitsch, A. (2002) Effects of transgenic glufosinate-tolerant oilseed rape (Brassicanapus) and the associated herbicide application on eubacterial and Pseudomonas communities in the rhizosphere. *FEMS Microbiology Ecology* **41**: 181–90.

Hai, B., Diallo, N.H., Sall, S., et al. (2009) Quantification of key genes steering the microbial nitrogen cycle in the rhizosphere of sorghum cultivars in tropical agroecosystems. *Applied and Environmental Microbiology* **75**: 4993–5000.

Hallin, S., Jones, C.M., Schloter, M., & Philippot, L. (2009) Relationship between N-cycling communities and ecosystem functioning in a 50-year-old fertilization experiment. *The ISME Journal* **3**: 597–605.

Hao, Y., Winans, S.C., Glick, B.R., & Charle, T.C. (2010) Identification and characterization of new LuxR/LuxI-type quorum sensing systems from metagenomic libraries. *Environmental Microbiology* **12**: 105–17.

He, Z., Deng, Y., Van Nostrand, J.D., et al. (2010) GeoChip 3.0 as a high-throughput tool for analyzing microbial community composition, structure and functional activity. *The ISME Journal* **4**: 1167–79.

He, Z., Gentry, T.J., Schadt, C.W., et al. (2007) GeoChip: a comprehensive microarray for investigating biogeochemical, ecological and environmental processes. *The ISME Journal* **1**: 67–77.

Hiller, N.L., Janto, B., Hogg, J.S., et al. (2007) Comparative genomic analyses of seventeen Streptococcus pneumoniae strains: insights into the pneumococcal supragenome. *Journal of Bacteriology* **189**: 8186–95.

Hol, W.H.G., De Boer, W., Termorshuizen, A.J., et al. (2010) Reduction of rare soil microbes modifies plant–herbivore interactions. *Ecology Letters* **13**: 292–301.

Hontzeas, N., Richardson, A.O., Belimov, A., Safronova, V., Abu-Omar, M.M., & Glick, B.R. (2005) Evidence for horizontal transfer of 1-aminocyclopropane-1-carboxylate deaminase genes. *Applied and Environmental Microbiology* **71**: 7556–8.

Horner-Devine, M.C., Lage, M., Hughes, J.B., & Bohannan, B.J. (2004) A taxa-area relationship for bacteria. *Nature* **432**: 750–3.

Huang, J.J., Han, J., Zhang, L., & Leadbetter, J.R. (2003) Utilization of acyl-homoserine lactone quorum signals for growth by a soil pseudomonad and *Pseudomonas aeruginosa* PAO1. *Applied and Environmental Microbiology* **69**: 5941–9.

Humbert, S., Tarnawski, S., Fromin, N., Mallet, M.P., Aragno, M., & Zopfi, J. (2010) Molecular detection of anammox bacteria in terrestrial ecosystems: distribution and diversity. *The ISME Journal* **4**(3): 450–4.

Jackson, L.E., Burger, M., & Cavagnaro, T.R. (2008) Roots, nitrogen transformations, and ecosystem services. *Annual Review of Plant Biology* **59**: 341–63.

Junca, H., & Pieper, D.H. (2004) Functional gene diversity analysis in BTEX contaminated soils by means of PCR-SSCP DNA fingerprinting: comparative diversity assessment against bacterial isolates and PCR-DNA clone libraries. *Environmental Microbiology* **6**: 95–110.

Kandeler, E., Deiglmayr, K., Tscherko, D., Bru, D., & Philippot, L. (2006) Abundance of narG, nirS, nirK, and nosZ genes of denitrifying bacteria during primary successions of a glacier foreland. *Applied and Environmental Microbiology* **72**: 5957–62.

Kane, M.D., Jatkoe, T.A., Stumpf, C.R., Lu, J., Thomas, J.D., & Madore, S.J. (2000) Assessment of the sensitivity and specificity of oligonucleotide (50mer) microarrays. *Nucleic Acids Research* **28**: 4552–7.

Keller, L., & Surette, M.G. (2006) Communication in bacteria: an ecological and evolutionary perspective. *Nature Reviews Microbiology* **4**: 249–58.

Kostic, T., Weilharter, A., Rubino, S., et al. (2007) A microbial diagnostic microarray technique for the sensitive detection and identification of pathogenic bacteria in a background of nonpathogens. *Analytical Biochemistry* **360**: 244–54.

Kyrpides, N.C. (2009) Fifteen years of microbial genomics: meeting the challenges and fulfilling the dream. *Nature Biotechnology* **27**: 627–32.

Lemos L.N., Fulthorpe R.R., Triplett E.W., Roesch L.F.W. (2011) Rethinking microbial diversity analysis in the high throughput sequencing era. *Journal of Microbiological Methods* **86**(1): 42–51.

Lesic, B., Lepine, F., Deziel, E., et al. (2007) Inhibitors of pathogen intercellular signals as selective anti-infective compounds. *PLoS Pathogens* **3**(9): 1229–39.

Liu, W.T., Marsh, T.L., Cheng, H., & Forney, L.J. (1997) Characterization of microbial diversity by determining terminal restriction fragment length polymorphisms of genes encoding 16S rRNA. *Applied and Environmental Microbiology* **63**: 4516–22.

Loreau, M., Naeem, S., Inchausti, P., et al. (2001) Biodiversity and ecosystem functioning: current knowledge and future challenges. *Science* **294**: 804–8.

Mathesius, U., Mulders, S., Gao, M., Teplitski, M., Caetano-Anollés, G., Rolfe, B.G., & Bauer, W.D. (2003) From the Cover: Extensive and specific responses of a eukaryote to bacterial quorum-sensing signals. *Proceedings of the National Academy of Sciences of the United States of America* **100**: 1444–9.

Mavrodi, D.V., Peever, T.L., Mavrodi, O.V., et al. (2010) Diversity and evolution of the phenazine biosynthesis pathway. *Applied and Environmental Microbiology* **76**: 866–79.

Mavrodi, O.V., Mavrodi, D.V., Thomashow, L.S., & Weller, D.M. (2007) Quantification of 2,4-diacetylphloroglucinol-producing Pseudomonas fluorescens strains in the plant rhizosphere by real-time PCR. *Applied and Environmental Microbiology* **73**: 5531–8.

McDonald, I.R., Bodrossy, L., Chen, Y., & Murrell, J.C. (2008) Molecular ecology techniques for the study of aerobic methanotrophs. *Applied and Environmental Microbiology* **74**: 1305–15.

McDonald, I.R., & Murrell, J.C. (1997) The particulate methane monooxygenase gene pmoA and its use as a functional gene probe for methanotrophs. *FEMS Microbiology Letters* **156**: 205–10.

Mendum, T.A., Sockett, R.E., Hirsch, P.R. (1999) Use of molecular and isotopic techniques to monitor the response of autotrophic ammonia-oxidizing populations of the beta subdivision of the class *Proteobacteria* in arable soils to nitrogen fertilizer. *Applied and Environmental Microbiology* **65**: 4155–62.

Millennium Ecosystem Assessment (MA) (2005) *Ecosystems and Human Well-being: Synthesis*. Island Press, Washington, DC.

Morales, S.E., Cosart, T., Holben, W.E. (2010) Bacterial gene abundances as indicators of greenhouse gas emission in soils. *The ISME Journal* **4**: 799–808.

Mosier, A.R., & Parkin, T. (2007) Gaseous emissions (CO_2, CH_4, N_2O, and NO) from diverse agricultural production systems. In: G. Benckiser and S. Schnell (eds.) *Biodiversity in agricultural production systems*, pp. 317–48. CRC Press/Taylor and Francis, Boca Raton, FL.

Muyzer, G., & Smalla, K. (1998) Application of denaturing gradient gel electrophoresis (DGGE) and temperature gradient gel electrophoresis (TGGE) in microbial ecology. *Antonie Van Leeuwenhoek* **73**: 127–41.

Muyzer, G., Teske, A., Wirsen, C.O., & Jannasch, H.W. (1995) Phylogenetic relationships of Thiomicrospira species and their identification in deep-sea hydrothermal vent samples by denaturing gradient gel electrophoresis of 16S rDNA fragments. *Archives of Microbiology* **164**: 165–72.

Neufeld, J.D., Chen, Y., Dumont, M.G., & Murrell, J.C. (2008) Marine methylotrophs revealed by stable-isotope probing, multiple displacement amplification and metagenomics. *Environmental Microbiology* **10**: 1526–35.

Nicol, G.W., Leininger, S., Schleper, C., & Prosser, J.I. (2008) The influence of soil pH on the diversity, abundance and transcriptional activity of ammonia-oxidizing archaea and bacteria. *Environmental Microbiology* **10**: 2966–78.

Norton, J.M., Alzerreca, J.J., Suwa, Y., & Klotz, M.G. (2002) Diversity of ammonia monooxygenase operon in autotrophic ammonia-oxidizing bacteria. *Archives of Microbiology* **177**: 139–49.

Pelz O., Tesar M., Wittich R.M., Moore E.R., Timmis K.N., & Abraham W.R. (1999) Towards elucidation of microbial community metabolic pathways: unraveling the network of carbon sharing in a pollutant-degrading bacterial consortium by immunocapture and isotopic ratio mass spectrometry. *Environmental Microbiology* **1**: 167–74.

Philippot, L. (2002) Denitrifying genes in bacterial and archaeal genomes. *Biochimica et Biophysica Acta* **1577**: 355–76.

Prell, J., & Poole, P. (2006) Metabolic changes of rhizobia in legume nodules. *Trends in Microbiology* **14**: 161–8.

Raaijmakers, J.M., & Weller, D.M. (1998) Natural plant protection by 2,4-diacetylphloroglucinol-producing *Pseudomonas* spp. in take-all decline soils. *Molecular Plant-Microbe Interactions* **11**(2): 144–52.

Reinhardt, D. (2007) Programming good relations—development of the arbuscular mycorrhizal symbiosis. *Current Opinion in Plant Biology* **10**: 98–105.

Rhee, S., Liu, X., Wu, L., Chong, S., Wan, X., & Zhou, J. (2004) Detection of genes involved in biodegradation and biotransformation in microbial communities by using 50-mer oligonucleotide microarrays. *Applied and Environmental Microbiology* **70**: 4303–17.

Riaz, K., Elmerich, C., Moneira, D., Raffoux, A., Dessaux, Y., & Faure, D. (2008) A metagenomic analysis of soil bacteria extends the diversity of quorum-quenching lactonases. *Environmental Microbiology* **10**: 560–70.

Rotthauwe, J.H., Witzel, K., & Liesack, W. (1997) The ammonia monooxygenase structural gene amoA as a functional marker: molecular fine-scale analysis of natural ammonia-oxidizing populations. *Applied and Environmental Microbiology* **63**: 4704–12.

Russell, J.B. (2007) The energy spilling reactions of bacteria and other organisms. *Journal of Molecular Microbiology and Biotechnology* **13**: 1–11.

Ryu, C.M., Farag, M.A., Hu, C.H., Reddy, M.S., Kloepper, J.W., & Paré, P.W. (2004) Bacterial volatiles induce systemic resistance in Arabidopsis. *Plant Physiology* **134**: 1017–26.

Sanger, F., Air, G.M., Barrell, B.G., et al. (1977) DNA sequencing with chain-terminating inhibitors. *Proceedings of the National Academy of Sciences of the United States of America* **74**: 5463–7.

Schauss, K., Focks, A., Leininger, S., et al. (2009) Dynamics and functional relevance of ammonia-oxidizing archaea in two agricultural soils. *Environmental Microbiology* **11**: 446–56.

Schipper, C., Hornung, C., Bijtenhoorn, P., Quitschau, M., Grond, S., & Streit, W.R. (2009) Metagenome-derived clones encoding for two novel lactonase family proteins involved in biofilm inhibition in *Pseudomonas aeruginosa*. *Applied and Environmental Microbiology* **7**(1): 224–33.

Schwieger, F., & Tebbe, C.C. (1998) A new approach to utilize PCR-single-strand-conformation polymorphism for 16S rRNA gene-based microbial community analysis. *Applied and Environmental Microbiology* **64**: 4870–6.

Sharma, S., Aneja, M.K., Mayer, J., Schloter, M., & Munch, J.C. (2004) RNA fingerprinting of microbial community in the rhizosphere soil of grain legumes. *FEMS Microbiology Letters* **240**: 181–6.

Sharma, S., Szele, Z., Schilling, R., Munch, J.C., & Schloter, M. (2006) Influence of freeze-thaw stress on the structure and function of microbial communities and denitrifying populations in soil. *Applied and Environmental Microbiology* **72**: 2148–54.

Shen, J.P., Zhang, L.M., Zhu, Y.J.Z, & He, J.Z. (2008) Abundance and composition of ammonia-oxidizing bacteria and ammonia-oxidizing archaea communities of an alkaline sandy loam. *Environmental Microbiology* **10**: 1601–11.

Stres, B., Danevcic, T., Pal, L., et al. (2008) Influence of temperature and soil water content on bacterial, archaeal and denitrifying microbial communities in drained fen grassland soil microcosms. *FEMS Microbiology Ecology* **66**: 110–22.

Strickland, M.S., Lauber, C., Fierer, N., & Bradford, M.A. (2009) Testing the functional significance of microbial community composition. *Ecology* **90**: 441–51.

Sunnucks, P., Wilson, A.C., Beheregaray, L.B., Zenger, K., French, J., & Taylor, A.C. (2000) SSCP is not so difficult: the application and utility of single-stranded conformation polymorphism in evolutionary biology and molecular ecology. *Molecular Ecology* **9**: 1699–710.

Szukics, U., Abell, G.C.J., Hödl, V., et al. (2010) Nitrifiers and denitrifiers respond rapidly to changed moisture and increasing temperature in a pristine forest soil. *FEMS Microbiology Ecology* **72**(3): 395–406.

Tautz, D., Ellegren, H., & Weigel, D. (2010) Next generation molecular ecology. *Molecular Ecology* **19**: 1–3.

Tavormina, P.L., Ussler, W. 3rd, Joye, S.B., Harrison, B.K., & Orphan, V.J. (2010) Distributions of putative aerobic methanotrophs in diverse pelagic marine environments. *The ISME Journal* **4**: 700–10.

Thomashow, L.S., & Weller, D.M. (1988) Role of a phenazine antibiotic from *Pseudomonas fluorescens* in biological control of *Gaeumannomyces graminis* var. *tritici*. *Journal of Bacteriology* **170**: 3499–508.

Torsvik, V., Daae, F.L., Sandaa, R.A., Ovreas, L. (1998) Novel techniques for analysing microbial diversity in natural and perturbed environments. *Journal of Biotechnology* **64**: 53–62.

Treusch, A.H., Leininger, S., Kletzin, A., Schuster, S.C., Klenk, H.P., & Schleper, C. (2005) Novel genes for nitrite reductase and Amo-related proteins indicate a role of uncultivated mesophilic crenarchaeota in nitrogen cycling. *Environmental Microbiology* **7**: 1985–95.

Tringe, S.G., von Mering, C., Kobayashi, A., et al. (2005) Comparative metagenomics of microbial communities. *Science* **308**: 554–7.

UN Millennium Project (2005) *UN Millennium Report, Investing in Development: A Practical Plan to Achieve the Millennium Development Goals*. UN Development Programme, New York.

Urich, T., Lanzen, A., Qi, J., Huson, D.H., Schleper, C., & Schuster, S.C. (2008) Simultaneous assessment of soil microbial community structure and function through analysis of the meta-transcriptome. *PLoS ONE* **3**: e2527.

Vázquez, S., Nogales, B., Ruberto, L., et al. (2009) Bacterial community dynamics during bioremediation of diesel oil-contaminated Antarctic soil. *Microbial Ecology* **57**: 598–610.

von Rad, U., Klein, I., Dobrev, P.I., et al. (2008) Response of *Arabidopsis thaliana* to N-hexanoyl-DL-homoserine-lactone, a bacterial quorum sensing molecule produced in the rhizosphere. *Planta* **229**: 73–85.

Wall, D.H., Bardgett, R.D., & Kelly, E. (2010) Biodiversity in the dark. *Nature Geoscience* **3**(5): 297–8.

Warnecke, F., Luginbuhl, P., Ivanova, N., et al. (2007) Metagenomic and functional analysis of hindgut microbiota of a wood-feeding higher termite. *Nature* **450**: 560–5.

Williamson, L.L., Borlee, B.R., Schloss, P.D., Guan, C., Allen, H.K., & Handelsman, J. (2005) Intracellular screen to identify metagenomic clones that induce or inhibit a quorum-sensing biosensor. *Applied and Environmental Microbiology* **71**: 6335–44.

Woese, C.R. (1987) Bacterial evolution. *Microbiological Reviews* **51**: 221–71.

Wolsing, M., & Priemé, A. (2004) Observation of high seasonal variation in community structure of denitrifying bacteria in arable soil receiving artificial fertilizer and cattle manure by determining T-RFLP of nir gene fragments. *FEMS Microbiology Ecology* **48**: 261–71.

Yachi, S., & Loreau, M. (1999) Biodiversity and ecosystem productivity in a fluctuating environment: The insurance hypothesis. *Ecology* **96**: 1463–8.

Yergeau, E., Kang, S., He, Z., Zhou, J., & Kowalchuk, G.A. (2007) Functional microarray analysis of nitrogen and carbon cycling genes across an Antarctic latitudinal transect. *The ISME Journal* **1**: 163–79.

Yilmaz, P., Gilbert, J.A., Knight, R., et al. (2011) The genomic standards consortium: bringing standards to life for microbial ecology. *The ISME Journal* **5**: 1565–7.

Zagordi, O., Geyrhofer, L., Roth, V., Beerenwinkel, N. (2010) Deep sequencing of a genetically heterogeneous sample: local haplotype reconstruction and read error correction. *Journal of Computational Biology* **17**(3): 417–28.

Zehr, J.P., Jenkins, B.D., Short, S.M., Steward, G.F. (2003) Nitrogenase gene diversity and microbial community structure: a cross-system comparison. *Environmental Microbiology* **5**: 539–54.

Zhang, H., Xie, X., Kim, M.S., Kornyeyev, D.A., Holaday, S., & Paré, P.W. (2008) Soil bacteria augment Arabidopsis photosynthesis by decreasing glucose sensing and abscisic acid levels in planta. *The Plant Journal* **56**: 264–73.

Zhang, Y., Zhang, X., Liu, X., *et al.* (2007) Microarray-based analysis of changes in diversity of microbial genes involved in organic carbon decomposition following land use/cover changes. *FEMS Microbiology Letters* **266**: 144–51.

Zhou, J. (2003) Microarrays for bacterial detection and microbial community analysis. *Current Opinions in Microbiology* **6**: 288–94.

Zhu, G., Jetten, M.S., Kuschk, P., Ettwig, K.F., & Yin C. (2010) Potential roles of anaerobic ammonium and methane oxidation in the nitrogen cycle of wetland ecosystems. *Applied Microbiology and Biotechnology* **86**(4): 1043–55.

CHAPTER 2.2

From Genes to Ecosystems: Plant Genetics as a Link between Above- and Belowground Processes

Jennifer A. Schweitzer, Michael D. Madritch,
Emmi Felker-Quinn, and Joseph K. Bailey

2.2.1 Introduction

This chapter addresses several key issues regarding the importance of a "genes to ecosystem" approach in considering plant–soil linkages. Firstly, we examine how plant functional traits bridge plant species interactions with soil community dynamics. Secondly, we consider the role of plant genetic variation on soil communities and then examine some of the mechanisms by which plant genetic variation affects ecosystem processes. We subsequently explore how plant–soil feedbacks may be strong evolutionary drivers of change in plant functional traits at local and landscape scales, and conclude by considering some key directions for further research. Taken together, this chapter highlights the genetic linkages between plants and soils (i.e. "the extended phenotype") that may have important, but hitherto little appreciated, evolutionary implications.

All plant species express phenotypic variation in morphological, physiological, and chemical traits which in turn drive belowground processes. Plant phenotypic variation is driven, in part, by genetic variation, and ecologists are increasingly aware that genetic variation within a species merits much more attention than has historically been accorded (Whitham et al. 2006; Johnson & Stinchcombe 2007; Hughes et al. 2008; Bailey et al. 2009). Intraspecific trait variation in plants influences aboveground processes, including structuring foliar arthropod communities (Wimp et al. 2004; Johnson & Agrawal 2005; Crutsinger et al. 2006;

Keith et al. 2010), providing resistance and resilience to environmental stress and disturbance (Hughes & Stachowicz 2004; Reusch et al. 2005; Hughes et al. 2009), and promoting resistance to invasion by exotic species (Crutsinger et al. 2008c; Velland et al. 2010). While the role of genetic variation in primary producers has been shown to impact a variety of aboveground response variables, the relationship between plant genetic variation and belowground processes has been less well explored. Recent work has shown that plant genetic variation is particularly important to consider from a belowground perspective as well, as genetically based plant–soil linkages can have strong ecological and evolutionary consequences (Lankau & Strauss 2007; Lankau et al. 2010; Pregitzer et al. 2010; Felker-Quinn et al. 2011; Madritch & Lindroth 2011). Given the long-term legacy effects of plants on soils, it is essential to understand how plant genetic variation influences belowground processes and the supporting services that ecosystems provide (Daily 1997).

Plants influence soil community composition and ecosystem processes in both agricultural and natural systems (Hobbie 1992; Bever et al. 1997; Hooper et al. 2000; Diab el Arab et al. 2001; Wardle et al. 2004; Wardle 2006). Most of this work, until recently, has been focused at the level of plant functional group or species. Variation in the quantity and quality of plant inputs to the soil (both above- and belowground) influences substrate availability for the soil community, including mutualists, root herbivores, pathogens, and decomposers and their activities, as well as soil food web interactions (Paul & Clark 1996; Wardle et al. 2003, 2004; Vandenkoornhuyse et al. 2003; De Deyn et al. 2004; Kang & Mills 2004; Wardle 2006; Horwath 2007). For example, many studies have documented that soil communities, litter decomposition, and mineralization processes differ in soils beneath dominant tree species (Hobbie 1992; Binkley & Menyailo 2005; Wardle 2006). These differences are often due to species-level variation in traits such as plant growth rate, leaf and root chemistry or production, root exudation or abiotic effects due to variation in canopy or rooting structure (Bever et al. 1996; Priha et al. 2001; Bartelt-Ryser et al. 2005; Grayston & Prescott 2005). An appreciation of the links between variation in plant traits and belowground processes in terrestrial ecosystems (Hobbie 1992; Binkley & Giardina 1998; Hooper et al. 2000; Wardle et al. 2004) suggests that plant intraspecific genetic diversity could influence the belowground community and associated ecosystem processes based on the same mechanisms that structure plant–soil interactions at the species level (Zinke 1962; Bever et al. 1996; Rhoades 1997).

Plant–soil linkages may feedback to influence many aspects of plant communities, including invasibility, plant competitive interactions, and successional dynamics (van der Putten et al. 1993, 2001; Casper & Castelli 2003; Reynolds et al. 2003; Bonkowski & Roy 2005; Kardol et al. 2007; Rout & Callaway 2009; de la Peña et al. 2010), and may also have evolutionary implications. For example, the net effects of diverse components of the soil community that either interact with the living plant (including root herbivores, pathogens, mutualists) or plant detritus (heterotrophic decomposers) can result in either positive or negative feedback to plant performance or persistence and feedbacks that can vary through time (Bezemer et al. 2006; Kardol et al. 2006; Kulmatiski et al. 2008; Diez et al. 2010). Negative plant–soil feedbacks from accumulation of pathogens or herbivores prevent species from persisting at fixed locations or at high abundances and also promote species co-occurrence (Bever 1994, 2003; Bever et al. 1997; Klironomos 2002; Bonanomi et al. 2005; Diez et al. 2010) while positive plant–soil feedbacks are mechanisms for persistence or local adaptation (Klironomos 2002; Calloway et al. 2004; Johnson et al. 2010; Mangan et al. 2010). Feedbacks from plants in either direction (negatively or positively) may have important implications for selection and subsequent evolutionary dynamics, which has received little attention in natural systems to date (see van der Putten et al. 2001). Moreover, recent advances demonstrate (at both the plant species and clone level) "home-field advantage" for decomposer communities and the processes they mediate and show that plant (species) specificity may exist for processes such as leaf litter decomposition (Ayres et al. 2006, 2009; Strickland et al. 2009; Madritch & Lindroth 2011; but see St. John et al. 2011). Together the plant–soil feedback and "home-field advantage" literature indicate that evolutionary consequences may be predicted based on the linkages between plants and soils.

While still nascent, studies of the relationship between plant genetic variation and soil communities and processes in natural systems often show tight connections. The expectation would be that plant-associated components of the soil community associated with living plant roots would create the tightest linkage (e.g. root pathogens or herbivores, fungal mutualists, as in agricultural systems) to influence soil processes and plant–soil feedbacks. While this indeed may be the case, the studies to date in natural systems have focused on how genetically-based variation in plants affects soil decomposer communities and the processes they mediate (Schweitzer et al. 2004, 2008a, 2011a; Madritch & Hunter 2005; Madritch et al. 2006, 2009; Madritch & Lindroth 2011). Trait variation at the level of plant phenotype, genotype, genotypic variation, population genetic variation, and genetic divergence (via local adaptation within a species) have all been examined to determine their effects on aspects of the soil decomposer community or on soil processes (see Table 2.2.1 for definitions of genetic terms). These studies have found that intraspecific variation of plant traits can cause decomposer communities to change their composition and/or activity, affect local soil processes, interact with the environment (both abiotic and biotic), and impact plant–soil feedbacks that in turn affect plant performance traits at multiple scales.

Table 2.2.1 Genetic terms used in the genetic-based plant–soil linkage literature

Phenotype
observed value of a trait for an individual that is a composite of genetic and environment effects

Genotype
genetic individual

Genotypic diversity
number of genotypes within a population; manipulations of clonal genotypes or sib-families in mixture from low to high

Genetic variance (V_G)
the variance in a phenotypic trait among individuals due to genetic differences

Genetic divergence
genetic differentiation between populations within a species due to natural selection or drift; can lead to speciation

(From Lynch & Walsh 1998; Connor & Hartl 2004, Hughes et al. 2008.)

2.2.2 The role of plant functional traits in bridging species interactions with soil community dynamics

Functional traits in plants vary both among and within species, and influence associated communities and soil food web dynamics, as well as ecosystem processes that soil communities mediate. Plant traits can directly influence soil processes by altering both biotic (e.g. resource quality and quantity for root-associated organisms) and abiotic conditions (e.g. soil moisture, temperature, and humidity). Moreover, plant traits can have indirect effects on associated communities and their activities by altering conditions that impact the physiology of soil decomposers as well as changing the availability of carbon (C) substrates (and other mineral nutrients). While these direct and indirect links vary in their ability to feedback positively or negatively to affect the fitness of the plant that produced the trait(s) (Binkley & Giardina 1998), the linkage between specific plant traits, soil communities and soil processes indicate that plant soil feedbacks would be predicted to be an important factor affecting genetic variation in plant functional traits (Fig. 2.2.1).

A significant body of work has demonstrated the utility of using plant functional traits in ecology (McGill et al. 2006; Westoby & Wright 2006; Cornwell et al. 2008; Hillebrand & Matthiessen 2009; de Bello et al. 2010). This may be important for determining aspects of the soil community or ecosystem-level processes (e.g. leaf litter decomposition). Traits are manifested as the phenotype upon which selection acts, and these vary with the individual. Beginning with plant communities (McGill et al. 2006), ecologists have come to recognize that plant traits, not necessarily species identity, are the major determinants of community and ecosystem consequences (Wright et al. 2004; Cornwell et al. 2008). Such trait-based approaches are applicable to any level of biodiversity and often focus on variation at the level of the individual organism to elucidate the effects of biodiversity on communities and ecosystems (Petchey & Gaston 2002, 2006). For instance, several authors include intraspecific variation into their estimates of functional diversity for plant species and its effect on ecosystem processes that when accounted for may change the importance of the species effect (Cianciaruso et al. 2009; Fajardo & Piper 2011).

Belowground communities (from herbivores to pathogens to decomposers) are amenable to trait-based approaches, in part, because of their tight associations with plants. The microbial decomposer component of the soil community is difficult to assess in the realm of typical richness indices as microbes often reproduce asexually and swap large functionally important sections of DNA, making "species" an arbitrary definition in the microbial world (Ayala et al. 2000). Consequently, microbial ecologists are beginning to focus on the functional traits of belowground communities as their defining characteristic using molecular tools that inform on the functionality of microbial genes (Brussard et al. Chapter 1.3; Hackl et al. Chapter 2.1, this volume). Green et al. (2008) suggest building on the functional trait work developed in plant communities, such as the universal "leaf economics spectrum" described by Wright et al. (2004), which uses six traits to explain global patterns of plant nutrient cycling rates, viz. leaf mass area, photosynthetic rate, leaf nitrogen (N), leaf phosphorus (P), dark respiration rate, and leaf lifespan. Instead of relying on traditionally-defined functional groups, variation along a continuous spectrum of important traits better defines an organism's functional role in

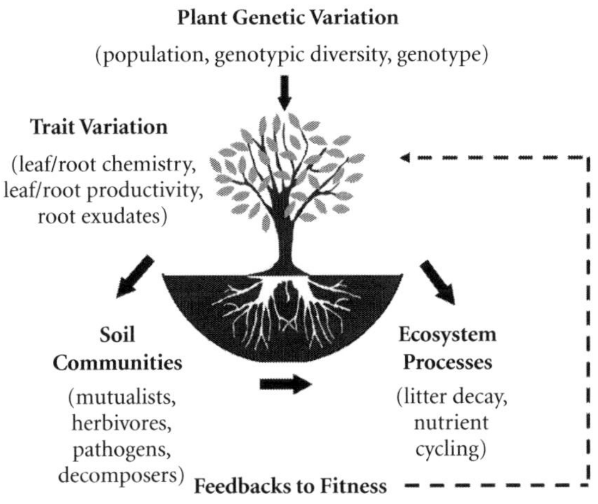

Figure 2.2.1 Schematic illustrating the linkages between plant genetic diversity and soil processes, with evolutionary implications. It demonstrates how plant genetic diversity influences the expression of plant traits that can directly and indirectly influence soil communities and soil processes. Significant variation in soil communities and processes may also feed back to influence the fitness and performance of plant offspring (positively or negatively) to influence population genetic variance. Genetic-based, plant–soil feedback may then be a little-understood mechanism in a larger selective arena that can affect plant population genetics.

an ecosystem. A major challenge in microbial ecology lies in determining which plant functional traits are most important to microbial community dynamics and belowground nutrient cycling phenomena (Green et al. 2008; Hackl et al. Chapter 2.1, this volume).

The influence of plant functional traits on microbial processes is well-documented, as variation in leaf chemistry partially defines a plant's effect on above- and belowground dynamics. The strong influence of plant structures on belowground processes (Cadisch & Giller 1997) necessitates that the ecology of individual plants has a large influence on the ecology of belowground systems (Bardgett 2002). For instance, the traits that make some plants successful invaders, such as high specific leaf area, rapid growth rate, and elevated nutrient concentrations, often have the effect of increasing decomposition and belowground nutrient cycling rates (Ehrenfeld et al. 2001; Allison & Vitousek 2004; Vila et al. 2011). Globally, variation in functional traits explains more variation in decomposition rates than do state factors. That is, across global scales, functional trait-driven variation in leaf litter decomposition exceeds climate-driven variation (Cornwell et al. 2008). In short, trait-based approaches use direct measurements of expressed phenotypes as a metric with which to explain community and ecosystem dynamics (Diaz & Cabido 2001; Cornwell & Ackerly 2009). Specific leaf area and leaf N concentration tend to be the two most important leaf-related traits affecting belowground processes (Wright et al. 2004; Cornwell et al. 2008), especially decomposition and nutrient release. Despite the recent advances linking plant functional traits to belowground decomposition processes, relatively little progress has been made towards identifying the functional traits of belowground organisms that are most important to a diverse array of soil community dynamics (i.e. root herbivores, pathogens, or mutualists) or biogeochemical processes (Green et al. 2008). In part, this is due to the inability to identify pathogenic or mutualistic interactions through a functional gene approach (Hackl et al. Chapter 2.1, this volume). Nonetheless, advances in trait-based ecology seek to explain belowground communities and processes using fine-scale biotic information.

2.2.3 The role of plant genetic variation on soil communities

The soil community is notoriously difficult to quantify or understand in a holistic manner. Nonetheless, the composition and/or the activity of the soil community (notably, the heterotrophic decomposer component) have been shown to respond to plant genetic variation. While plant-associated components of the soil community, such as root feeders, symbiotic organisms, and pathogens, may be expected to be the most sensitive to genetic variation in plants, in natural systems the decomposer community has been most examined as it relates to plant trait variation ranging in the scale of variation from phenotype to genotype to genotypic variation to population genetic variance. For example, short-term responses (<3 years) of litter phenotype transplant treatments had little effect on soil bacterial communities (Madritch & Hunter 2005). However, longer-term common garden and field experiments have shown that soil microbial community composition or enzymatic activity can vary based on phenotypic variation in plants. Aspen (*Populus tremuloides*) clones interact with nutrient availability to influence the activity of six extracellular enzymes (including C, protein, and polyphenol aromatic degrading enzymes) as well as rates of soil respiration in litter microcosms (Fig. 2.2.2a, data from the low nutrient treatments only; Madritch *et al.* 2007). Similarly, Schweitzer *et al.* (2008a) found genetic variation in microbial community composition in soils beneath randomized, replicate copies (n = 3–5) of five different genotypes of *Populus angustifolia* after 16 years of growth and litter inputs to their associated soils. Each genotype, regardless of position in the common garden, had a similar microbial community in the soils directly beneath each tree, indicating a significant influence of the traits of the tree on soil microorganisms (Fig. 2.2.2b). Similarly, at the population level, forest stands with intermediate gene diversity (i.e. heterozygosity) show a marked shift in soil microbial community composition and extracellular enzyme activity from high or low gene diversity stands, that was correlated with both plant productivity and plant chemistry (Schweitzer *et al.* 2011b). In contrast, the saprophytic fungal community on wood samples (characterized on the basis of fruiting bodies) decomposed in the field for 24 months showed no effects of within-species genetic divergence in *Eucalyptus globulus*, as local properties of wood had a larger effect on fungal composition and richness than did origin of tree provenance (i.e. race) during the early stages of wood decay (Barbour *et al.* 2009a).

The composition and activity of belowground microbial communities track fine-scale variation in overstory plants. For example, in a 3-year manipulative experiment where aspen litter was reciprocally placed on the forest floor across multiple aspen clones, Madritch and Lindroth (2011) demonstrated that both microbial community composition and extracellular enzyme activity shifted from one resembling the native community to one that was more compositionally similar to the transplanted community. Plots that matched litter genotypes with native soil communities experienced faster litter decay and lost less inorganic soil N than did plots where the litter genotype was foreign to the soil community. These microbial data overall suggest that variation in tree traits, in many cases in terms of plant secondary chemistry, partially determines the community composition or activity of the soil microbial community in both common garden and field studies.

Despite the effects of plant genetic variation or genotype on soil microbial communities, a range of responses to variation in plant traits in other trophic levels within the leaf litter or soil faunal community are apparent. For example, in a leaf-litter amendment study in a common environment Madritch and Hunter (2005) found no effect of nine different oak (*Quercus laevis*) litter phenotypes on the soil micro-arthropod community composition (Acari, Collembola, and other micro-arthropod groups). Similarly, Crutsinger *et al.* (2008b) found that litter-based micro-arthropod communities (Acari, Collembola) extracted from litterbags of *Solidago altissima*, decomposed in a common environment, showed weak effects of plant genotype, and no effects of genotypic diversity, compared to foliar herbivores (and other arthropod guilds) which showed strong responses to both genotype and genotypic diversity (Crutsinger *et al.* 2006). However, the genotypic diversity effects aboveground varied over time and across spatial scales (Crutsinger *et al.* 2008a, 2009). In contrast, leaf litter of two genetically diverged (i.e. locally-adapted) populations of *Eucalyptus globulus* decomposed in a common environment indicated significant differences in richness, abundance, and community

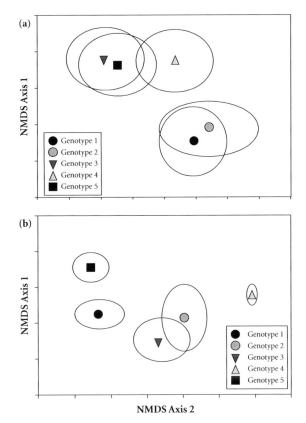

Figure 2.2.2 Experimental results indicate the extracellular enzyme activity or the community composition of soil microbial communities varies beneath trees of different genotypes in common environments. Each symbol (surrounded by ellipses representing 1 ± standard error of the mean) represents the mean activity of six extracellular enzymes in soils associated with *Populus tremuloides* genotypes (a) or the mean microbial community composition in soils associated with *P. angustifolia* (b) genotypes (Bray–Curtis dissimilarity; n = 3–5 replicates/genotype). As this is a distance measure, the non-metric multidimensional scaling (NMDS) axes are unitless. (Data are from Madritch et al. 2007 and Schweitzer et al. 2008a.)

composition (including Acari, Collembola, and other macro-arthropods) after 4 and 8 months of litter decay (Barbour et al. 2009b). These data suggest that the soil micro-arthropod community may respond to plant genetic variation.

2.2.4 The role of plant genetic variation on ecosystem processes

Plants influence soil processes both directly and indirectly through inputs of organic matter that affect local soil communities and the processes they mediate. Within-species variance has been shown to impact soil processes from litter decomposition to total pools and fluxes of soil nutrients, often reflecting concomitant changes in soil communities. As with soil communities, soil processes related to plant genetic variation interact with environmental variation and show a continuum of plant–soil linkages from weak to strong.

Variation at the level of phenotype or genotype, but not genotypic diversity per se, often results in changes in rates of leaf litter decomposition. Litter phenotype or genotype across multiple plant species affects rates of leaf litter mass loss during early (but not always later) stages of decay when labile constituents of the litter are leached or degraded, as well as affecting nutrient release over time (Madritch & Hunter 2002, 2005; Silfer et al. 2007).

For example, genotypes of *Populus angustifolia* × *P. fremontii* tree types differed in decay by 18% as well as demonstrated strong differences in N and P release (Fig. 2.2.3a). The differences in decay were correlated with the concentration of condensed tannin in the foliage (Schweitzer *et al.* 2005). Genotypes of *Solidago altissma* were found to vary in decomposition rate constants by 49%, but this variation was not as great as that among three species of *Solidago* (Crutsinger *et al.* 2009), indicating that within-species variation in traits may not equal trait variation among species.

While the effects of genotypes in common gardens indicate that genotypic variation can influence rates of litter decay and nutrient release, as would be predicted, genotypic interactions with

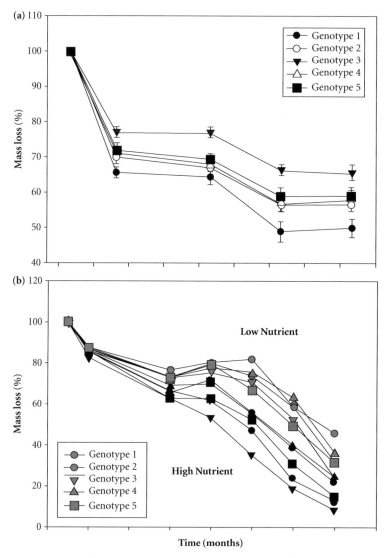

Figure 2.2.3 Experimental results indicate differences in mass loss among genotypes with and without abiotic environmental variation. Genotypes of *Populus angustifolia* decompose at different rates after 24 months in the field (a). Similarly, genotypes of *P. tremuloides* grown under high (black symbols) or low (gray symbols) nutrient treatments decompose at different rates after 13 months in the field (b). (Data are modified from Schweitzer *et al.* 2005 and Madritch *et al.* 2006.)

abiotic and biotic variation also occur. For example, genotypes of aspen grown in high- and low-nutrient environments demonstrate different patterns of foliar chemistry that are influenced by genotype, nutrient treatment, and their interactions over time. This interactive effect of genotype and nutrient treatment on litter chemistry results in genotype × nutrient treatment interactions in the decay of the same aspen genotypes (Fig. 2.2.3B), as well as the release of N, C, and sulphur (S) from decomposing leaf litter (Madritch et al. 2006). When these genotypes were mixed into genotypic diversity treatments to decompose, significant but weak effects of genotypic diversity (three and five genotype mixtures) on decay were found. Fertilization had a much larger relative effect than did genotype for all response variables, indicating that the nutrient availability aspect of the abiotic environment was more important to litter decay than was plant genotype in this study. Therefore, environmental variation interacts with genotype to structure microbially-mediated processes such as litter decomposition, indicating the relative importance of plant genetic variance.

Variation in net rates of N mineralization, net nitrification, or soil nutrient availability can occur in soils associated with plants of varying genetic variation. Individual replicated genotypes from two species of *Populus* (as well as their F_1 hybrids) grown in a common environment indicate that tree genotypes can demonstrate larger variation in annual net rates of N mineralization in their associated soils than the variation between species (Schweitzer et al. 2011a), however, as with Crutsinger et al. (2009), the range of variance sampled within and across species will determine how general this pattern may be. Moreover, genotype-specific annual rates of N mineralization indicate that processes such as these can have a genetic basis, which indicates that heritable plant traits can repeatedly influence their associated soils. Soil ammonium availability can also vary by litter phenotype treatments (Madritch & Hunter 2002, 2005), although neither soil ammonium nor nitrate change with plant litter treatments of increasing phenotypic diversity (Madritch & Hunter 2005). This then suggests that phenotype composition is more important than phenotypic diversity per se. At the population level, soil ammonium is correlated with intermediate gene diversity across a gradient of population gene diversity (Schweitzer et al. 2011b). While the rates of soil nutrient turnover may vary by plant genetic variation, nutrient uptake also has been shown to vary by plant genotype (Hughes et al. 2009) which may also influence overall pools of soil nutrients.

In both common garden and field environments, pools of soil C and N have been shown to vary by plant genotype. Schweitzer et al. (2008a) found variation in pools of microbial N and microbial C among 5–8 genotypes each of *P. fremontii*, *P. angustifolia*, and their hybrids in 16-year-old common garden. No differences were found in total soil C or N (Schweitzer et al. 2011a). At the level of individual plant clone across a landscape, Madritch et al. (2009) found that individual clones of aspen (from 12–40 m^2) in the field varied in foliar chemistry that influenced the extracellular enzyme activity of the soil microbial community (relative to adjacent, non-aspen soils). Variation in enzyme activity consequently influenced total pools of soil C and N. These data show linkages from the plant through the microbial community to soil nutrient pools and indicate that plant chemistry, soil microbial communities, and soil processes are all correlated.

2.2.5 The evolutionary implications of plant–soil linkages

While plants can influence many aspects of soils, including soil decomposer communities and the processes they mediate, plant–soil feedbacks in natural systems represent an emerging area of research linking ecology and evolution. Soils and underlying parent material can determine the distribution, population genetic structure and evolution of plant species at large scales (Ellis & Weis 2006; Fierer & Jackson 2006; Alvarez et al. 2009). Simultaneously, especially at local scales, plants can influence soil abiotic and biotic properties that feed back to impact plant diversity and succession, persistence of invasive species and overall soil fertility (Ehrenfeld et al. 2005; Kardol et al. 2006, 2007; Kulmatiski et al. 2008; Mangan et al. 2010). Whether directly or indirectly, plant functional traits can create conditions that affect soil communities (i.e. their composition or activity) and decomposition processes that soil communities at least partly regulate (i.e. nutrient depolymerization or mineralization processes). The data to date, described earlier, demonstrate that genetic variation in functional plant traits can influence soil communities and the processes in soil they mediate (i.e. create an "extended phenotype"; Whitham et al. 2003). Less is known about how soils may impact the evolutionary dynamics of plant functional traits. For soils to be important factors driving the evolution of functional plant traits, several conditions must be met: there must be genetic variation in the plant population for soils to influence, and plants must respond to such selective forces imposed by soils (Brady et al. 2005). Emerging examples linking ecological and evolutionary dynamics provides compelling evidence that an evolutionary response of plants to soils in the form of local adaptation and population-level genetic divergence is possible, and that soil communities may have large impacts on the expression of plant genetic variation.

Recent studies examining "home-field advantage" for a range of community and ecosystem responses in specific "home," or native conditions, versus "away" conditions are experimental, mechanistic tests that indicate local adaptation. For example, Johnson et al. (2010) found that Andropogon ecotypes adapt to their local soil via indigenous arbuscular mycorrhizal fungal communities such that mycorrhizal exchange of the most limiting resource is maximized. Because plants have evolved in response to a host of biotic and abiotic soil factors, these results indicate that "soils" act as agents of selection.

In addition to the evolutionary response of plants to arbuscular mycorrhizal fungal communities described in Johnson et al. (2010), there are many potential mechanisms by which "soils" may act as agents of selection. It has been proposed that plants are part of a multitrophic selection arena belowground therefore, as with any evolutionary dynamic, the selective impact and evolutionary response of the interacting species may vary in space, time, and context (sensu van der Putten et al. 2001; Thompson 2005). Indeed, studies indicating the importance of "home-field advantage" in litter decomposition (Gholz et al. 2000; Ayres et al. 2009; Strickland et al. 2009; Madritch & Lindroth 2011; St. John et al. 2011) show that decomposer communities more efficiently utilize litter nutrients from their "home" than when in "away" conditions. These "home-field advantage" studies provide mechanistic examples of how local litter alters decomposer communities and may act as an important selective force.

When the mechanisms of "home-field" effects are combined with the fitness and performance consequences of feedbacks, the net effect is a plant soil interaction with evolutionary consequences, even if the specific selection gradients are unknown. For example, Populus spp. in the western USA vary in plant phytochemical traits that influence rates of leaf litter decay, soil microbial communities, and rates of soil net N mineralization and nutrient availability (Schweitzer et al. 2004, 2008b; Rehill et al. 2006; Fischer et al. 2010). In a greenhouse experiment, Pregitzer et al. (2010) planted seedlings from 20 randomly collected P. angustifolia genetic families in soils conditioned by various Populus species in the field and measured subsequent survival and performance. Even though P. angustifolia soils were less fertile overall, P. angustifolia seedlings grown in P. angustifolia-conditioned soils were twice as likely to survive, grew 24% taller, had 27% more leaves, and 29% greater aboveground biomass than P. angustifolia seedlings grown in non-native P. fremontii or hybrid soils. Increased survival resulted in higher trait variation among seedlings in native soils compared to seedlings grown in non-native "away" soils. Soil microbial biomass explained more of the variation in seedling performance than soil texture, pH, or nutrient availability, suggesting microbial interactions and feedbacks between plants, soils, and associated microorganisms. Overall, these data suggest that a positive soil feedback helps maintain genetic variance in P. angustifolia seedlings, although the specific biotic components of the soil community, and their interactions, that influenced this pattern are unknown.

This framework may be expanded to broad geographic scales, as well as across populations where soils may also act as agents of selection influencing the fitness and performance of plants. Felker-Quinn et al. (2011) found that population-level genetic

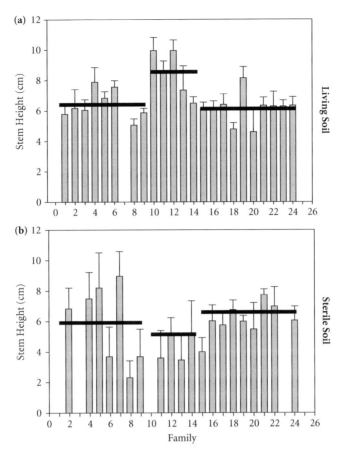

Figure 2.2.4 Experimental results indicate the importance of the soil community in expressing family and population-level variance in tree seedling height. The top panel (a) indicates family and population responses when seedlings were grown in living soil while the bottom panel (b) indicates that the range of genetic variation due to family or population is reduced without the soil community (i.e. in sterile soils). Each bar represents the mean height of an individual family, while the black horizontal bars indicate population level means from each of the three populations of *Ailanthus altissima*. (Data are modified from Felker-Quinn et al. 2011.)

variation of three geographically distinct populations of *Ailanthus altissima* resulted in genetic families that vary in growth patterns and phenotypic traits, as determined by growing families of trees from each population in their local and "away" soil. When the seedlings from each family and population were reciprocally grown in soils from each population, with both a live soil and sterile soil treatments, they found that feedbacks from population origin differentially influenced the performance of seedlings. Two populations showed positive feedbacks, whereby seedling performance was enhanced in "home" soils rather than "away" soils, while one population performed better in "away" soils, indicating negative feedback effects. Moreover, population genetic variance was found when the plants were grown in a live control soil (with an intact soil community), but there was little among-population genetic variance when the seedlings were grown in a sterile control soil for a range of traits (Fig. 2.2.4). These results show that the expression of additive and population level genetic variation depends upon the presence of the soil biotic community and that the interpretation of quantitative evolutionary divergence in plants depends upon the presence of the soil biotic community.

The literature that is based on aboveground systems has commonly found that herbivores and tritrophic interactions provide important selective gradients for plants (Fritz & Simms 1992; Hunter &

Price 1992; Thompson 2005). However, the soil community is also emerging as an important factor in mediating the evolutionary dynamics of plant functional traits. Both Pregitzer et al. (2010) and Felker-Quinn et al. (2011) show the importance of within-species genetic variance on plant–soil feedback that is mediated by as yet unknown aspects of the soil community to influence the selection of plant traits in future generations. These data which are some of the first in natural systems to show that genetically-based functional variation in plant traits can influence soil communities and that soil communities affected by functional plant traits can act as agents of selection for aboveground traits through feedbacks. As such, they implicate the soil community as an important regulator of both belowground processes as well as aboveground genetic variation in these systems.

2.2.6 Conclusions and future directions

What do we really know so far? As many of the discussed examples indicate, ecological interactions between plants and soils can result in specific community and ecosystem responses that may feed back to influence trait variation within and among plant populations. Specifically, the studies reviewed here indicate that: 1) plant genetic variance results in shifts in plant phenotypic traits; 2) plant genetic variance impacts decomposer communities and the activity of extracellular enzymes that mediate organic matter dynamics; 3) plant genetic variance impacts aspects of soil C and N cycling; 4) genetic-based "extended phenotype" feedbacks influence plant trait variance (through the soil community), with evolutionary consequences.

Plant traits, at the intraspecific level, may result in predictable impacts on soil communities and aspects of C and N cycling. Plant chemistry, in particular, may constitute potent plant functional traits that influence decomposer soil communities, as indicated by the importance of plant foliar N, lignin, and polyphenol content for many of the earlier examples (*sensu* Grayston et al. 1998; Schweitzer et al. 2005, 2008b, 2011b; Madritch et al. 2006, 2009). Soils change in response to genetic variation in plant traits, directly or indirectly by influencing heterotrophic microbial communities, microbial activities, and soil food web dynamics.

Table 2.2.2 indicates the range of studies and relative strength of responses of genetic-based plant–soil linkages to date. It demonstrates that phenotypic variation occurs across a continuum of genetic variation, from genotype to population-level divergence, which can influence variation in belowground decomposer communities and biogeochemistry, to various degrees. More work is necessary to understand the detailed mechanisms of these linkages including the roles of roots and root-associated communities, root exudates, and genetic-based interactions between C, N, and phosphorus cycles.

While the initial work indicating the importance of genetic-based plant–soil linkages has been done in common environments with limited environmental variation, experiments that incorporate environmental variation (e.g. nutrient addition) in natural systems also indicate that plant genetic factors can drive variation in belowground processes (Madritch et al. 2006, 2009; Madritch & Lindroth 2011; Schweitzer et al. 2011b). However, it is largely unknown to what extent plant genetic factors interact with large-scale environmental heterogeneity to influence the distribution and activity of soil communities and their resulting nutrient dynamics. Moreover, understanding the importance of plant genetic variance in the context of whole plant community dynamics (i.e. the biotic environment) remains unresolved.

Recent data suggest that genetic-based, plant–soil feedbacks may be an important factor in maintaining plant genetic variation across landscapes. The few studies to date on this topic suggest that positive and negative feedbacks may be an important mechanism for determining population genetic variation. Plant–soil linkage studies may represent a holistic way to assess the evolutionary consequences of multiple interacting feedback loops as plants and soils are players in a multitrophic selection arena (*sensu* van der Putten et al. 2001). While the "extended phenotype" of plant genetic variance may feed back to influence subsequent population-level or trait variance (Fig. 2.2.1), other above- and belowground interacting factors, such as herbivory, tritrophic interactions, and plant–plant interactions, are also acting as important selective agents that

Table 2.2.2 Range of studies and the relative responses, to date, examining the effects of plant genetic variation on aspects of soil communities, soil nutrient pools, and fluxes. Empty cells indicate knowledge gaps

	Genotype/individual	Genotypic diversity	Population genetic variance	Genetic divergence (provenance)
Biomass/net primary productivity	Strong	Strong		
Litter decomposition	Moderate	Weak		
Litter nutrient release	Moderate	Weak		
Soil invertebrates	Strong	Weak		Moderate
Microbial community composition	Weak–moderate			None
Extracellular enzyme activity	Strong		Moderate	
Microbial nutrient pools	None–weak		None–weak	
Soil respiration	None–moderate			
Total soil nitrogen/carbon	None	None	None	
Net nitrogen mineralization	Strong			
Nitrogen availability	Strong		Moderate	Moderate

together will shape the genetic landscape of each interaction. These data along with the strong evidence demonstrating the role of soil communities and plant–soil feedback in mediating plant competitive and successionary dynamics (van der Putten et al. 1993; Aerts 1999; Bever 2003; Reynolds et al. 2003; Bonkowski & Roy 2005; Kardol et al. 2006), indicate an important but still little-appreciated role of plant–soil feedback in mediating landscape patterns of ecology and evolution, which may ultimately provide support for the role of adaptive evolution in ecosystem ecology (Matthews et al. 2011; Schoener 2011;).

References

Aerts, R. (1999) Interspecific competition in natural plant communities: mechanisms, trade-offs and plant-soil feedbacks. *Journal of Experimental Botany* **50**: 29–37.

Allison, S.D., & Vitousek, P.M. (2004) Rapid nutrient cycling in leaf litter from invasive plants in Hawai'i. *Oecologia* **141**: 612–19.

Alvarez, N., Theil-Egenter, C., Tribsch, A., et al. (2009) History or ecology: Substrate type as a major driver of spatial genetic structure in Alpine plants. *Ecology Letters* **12**: 632–40.

Ayala, F.J., Fitch, W.M., & Clegg, M.T. (2000) Variation and evolution in plants and microorganisms: towards a new synthesis 50 years after Stebbins. *Proceedings of the National Academy of Sciences of the United States of America* **97**: 6941–4.

Ayres, E., Dromph, K.M., & Bardgett, R.D. (2006) Do plant species encourage soil biota that specialise in the rapid decomposition of their litter? *Soil Biology and Biochemistry* **38**: 183–6.

Ayres, E., Steltzer, H., Simmons, B.L., et al. (2009) Home-field advantage accelerates leaf litter decomposition in forests. *Soil Biology and Biochemistry* **41**: 606–10.

Bailey, J.K., Schweitzer, J.A., Úbeda, F., et al. (2009) From genes to ecosystems: A synthesis of the effects of plant genetic variation across levels of organization. *Philosophical Transaction of the Royal Society B* **364**: 1607–16.

Barbour, R.C., Baker, S.C., O'Reilly-Wapstra, J.M., Harvest, T.M., & Potts, B.M. (2009a) A footprint of tree-genetics on the biota of the forest floor. *Oikos* **118**: 1917–23.

Barbour, R.C., Storer, M.J., & Potts, B.M. (2009b) Relative importance of tree genetics and microhabitat on macrofungal biodiversity on coarse woody debris. *Oecologia* **160**: 335–42.

Bardgett, R. (2002) Causes and consequences of biological diversity in soil. *Zoology* **105**: 367–74.

Bartelt-Ryser, J., Joshi, J., Schmid, B., Brandl, H., & Balser, T. (2005) Soil feedbacks of plant diversity on soil microbial communities and subsequent plant growth. *Perspectives in Plant Ecology, Evolution and Systematics* **7**: 27–49.

Bever, J.D. (1994) Feedback between plants and their soil communities in an old field community. *Ecology* **75**: 1965–77.

Bever, J.D. (2003) Soil community feedback and the coexistence of competitors: conceptual frameworks and empirical tests. *New Phytologist* **157**: 465–73.

Bever, J.D., Morton, J., Antonovics, J., & Schultz, P.A. (1996) Host-dependent sporulation and species diversity of mycorrhizal fungi in a mown grassland. *Journal of Ecology* **75**: 1965–77.

Bever, J.D., Westover, K.M., & Antonovics, J. (1997) Incorporating the soil community into plant population dynamics: the utility of the feedback approach. *Journal of Ecology* **85**: 561–73.

Bezemer, T.M., Lawson, C.S., Hedlund, K., *et al.* (2006) Plant species and functional group effects on abiotic and microbial soil properties and plant–soil feedback responses in two grasslands. *Journal of Ecology* **94**: 893–904.

Binkley, D., & Giardina, C. (1998) Why do tree species affect soils? The warp and woof of tree–soil interactions. *Soil Biology and Biochemistry* **42**: 89–106.

Binkley, D., & Menyailo, O. (2005) *Tree Species Effects on Soils: Implications for Global Change*. NATO Science Series, Springer, Dordrecht.

Bonanomi, G., Giannino, F., & Mazzoleni, S. (2005) Negative plant–soil feedback and species coexistence. *Oikos* **111**: 311–21.

Bonkowski, M., & Roy, J. (2005) Soil microbial diversity and soil functioning affect competition among grasses in experimental microcosms. *Oecologia* **143**: 232–40.

Brady, K.U., Kruckeberg, A.R., & Bradshaw Jr., H.D. (2005) Evolutionary ecology of plant adaptation to serpentine soils. *Annual Review of Ecology, Evolution and Systematics* **36**: 243–66.

Cadisch, G., & Giller, K.E. (1997) *Driven by Nature: Plant litter quality and decomposition*. CAB International, Wallingford.

Calloway, R.M., Thelen, G.C., Rodriguez, A., & Holben, W.E. (2004) Soil biota and exotic plant invasion. *Nature* **427**: 731–3.

Casper, B.B., & Castelli, J.P. (2003) Evaluating plant–soil feedback together with competition in a serpentine grassland. *Ecology Letters* **10**: 394-300.

Cianciaruso, M.V., Batalha, M.A., Gaston, K.J., & Petchey, O.L. (2009) Including intraspecific variability in functional diversity. *Ecology* **90**: 81–9.

Connor, J.K., & Hartl, D.L. (2004) *A Primer of Ecological Genetics*. Sinauer, Sunderland.

Cornwell, W.K., & Ackerly, D.D. (2009) Community assembly and shifts in plant trait distributions across an environmental gradient in coastal California. *Ecological Monographs* **79**: 109–26.

Cornwell, W.K., Cornelissen, J.H.C., Amantangelo, K., *et al.* (2008) Plant species traits are the predominant control on litter decomposition rates within biomes worldwide. *Ecology Letters* **11**: 1065–71.

Crutsinger, G.M., Collins, M.D., Fordyce, J.A., Gompert, Z., Nice, C.C., & Sanders, N.J. (2006) Plant genotypic diversity predicts community structure and governs an ecosystem process. *Science* **313**: 966–8.

Crutsinger, G.M., Collins, M.D., Fordyce, J.A., & Sanders, N.J. (2008a) Temporal dynamics in non-additive responses of arthropods to host-plant genotypic diversity. *Oikos* **117**: 255–64.

Crutsinger, G.M., Reynolds, N., Sanders, N.J., & Classen, A.T. (2008b) Disparate effects of host plant genotypic diversity on above- and belowground communities. *Oecologia* **158**: 65–75.

Crutsinger, G.M., Sanders, N.J., & Classen, A.T. (2009) Contrasting intra- and interspecific variation on litter dynamics. *Basic and Applied Ecology* **10**: 535–43.

Crutsinger, G.M., Souza, L., & Sanders, N.J. (2008c) Intraspecic diversity and dominant genotypes resist plant invasions. *Ecology Letters* **11**: 16–23.

Daily, G.C. (eds.) (1997) *Nature's services: societal dependence on natural ecosystems*. Island Press, Washington DC.

de Bello, F., Lovorel, S., Albert, C.H., *et al.* (2010) Quantifying the relevance of intraspecific trait variability for functional diversity. *Methods in Ecology and Evolution* **2**: 163–74.

De Deyn, G.B., Raaijakers, C.E., van Ruijven, J., Berendse, F., van der Putten, W.H. (2004) Plant species identity and diversity effects on different trophic levels of nematodes in the soil food web. *Oikos* **106**: 576–86.

de la Peña, E., de Clercq, N., Bonte, D., Roiloa, S., Rodríguez-Echeverría, S., & Freitas, H. (2010) Plant-soil feedback as a mechanism of invasion by *Carpobrotus edulis*. *Biological Invasions* **12**: 3637–48.

Diab el Arab, H.G., Vilich, V., & Sikora, R.A. (2001) The use of phospholipid fatty acid analysis (PFLA) in determination of rhizosphere specific microbial communities (RSMC) of two wheat cultivars. *Plant and Soil* **228**: 291–7.

Diaz, S. & Cabido, M. (2001) Vive la difference: plant functional diversity matters to ecosystem processes. *Trends in Ecology and Evolution* **16**: 646–55.

Diez, J.M., Dickie, I., Edwards, G., Hulme, P.E., Sullivan, J.J., & Duncan, R.P. (2010) Negative soil feedbacks accumulate over time for non-native plant species. *Ecology Letters* **13**: 803–9.

Ehrenfeld, J.G., Kourtev, P., Huang, W.Z. (2001) Changes in soil functions following invasions of exotic understory plants in deciduous forests. *Ecological Applications* **11**: 1287–300.

Ehrenfeld, J.G., Ravit, B., & Elgersma, K. (2005) Feedback in the plant-soil system. *Annual Review of Environmental Restoration* **30**: 75–115.

Ellis, A.G., & Weis, A.E. (2006) Coexistence and differentiation of "flowering stones": the role of local adaptation to soil microenvironment. *Journal of Ecology* **94**: 322–35.

Fajardo, A., & Piper, F.I. (2011) Intraspecific trait variation and covariation in a widespread tree species (*Nothofagus pumilio*) in southern Chile. *New Phytologist* **189**: 259–71.

Felker-Quinn, E., Bailey, J.K., & Schweitzer, J.A. (2011) Geographic mosaics of plant-soil feedbacks in a highly invasive plant: implications for invasion success. *Ecology* **92**: 1208–14.

Fierer, N., & Jackson, R., (2006) The diversity and biogeography of soil bacterial communities. *Proceedings of the National Academy of Sciences of the United States of America* **103**: 626–31.

Fischer, D.G., Hart, S.C., Schweitzer, J.A., Selmants, P.C., & Whitham, T.G. (2010) Soil nitrogen availability varies with plant genetics across diverse river drainages. *Plant and Soil* **331**: 391–400.

Fritz, R.S., & Simms, E.L. (eds.) (1992) *Plant resistance to herbivores and pathogens*. University of Chicago Press, Chicago, IL.

Gholz, H.L., Wedin, D.A., Smitherman, S.M., et al. (2000) Longterm dynamics of pine and hardwood litter in contrasting environments: toward a global model of decomposition. *Ecosystems* **6**: 751–65.

Grayston, S.J., & Prescott, C.E. (2005) Microbial communities in forest floors under four tree species in coastal British Columbia. *Soil Biology and Biochemistry* **37**: 1157–67.

Grayston, S.J., Wang, S., Campbell, C.D., & Edwards, A.C. (1998) Selective influence of plant species on microbial diversity in the rhizosphere. *Soil Biology and Biochemistry* **30**: 369–78.

Green, J.L., Bohannan, B.J.M., & Whitaker, R.J. (2008) Microbial biogeography: from taxonomy to traits. *Science* **320**: 1039–43.

Hillebrand, H., & Matthiessen, B. (2009) Biodiversity in a complex world: consolidation and progress in functional biodiversity research. *Ecology Letters* **12**: 1405–19.

Hobbie, S.E. (1992) Effects of plant species on nutrient cycling. *Trends in Ecology & Evolution* **7**: 336–9.

Hooper, D.U., Bignell, D.E., Brown, V.K. et al. (2000) Interactions between above- and belowground biodiversity in terrestrial ecosystems: Patterns, mechanisms, and feedbacks. *BioScience* **50**: 1049–61.

Horwath, W. (2007) Carbon cycling and formation of soil organic matter. In: E.A. Paul (ed.) *Soil microbiology, ecology and biochemistry*, (3rd edn.), pp. 303–37. Academic Press, Amsterdam.

Hughes, A.R., Inouye, B., Johnson, M.T.J., Underwood, N., & Vellend, M. (2008) Ecological consequences of genetic diversity. *Ecology Letters* **11**: 609–23.

Hughes, A.R., & Stachowicz, J.J. (2004) Genetic diversity enhances the resistance of a seagrass ecosystem to disturbance. *Proceedings of the National Academy of Sciences of the United States of America* **101**: 8998–9002.

Hughes, A.R., Stachowicz, J.J., & Williams, S.L. (2009) Morphological and physiological variation among seagrass (*Zostera marina*) genotypes. *Oecologia* **159**: 725–33.

Hunter, M.D., & Price, P.W. (1992) Playing chutes and ladders: heterogeneity and the relative roles of bottom-up and top-down forces in natural communities. *Ecology* **73**: 724–32.

Johnson, M.T.J., & Agrawal, A.A. (2005) Plant genotype and environment interact to shape a diverse arthropod community on evening primrose (*Oenothera biennis*). *Ecology* **86**: 874–85.

Johnson, M.T.J., & Stinchcombe, J.R. (2007) An emerging synthesis between community ecology and evolutionary biology. *Trends in Ecology & Evolution* **22**: 250–7.

Johnson, N.C., Wilson, G.W.T., Bowker, M.A., Wilson, J.A., Miller, R.M. (2010) Resource limitation is a driver of local adaptation in mycorrhizal symbioses. *Proceedings of the National Academy of Sciences of the United States of America* **107**: 2093–8.

Kang, S., & Mills, A. (2004) Soil microbial community structure changes following disturbance of the overlying plant community. *Soil Science* **169**: 55–65.

Kardol, P., Bezemer, M.T., & van der Putten, W.H. (2006) Temporal variation in plant-soil feedbacks controls succession. *Ecology Letters* **9**: 1080–8.

Kardol P., Cornips, N.J., van Kempen, M.M.L., Tanja Bakx-Schotman, J.M., & van der Putten, W.H. (2007) Microbe-mediated plant-soil feedback causes historical contingency effects in plant community assembly. *Ecological Monographs* **77**:147–62.

Keith, A.R., Bailey, J.K., & Whitham, T.G. (2010) A genetic basis to community repeatability and stability. *Ecology* **91**: 3398–406.

Klironomos, J.N. (2002) Feedback with soil biota contributes to plant rarity and invasiveness in communities. *Nature* **417**: 67–70.

Kulmatiski, A., Beard, K.H., Stevens, J., & Cobbold, S.M. (2008) Plant-soil feedbacks: a meta-analytical review. *Ecology Letters* **11**: 980–92.

Lankau, R.A., Nuzzo, V., Spyreas, G., & Davis, A.S. (2010) Evolutionary limits ameliorate the negative impact of an invasive plant. *Proceedings of the National Academy of Sciences of the United States of America* **36**: 15362–7.

Lankau, R.A., & Strauss, S.Y. (2007) Mutual feedbacks maintain both genetic and species diversity in a plant community. *Science* **317**: 1561–3.

Lynch, M., & Walsh, B. (1998) *Genetics and analysis of quantitative traits*. Sinauer, Sunderland, MA.

Madritch, M.D., Donaldson, J.R., & Lindroth, R.L. (2006) Genetic identity of *Populus tremuloides* litter influences decomposition and nutrient release in a mixed forest stand. *Ecosystems* **9**: 528–37.

Madritch, M.D., Donaldson, J.R., & Lindroth, R.L. (2007) Canopy herbivory mediates the influence of plant genotype on soil processes through frass deposition. *Soil Biology and Biochemistry* **39**: 1192–201.

Madritch, M.D., Greene, S.L., & Lindroth, R.L. (2009) Genetic mosaics of ecosystem functioning across aspen-dominated landscapes. *Oecologia* **160**: 119–27.

Madritch, M.D., & Hunter, M.D. (2002) Phenotypic diversity influences ecosystem functioning in an oak sandhills community. *Ecology* **83**: 2084–90.

Madritch, M.D., & Hunter, M.D. (2005) Phenotypic variation in oak litter influences short- and long-term nutrient cycling through litter chemistry. *Soil Biology and Biochemistry* **37**: 319–27.

Madritch, M.D., & Lindroth, R.L. (2011) Soil microbial communities adapt to genetic variation in leaf litter inputs. *Oikos* **120**: 1696–704.

Mangan, S.A., Schnitzer S.A., Herre, E.A., *et al.* (2010) Negative plant-soil feedback predicts tree species relative abundance in a tropical forest. *Nature* **466**: 698–9.

Matthews, B., Narwani, A., Hausch, S., *et al.* (2011) Toward an integration of evolutionary biology and ecosystem science. *Ecology Letters* **14**: 690–701.

McGill, B.J., Enquist, B.J., Weiher, E., & Westoby, M. (2006) Rebuilding community ecology from functional traits. *Trends in Ecology & Evolution* **21**: 178–85.

Paul, E.A., & Clark, F.E. (1996) *Soil microbiology and biochemistry*. Academic Press, San Diego CA.

Petchey, O.L., & Gaston, K.J. (2002) Functional diversity (FD), species richness and community composition. *Ecology Letters* **5**: 402–11.

Petchey, O.L., & Gaston, K.J. (2006) Functional diversity: back to basics and looking forward. *Ecology Letters* **9**: 741–58.

Pregitzer, C.P., Bailey, J.K., Hart, S.C., & Schweitzer, J.A. (2010) Soils as agents of selection: feedbacks between plants and soils alter seedling survival and performance. *Evolutionary Ecology* **24**: 1045–9.

Priha, O., Grayston, S.J., Hiukka, R., Pennanen, T., & Smolander, A. (2001) Microbial community structure and characteristics of the organic matter in soils under *Pinus sylvestris*, *Picea abies* and *Betula pendula* at two forest sites. *Biology and Fertility of Soils* **33**: 17–24.

Rehill, B.J., Whitham, T.G., Martinsen, G.D., Schweitzer, J.A., Bailey, J.K. and Lindroth, R.L. (2006) Developmental trajectories in cottonwood phytochemistry. *Journal of Chemical Ecology* **32**: 2269–85.

Reynolds, H.L., Packer, A., Bever, J.D., & Clay, K. (2003) Grassroots ecology: Plant-microbe interactions as drivers of plant community structure and dynamics. *Ecology* **84**: 2281–91.

Reusch, T.B.H., Ehlers, A., Hämmereli, I., & Worm, B. (2005) Ecosystem recovery after climatic extremes enhanced by genotypic diversity. *Proceedings of the National Academies of Science* **102**: 2826–31.

Rhoades, C.C. (1997) Single-tree influences on soils properties in agroforestry: lessons from natural forestry and savannah ecosystems. *Agroforestry Systems* **35**: 71–94.

Rout, M.E., & Callaway, R.M. (2009) An invasive plant paradox. *Science* **324**:734–5.

Schoener, T.W. (2011) The newest synthesis: understanding the interplay of evolutionary and ecological dynamics. *Science* **331**: 426–9.

Schweitzer, J.A., Bailey, J.K., Fischer, D.G., *et al.* (2008a) Soil microorganism-plant Interactions: heritable relationship between plant genotype and associated microorganisms. *Ecology* **89**: 773–81.

Schweitzer, J.A., Bailey, J.K., Fischer, D.G., LeRoy, C.J., Whitham, T.G., Hart, S.C. (2011a) Functional and heritable consequences of plant genotype on community composition and ecosystem processes. In: Ohgushi T, Schmitz O, Holt R (eds.) *Ecology and Evolution of Trait-mediated Indirect Interactions: Linking Evolution, Community, and Ecosystem*. British Ecological Society (in press).

Schweitzer, J.A., Bailey, J.K., Hart, S.C., Wimp, G.M., Chapman, S.C., & Whitham, T.G. (2005) The interaction of plant genotype and herbivory decelerate leaf litter decomposition and alter nutrient dynamics. *Oikos* **110**: 133–45.

Schweitzer, J.A., Bailey, J.K., Rehill, B.J., Hart, S.C., Lindroth, R.L., & Whitham, T.G. (2004) Genetically based trait in dominant tree affects ecosystem processes. *Ecology Letters* **7**: 127–34.

Schweitzer, J.A., Fischer, D.G., Rehill, B.J., *et al.* (2011b) Forest gene diversity influences the composition and function of soil microbial communities. *Population Ecology* **53**: 35–46.

Schweitzer, J.A., Madritch, M.D., Bailey, J.K., *et al.* (2008b) From genes to ecosystems: the genetic basis of condensed tannins and their role in nutrient regulation in a *Populus* model system. *Ecosystems* **11**: 1005–20.

Silfer, T., Roininen, H., Oksanen, E., & Rousi, M. (2007) Genetic and environmental determinants of silver birch growth and herbivore resistance. *Forest Ecology and Management* **257**: 2145–9.

St. John, M.G., Orwin, K.H., & Dickie, I.A. (2011) No "home" versus "away" effects of decomposition found in a grassland forest reciprocal litter transplant study. *Soil Biology and Biochemistry* **43**: 1482–9.

Strickland, M.S., Lauber, C., Fierer, N., & Bradford, M.A. (2009) Testing the functional significance of microbial community composition. *Ecology* **90**: 441–51.

Thompson, J.N. (2005) *The geographic mosaic of coevolution*. University of Chicago Press: Chicago, IL.

Vandenkoornhuyse, P., Ridgway, K.P., Watson, I.J., Fitter, A.H., & Young, J.P.W. (2003) Co-existing grass species have distinctive arbuscular mycorrhizal communities. *Molecular Ecology* **12**: 3085–95.

van der Putten, W.H., Van Dijk, C., & Peters, B.A.M. (1993) Plant–specific soil-borne diseases contribute to succession in foredune vegetation. *Nature* **362**: 53–6.

van der Putten, W.H., Vet, L.E.M., Harvey, J.A., & Wäckers, F.L. (2001) Linking above- and belowground multitrophic interactions of plants, herbivores, pathogens, and their antagonists. *Trends Ecology & Evolution* **16**: 547–54.

Velland, M., Drummond, E.B.M., & Tomimatsu, H. (2010) Effects of genotypic diversity on the invasiveness and invasibility of plant populations. *Oecologia* **162**: 371–81.

Vila, M., Espinar, J.L., Hejda, M., *et al.* (2011) Ecological impacts of invasive alien plants: a meta-analysis of their effects on species, communities and ecosystems. *Ecology Letters* **14**: 702–8.

Wardle, D.A. (2006) The influence of biotic interactions on soil biodiversity. *Ecology Letters* **9**: 870–86.

Wardle, D.A., Bardgett, R.D., Klironomos, J.N., Setälä, H., van der Putten, W.H., & Wall, D.H. (2004) Ecological linkages between aboveground and belowground biota. *Science* **304**: 1629–33.

Wardle, D.A., Yeates, G.W., Williamson W., & Bonner, K.I. (2003) The response of a three trophic level soil food web to the identity and diversity of plant species and functional groups. *Oikos* **102**: 45–56.

Westoby, M., & Wright, I. J. (2006) Land-plant ecology on the basis of functional traits. *Trends in Ecology & Evolution* **21**: 261–8.

Whitham, T.G., Bailey, J.K., Schweitzer, J.A., *et al.* (2006) A framework for community and ecosystem genetics: From genes to ecosystems. *Nature Reviews Genetics* **7**: 510–23.

Whitham, T.G., Martinsen, G.D., Young, W., *et al.* (2003) Community and ecosystem genetics: A consequence of the extended phenotype. *Ecology* **84**: 559–73.

Wimp, G.M., Young, W.P., Woolbright, S.A., Martinsen, G.D., Keim, P., & Whitham, T.G. (2004) Conserving plant genetic diversity for dependent animal communities. *Ecology Letters* **7**: 776–80.

Wright, I.J., Reich, P.B., Westoby, M., *et al.* (2004) The worldwide leaf economics spectrum. *Nature* **428**: 821–7.

Zinke, P.J. (1962) The pattern of individual forest trees on soil properties. *Ecology* **43**: 130–3.

CHAPTER 2.3

Delivery of Soil Ecosystem Services: From Gaia to Genes

Katarina Hedlund and Jim Harris

2.3.1 Introduction

A number of threats to soils and human well-being have been identified arising from climate change and conversion and intensive use of land, which will compromise our ability to meet current and future demands for provisioning of food, fibre, clean air, and water (Barrios 2007). We present two diverse approaches for consideration that will assist in the development of management strategies for addressing these issues.

Firstly, the Millennium Ecosystem Assessment (MEA) which was initiated by the United Nations (UN) Convention on Biological Diversity and several other international organizations, illustrate how human well-being depends on the state and the use of the Earth system and its natural biodiversity. The MEA has stated that ecosystem services are dependent on functions that are provided by biodiversity both globally and locally. Furthermore, in terrestrial systems these are underpinned by the soil ecosystem. Soils and soil biodiversity are critical to delivering supporting services and the provisioning, regulating, and cultural services which flow from them.

Secondly, the Gaia hypothesis considers the Earth as if it were a single organism where biological and abiotic components interact in a homeostatic mechanism maintaining equilibrium to support favourable conditions where life can exist (Lovelock 1979). The development and maintenance of feedback between and within the abiotic and biotic components of this mechanism are critical to its effective and sustained function. Both the MEA and Gaian frameworks embrace a global concept of delivery of ecosystem services, but these are poorly-developed with regard to the role of soil, especially soil biota. This chapter describes the emerging framework on ecosystem service delivery as we explore and emphasise soil and soil biodiversity as critical components of natural capital that deliver ecosystem services and which have global implications for international policy and decisions in the face of global change.

In terrestrial systems the majority of ecosystem services arise from soil functions which are dependent to a greater or lesser extent on the interactions between organisms, organic and mineral fractions of soil (Kibblewhite *et al.* 2008). Soil organisms that feed on detritus affect the structure and composition of plant communities through nutrient supply, whereas herbivores, pathogens and symbionts affect plant communities directly through their interaction with living plant roots (Wardle *et al.* 2004). These functions can, according to the MEA, be defined as supporting services and include functions that involve soil structure formation, biotic resistance to pathogens and regulation of the aboveground community composition and productivity. Therefore, these supporting services are pivotal for developing and maintaining the soil pore network which provides the physical framework in and through which soil processes occur, regulating gas exchange, hydrological processes and organism dynamics (e.g. Young & Ritz 2005; Rillig *et al.* 2010). Biotic resistance can control and modulate pests and pathogen attacks (Garbeva *et al.* 2004). Soil biodiversity can also support aboveground diversity to conserve nature (Van der Heijden *et al.* 1998). Soils and soil biota also provide regulating ecosystem services, which are soil functions that influence carbon (C) and nitrogen (N) cycling and which relate

Soil Ecology and Ecosystem Services. First Edition. Edited by Diana H. Wall *et al.*
© 2012 Oxford University Press. Published 2012 by Oxford University Press.

to yield and quality (Marschner & Rengel 2007; Osler & Sommerkorn 2007) while simultaneously retaining C and N within the soils. Gas and trace gas emissions from the soil are prevented, thus mitigating climate change.

2.3.2 Ecosystem services delivery and Gaia theory

The optimization of provisioning services such as crop production has led to global changes in land-use and losses of biodiversity that tend to counteract ecosystem services such as water regulation, nutrient retention, C sequestration, and biotic resistance (Foley et al. 2005; Fig. 2.3.1A). In a world without human society, ecosystems are multifunctional as a result of feedback and evolutionary mechanisms (Fig. 2.3.1B). This means that in natural communities, ecosystem functioning is generally high at several of the identified services that we now utilize. This essentially reflects a system where society has operated outside of the "normal" constraints, or carrying capacity, of the Earth system, and has been able to do so by means of harnessing fossil fuel inputs—for transport, cultivation, fertilisers, pesticides, and other manufactured goods. To assess the state of the ecosystem and the ecosystem goods and service flows in response to management interventions, we need to measure not just those services and goods, but also the material and energy gains and losses from the system.

It is possible and pertinent to consider James Lovelock's Gaia hypothesis to optimise the production of multiple ecosystem services (Lovelock 1979).

This proposes that the Earth functions as if it were a single organism, with a physiological mechanism for homeostasis reacting to ever increasing insolation. These are founded upon feedback mechanisms between biological and abiotic components of the Earth system such that dynamic equilibria are maintained, for example through cloud formation in response to warming, oceanic drawdown of carbon dioxide, and the changing albedo from grassland to forests (Lovelock 2003). Evidence for feedback mechanisms are found scattered throughout geologic history, much of which was dominated by life in seas. The development of oxygen in the atmosphere (Karhu & Holland 1996) preceded and facilitated colonization of terrestrial areas, and saw the beginning of soil as a living system, driven by primary producers.

The removal of huge quantities of CO_2 from the atmosphere occurred in the Devonian Period and contributed to the development of deep-rooted plants which accelerated soil formation by mechanical effects, including crack formation, deep injection of C and allowing the colonization of soil micro-organisms, chemically altering the environment (Hinsinger et al. 2009). Recent work suggests that these key weathering processes were driven by a combination of plant roots and mycorrhizal fungi, which co-evolved with the development of land plants, the processes of which continue until today (Taylor et al. 2009). Van Breemen (1993) suggested that feedback mechanisms involving primary producers and decomposers develop soil properties favouring net primary productivity, and maintain a supply of water essential for net

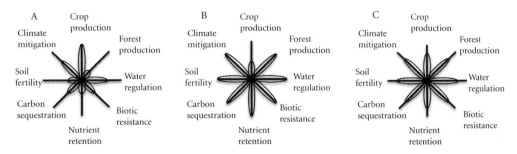

Figure 2.3.1 "Flower diagrams" indicating the delivery of ecosystem services on each axis of a particular service at different types of land use: a) crop production, b) natural ecosystem, c) cropland with integrated management of ecosystem service production of multiple services. (Modified from Foley et al. 2005.)

primary productivity. However, van Breemen preferred to couch these observations in terms of evolution, and felt they were not sufficient to explain the development of a Gaian mechanism as a whole. The main thrust of this argument was extended by van Breemen and Finzi (1998) who pointed out that there was strong evidence for plant-soil feedbacks in many ecosystems which focused on physical soil factors such as keeping soils moist and accumulating C and so, influencing ecosystem functions.

While these observations focused on abiotic soil factors, at the same time the concept of biotic feedback between soils and plants was being developed. Van der Putten and Peters (1997) demonstrated the control over plant communities which was exerted by specific growth-depressing soil organisms and with impacts on inter-specific competition. This identified a further feedback loop, potentially producing a more sensitive response to changing conditions and feedback to the wider system. Diaz *et al.* (1993) identified a mechanism where increased carbon dioxide caused increased root exudation, an increase in the soil microbial biomass N, leading to the inhibition of further plant biomass production through plant roots and being outcompeted for soil N by the microbial biomass. Bever *et al.* (1997) demonstrated, by means of a computer model, how negative feedback (a prerequisite in the Gaian model) between the soil community and plant community might maintain plant species diversity. Since diversity in the genotype is important for providing a wide as possible potential response to changing conditions (insurance hypothesis), these interactions could be considered important in the overall homeostatic mechanism.

Singh *et al.* (2010) have reviewed the ways in which microbially-mediated processes involved in feedback loops are likely respond to climate change, and how these could be manipulated to provide mitigation of these effects by reducing atmospheric greenhouse gas emissions from terrestrial systems. They draw out an important consideration of shifts in communities: if the soil community is unchanged, then responses to climate will be an increase in process rate, but with the same underlying behaviour and regulators. However, if the community shifts, we may see a threshold effect where the response shifts to a different trajectory, making prediction problematic as the underlying behaviours and regulators will have changed. Xu *et al.* (2011) commented on and agreed with this review, however, proposed considering the feedback mechanism by illustrating the interaction between inorganic N dynamics, microbial communities, soil factors, and the response of heterotrophic respiration under warming in high latitude systems. As an example, increases in soil inorganic N may cause greater production of cellulases, but block the production of ligninases (Carreiro *et al.* 2000), leading to an increased turnover in labile C, and a decreased turnover of recalcitrant C, with the implication for soil C storage and organic matter stabilization. These mechanisms have been barely researched, and certainly not included in global climate models—those that do incorporate soil compartments tend to be simplistic models with no inclusion of the effects of shifts in community composition. Phillips (2009) has placed soils at the centre of Gaian regulatory mechanism, acting as extended composite phenotypes and expressing the effects of genes through the effects of organisms. This extends the notion of Earth as set of tightly-coupled, densely interwoven systems (Fig. 2.3.2), and consistent with Gaian feedback and control.

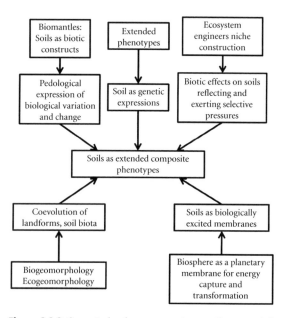

Figure 2.3.2 Conceptual pathways converging on soils as extended phenotypes. (Modified from Phillips 2009.)

For the Gaia hypothesis to be considered as part of the ecosystem services concept, the interactions among services, or feedback among organisms, need to be considered in schemes designed to provide management information, and at a scale and resolution appropriate to the scale of management intervention. This can be visualized in the flower diagrams (Fig. 2.3.1C) where the future management of resources needs to account for a production of multiple ecosystem services in order to optimise the use of global and local resources.

Moving forward to understanding how multiple ecosystem services and Gaia-scale feedbacks can be optimized by management or conservation activities, we first need to know at what biological levels (i.e. genes to ecosystems) key functions delivering ecosystem services are produced and how feedbacks among organisms influence the total sum of such actions. Secondly, we need to understand what spatial and temporal scales the organisms are acting upon, so that it is possible to manage actions in optimizing the delivery of ecosystem services.

2.3.3 At what biological levels are soil ecosystem services produced?

The delivery of ecosystem services are produced by direct (root-feeding, symbiosis) or indirect (decomposition-related) interactions among soil organisms and plants (Wardle et al. 2004). We need to know the identity behind the key interactions which result in the service produced to better understand and manage the delivery of services. Knowledge of the specificity behind functions and interactions is emerging, both at the gene level as well as at the community level thanks to the contemporary tools which are used to resolve the genetic identity of organisms and their actions. Here, these interactions will be discussed while focussing in particular on direct interactions, as these are often overlooked in Gaia-type nutrient and energy flux feedbacks.

2.3.3.1 Gene level

Characterizing the diversity and expression of genes directly involved in the functioning of ecosystems has been one approach to understanding the functional importance of interactions among soil organisms on the level of genes. For example mycorrhizal symbiosis is involved in the delivery of provisioning services as timber growth. Molecular investigations of ectomycorrhizal fungi (Le Quere et al. 2004; Martin et al. 2008) or *Frankia* strains (Oakley et al. 2004) show that variation in the host selection of one symbiont species is associated with differences in gene expression within a species at the local scale. This means that all individuals of a fungal species do not always form symbiosis with a particular plant species, but rather the fungal individuals that have specific patterns of gene expression. The development of tools within functional metagenomics, including large-scale sequencing and microarray analyses, can now be adapted for measuring the diversity and expression of functional genes in natural environments (Wright et al. 2005; Dinsdale et al. 2008; McGuire et al. 2010). These new techniques, which are further explained in chapter 2.1 by Hackl et al. (this volume), may help us determine interactions that we cannot detect at a phylogenetic level, but only at a functional-gene level, as for example known for symbiont-host plant recognition systems.

2.3.3.2 Community level

Other ecosystem services, such as biotic resistance, can be delivered by a number of different groups of organisms interacting towards protection against pathogens or predators, through a multi-trophic interaction among hosts, enemies and their antagonists. This has been suggested as a mechanism when functional complementarity among arbuscular mycorrhizal (AM) fungal species is an important driver for plant resistance to root pathogens rather than species-specific interactions (Wehner et al. 2010; Maherali & Klironomos 2007). In an agricultural context there are specific bacterial communities related to different varieties of wheat (Germida & Siciliano 2001) and a high diversity of the rhizosphere community can thus both support the presence of growth supporting bacteria as well as suppress the infection by plant pathogens (Mazzola & Gu 2002; Matos et al. 2005). From natural ecosystems, it is known that grassland plant species develop species-specific rhizosphere communities (Bezemer et al. 2010), which would also suggest that specific interactions can play a role in plant protection against soil-borne enemies. While being specific, these interactions can involve relatively large consortia of microbial taxa (Mendes et al. 2011). With techniques such as monoclonal antibody-based detection systems, and DNA sequencing to detect species diversity of microorganisms and activity of specific plant pathogens, it is possible to monitor both the development of the pathogen within the plant as well as an AM-fungus infecting its host and preventing a number of plant crops from being infected by fungal pathogens such as *Fusarium* and *Pythium* (Arya et al. 2010; Mohandas et al. 2010).

2.3.3.3 Ecosystem level

At the ecosystem level, the ecosystem service of biotic resistance can be regulated by indirect interactions among organisms as these are members of the food web of the soil community. Evidence using molecular detection methods has revealed that many presumed pathogenic microorganism species can have different trophic relationships with a plant at different periods of their life cycle (Huang et al. 2009). Microbial species may shift among symbiotic, mutualistic and parasitic interactions with their plant host, thus inappropriate timing of use of pesticides in normal agricultural management may also reduce the symbiotic phases of the pathogen (Newton et al. 2010). Therefore, knowledge of trophic expressions and of the temporal shifts of pathogen activity will greatly improve the provisioning of ecosystem services. Newton et al. (2010) suggest that crop management schemes should rather promote the complex heterogeneity of soil interactions that yields enhanced crop function, by tilting the balance towards beneficial organisms than focusing on the pathogenic stages of the soil pathogens.

Disease suppression of the root pathogen *Rhizoctonia*, has in a metagenomic study by Mendes et al. (2011) not been explained by a specific soil bacterial group, rather by the composition of the entire community. This effect can be caused by direct interaction among bacteria competing for resources where species-specific interactions trigger secondary metabolite production that acts as anti-microbial cues within the soil community (Garbeva et al. 2011). With such microarray-based analyses of transcribed responses by species involved in microbial interactions, we can now see the potential of how complex interactions may trigger indirect responses favouring biotic resistance in some soil communities but not in others.

Symbionts such as AM fungi can also influence remediation and restoration activities of soils, in order to increase delivery of multiple ecosystem services such as plant production and detoxification. The AM fungal symbioses can increase plant recovery and yield lower amounts of metals in the plant tissue at restoration of bauxite sites (Babu & Reddy 2011). However, it is the composition of AM fungal species communities that is the key to restoring the plant communities, as it is the outcome of the host plant-symbiont interaction together with plant competition that determines the plant community composition (Ji et al. 2010).

2.3.4 At what spatial scales can we describe and quantify soil ecosystem services?

For us to use the ecosystem services concept for promoting a sustainable use of soils, we need to describe and quantify the processes responsible for determining the ecosystem services, and also understand how they act across spatial scales. The production of soil ecosystem services is dependent on the underlying diversity present in a particular field, whereas this diversity at least in part, depends on the dispersal ability of species from other areas in the surrounding landscape. For services produced by aboveground organisms that we can readily observe, such as pollinators and natural enemies, the dispersal ability of organisms producing the services is the key to whether they can decolonize fields after disturbance from agricultural practices such as pesticide application or tillage. This may result in source-sink relationships between local and regional scales, which can maintain populations and species that otherwise, are rare and vulnerable to disturbance. This may enable the communities to be dynamic in space and time when

Table 2.3.1 Soil ecosystem service producers and their influence at different scales in production landscapes (adapted from Zhang et al. 2007, with permission from Elsevier)

Ecosystem service	Field	Landscape/regions	Global
Carbon sequestration	Plants and soil organisms sequester carbon (Hedlund 2002)	Plants and soil organisms provides multiple ecosystem services (Raudsepp-Hearne et al. 2010)	Plant and soil organisms promote carbon sinks (Lal 2010)
Nutrient retention	Soil microorganisms and plants (Fierer et al. 2009)	Plant cover and immobilization of nutrients by soil microorganisms (Carpenter et al. 1998)	Plants and soil organisms influence on nitrogen cycle (Robertson & Vitousek 2009)
Resistance to pests and invasive species	Soil community diversity with predators and microbial antagonists (Lucas 2011)	Natural enemies of plant pests as spiders, carabids (Schmidt et al. 2008)	Biological invasions due to release from net negative plant–soil feedback (Van der Putten et al. 2007)
Soil structure	AM fungi Earthworms Microorganisms (Barrios 2007; Rillig et al. 2010).	Erosion control by increased soil carbon and plant cover (Lal 2010)	

responding to land-use or climate change (Hedlund et al. 2004; De Deyn & Van der Putten 2005). We tend to scale the soil ecosystem service delivery according to a non-dynamic view of the soil communities and thus their production of services strongly depends on the present land-use management (Zhang et al. 2007).

On a field scale we can quantify and monitor the activity of organisms and the resulting delivery of ecosystem services as both plants and soil organisms through their biomass nutrient retention and C in the soil (Table 2.3.1; Hedlund 2002; Fierer et al. 2009). Even though we lack information on explicit spatial scales at which soil ecosystem services are produced, we know that microorganisms vary in abundance and diversity over spatial scales similar to larger organisms and that there is growing evidence that microbial species distributions are non-random (Martiny et al. 2006). Dispersal propagules of bacteria, fungi, protozoa, and nematodes have been found thousands of metres from their sources, which demonstrate their ability to disperse passively in space (Hovmoller et al. 2002; Finlay 2002). Spores from wood-decaying fungi can disperse over distances encompassing regional scales (Edman & Jonsson 2001). With an increased knowledge of biogeography of microorganisms (e.g. Green & Bohannan 2006) we may find other ways to predict responses to environmental change by microorganisms.

Suppression of pests in the soil by direct biotic resistance towards root pathogens will rather act on a field scale as specific communities will counteract the pathogens (Lucas 2011; Mendes et al. 2011). Such field scale effects can also produce biotic resistance to above-ground pests and invasive species (Van der Putten and Peters 1997; Schmidt et al. 2008). However, certain ecosystem services are produced by organisms that are sensitive to management and can not be replaced in space or time by others as their disperal abilities and recolonization rates are low. Such species groups include AM fungi and earthworms that can require recolonisation when for example they do not survive intensive soil tillage (Helgason et al. 1998; Gormsen et al. 2004). These organisms produce ecosystem services such as soil aggregate formation and thus increase soil porosity (Barrios 2007; Rillig et al. 2010). Though on a landscape scale these organisms deliver also more general services, for example erosion control by forming soil aggregates (Lal 2010).

2.3.5 Use of soil ecosystem services in a policy context

The framework of ecosystem services has been further taken into the decision making process by using ecosystem services from natural capital as worth conserving and thus fulfilling the intentions of the MEA framework. Daily et al. (2009)

have proposed the use of a decision-loop that is based upon using ecosystem services as a way of accounting for natural capital by valuing such services in decisions of how to manage biodiversity. This decision loop can be iterated to test different land-use actions or conservation policies, both on a global but also on a local scale (Fig. 2.3.3). With this framework it is possible to understand positive and negative feedbacks on biodiversity, ecosystem services from policy and management decisions. In another recent framework, the Economics of Ecosystems and Biodiversity program (TEEB—initiated by the United Nations Environment Programme (UNEP)), ecosystem services have been studied by a tiered approach where values of biodiversity are recognized, demonstrated and captured (Sukhdev et al. 2010). A number of case studies included in the TEEB report focus on the need for decision makers to assess and communicate the role of biodiversity and ecosystem services to achieve human well-being.

TEEB and MEA frameworks recognize the same process for using the decision loop (Fig. 2.3.3). We can identify gaps of knowledge and iterate the loop with different actions and on different scales. We first need to identify how land-use and climate change impact soil biodiversity as we work our way through the loop. Major threats to soil biodiversity are identified as a loss of organic matter, soil erosion, compaction, and salinization in Europe as well as in other parts of the world (Jeffery & Gardi 2010; Lal 2010). When threats and possible activities of mitigation are considered, the delivery of ecosystem services has to be quantified in order to determine the trade-offs among individual services. Management actions will act on several biological levels and spatial scales (Zhang et al. 2007). Within a soil community there are multiple interactions performing functions as decomposition of C and N, competitive interactions making pest or invasive species more or less prone to invade the soil community and predator-prey interactions that control species abundances. Direct as well as indirect interactions among soil organisms may result in non-intuitive relationships between land-use and ecosystem service provisioning, especially if interacting organisms are affected by processes at different spatial and temporal scales (Kremen 2005). For example, species providing different services may be linked in food webs, and a species may also provide more than one service (Gamfeldt et al. 2008). Trade-offs between specific services will determine the outcome of the multiple interactions (Raudsepp-Hearne et al. 2010); and to estimate multiple soil ecosystem services, the concept and theories of soil food web interactions can be used to quantify the delivery of ecosystem services or even the stability of the services (de Ruiter et al. 1994; Neutel et al. 2002).

The appreciation of ecosystem services in the decision-loop framework can be achieved with an economic valuation of the services to optimise the

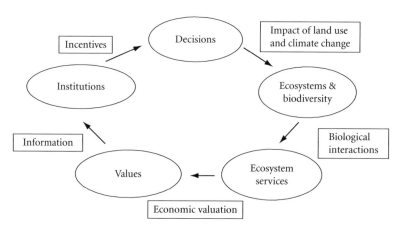

Figure 2.3.3 A decision loop which can be used for policy development accounting for ecosystem services when taking actions and decision on natural capital. (Modified from Daily et al. 2009, with permission from the Ecological Society of America.)

efforts for conservation or for mitigating biodiversity and soil degradation (Tallis & Polasky 2009). Recently, soil ecosystem services have been a focus as soils have been regarded as natural capital that can be valued and protected by its own value of providing regulating and supporting services (Dominati et al. 2010). The valuation of ecosystem services can be used in directing policies towards conserving soil biodiversity and at the same time optimising provisioning ecosystem services. The Natural Capital Project (Tallis et al. 2010) has recently developed a tool "Integrated valuation of ecosystem services and trade-offs" (InVEST; Kareiva et al. 2011) that incorporates natural capital into the decision making process. This tool can be utilized to determine land-use or climate change actions which can be iterated through the decision loop and which can estimate the consequences of provisioning services as well as conservation efforts, on a regional scale (Nelson et al. 2008) as well as on a global scale (Nelson et al. 2010). Soil ecosystem services are not explicitly valued; only regulating services promoting soil C sequestration. Both the TEEB report and the InVEST tool are examples of how values of ecosystem services can be used to adapt policies for a more sustainable use of natural resources. By iterating the decision loop with different scenarios of management or decision, and quantifying services delivered at gene and ecosystem levels, both positive and negative feedbacks of the delivery of ecosystem services can be identified.

The appreciation of soils in regulating services as water purification, C sequestration and climate regulation, have received a much wider appreciation for soils as a natural capital through estimating costs of losses of services (Adhikari & Nadella 2011). The costs of pathogen outbreaks in agriculture may have even larger costs to society as they include the reduction of ecosystem services by intensification of agriculture and forestry. The natural capital of mitigating climate change by sequestration of C in soils is suggested to become a major driver for promoting soil ecosystem services as a natural capital (Lal 2010). However the way forward from quantification of soil ecosystem services to decisions by valuing soil as natural capital is still open. This development will require further exploration of quantification of the important soil ecosystem services and their inter-relationships.

2.3.6 Conclusions

In order to provide an effective understanding of the impacts of different threats to soils and how to promote sustainable soil use, we need to understand the relationships between the diversity of soil organisms from the levels of genes to communities, and their concomitant delivery of ecosystem services. To promote sustainable use of soils we have to understand at what scales, in space and time, ecosystem services can be quantified and managed.

The ability of soil organisms to secure the delivery of services by either overcoming disturbances in space or time is unique among terrestrial organisms, though we lack knowledge of the tipping points in biodiversity in which sustainable delivery of services is degraded. Targeting trade-offs among multiple ecosystem services which are dependent on soils, are more likely to result in win-win interventions than pursuing interventions which are focused on conservation alone (Diaz et al. 2009). This is because the delivery of different services is often linked to the same ecosystem functions and feedback mechanisms. To effectively determine trade-offs between services, the scales at which the soil ecosystem services are acting need to be elucidated. To make such interventions a reality, we need to enhance the appreciation of the role of soils and soil biodiversity in the policy arena and demonstrate the complete dependence of ecosystem service delivery on the natural capital which is found in soil and the need to conserve and restore it.

Both the MEA and Gaian frameworks embrace a global concept of delivery of ecosystem services but these have been poorly developed with regard to the role of soil and soil biota in particular. Soil ecosystem services delivery may act locally on a micro scale in soil aggregates and with a high resolution of specific functions while having global consequences for C and N fluxes. These global consequences and especially changes in global atmospheric composition and climate conditions have major implications for international policy. Hence policy decisions depend on accurate predictions on consequences of ongoing and alternative policies, which require adequate insight in the functioning of ecosystems. That is where gene-based and Gaian approaches cross paths. Knowledge at both the gene and the global level of delivery of services and the key intermediary stages will assist us in developing management aimed at promoting sustainable soil use in agriculture or forestry, as well as mitigate global climate change by enhancing regulation services. Despite the importance of soil organisms in the release and turnover of C, climate models do not include mechanisms of the detailed dynamics of soil systems in time and space, and at different scales—without this, the models will always be partial and inaccurate. In most future global development scenarios an even more intensive production is envisaged which will inevitably impair C storage, nutrient retention, and soil structure formation, which threaten fertility and sustainable use of soils (Lal 2010). There is a growing awareness in several sectors of the potential for conserving and restoring natural capital of soil to promote sustainable production (Bennett et al. 2010; Gardi et al. 2009) though the use of biotic interactions to direct soil ecosystem services is yet to be developed and transferred.

Although the MEA has provided a framework within which to guide our interventions to conserve and restore the natural capital of our global ecosystem, this is not enough—we have to understand the need to re-instate the feedback loops on which the Gaian homeostatic system depends. That is why we have sought to bring these two approaches together in this chapter. In this way we may determine how widespread land conversion, and the potential for ecological and evolutionary feedback responses to a changing climate, may influence the sustainability and quality of human life on Earth. More importantly this provides guidance as to how we might manage our natural soil capital for a more sustainable delivery of ecosystem services on a global scale.

References

Adhikari, B., & Nadella, K. (2011) Ecological economics of soil erosion: a review of the current state of knowledge. *Annals of the New York Academy of Sciences* **1219**: 134–52.

Arya, A., Arya, C., & Misra, R. (2010) Mechanism of action in arbuscular mycorrhizal symbionts to control fungal diseases. In: A. Arya & A.E. Perello (eds.) *Management of fungal plant pathogens*, pp. 171–82. CABI, Wallingford.

Babu, A.G., & Reddy, M.S. (2011) Influence of arbuscular mycorrhizal fungi on the growth and nutrient status of Bermuda grass grown in alkaline bauxite processing residue. *Environmental Pollution* **159**: 25–9.

Barrios, E. (2007) Soil biota, ecosystem services and land productivity. *Ecological Economics* **64**: 269–85.

Bennett, L.T., Mele, P.M., Annett, S., & Kasel, S. (2010) Examining links between soil management, soil health, and public benefits in agricultural landscapes: An Australian perspective. *Agriculture Ecosystems and Environment* **139**: 1–12.

Bever, J.D., Westover, K.M., & Antonovics, J. (1997) Incorporating the soil community into plant population dynamics: the utility of the feedback approach. *Journal of Ecology* **85**: 561–73.

Bezemer, T.M., Fountain, M.T., Barea, J.M., et al. (2010) Divergent composition but similar function of soil food webs beneath individual plats: plant species and community effects. *Ecology* **91**: 3027–36.

Carpenter, S.R., Caraco, N.F., Correll, D.L., Howarth, R.W., Sharpley, A.N., & Smith, V.H. (1998) Nonpoint pollution of surface waters with phosphorus and nitrogen. *Ecological Applications* **8**: 559–68.

Carreiro, M., Sinsabaugh, R.L., Repert, D.A., & Parkhurst, D.F. (2000) Microbial enzyme shifts explain litter decay responses to simulated nitrogen deposition. *Ecology* **81**: 2359–65.

Daily, G.C., Polasky, S., Goldstein, J., et al. (2009) Ecosystem services in decision making: time to deliver. *Frontiers in Ecology and the Environment* **7**: 21–8.

De Deyn, G.B., & Van der Putten, W.H. (2005) Linking aboveground and belowground diversity. *Trends in Ecology & Evolution* **20**: 625–33.

De Ruiter, P.C., Neutel, A.M., & Moore, J.C. (1994) Modelling food webs and nutrient cycling in agro-ecosystems. *Trends in Ecology & Evolution* **9**: 378–83.

Diaz, S., Grime, J.P., McPherson, E.F., & Harris, J.A. (1993) Evidence of a feedback mechanism limiting plant response to elevated carbon dioxide. *Nature* **364**: 616–17.

Diaz, S., Hector, A., & Wardle, D.A. (2009) Biodiversity in forest carbon sequestration initiatives: not just a side benefit. *Current Opinion in Environmental Sustainability* **1**: 55–60.

Dinsdale, E.A., Edwards, R.A., Hall, D., et al. (2008) Functional metagenomic profiling of nine biomes. *Nature* **452**: 629–32.

Dominati, E., Patterson, M., & Mackay, A. (2010) A framework for classifying and quantifying the natural capital and ecosystem services of soils. *Ecological Economics* **69**: 1858–68.

Edman, M., & Jonsson, B.G. (2001) Spatial pattern of downed logs and wood-decaying fungi in an old-growth *Picea abies* forest. *Journal of Vegetation Science* **12**: 609–20.

Fierer, N., Carney, K.M., Horner-Devine, M.C., & Megonigal, J.P. (2009) The biogeography of ammonia- oxidizing bacterial communities in soil. *Microbial Ecology* **58**: 435–45.

Finlay, B.J. (2002) Global dispersal of free-living microbial eukaryote species. *Science* **296**: 1061–3.

Foley, J.A., DeFries, R., Asner, G.P., *et al.* (2005) Global consequences of land use. *Science* **309**: 570–4.

Gamfeldt, L., Hillebrand, H., & Jonsson, P.R. (2008) Multiple functions increase the importance of biodiversity for overall ecosystem functioning. *Ecology* **89**: 1223–31.

Garbeva, P., Hol, W.H.G., Termorshuzen, A.J., Kowalchuk, G.A., & De Boer, W. (2011) Fungistasis and general soil biostasis – A new synthesis. *Soil Biology & Biochemistry* **43**: 469–77.

Garbeva, P., Van Veen, J.A., & van Elsas, J.D. (2004) Microbial diversity in soil: Selection of microbial populations by plant and soil type and implications for disease suppressiveness. *Annual Review of Phytopathology* **42**: 243–70.

Gardi, C., Montanarella, L., Arrouays, D., *et al.* (2009) Soil biodiversity monitoring in Europe: ongoing activities and challenges. *European Journal of Soil Science* **5**: 807–19.

Germida, J.J., & Siciliano, S.D. (2001) Taxonomic diversity of bacteria associated with the roots of modern, recent and ancient wheat cultivars. *Biology and Fertility of Soils* **33**: 410–15.

Gormsen, D., Hedlund, K., Korthals, G.W., *et al.* (2004) Management of plant communities on set-aside land and its effects on earthworm communities. *Journal of European Soil Ecology* **40**: 123–8.

Green, J., & Bohannan, B.J.M. (2006) Spatial scaling of microbial biodiversity. *Trends in Ecology & Evolution* **21**: 501–9.

Hedlund, K. (2002) Soil microbial community structure in relation to plant diversity management on former agricultural land. *Soil Biology & Biochemistry* **34**: 1299–307.

Hedlund, K., Griffith, B., Christensen, S., *et al.* (2004) Trophic interactions in changing landscapes: responses of soil food webs. *Basic and Applied Ecology* **5**: 495–503.

Helgason, T., Daniell, T.J., Husband, R., *et al.* (1998) Ploughing up the wood-wide web? *Nature* **394**: 431.

Hinsinger, P., Bengough, A.G., Vetterlein, D., & Young, I.M. (2009) Rhizosphere: biophysics, biogeochemistry and ecological relevance. *Plant and Soil* **321**: 117–52.

Hovmoller, M.S., Justesen, A.F., & Brown, J.K.M. (2002) Clonality and long-distance migration of *Puccinia striiformis* f. sp *tritici* in north-west Europe. *Plant Pathology* **51**: 24–32.

Huang, Y.J., Pirie, E.J., Evans, N., Deloume, R., King, G.J., & Fitt, B.D.L. (2009) Quantitative resistance to symptomless growth of Leptosphaeria maculans (phoma stem canker) in Brassica napus (oilseed rape). *Plant Pathology* **58**: 314–23.

Jeffery, S., & Gardi, C. (2010) Soil biodiversity under threat—a review. *Acta Societatis Zoologicae Bohemicae* **74**: 7–12.

Ji, B.M., Bentivenga, S.P., & Casper, B.B. (2010) Evidence for ecological matching of whole AM fungal communities to the local plant-soil environment. *Ecology* **91**: 3037–46.

Kareiva, P., Tallis, H., Ricketts, T.H., Daily, G.C., & Polasky, S. (2011) *Natural Capital: Theory and Practice of Mapping Ecosystem Services*. Oxford University Press, Oxford.

Karhu, J.A., & Holland, H.D. (1996) Carbon isotopes and the rise of atmospheric oxygen. *Geology* **24**: 867–70.

Kibblewhite, M.G, Ritz, K., & Swift, M.J. (2008) Soil health in agricultural systems. *Philosophical Transactions of the Royal Society, London B* **363**: 685–701.

Kremen, C. (2005) Managing ecosystem services: what do we need to know about their ecology? *Ecology Letters* **8**: 468–79.

Lal, R. (2010) Managing soils and ecosystems for mitigating anthropogenic carbon emissions and advancing global food security. *Bioscience* **60**: 708–21.

Le Quere, A., Schutzendubel, A., Rajashekar, B., *et al.* (2004) Divergence in gene expression related to variation in host specificity of an ectomycorrhizal fungus. *Molecular Ecology* **13**: 3809–19.

Lovelock, J.E. (1979) *Gaia: A new look at life on Earth*. Oxford University Press, Oxford.

Lovelock, J.E. (2003) The Living Earth. *Nature* **426**: 769–70.

Lucas, J.A. (2011) Advances in plant disease and pest management. *Journal of Agricultural Science* **149**: 91–114.

Maherali, H., & Klironomos, J. (2007) Influence of phylogeny on fungal community assembly and ecosystem functioning. *Science* **316**: 1746–8.

Marschner, P., & Rengel, Z. (2007) *Nutrient Cycling in Terrestrial Ecosystems*. Springer, Berlin.

Matos, A., Kerkhof, L., & Garland, J.L. (2005) Effects of microbial community diversity on the survival of *Pseu-*

domonas aeruginosa in the wheat rhizosphere. *Microbial Ecology* **49**: 257–64.

Martin, F., Aerts, A., Ahrén, D., et al. (2008) The genome of *Laccaria bicolor* provides insights into mycorrhizal symbiosis. *Nature* **452**: 88.

Martiny, J.B.H., Bohannan, B.J.M., Brown, J.H., et al. (2006) Microbial biogeography: Putting microbes on the map. *Nature Reviews Microbiology*, **4**: 102–12.

Mazzola, M., & Gu, Y.H. (2002) Wheat genotype-specific induction of soil microbial communities suppressive to disease incited by Rhizoctonia solani anastomosis group (AG)-5 and AG-8. *Phytopathology* **92**:1300–7.

McGuire, K.L., Bent, E., Borneman, J., et al. (2010) Functional diversity in resource use by fungi. *Ecology* **91**: 2324–32.

Mendes, R., Kruijt, M., de Bruijn, I., et al. (2011) Deciphering the rhizosphere microbiome for disease-suppressive bacteria. *Science* **332**: 1097–100.

Mohandas, S., Manjula, R., Rawal, R.D., et al. (2010) Evaluation of arbuscular mycorrhiza and other biocontrol agents in managing *Fusarium oxysporum* f. sp. *Cubense* infection in banana cv. Neypoovan. *Biocontrol Science and Technology* **20**: 165–81.

Nelson, E., Polasky, S., Lewis, D.J., et al. (2008) Efficiency of incentives to jointly increase carbon sequestration and species conservation on a landscape. *Proceedings of the National Academy of Sciences of the United States of America* **105**: 9471–6.

Nelson, E., Sander, H., Hawthorne, P., et al. (2010) Projecting global land-use change and its effect on ecosystem service provision and biodiversity with simple models. *PLoS One* **5**: e14327.

Neutel, A.M., Heesterbeek, J.A.P., & de Ruiter, P.C. (2002) Stability in real food webs: weak links in long loops. *Science* **296**: 1120–3.

Newton, A.C., Fitt, B.D.L., Atkins, S.D., Walters, D.R., & Daniell, T.J. (2010) Pathogenesis, parasitism and mutualism in the trophic space of microbe-plant interactions. *Trends in Microbiology* **18**: 365–73.

Oakley, B., North, M., Franklin, J.F., Hedlund, B.P., & Staley, J.T. (2004) Diversity and Distribution of *Frankia* Strains Symbiotic with *Ceanothus* in California. *Applied Environmental Microbiology* **70**: 6444–52.

Osler, G.H.R., & Sommerkorn, M. (2007) Toward a complete soil C and N cycle: Incorporating the soil fauna. *Ecology* **88**: 1611–21.

Phillips, J.D. (2009) Soils as extended composite phenotypes. *Geoderma* **149**: 143–51.

Raudsepp-Hearne, C., Peterson, G.D., & Bennett, E.M. (2010) Ecosystem service bundles for analyzing trade-offs in diverse landscapes. *Proceedings of the National Academy of Sciences of the United States of America* **107**: 5242–7.

Rillig, M.C., Mardatin, N.F., Leifheit, E.F., et al. (2010) Mycelium of arbuscular mycorrhizal fungi increases soil water repellency and is sufficient to maintain water-stable soil aggregates. *Soil Biology & Biochemistry* **42**:1189–91.

Robertson, G.P., & Vitousek, P.M. (2009) Nitrogen in agriculture: Balancing the cost of an essential resource. *Annual Review of Environment and Resources* **34**: 97–125.

Schmidt, M.H., Thies, C., Nentwig, W., & Tscharntke, T. (2008) Contrasting responses of arable spiders to the landscape matrix at different spatial scales. *Journal of Biogeography* **35**:157–66.

Singh, B.K., Bardgett, R.D., Smith, P., & Reay, D.S. (2010) Microorganism and climate change: terrestrial feedbacks and mitigation options. *Nature Reviews* **8**: 779–90.

Sukhdev, P., Wittmer, H., Schröter-Schlaack, C., & Nesshöver, C. (2010) *The Economics of Ecosystems and Biodiversity: Mainstreaming the Economics of Nature: A synthesis of the approach, conclusions and recommendations of TEEB*. United Nations Environment Programme.

Tallis, H., & Polasky, S. (2009) Mapping and valuing ecosystem services as and approach for conservation and natural-resource management. *Annals of the New York Academy of Sciences* **1162**: 265–83.

Tallis, H., Yukuan, W., Bin, F., et al. (2010) The Natural Capital Project. *Bulletin of the British Ecological Society 2010* **41**(1): 10–13.

Taylor, L.L., Leake, J.R., Quirk, J., Hardy, K., Banwart, S.A., & Beerling, D.J. (2009) Biological weathering and the long-term carbon cycle: integrating mycorrhizal evolution and function into the current paradigm. *Geobiology* **7**: 171–91.

Van Breemen, N. (1993) Soils as biotic constructs favouring net primary productivity. *Geoderma* **57**: 183–211.

Van Breemen, N., & Finzi, A.C. (1998) Plant–soil interactions: ecological aspects and evolutionary implications. *Biogeochemistry* **42**: 1–19.

Van der Heijden, M.G.A., Klironomos, J.N., Ursic, M., et al. (1998) Mycorrhizal fungal diversity determines plant biodiversity, ecosystem variability and productivity. *Nature* **396**: 69–72.

Van der Putten, W.H., Klironomos, J.N., & Wardle, D.A. (2007) Microbial ecology of biological invasions. *The ISME Journal* **1**: 28–37.

Van der Putten, W.H., & Peters, B.A.M. (1997) How soil-borne pathogens may affect plant competition. *Ecology* **78**: 1785–95.

Wardle, D.A., Bardgett, R.D., Klironomos, J.N., et al. (2004) Ecological linkages between aboveground and belowground biota. *Science* **304**: 1629–33.

Wehner, J., Antunes, P.M., Powell, J.R., Mazukatow, J., & Rillig, M.C. (2010) Plant pathogen protection by arbuscular mycorrhizas: A role for fungal diversity? *Pedobiologia* **53**: 197–201.

Wright, D.P., Johansson, T., Le Quere, A., Soderström, B., & Tunlid, A. (2005) Spatial patterns of gene expression in the extramatrical mycelium and mycorrhizal root tips formed by the ectomycorrhizal fungus Paxillus involutus in association with birch (*Betula pendula*) seedlings in soil microcosms. *New Phytologist* **167**: 579–96.

Xu, C.G., Liang, C., Wullschleger, S., Wilson, C., & McDowell, N. (2011) Importance of feedback loops between soil inorganic nitrogen and microbial communities in the heterotrophic soil respiration response to global warming. *Nature Reviews Microbiology* **8**: 779–90.

Young, I.M., & Ritz, K. (2005) The habitat of soil microbes. In: R.D. Bardgett, M.B. Usher & D.W. Hopkins (eds.) *Biological Diversity and Function in Soils*, pp. 31–43. Cambridge University Press, Cambridge.

Zhang, W., Ricketts, T.H., Kremer, C., Carney, K., & Swinton, S.M. (2007) Ecosystem services and dis-services to agriculture. *Ecological Economics* **64**: 253–60.

Synthesis

Wim H. van der Putten and Karl Ritz

In this section, the relationships between genes and ecosystem services have been explored and explained across three levels of scale. At the scale of microbes, in spite of the overwhelmingly new insights into the ecology and functioning of microbial communities, it is astonishing how little is known of consequences of human-induced global changes for the ecology and evolution of soil organisms. Such knowledge is needed to evaluate how management and policy actions work to affect the provisioning of ecosystem services on which human society depends. For example, relatively little is known about the effects of land-use transition, soil pollution, and global atmospheric and climate change on composition and functioning of soil microbial and faunal communities. Hackl and colleagues predict that the entire soil community composition may have been described in molecular terms by 2020. Such sequence-based techniques will provide an enormous volume of data, which will also require new informatics approaches to store, sort, process, and query such data. The infrastructure to handle such data is being developed, but it remains a significant challenge to realize such technology in a routine manner in order to address ecological questions. As a landmark, we may wish to take a kilogram of soil, appropriately sourced, determine the molecular profile within it, and interpret the actual and latent ecosystem functioning it does, and could, provide. When realizable at a reasonable cost, such an approach may help us to produce rapid and effective assessments of ecosystem services and of the consequences of management decisions for the sustainability of these services.

It is hitherto unknown if soil communities as a whole can become adapted to global changes (e.g. global warming). The first question is, to what extent can soil biota become locally adapted? Schweitzer and colleagues show that soil biota indeed can be locally adapted. Plant interactions with symbiotic mutualists or soil pathogens could have evolutionary potential, which is well known for resistance breeding. Interestingly, recent studies point out that such adaptation may also take place with respect to decomposition and thus provide plants with a so-called "home-field advantage." Ultimately, at the level of plant genotypes, there will be a trade-off between any home-field advantage, which would favor those plants that remain at a specific site, with the need to either physically escape from soil-borne pathogens by dispersal, or in time by selection against non-resistant plant genotypes. At the same time, plant adaptation to soil-borne biota will influence and be influenced by aboveground selection due to plant—aboveground herbivore–carnivore and symbiont interactions. The identification of "foundation species," which have a disproportionate role in community genetics, has called for awareness that there may be drivers and passengers in such complex communities. In other systems where there are less obvious foundation species, the question is, who is driving evolution and when? This chapter provides a number of exciting and stimulating cases that create awareness for a need of more work into the role of soil ecology in community genetics.

The study of aboveground–belowground coevolution will be particularly relevant (besides addressing consequences of global change) when examining the long-term population dynamics of introduced exotic species. These species, especially when having disproportionally negative effects on biodiversity, ecosystem processes, and ecosystem services, may need to be controlled for at least by minimizing the unwanted side effects. The opposite effect is visible in the case of maintaining crop performance. Crop species are usually grown in monocultures and any factor that controls crop abundance and yield will provide a cost on the delivery of ecosystem services.

As soil biota play an important role in controlling the abundance of plant species, release from native enemies will have strong impacts on plant invasiveness in new ranges, especially because symbiotic mutualistic soil biota could be more generalists than the enemies. These processes have received relatively little attention from the perspective of gene activities and addressing these processes from an evolutionary gene to network perspective will undoubtedly shed new light on the role of soil biota in driving such complex interactions and the potential for humans to influence them to their own benefit.

The Millennium Ecosystem Assessment has compiled an enormous amount of information from all around the world on how biodiversity loss may influence the functioning of ecosystems. However, it clearly falls short in terms of the extent to which is considers soil biota. The same applies to the role of soil ecology in Gaia theory. Climate models, management, and policy decisions are all quite devoid of soil ecological information. The chapter by Hedlund and Harris makes an interesting case as to why such knowledge is needed and how insights into soil ecology may influence decision-making by managers and policy-makers. Scientists that work on abiotic aspects of soil are well organized globally. However, soil ecologists may find another challenge here to become united and develop global soil ecology initiatives. The core message of the entire section is that belowground genes matter for the provisioning of ecosystem services at all scales, from nucleic acid molecules to our entire planet.

SECTION 3

Community Structure and Biotic Assemblages

SECTION EDITORS: **Donald R. Strong** and **Valerie Behan-Pelletier**

Introduction

Donald R. Strong and Valerie Behan-Pelletier

If "carbon is King and nitrogen is Queen" (Mancinelli 2003), then soil is the grandest, liveliest, and most powerful castle in the kingdom of life on Earth. Soil is uniquely earthen, as far as we know. A test question for introductory biology courses goes, "Supply a word that would make the following press release of the future correct, '———— has been discovered in Martian soil.'" All soil ecologists will quickly give a correct answer, and a large fraction of students will also answer correctly, if the course has had substantial content in soil ecology (the lectures will briefly note why biologists use regolith rather than soil for the fine stuff covering the surfaces of other planets). Sadly, many biology students probably finish their undergraduate years conflating soil with "dirt," as do most non-ecologically inclined students and most of the public. They are ignorant of the fact that the thin film of earth's soil is alive and at the same time more than just alive. The living soil acts as earth's lungs. It is among our most precious natural capital. The living soil contributes hugely to proximate matters of vital importance: agriculture, mariculture, fishing, pollution control, and water purification, retention, and cycling. On the broadest and longest-term horizons, communities of organisms in the soil are at the core of biogeochemical cycling at both local and global scales. Our living soil is one of the keys to the maintenance of biodiversity, ecosystem and biogeochemical processes, and life on earth, both on land and in the sea.

High diversity of soil organisms and soil environment is another poorly appreciated, yet vital fact. The diversity of animals, plants, and microbes all contribute to the vitality of soils and to the services provided for humanity. Each of these groups is very diverse in function, and their contributions to soil properties vary geographically.

In six chapters this section of the book explores patterns of biodiversity at scales, from rhizosphere, to soil samples, to plot, to landscape in vertical and horizontal separation. It covers the diversity among single taxa, of species assemblages, and across taxa in generating the harlequin patchiness of soil biota and functions at local scales. The influences of non-living physical and chemical properties, climate, as well as the weaving in of biotic factors in affecting the services provided by soil are a complementary topic of chapters in this section. The section goes on to consider how small-scale soil processes scale up to global patterns. One axis in scaling up is spatial, from tiny areas to larger and larger areas. Another axis is organismal, including larger and larger organisms. In the soil, organismal scale runs from bacteria, Archea, and tiny algae on the small end to animals and higher plants on the large end. Larger organisms do things that tiny organisms cannot, such as most of what falls under the rubric of ecological engineering.

These chapters probe local to global patterns for biota and ecosystem processes and services and consider the roles of different functional groups, functional equivalency, and species traits in ecosystem dynamics and biogeochemistry. Finally, they explore the promise of soil ecology in conservation, maintenance of biodiversity, and food for our hungry planet. Humans have done great harm to soils through pollution, erosion, salinization, and reductions of carbon, nutrients, and biota, among other insults. Paraphrasing Ed Wilson (1987), we humans need these soils and their biota, but they don't need us.

References

Mancinelli, R.L. (2003) What Good Is Nitrogen? In: L.J. Rothschild, & A.M. Lister (eds.) *Evolution on Planet Earth, The Impact of the Physical Environment*, pp. 25–34. London: Academic Press.

Wilson, E.O. (1987) The little things that run the world (the importance and conservation of invertebrates). *Conservation Biology* **1**: 344–6.

CHAPTER 3.1

Succession, Resource Processing, and Diversity in Detrital Food Webs

Justin Bastow

3.1.1 The surprising diversity of soil communities

The diversity of soil communities, and the difficulties this diversity poses for traditional ecological theory, has been a major theme of soil ecology for decades. Anderson (1975a) referred to the estimated thousand species of soil animals with seemingly unspecialized diets in a temperate woodland as "the enigma of soil animal diversity." Although our understanding of and appreciation for the biodiversity in soils have increased considerably in the last few decades (Wurst et al. Chapter 1.2, this volume), it remains difficult to reconcile with the theory of competitive exclusion, and its implication that no two species exploiting the same ecological niche can coexist.

Living and decomposing plants provide almost all of the carbon and energy that support soil food webs (but see Shimmel & Darley (1985) for an estimate of the algal contribution). The utilization of terrestrial plants is something soil organisms have in common with the other extraordinarily diverse group of organisms in the biosphere, the herbivorous insects. In the case of herbivorous insects, their utilization of terrestrial plants, and angiosperms in particular, appears to provide insight into their diversity. Most of the diverse lineages of insects radiated with angiosperms during the Cretaceous (such as Phytophaga, in the Coleoptera (Farrell 1998), and the Lepidoptera (Grimaldi 1999)) and the coevolution of insects and angiosperms is cited as a contributing factor in the diversity of both groups (Ehrlich & Raven 1964; Grimaldi 1999). Insect herbivores are often highly specialized in terms of the plant species that they feed on, and often the plant tissues as well. Although ecological theory predicts that no two species may coexist if they inhabit the same niche, an estimated 20,000 species of herbivorous insects can be found in a hectare of seasonal forest in Panama (Erwin 1982). The ability of large numbers of herbivorous insect species to coexist in most terrestrial ecosystems is likely related to the large number of angiosperm species as well as the diversity of ways in which insects have specialized in feeding on them (Daugherty 2009). It is therefore not surprising that the diversity of herbivorous insects is correlated with the diversity of angiosperms; both groups are most diverse in the tropics and the least diverse at high latitudes.

Some soil fauna, including many insect larvae, feed on plant roots (Brown & Gange 1990). Coevolution with their host plants may have led to specialization by these root-feeding soil animals, thereby allowing for greater coexistence and diversity in soils. Many soil organisms, however, feed on decomposing plant litter, and coevolution provides very little insight into their diversity. Decomposers and detritivores exert no direct selective pressure on the plants they feed on, eliminating the possibility for coevolution (Scheu & Setälä 2001). It is not surprising then, that the diversity of soil organisms does not seem to be correlated with the diversity of plants, as will be discussed more towards the end of this chapter.

Perhaps also because of the absence of coevolution, detritivores and decomposers appear to be much less specialized in terms of the plant species

Soil Ecology and Ecosystem Services. First Edition. Edited by Diana H. Wall *et al.*
© 2012 Oxford University Press. Published 2012 by Oxford University Press.

they utilize than herbivorous insects (Anderson 1975a). The widespread sowbug *Porcellio scaber*, for example, can be reared on leaf litter from temperate deciduous trees (Zimmer & Topp 1997), forbs, grasses, and leguminous shrubs as well as woody debris (Bastow unpublished data), and is a facultative herbivore (Paris & Sikora 1965), predator (Edney *et al.* 1974), and copraphage (Kautz *et al.* 2002). Among earthworms, there are clear geophages and detritivores (Curry & Schmidt 2007), but detritivores often exhibit no detectable preferences among litters from different plant species (Neilson & Boag 2003). Those preferences that are seen among detritivorous earthworms appear to be the result of differences in litter quality (e.g. carbon:nitrogen ratio or phenolic content; Hendriksen 1990), not of specialization on particular plant taxa.

Although comparisons of soil fauna under different plant species generally find distinct assemblages of fauna, these differences are driven by changes in the relative abundance of taxa; most soil fauna can be found under a wide range of plant species. For example, comparisons of soil mite assemblages in monocultures of four different tree species in three different families found that many mites were found under all four tree species, as well as in forested and meadow reference plots (Badejo & Tian 1999). Findings were similar for mites in plots of three different species of legume (Badejo *et al.* 2002). This suggests that the specialization on particular species and genera of plants that is found in herbivorous insects is rare or absent in detritivorous soil fauna. The seemingly general diets of soil organisms suggest that there is considerable overlap in resource use by detritivores and decomposers, putting the diversity of soil food webs further at odds with traditional niche theory (Maraun *et al.* 2003).

Soil communities are certainly not the only ones to have seemed more diverse than ecological theory predicted possible (e.g. plankton; Hutchison 1966), and ecologists have proposed a number of mechanisms for how consumers may coexist despite apparently competing for resources or habitat. Some of these mechanisms include predation on the competitively dominant consumer (Paine 1966), disturbance (Connell 1978), and spatial or temporal heterogeneity (Chesson 2000). With regards to soil diversity in particular, Giller (1996) proposed that favorable abiotic conditions, low resource competition, and the large range in body sizes of soil organisms may also help explain the "enigma."

Although all of these factors likely contribute to coexistence in soils, this chapter will focus on the explanation for soil diversity that Anderson offered: that detritivores and decomposers specialize on particular successional stages of their resource (1975b). As dead plant and animal matters decompose, they support successions of consumers, meaning that the consumers of fresh detritus are characteristically and predictably different from the consumers of more decomposed detritus. In addition to reducing their niche overlap with one another, organisms which utilize detritus at different points during their decomposition may interact with one another via the succession of their shared resource (i.e. processing chain interactions). These interactions between early and later successional consumers of detritus are frequently facilitative interactions, and the ubiquity and importance of positive species interactions is one of the features which distinguishes soil food webs from aboveground food webs (Lavelle *et al.* 1995; Wall & Moore 1999).

In this chapter I will review detrital succession in soils, some of the mechanisms that structure these successions, and some of the theoretical models for succession that may provide insight into soil food webs. Then I will discuss how this process relates to species interactions in soils and the extent to which it helps explain the diversity of soils at local and regional scales. Although Anderson's concept of successional specialization does not offer much insight into the diversity of root-feeding soil organisms, it relates to many of the unique interactions and processes of detrital food webs.

3.1.2 From litter and carrion to soil organic matter: detrital succession in soils

Less than a fifth of plant biomass is consumed by herbivores in terrestrial ecosystems (Cyr & Pace 1993). The remaining plant biomass becomes detritus when shed as senescent leaves and stems or following the death of the plant. Over the course of its decomposition, the particle size of the detritus is reduced (Hassall & Sutton 1978; Kheirallah 1990;

Martin 1991). The carbon: nitrogen ratio of decomposing plant matter generally decreases as nitrogen is immobilized and then, often, increases as nitrogen is mobilized. The organic matter is transported from the litter layer into the soil (Chamberlain et al. 2006; Frelich et al. 2006), its large organic molecules are broken into smaller ones and its small organic molecules are mineralized (Swift et al. 1979). Moore et al. (2004) describe this ontogeny of detritus as creating a "mosaic of resources" for consumers during the course of its decomposition. It may also be said to create a "mosaic of habitats," as its size, texture, moisture, and position in the soil change.

The predictable changes that litter undergoes during its decomposition are the result of the microbes and fauna that feed on it during this process. The assemblage of this food web undergoes succession as the detritus that supports the food web changes in terms of both its resource and habitat quality. Patterns of succession in both microbes and fauna have been documented for a variety of types of terrestrial detritus, and will be briefly summarized in this section. Mechanisms for these successions have been better elucidated in some systems than others, but changes in resource quality during decomposition are clearly of paramount importance for most detrital resources.

The microbial assemblage found on decomposing plant litter generally changes in composition and increases in diversity over time (Renvall 1995; Peters et al. 2000). The earliest "colonizers" of detritus are those microbes that colonize the plant tissues while they are still alive (phyllosphere or parasitic fungi and bacteria (Hudson 1968; Osono 2006)). While Gram-positive bacteria, including actinomycetes, dominate the assemblage on rice straw during the first 2 weeks of decomposition, Gram-negative bacteria and fungi become more abundant after the first 2 months (Cahyani et al. 2002). Nitrogen-fixing and ammonifying bacteria rapidly colonize decomposing litter, while nitrifying bacteria colonize later and utilize ammonia made available by the ammonifiers (Torres et al. 2005).

Fungi increase in abundance relative to bacteria during the decomposition of litter (reflected by changes in the nematode assemblage; Ferris & Matute 2003; Georgieva et al. 2005). Among saprotrophic fungi, early successional species (primarily Ascomycota and Zygomycota) utilize more labile or soluble carbon from the detritus, including sugars and cellulose, while later successional species (primarily Basidiomycota) utilize more recalcitrant carbon, especially lignin, chitin and tannins (Garrett 1951; Watson et al. 1974; Osono & Takeda 2001; Hanson et al. 2008). Sugar and cellulose-digesting fungi sometimes reoccur later during the decomposition, however, to utilize labile carbon made available by the lignin-degraders (Hudson 1968; Osono & Takeda 2001). Holmer and Stenlid (1997) found a correlation between competitive ability and successional sequence among wood-decomposing basidiomycetes, suggesting that competitive exclusion may be important in determining the succession of fungi within a broad functional group (wood-rotting basidiomycetes). Parasitic fungi that colonize living trees can also persist on fallen trees for decades, delaying the onset of the "typical" fungal succession (Renvall 1995). Saprotrophic fungi are consumed by other parasitic or saprotrophic fungi, which occur later during the succession and are often quite specific about which fungi they feed on (Niemelä et al. 1995).

The assemblage of soil fauna found on decomposing litter also changes during the decomposition process. As with microbes, changes in the resource quality of litter are clearly important for soil fauna, but changes in the habitat quality of the detritus and its location in the soil profile appear to also be of critical importance. Large detritivores that feed on litter on the soil surface ("litter transformers" or "shredders," such as isopods, millipedes, and epigeic and anecic earthworms) feed on detritus earlier in its decomposition than soil dwelling organisms that ingest soil organic matter (such as endogeic earthworms, and some collembolans and insect larvae). There is often a delay before microbivores colonize litter (Eitminavičiūtė et al. 1976; Lagerlöf & Andrén 1985), and then a shift from bacterivorous to fungivorous fauna, presumably following succession in the microbial assemblage (nematodes: Vreeken-Buijs & Brussaard 1996; Ferris & Matute 2003; Georgieva et al. 2005; mites: Vreeken-Buijs & Brussaard 1996). Predatory fauna, predictably, arrive on decomposing organic matter after their microbivorous prey (Wardle et al. 1995; Vreeken-Buijs & Brussaard 1996). Among collembolans, there is succession from surface-dwelling collembolans to soil-dwelling collembolans (Hasegawa & Takeda 1995; Takeda 1995), and from fungivores to detritivores which ingest soil organic matter (Hasegawa 1997). A number of studies have found turnover in the genera or species of oribatid mites, with some occurring in fresh litter and others in older, more aged litter (Anderson 1975b; Hasegawa 1997; Fagan et al. 2006), although the ecological differences between early and later successional oribatids are generally unclear. Santos and Whitford (1981) found a very predictable succession of prostigmatid and mesostigmatid mites in a desert soil food web. The earliest mites found on shrub litter, Tydeidae, fed on bacterivorous nematodes. Fungivorous Tarsonemidae or Pyemotidae mites peaked once 30–40% of the litter mass was lost, and predatory mesostigmatid mites arrived at 40–45% litter loss.

The relative timing of different groups of microarthropods, however, appears to be highly variable. Hasegawa (1997) found collembolans to be earlier colonizers of pine needles than mites, with median colonization times of 1.5 and 2.5 years, respectively, which were qualitatively similar to the findings of Irmler (2000) for both beech and mixed oak, hazel, and spruce litters. Wardle et al. (1995), however, found oribatid mites to be among the earliest colonizers of sawdust (peaking at day 0), with collembolans and predatory mites following after 100–200 days. Vreeken-Buijs and Brussaard (1996) found collembolans to colonize wheat residue after bacterivorous and nematophagous mites, and at about the same time as fungivorous and predatory mites, similar to Santos and Whitford's study of desert shrub litter (1981). Differences in litter quality, climate, or local species pools presumably explain these differences.

The abundance of soil organisms is also affected by seasonal changes in abiotic conditions, and some studies have inferred that observed turnover in soil organisms was seasonal rather than successional (Osler et al. 2004; Lindo & Winchester 2007). Several methods exist, however, for distinguishing between the effects of seasonality and detrital ontogeny (both resource and habitat changes). These include comparing the turnover of organisms on litter placed in the field in different seasons (Santos & Whitford 1981; Osler et al. 2004), comparing the assemblage of soil organisms on the same litter in the same seasons over multiple years (Anderson 1975b; Takeda 1995; Wardle et al. 1995; Irmler 2000), or comparing the assemblages of organisms on focal litter to the surrounding soil (Takeda 1995).

Aside from fungal successions on logs, the two most well-studied substrates for detrital succession are probably animal dung and carrion, whose study is aided by their fast decomposition (Carter et al. 2007) and the dominant role played by macroscopic arthropods (Payne 1965). In contrast to plant litter, which accounts for the vast majority of detritus in terrestrial ecosystems (Swift et al. 1979), carrion and dung are patchily distributed resources of much higher resource quality (i.e. higher nutrient and moisture contents, and less recalcitrant carbon) (Carter et al. 2007). This suggests that dispersal and competition may be more important in the detrital food webs of dung and carrion than of plant litter.

On mammal carcasses in temperate regions, the earliest insects to colonize are calliphorid, and sometimes sarcophagid, flies, which arrive within minutes and lay eggs (Payne 1965; Amendt et al. 2004). Other flies, histerid, silphid, and staphylinid beetles, and generalist scavengers, such as ants and yellow jackets, arrive during the first day or two, while anaerobic bacteria break down the carcass internally (Carter et al. 2007). A combination of the internal build-up of bacterial waste gasses and external feeding by newly hatched fly larvae ruptures the carcass (Carter et al. 2007), leading to a period of rapid tissue removal by insect larvae (Payne 1965) and aerobic microbial activity (Putnam 1978). Once most of the flesh is consumed, drier more recalcitrant tissues such as bones and hair are utilized by dermestid, clerid, and scarab beetles as well as generalist detritivores such as isopods, millipedes, and mites (Payne 1965; Amendt et al. 2004). Carrion decomposition can follow very different trajectories, however, if cold, winter conditions prevent insect colonization of the carrion (Putnam 1978), or if the carcass is buried, which, in the case of small mammals, can occur if it is found early by burying beetles.

Succession is also seen in the assemblages of beetles feeding and ovipositing on mammal dung. As dung ages the moisture content of the dung decreases, which changes its relative desirability to different species of beetles. Cattle dung is first used by "tunnelers," beetles which rapidly (within 24 hours) excavate underground tunnels and stock the tunnels with dung for their developing larvae (Doube et al. 1988; Menéndez & Gutiérrez 1999). The remaining, drier, dung is used by beetles which bury dung progressively over several days (Doube et al. 1988) or whose larvae develop in the unburied dung pats ("dwellers"; Menéndez & Gutiérrez 1999). Early successional beetles on large mammal dung in the tropics are generally "rollers," which roll away balls of fresh dung to stock their broods (Sabu et al. 2006). They are followed by a diversity of both tunnelers and dwellers. In cool, wet climates, where dung moisture can limit larval success, early successional beetles include tunnelers, dwellers who lay their eggs in soil near the dung pat, and dwellers who their lay eggs in silken egg cocoons, three strategies that reduce the exposure of eggs and larvae to moisture (Gittings & Giller 1998). Once the dung is drier, it is colonized by dwellers that lay eggs directly in the dung pat. Predatory beetles arrive a few days after the coprophages to prey on the larvae of dung beetles and flies (Menéndez & Gutiérrez 1999). Clearly dispersal abilities and the resource quality of dung are important in determining the sequence in which beetles utilize dung, and dung beetles are widely presumed to compete intensely for their ephemeral resource (see Finn & Gittings 2003 for a review of evidence for competition). Interestingly, some dung beetles also accelerate moisture loss from dung pats by burrowing into them (Sabu et al. 2006). This would affect the resource quality of the remaining dung, and may determine which beetles are subsequently able to colonize the dung pats.

Although there have been countless observational studies of consumer succession on decomposing organic matter in the last century, there have been far fewer experimental manipulations or other attempts to test mechanisms that might be structuring these successions. A wide variety of mechanisms have been inferred to determine the sequence with which consumers utilize detritus, including changes in the resource and habitat quality of the detritus and differences among consumers in their competitive abilities or their colonization or development rates. Theoretical models for succession and processing chain interactions, the topic of the next section, may be helpful in clarifying the range of mechanisms that can structure succession and the diversity of ways that consumers can interact during the course of succession.

3.1.3 Mechanisms and models for detrital succession

The patterns described earlier in microbial and faunal assemblages using terrestrial detritus suggest that succession is a dominant feature of detrital food webs, and critical in understanding soil communities. Modern ecology began with the study of succession; Cowles (1899) described the succession of vegetation on sand dunes, and studies of fungal succession on dung soon followed (Salmon & Massee 1901, as cited by Hudson 1968). Succession is often assumed to be structured by facilitative interactions; early successional species seem to modify the habitat or resource in ways that facilitate colonization by later species. Connell and Slatyer (1977) showed, however, that this "facilitation" model was only one of at least three models (Fig. 3.1.1) that could lead to a predictable succession of species. Whereas the facilitation model assumes that early successional species are the only ones capable of colonizing a new resource or habitat (step 2), early successional species may simply be the species that are most likely to colonize it on account of their superior dispersal or foraging abilities (as in both Connell & Slatyers' "tolerance" and "inhibition" models). Early successional species are assumed to make further recruitment by early successional species less likely (step 4), but may make colonization by later successional species easier (facilitation model), more difficult (inhibition model) or they may have no effect on later successional species (tolerance model) (step 3). Differences among species in lifespan and competitive ability, rather than facilitation, explain the transition from early successional species to later successional species in the tolerance and inhibition models (step 5).

Despite the paucity of experimental manipulations of detrital succession, examples can be found in the soils literature that seem to fit steps from all of Connell and Slatyer's models. Consumers in the detrital food web often improve the resource quality

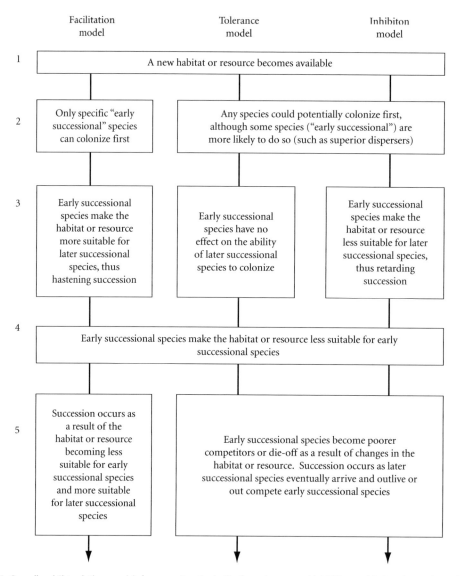

Figure 3.1.1 Connell and Slatyer's three models for succession: the facilitation, tolerance, and inhibition models (in columns, from left to right). Each model describes a sequence of events and interactions that occur following the availability of a new habitat or resource (step 1, at the top). Steps, described in the text, are numbered on the left. Steps 1 and 4 are the same in all three models, while steps 2 and 5 are the same in the tolerance and inhibition models. (Redrawn and modified from Connell & Slatyer 1977, with permission from The University of Chicago Press.)

of detritus for subsequent consumers (step 3 in the facilitative model): detritivores ("litter transformers" or "shredders") facilitate microbial utilization of litter by reducing its particle size (i.e. comminution; Hassall & Sutton 1978; Kheirallah 1990; Martin 1991) and microbial conditioning of litter improves its resource quality for detritivores (Daniel et al. 1997). The inhibition model, by contrast, would seem to describe the situation of parasitic fungi which invade a living tree and, following the tree's death, suppress colonization by saprotrophic fungi for years or decades (Renvall 1995), or when burying beetles prevent further colonization of small mammal carcasses by other insects. It is difficult to find unambiguous examples in the soil literature that fit the tolerance model, although the succession from primary sugar and cellulose-digesting fungi to lignin-degrading fungi might fit this model.

None of Connell and Slatyer's models seems to describe the initial stages of detrital succession on plant litter very well, however. Although detritivores and microbes often seem to facilitate one another, for example, this facilitation is not strictly necessary for either microbes or litter-feeding detritivores to utilize litter (as in step 2 of the facilitation model). Additionally, plant litter is a ubiquitous resource in many terrestrial ecosystems, so dispersal ability is unlikely to determine which consumers are early successional (as in step 2 of the tolerance and inhibition models). Abundant plant litter would also suggest that resource competition is relatively weak in litter-based food webs, and that competition is unlikely to explain the later stages of succession (Scheu & Schulz 1996). Carrion and dung based food webs are more likely to fit the tolerance or inhibition models, because dispersal and competition are more important on such patchily distributed, ephemeral resources. Logs, from the point of view of microbes, are also patchily distributed, and may fit the tolerance or inhibition models better than other types of plant detritus; Holmer and Stenlid (1997) argue that early successional fungi on logs are better dispersers and found that later successional fungi are better competitors.

One difference between traditional ecological succession and detrital succession is that traditional succession describes the sequence of species inhabiting a location in space ("seral" succession), whereas detrital succession describes the sequence of species utilizing a resource ("substratum" or "resource" succession). Although Connell and Slatyer include examples of detrital succession (which they term "heterotrophic succession" and regard as examples of their facilitation model), their models were primarily intended to describe seral succession. One important difference between succession in a habitat and succession on a resource is that occupants of a habitat may leave the habitat unchanged, whereas consumers of a resource are inevitably depleting the resource, regardless of how else they are modifying it. Models which explicitly incorporate resource dynamics may, therefore, be able to provide more mechanistic insight into when early successional consumers are likely to facilitate or inhibit later successional consumers.

Incorporating resource quality is both the difficult and crucial step in modeling food webs based on plant litter. Decomposers and detritivores appear to be much more limited by resource quality than resource quantity, in that experimental demonstrations of resource limitation in soil food webs typically involve the addition of higher quality resources, such as sugar (Bååth *et al.* 1978; Jonasson *et al.* 1996; Mikola & Setälä 1998; Magill & Aber 2000; Nieminen & Setälä 2001) or nutrient rich organic matter (Aescht & Foissner 1992; Tian *et al.* 1993; Chen & Wise 1999; Ruess *et al.* 2002; Jaffee 2004). The difficulty in incorporating resource quality into food web models, however, is that so many different factors contribute to resource quality, including the concentrations of numerous nutrients, the chemical forms of carbon and nutrients, the moisture content and the particle size of the resource. Consumers within the same food web differ in their sensitivity to different aspects of resource quality, and also affect the resource quality of the remaining detritus as a result of their feeding. Traditional ecological models for species interactions, which focus on resource quantity, are of little use in understanding detrital succession.

Heard (1994) modeled the interaction between two consumers that utilize the same resource, but do so when the resource is at two different successional stages (an "upstream condition" and a "downstream condition" *sensu* Heard; Fig. 3.1.2). The resource in Heard's model is donor-controlled, meaning that the rate with which it enters the system is not affected by anything within the system. Detrital food webs are regarded as donor-controlled on short timescales, because neither decomposers nor the standing biomass of detritus affect the rate with which new detritus is produced. The resource in Heard's model then passes through the two successional stages in a unidirectional manner. The resource can be processed through the feeding behavior of consumers ("consumer-dependent processing") or through consumer-independent mechanisms. Early successional consumers ("upstream consumers") can affect the dynamics of later successional consumers ("downstream consumers") if they affect either the rate or the efficiency with which the resource is processed.

At equilibrium, early successional consumers will either positively or negatively affect later successional consumers depending on whether consumer-dependent processing is more or less efficient than consumer-independent processing (Heard 1994). The efficiency with which a consumer processes a resource is generally inversely related to their feeding and metabolic efficiency; a detritivore, for example, that ingests most of the litter that it fragments and assimilates most of the litter it ingests will produce very little in the way of feces or incidentally fragmented litter. Thus "sloppy" or inefficient consumers facilitate later successional consumers, while highly efficient consumers preempt their access to the resource. At shorter time scales (non-equilibrium), consumers generally facilitate later successional consumers by increasing the rate with which the resource is processed, regardless of the efficiency with which they process the resource (Heard 1995).

Although formulated as a general model of consumer interaction via resource processing, Heard's "processing chain" model is particularly applicable to interactions among aquatic detritivorous insects, the subject of Heard's empirical research. Functional groups of aquatic detritivores are defined largely by the particle size of organic matter they feed on, and litter processed by abiotic forces (tumbling in the water column) is widely presumed to reach a condition similar to that of litter processed by "shredders" (aquatic detritivores that feed on intact leaves and produce fine particulate organic matter). "Consumer-independent processing" is perhaps harder to define in terrestrial systems, where physical forces, such as leaching and photodegradation, are less important in decomposition and do not produce detritus similar in quality or condition to detritus processed by organisms. Nonetheless, Heard's model may apply well to the interactions between particular groups of consumers in the detrital food web (Table 3.1.1), and also demonstrates some general, but previously unnoticed, features of resource processing interactions. As with succession, resource processing interactions have generally been presumed to be facilitative, but Heard's model shows that early successional consumers may also negatively affect later successional consumers ("resource pre-emption" *sensu* Heard). The rate of resource processing and the efficiency of resource processing

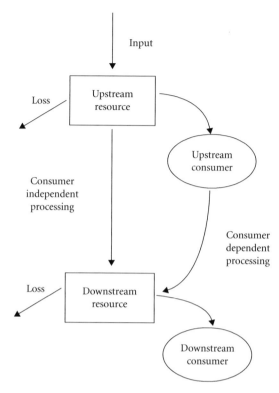

Figure 3.1.2 Heard's processing chain model describing the effects of an early successional consumer on a later successional consumer (upstream and downstream consumers, respectively, *sensu* Heard). The consumers feed on the same resource, but do so when the resource is in two different conditions or stages (upstream and downstream resources). The resource enters the system in the upstream condition at a fixed rate ("Input"), and can be processed into the downstream condition either through the action of the upstream consumer or through consumer independent processing. The resource is lost from the system from both conditions at fixed rates ("Loss"). (Redrawn and modified from Heard 1994.)

are both important in determining whether early successional consumers positively or negatively affect later successional consumers, but these factors operate on different timescales. The rate of resource processing is more important in determining the nature of the interaction in the short term, while the efficiency of resource processing is important at equilibrium. The apparent ubiquity of facilitative interactions may indicate that short-term, transient dynamics are more important than equilibrium conditions in understanding interactions in soils.

Heard's general model can be made modified to more precisely describe particular components of

Table 3.1.1 Examples of interactions from the soil literature that may be described by Heard's (1994) processing chain model

Upstream resource	Downstream resource	Upstream consumer	Downstream consumer	Possible examples
Litter	Soil organic matter	Isopods	Soil microbes	Hanlon & Anderson 1980, Hassall et al. 1987, Kayang et al. 1996, Kautz and Topp 2000, Hättenschwiler & Bretscher 2001, Zimmer et al. 2005
Litter	Soil organic matter	Litter feeding earthworms	Soil microbes	Zimmer et al. 2005
Litter	Soil organic matter	Litter feeding earthworms	Soil feeding earthworms	Briones et al. 2005
Litter	Microbes	Isopods	Microbivorous nematodes	Bastow 2007
Proteins and other nitrogenous organic molecules	Ammonia	Ammonifying bacteria	Nitrifying bacteria	Torres et al. 2005
Lignin bound cellulose and sugars	Cellulose and sugars	Lignin-degrading fungi	Sugar, cellulose fungi	Osono & Takeda 2001
Wet dung	Dry dung	"Tunneling" dung beetles	"Dwelling" dung beetles	Sabu et al. 2006

the soil food web. Moore et al. (2004), for example, used a model similar to Heard's to demonstrate the complexity of consumer interactions that can arise from detrital ontogeny (Fig. 3.1.3). Moore et al.'s model includes two resource conditions (recalcitrant and labile detritus) and two consumers (fungi and bacteria). As in Heard's model, the resource moves between conditions in a single direction (from recalcitrant to labile), but in contrast to Heard's, resources can enter the system in either condition, and the resources remain in the system until consumed. Fungi in Moore et al.'s model can feed on detritus in both conditions, while bacteria can only feed on labile detritus. Both consumers contribute to both resource pools via mortality. Although their analysis of the model was limited to just a few of its parameters, Moore et al. showed that coexistence of bacteria and fungi was possible under biologically realistic conditions (i.e. bacteria consume labile detritus more rapidly and assimilate it more efficiently than fungi do) and that the balance between bacteria and fungi is determined by the quality of detrital inputs and the rate at which fungi process recalcitrant detritus. Bacteria and fungi competed, but bacteria were also facilitated by, and often dependent on, fungi.

Heard models resources as occurring in two discrete states or conditions (upstream and downstream). It is therefore most appropriate in describing upstream consumers which have a relatively large and discontinuous effect on the quality of the resource, such as when detritivores convert litter into feces, triggering what Scheu and Setälä (2001) term a "feces cascade" in the microbial and microbivorous components of the food web. The response of detritivores to nitrogen immobilization in litter by microbes, by contrast, would be best modeled using a continuous description of resource quality, because detritivores would perceive nitrogen immobilization as a continuous decrease in the carbon: nitrogen ratio of the litter. It may generally be true that models with discrete resource states are better for describing how large organisms affect the resource quality perceived by smaller organisms, while models with continuously varying resource quality are better for describing how small organisms affect the resource quality perceived by larger organisms.

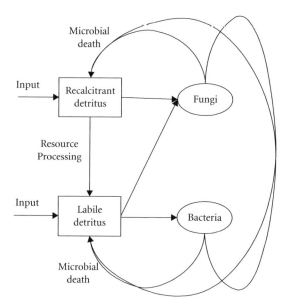

Figure 3.1.3 Moore et al.'s detrital ontogeny model describing the interactions between fungi and bacteria. The resource, detritus, occurs in two conditions, recalcitrant and labile. Detritus enters the system at fixed rates, but can enter the system in either condition. Fungi feed on detritus in both conditions and control the rate with which recalcitrant detritus becomes labile. Bacteria feed on detritus in the labile condition only. Fungi and bacteria contribute to both resource pools through their own mortality. The resource is never lost from the system (except through the metabolic inefficiencies of bacteria and fungi). (Redrawn and modified from Moore et al. 2004, with permission from John Wiley & Sons.)

Regardless of its realism or literal applicability, Heard's processing chain model demonstrates the need for experimental tests of the interactions between consumers inferred to interact with one another via detrital ontogeny. Consumers that feed on the same detritus at different points in its decomposition, while it is in different resource conditions or states, may positively or negatively affect one another, and the factors which determine the nature of this interaction (rates and efficiencies of resource processing) are difficult to intuit.

3.1.4 Can successional specialization explain coexistence and the diversity in soils?

Although Anderson proposed successional specialization as an explanation for the coexistence of the extraordinary diversity of soil organisms in a temperate woodland, the results of his attempts to document this succession in soil fauna were fairly mixed (1975b). Although the composition of oribatid mites changed during the course of decomposition in his litterbags, and enchytraeids only arrived in litterbags after a full year, other groups, such as Collembola and insect larvae, did not seem to differentiate between fresh and aged litter. Anderson concluded that detrital ontogeny was just one factor, along with habitat complexity, that allowed for the coexistence of so many soil fauna. Although it is clear that there is succession in the food web supported by decomposing organic matter, and that many interactions among soil organisms can only be understood in light of detrital ontogeny, it remains difficult to say how much detrital succession contributes to soil diversity.

It is clearer, after three decades, however, how some groups of soil organisms partition detritus as a resource. Microbes, for example, appear to exhibit "enzymatic specialization," meaning that they differ in their ability to degrade particular organic compounds. This is particularly well studied in fungi, for whom the sugar-, cellulose-, and lignin-decomposing fungi categories (Garrett 1951; Watson et al. 1974) still appear useful, but have been expanded on in terms of both the number of different substrates (Kjøller & Struwe 2002; Hanson et al. 2008) and the inclusion of non-culturable fungi (Hanson et al. 2008). Such enzymatic specialization translates into successional specialization in nature because the relative abundance of different organic compounds in detritus changes during decomposition, with the labile carbon being depleted much earlier than the recalcitrant carbon (Kjøller & Struwe 2002). A great deal of fungal diversity remains unexplained by our current knowledge of their enzymatic specialization, however. Hanson et al., for example, found 11 soil fungi that responded to particular carbon compounds (glycine, sucrose, cellulose, lignin, or tannin-protein), but this accounted for <10% of the fungi they isolated. Buée et al. (2009) found approximately 1000 soil fungi in 4-g samples of forest soil. Although this estimate includes mycorrhizal and parasitic fungi, it is likely that forests contain hundreds, and possibly thousands, of saprophytic fungi whose roles we are just beginning to understand. Soil bacteria likely also differ in terms of their abilities to degrade particular organic compounds, although bacteria as a whole seem to utilize more labile substrates than soil fungi. Some soil bacteria are also enzymatically specialized on particular conversions of nitrogen (e.g. ammonifiers, nitrifiers, denitrifiers). Because the nitrogen in decaying organic matter undergoes these conversions in a sequence, bacteria involved in nitrogen cycling can be thought of as having a specialized form of successional specialization (Torres et al. 2005).

Successional specialization has generally been harder to apply to the diversity of soil fauna. Although there are clear differences in both resource and habitat use by large, surface-feeding detritivores and smaller, soil-dwelling microbivores and predators, each of these groups contains a surprising number of species. The diversity of microbivores, in particular, has been difficult to explain. In laboratory feeding trials, fungivores and detritivores often show very little discrimination between fungi, can survive on a wide variety of foods, and, to the extent that they do have detectable preferences, often prefer to eat the same things (Maraun et al. 2003). There is evidence from stable isotope studies, however, that such results do not accurately reflect the behavior of these animals in the field. Stable isotope analysis uses the ratios of stable, naturally occurring isotopes in the tissues of organisms to infer the trophic relationships in communities. The ratio of ^{15}N to ^{14}N increases with each trophic transfer, so it is often used to determine the number of trophic exchanges between a consumer and a resource at the base of the food web. Although oribatid mites are often presumed to be generalist fungivores and detritivores, Schneider et al. (2004) used nitrogen stable isotope ratios to show that oribatid mites in a temperate forest include at least four distinct trophic groups. Schneider et al. interpreted these four groups as feeding on 1) lichens and algae, 2) litter, 3) primarily fungi, and 4) animals and fungi. Organic matter becomes enriched with ^{15}N over time (Hyodo et al. 2008) because microbes preferentially mineralize the lighter isotope. This pattern can be seen for both $^{15}N:^{14}N$ and $^{13}C:^{12}C$ ratios by looking down soil profiles (Dijkstra et al. 2006). Therefore, differences among mites in nitrogen stable isotope composition may also reflect successional specialization among mites. For example, two species of mites which both feed on the same species of fungi may acquire distinct stable isotope ratios if one of the species feeds at the soil surface on fungi utilizing fresh litter and the other feeds deeper in the soil on fungi utilizing more aged litter. Pollierer *et al.* (2009) were thus able to use stable carbon and nitrogen ratios to distinguish between animals feeding on litter, recent belowground carbon (indirectly, via fungi) and older belowground carbon. Although the use of stable isotope analysis in soil food webs is still relatively new, it is proving useful for distinguishing between carbon from roots and from litter, as well as distinguishing between younger and older organic matter. Both of these distinctions are important for connecting soil organisms and food webs to detrital ontogeny, and it is clear already that there is greater trophic specialization in soil fauna than had been detected by feeding trials, gut analysis and morphological studies.

Tissue concentrations of the radioactive isotope ^{14}C have also been used to determine the age of organic matter soil organisms assimilate (Briones *et al.* 2005). This technique compares tissue ^{14}C to records of atmospheric ^{14}C concentrations, which became elevated as a result of atmospheric nuclear testing and have been decreasing since the Partial Test Ban Treaty of 1963. Epigeic earthworms, according to this technique, are comprised of carbon that was fixed by photosynthesis 0–4 years earlier, while endogeic earthworms are comprised of 5–9-year-old carbon (Briones *et al.* 2005; Hyodo *et al.* 2008). Similarly, grass feeding termites feed on young organic matter (2 years old), while soil and wood feeding termites feed on considerably older organic matter (7–12 and 8–21 years old, respectively, Hyodo *et al.* 2008). These functional groups of fauna were already known to feed on quite distinct ages of organic matter, but these studies show that ^{14}C can be used to very precisely relate organisms to the ontogeny of their food source. It would be interesting to apply this technique in the future to soil fauna with more ambiguous feeding behaviors.

In addition to allowing for the coexistence of species that would otherwise compete, successional specialization by soil microbes or fauna may cause species to have synergistic effects on ecosystem functioning (a topic reviewed by Wurst *et al.* Chapter 1.2, this volume). Whereas unspecialized decomposers and detritivores would be redundant in terms of their effects on decomposition rates or nutrient cycling, when decomposers and detritivores are specialized on detritus at particular stages during its decomposition there is a positive relationship between their diversity and the rates of these processes. Tiunov and Scheu (2005), for example, found a positive relationship between fungal diversity and decomposition rate, which they attributed to facilitative interactions between sugar and cellulose-degrading fungi and, more generally, to the fact that resource processing by decomposers creates resources for other decomposers. Evidence for a relationship between diversity and ecosystem functioning in soil is quite mixed, however, (Wurst *et al.* Chapter 1.2, this volume), suggesting that there is redundancy among consumers of detritus even at particular stages in its decomposition. Experiments may also fail to find a positive relationship between diversity and decomposition processes, however, if they focus on a single stage in detrital ontogeny, such as litter mass loss or carbon mineralization. Such experiments would artificially exclude successional specialization of soil organisms, and may make these specialized consumers appear redundant. A full appreciation for the effects of diversity on decomposition and nutrient cycling, therefore, requires that such experiments include the full ontogeny of detritus, either sequentially or simultaneously.

3.1.5 Latitudinal gradients in soil diversity: detrital food webs thwart ecology's oldest pattern

Terrestrial plants and most groups of aboveground animals are known to be far more diverse in the tropics than in temperate regions. The inverse relationship between species diversity and latitude for these well known groups of organisms is perhaps ecology's oldest pattern (Hawkins 2001, as cited by De Deyn & Van der Putten (2005)). This pattern appears not to hold true, however, for soil organisms (Wardle 2002; De Deyn & Van der Putten 2005). Although many groups of soil organisms are poorly sampled, soil bacteria (Fierer & Jackson 2006), mycorrhizal fungi (Allen *et al.* 1995; Tedersoo & Nara

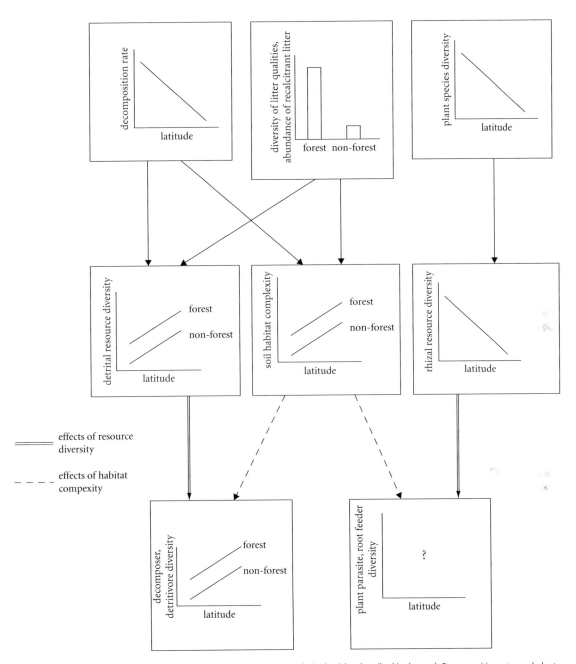

Figure 3.1.4 A conceptual model for the diversity of soil organisms across latitudes (also described in the text). Decomposition rates and plant species diversity both decline with increasing latitude (i.e. moving away from the equator), while the diversity of litter and the abundance of recalcitrant litter inputs is higher in forests than non-forests across all latitudes (top row). Decomposition rates, relative to the rate of litter inputs, are inversely related to the diversity of detrital resources present in a community at a single point in time, while the diversity and recalcitrance of litter inputs are positively related to the diversity of detrital resources (middle row). These two factors affect soil habitat complexity in a similar manner. Plant species diversity is positively related to the diversity of rhizal inputs. The diversity of decomposers and detritivores is determined by the diversity of detrital resources and the complexity of the soil habitat, both of which suggest that diversity should be higher in forests than non-forests and should increase with increasing latitude (bottom row). The diversity of plant parasites and root-feeding microbes and fauna is determined by soil habitat complexity and rhizal resource diversity, but these two factors are inversely related to one another, so the resulting pattern of rhizal food web diversity is ambiguous.

2010) and soil nematodes (Boag & Yeates 1998) appear to be most diverse at intermediate latitudes, while oribatid mites (Maraun et al. 2007) and earthworms (Lavelle et al. 1995) have similar diversities across low and intermediate latitudes. Temperate forests appear to have especially diverse soil communities (Boag & Yeates 1998; Maraun et al. 2007; Tedersoo & Nara 2010), although soil bacteria are surprisingly diverse in arid grasslands and deserts (Fierer & Jackson 2006).

There are numerous explanations for the latitudinal gradients in diversity seen in aboveground organisms, but for many animals, including insects, the diversity of plants is likely a major factor. Diverse assemblages of plants provide herbivores with a diversity of resources and all aboveground animals with a diversity of microhabitats. If decomposers and detritivores specialize on detritus at particular stages in its decomposition, rather than on detritus of particular taxa of plants, a diversity of plant species would not translate into a diversity of resources for the detrital food web. The habitat for soil organisms is comprised of both live plant roots and decomposing organic matter (the litter layer and soil organic matter). Although the diverse plant assemblages of the tropics may contribute to a diverse soil habitat in terms of live plant roots, the fast decomposition rates found in the tropics (Swift et al. 1979) would prevent the development of thick, stratified litter layers that contribute to soil habitat complexity in temperate regions (Wardle 2002). It is perhaps not surprising, then, that soil organisms do not consistently exhibit the tropical peak in diversity found in other groups.

If successional specialization in the detrital food web affects latitudinal patterns in soil diversity, the diversity of resources available to decomposers and detritivores would likely depend on two factors: the quality of detrital inputs and the decomposition rate. Because microbes seem to specialize on carbon molecules of varying lability, a diversity of litter qualities (e.g. herbaceous plant litter as well foliar, fine and coarse woody tree litter) is probably important for detrital food web diversity. Recalcitrant, difficult to decompose substrates support higher diversities of microbes, because a greater diversity of enzymes are required to break down such substrates (Hanson et al. 2008), so it is likely woody detritus is especially important for detrital diversity. Slow decomposition rates, relative to the frequency of litter inputs, would lead to litter being present at a variety of stages of decomposition, allowing for greater specialization by the detrital food web on particular stages of decomposition. Taken together, these two factors would predict that the diversity of decomposers and detritivores would be higher in forests than non-forests and would increase with increasing latitude, because decomposition rates tend to be slower at higher latitude (Fig. 3.1.4). Of course, these same factors would also lead to a more complex and heterogeneous soil habitat, which is also thought to contribute to the diversity of soil organisms. It is perhaps not surprising that it is difficult to disentangle the role of resource and habitat diversity in soil species diversity, given that many soil organisms live on or in their food.

Root-feeding soil organisms, such as plant-parasitic nematodes and mites, would not necessarily follow this pattern, however, because they do not exhibit successional specialization. If resource diversity is the most important factor controlling soil diversity, detrital and "rhizal" (i.e. root-feeding) food webs would, therefore be predicted to exhibit different latitudinal patterns in diversity. If abiotic conditions or habitat complexity, on the other hand, are the most important factors, detrital and rhizal food webs would be predicted to have similar latitudinal patterns in diversity. It may be difficult to assign members of groups, such as rhizosphere bacteria and their consumers, to detrital or rhizal food webs, but this would be a valuable comparison to make in groups where detrital and rhizal food webs are more distinct.

3.1.6 Future directions in understanding detrital succession

This overview of the patterns and mechanisms of succession during decomposition in soil food webs suggests a number of interesting areas for future research. Despite the enormous number of observational studies of succession on decomposing litter, carrion, and dung, there have been very few experimental tests of processing chain interactions among detritivores and decomposers. Many studies selectively exclude particular components of the detrital food web to determine the effects these organisms have on decomposition or nutrient cycling, such as when larger fauna are excluded from litterbags. These studies rarely measure the effects of such exclusion on other members of the detrital food web, however. Knowing how species affect ecosystem processes without knowing how they affect one another, however, limits our ability to understand how these processes are controlled by complex, diverse natural food webs. For example, it is likely that epigeic and endogeic earthworms in a community feed on the same decomposing organic matter, but epigeic earthworms do so when that organic matter is in the litter layer, while endogeic earthworms feed on it after it has been processed into soil organic matter. Knowing how epigeic earthworms, through their processing of litter, affect the dynamics of endogeic earthworms may be helpful in putting together information about how each group affects nutrient cycling into a complete understanding of earthworms and nutrient cycling. Some examples of manipulative experiments of consumer interactions during detrital succession are the studies of fungal interactions during the decay of wood (Holmer & Stenlid 1997; Folman *et al.* 2008) and the interactions between large detritivores and microbes (Hanlon & Anderson 1980; Kayang *et al.* 1996; Daniel *et al.* 1997).

The increasing use of isotope analysis in soil studies is one of the most exciting developments in the last decade; these techniques are enormously helpful in tracing energy and nutrient through soils without having to first dismember the natural soil community beyond recognition. Two areas of isotope research are particularly relevant to detrital food webs. The first is figuring out how much carbon different members of the soil community acquire from decomposing organic matter and how much they acquire from live plant roots. Historically, soil ecologists regarded fauna that were not directly observed feeding on roots as being part of the detrital food web. Recent isotope studies, however, have found a surprising amount of rhizal carbon throughout the soil food web (Albers *et al.* 2006; Pollierer *et al.* 2007; but see Elfstrand *et al.* 2008 for a counter example). The nature of interactions in the "detrital" food web will obviously be quite different if these microbes and fauna acquire the bulk of their energy, directly or indirectly, from live plant roots and their exudates, and only rely on decomposing organic matter for nutrients and habitat. Experiments distinguishing between rhizal and detrital carbon are difficult to perform in natural communities, however, and interpreting the results is often complicated by the presence of soil organic matter of ambiguous origin. Scheunemann *et al.* (2010), for example, found that earthworms and collembolans in an agricultural field acquired most of their carbon from soil organic matter which was decades old. The importance of rhizal and detrital carbon for soil organisms is also likely to differ among ecosystems.

The second application of isotope analysis that is relevant to understanding detrital food webs, of course, is the finding that soil fauna which appear in lab studies to have similar diets nonetheless have distinct stable isotope ratios in the field (Schneider *et al.* 2004). Resolving this discrepancy, likely through isotope studies that include a more detailed analysis of possible food sources (including microbes), may be very helpful in understanding the coexistence of soil fauna.

Finally, experimentally distinguishing between the role of detritus as a resource and the role of detritus as a habitat may be quite helpful in understanding soil food webs. It is often unclear whether detrital succession is structured by the changing resource quality of detritus or the changing habitat quality. One obvious approach for separating resource and habitat effects is the use of artificial substrates, which mimic the abiotic properties of detritus yet provide no nourishment. Fagan *et al.* (2006), for example, used litterbags filled with styrofoam chips to determine which mites colonize litterbags for the habitat they provided, as opposed to the resource. Distinguishing these two factors is important in understanding species interactions among soil organisms. Heard's processing chain model describes resource-mediated interactions, which can shift between competition and facilitation depending on the processing rates and efficiencies of upstream consumers. Metabolic requirements limit the number and degree to which organisms can benefit from resource-mediated interactions, because all consumers of the resource need to acquire sufficient energy from a finite supply. If early successional species primarily affect later successional species by modifying the habitat quality of detritus, resource competition is less important and facilitation more likely. Detritivores may have widespread effects on soil microbes and fauna, including those which acquire most of their carbon from live plant roots, when they modify the habitat qualities of detritus or the surrounding soil. Habitat-mediated interactions account for much of the literature on facilitative interactions in marine ecosystems (Stachowicz 2001), and these interactions may be widespread in soils as well (Scheu & Setälä 2001).

References

Aescht, E., & Foissner, W. (1992) Effects of mineral and organic fertilisers on the microfauna in a high-altitude reafforestation trial. *Biology and Fertility of Soils* **13**: 17–24.

Albers, D., Schaefer, M., & Scheu, S. (2006) Incorporation of plant carbon into the soil animal food web of an arable system. *Ecology* **87**: 235–45.

Allen, E.B., Allen, M.F., Helm, D.J., Trappe, J.M., Molina, R., & Rincon, E. (1995) Patterns and regulation of mycorrhizal plant and fungal diversity. *Plant and Soil* **170**: 47–62.

Amendt, J., Krettek R., & Zehner R. (2004) Forensic entomology. *Naturwissenschaften* **91**: 51–65.

Anderson, J.M. (1975a) The enigma of soil animal diversity. In: J. Vanek (ed.) *Progress in soil zoology. Proceedings of the 5th international colloquium on soil zoology*, Prague, 17–22 September. W. Junk, The Hague.

Anderson, J.M. (1975b) Succession, diversity and trophic relationships of some soil animals in decomposing leaf litter. *Journal of Animal Ecology* **44**: 475–95.

Bååth, E., Lohm, U., Lundgren, B., Rosswall, T., Söderström, B., & Wirén, A. (1978) The effect of nitrogen and carbon supply on the development of soil organism populations and pine seedlings: a microcosm experiment. *Oikos* **31**: 153–63.

Badejo, M.A., Espindola, J.A.A., Guerra, J.G.M., De Aquino, A.M., & Correa, M.E.F. (2002) Soil oribatid mite communities under three species of legumes in an ultisol in Brazil. *Experimental and Applied Acarology* **27**: 283–96.

Badejo, M.A., & Tian, G. (1999) Abundance of soil mites under four agroforestry tree species with contrasting litter quality. *Biology and Fertility of Soils* **30**: 107–12.

Bastow, J.L. (2007) Processing chain interactions in a grassland detrital food web: The role of *Porcellio scaber* in the California coastal prairie. Ph.D. University of California, Davis.

Boag, B., & Yeates, G.W. (1998) Soil nematode biodiversity in terrestrial ecosystems. *Biodiversity and Conservation* **7**: 617–30.

Briones, M.I., Garnett, M.H., & Piearce, T.G. (2005) Earthworm ecological groupings based on ^{14}C analysis. *Soil Biology and Biochemistry* **37**: 2145–9.

Brown, V.K., & Gange, A.C. (1990) Insect herbivory below ground. *Advances in Ecological Research* **20**: 1–58.

Buée, M., Reich, M., Murat, C., Morin, E., Nilsson, R.H., Uroz, S., & Martin, F. (2009) 454 Pyrosequencing analyses of forest soils reveal an unexpectedly high fungal diversity. *New Phytologist* **184**: 449–56.

Cahyani, V.R., Watanabe, A., Matsuya, K., Asakawa, S., & Kimura, M. (2002) Succession of microbiota estimated by phospholipid fatty acid analysis and changes in organic constituents during the composting process of rice straw. *Soil Science and Plant Nutrition* **48**: 735–43.

Carter, D.O., Yellowlees, D., & Tibbett, M. (2007) Cadaver decomposition in terrestrial ecosystems. *Naturwissenschaften* **94**: 12–24.

Chamberlain, P.M., McNamara, N.P., Chaplow, J., Stott, A.W., & Black, H.I.J. (2006) Translocation of surface litter carbon into soil by Collembola. *Soil Biology and Biochemistry* **38**: 2655–64.

Chen, B., & Wise, D.H. (1999) Bottom-up limitation of predaceous arthropods in a detritus-based terrestrial food web. *Ecology* **80**: 761–72.

Chesson, P. (2000) General theory of competitive coexistence in spatially-varying environments. *Theoretical Population Biology* **58**: 211–37.

Connell, J.H. (1978) Diversity in tropical rainforests and coral reefs. *Science* **99**: 1302–10.

Connell, J.H., & Slatyer, R.O. (1977) Mechanisms of succession in natural communities and their role in community stability and organization. *American Naturalist* **111**: 1119–44.

Cowles, H.C. (1899) The ecological relations of the vegetation on the sand dunes of Lake Michigan. *The Botanical Gazette* **27**: 95–391.

Curry, J.P., & Schmidt, O. (2007) The feeding ecology of earthworms—a review. *Pedobiologia* **50**: 463–77.

Cyr, H., & Pace, M.L. (1993) Magnitude and patterns of herbivory in aquatic and terrestrial ecosystems. *Nature* **361**: 148–50.

Daniel, O., Schönholzer, F., Ehlers, S., & Zeyer, J. (1997) Microbial conditioning of leaf litter and feeding by the wood-louse *Porcellio scaber*. *Pedobiologia* **41**: 397–401.

Daugherty, M.P. (2009) Specialized feeding modes promote coexistence of competing herbivores: insights from a metabolic pool model. *Environmental Entomology* **38**: 667–76.

De Deyn, G.B., & Van der Putten, W.H. (2005) Linking aboveground and belowground diversity. *Trends in Ecology & Evolution* **20**: 625–33.

Dijkstra, P., Ishizu, A., Doucett, R., et al. (2006) ^{13}C and ^{15}N natural abundance of soil microbial biomass. *Soil Biology and Biochemistry* **38**: 3257–66.

Doube, B.M., Giller, P.S., & Moola, F. (1988) Dung burial strategies in some South African coprine and onitine dung beetles (Scarabaeidae: Scarabaeinae). *Ecological Entomology* **13**: 251–61.

Edney, E.B., Allen, W., & McFarlane, J. (1974) Predation by terrestrial isopods. *Ecology* **55**: 428–33.

Ehrlich, P.R., & Raven, P.H. (1964) Butterflies and plants: a study in coevolution. *Evolution* **18**: 586–608.

Eitminavičiūtė, I., Bagdanavičienė, Z., Kadytė, B., Lazaukienė, L., & Sukackienė, I. (1976) Characteristic successions of microorganisms and soil invertebrates in the decomposition process of straw and lupine. *Pedobiologia* **16**: 106–15.

Elfstrand, S., Lagerlöf, J., Hedlund, K., & Mårtensson, A. (2008) Carbon routes from decomposing plant residues and living roots into soil food webs assessed with ^{13}C. *Soil Biology and Biochemistry* **40**: 2530–9.

Erwin, T.L. (1982) Tropical forests: their richness in Coleoptera and other Arthropod species. *The Coleopterists Bulletin* **36**: 74–5.

Fagan, L.L., Didham, R.K., Winchester, N.N., et al. (2006) An experimental assessment of biodiversity and species turnover in terrestrial vs canopy leaf litter. *Oecologia* **147**: 335–47.

Farrell, B.D. (1998) "Inordinate fondness" explained: why are there so many beetles? *Science* **281**: 555–9.

Ferris, H., & Matute, M.M. (2003) Structural and functional succession in the nematode fauna of a soil food web. *Applied Soil Ecology* **23**: 93–110.

Fierer, N., & Jackson, R.B. (2006) The diversity and biogeography of soil bacterial communities. *Proceedings of the National Academy of Sciences of the United States of America* **103**: 626–31.

Finn, J.A., & Gittings, T. (2003) A review of competition in north temperate dung beetle communities. *Ecological Entomology* **28**: 1–13.

Folman, L.B., Klein Gunnewiek, P.J.A., Boddy, L., & de Boer, W. (2008) Impact of white-rot fungi on numbers and community composition of bacteria colonizing beech wood from forest soil. *FEMS Microbial Ecology* **63**: 181–91.

Frelich, L.E., Hale, C.M., Scheu, S., et al. (2006) Earthworm invasion into previously earthworm-free temperate and boreal forests. *Biological Invasions* **8**: 1235–45.

Garrett, S.D. (1951) Ecological groups of soil fungi: a survey of substrate relationships. *New Phytologist* **50**: 149–66.

Georgieva, S.S., Christensen, S., & Stevnbak, K. (2005) Nematode succession and microfauna-microorganism interactions during root residue decomposition. *Soil Biology and Biochemistry* **37**: 1763–74.

Giller, P.S. (1996) The diversity of soil communities, the "poor man's tropical rainforest." *Biodiversity and Conservation* **5**: 135–68.

Gittings, T., & Giller, P.S. (1998) Resource quality and the colonisation and succession of coprophagous dung beetles. *Ecography* **21**: 581–92.

Grimaldi, D. (1999) The co-radiations of pollinating insects and angiosperms in the Cretaceous. *Annals of the Missouri Botanical Garden* **86**: 373–406.

Hanlon, R.D.G., & Anderson, J.M. (1980) Influence of macroarthropod feeding activities on microflora in decomposing oak leaves. *Soil Biology and Biochemistry* **12**: 255–61.

Hanson, C.A., Allison, S.D., Bradford, M.A., Wallenstein, M.D., & Treseder, K.K. (2008) Fungal taxa target different carbon sources in forest soils. *Ecosystems* **11**: 1157–67.

Hasegawa, M. (1997) Changes in Collembola and Cryptostigmata communities during the decomposition of pine needles. *Pedobiologia* **41**: 225–41.

Hasegawa, M., & Takeda, H. (1995) Changes in feeding attributes of four collembolan populations during the decomposition process of pine needles. *Pedobiologia* **39**: 155–69.

Hassall, M., & Sutton, S.L. (1978) The role of isopods as decomposers in a dune grassland ecosystem. *Science Proceedings of the Royal Dublin Society, Series A* **6**: 117–27.

Hassall, M., Turner, J.G., & Rands, M.R.W. (1987) Effects of terrestrial isopods on the decomposition of woodland leaf litter. *Oecologia* **72**: 597–604.

Hättenschwiler, S., & Bretscher, D. (2001) Isopod effects on decomposition of litter produced under elevated CO_2, N deposition and different soil types. *Global Change Biology* **7**: 565–79.

Hawkins, B.A. (2001) Ecology's oldest pattern? *Trends in Ecology & Evolution* **16**(8):470.

Heard, S.B. (1994) Processing chain ecology: resource condition and interspecific interactions. *Journal of Animal Ecology* **63**: 451–64.

Heard, S.B. (1995) Short-term dynamics of processing chain systems. *Ecological Modelling* **80**: 57–68.

Hendriksen, N.B. (1990) Leaf litter selection by detritivore and geophagous earthworms. *Biology and Fertility of Soils* **10**: 17–21.

Holmer, L., & Stenlid J. (1997) Competitive hierarchies of wood decomposing basidiomycetes in artificial systems based on variable inoculum sizes. *Oikos* **79**: 77–84.

Hudson, H.J. (1968) The ecology of fungi on plant remains above the soil. *New Phytologist* **67**: 837–74.

Hutchison, G.E. (1966) The paradox of the plankton. *American Naturalist* **95**: 137–45.

Hyodo, F., Tayasu, I., Konate, S., Tondoh, J.E., Lavelle, P., & Wada, E. (2008) Gradual enrichment of ^{15}N with humification of diets in a below-ground food web: relationship between ^{15}N and diet age determined using ^{14}C. *Functional Ecology* **22**: 516–22.

Irmler, U. (2000) Changes in the fauna and its contribution to mass loss and N release during leaf litter decomposition in two deciduous forests. *Pedobiologia* **44**: 105–18.

Jaffee, B.A. (2004) Do organic amendments enhance the nematode-trapping fungi *Dactylellina haptotyla* and *Arthrobotrys oligospora*? *Journal of Nematology* **36**: 267–75.

Jonasson, S., Michelsen, A., Schmidt, I.K., Nielsen, E.V., & Callaghan, T.V. (1996) Microbial biomass C, N and P in two arctic soils and responses to addition of NPK fertilizer and sugar: implication for plant nutrient uptake. *Oecologia* **106**: 507–15.

Kautz, G., & Topp, W. (2000) Acquisition of microbial communities and enhanced nutrient availablility of soil nutrients by the isopod *Porcellio scaber* (Latr.)(Isopoda: Oniscidea). *Biology and Fertility of Soils* **31**: 102–7.

Kautz, G., Zimmer, M., & Topp, W. (2002) Does *Porcellio scaber* (Isopoda: Oniscidea) gain from copraphagy? *Soil Biology and Biochemistry* **34**: 1253–9.

Kayang, H., Sharma, G.D., & Mishra, R.R. (1996) The influence of isopod grazing on microbial dynamics in decomposing leaf litter of *Alnus nepalensis* D. Don. *European Journal of Soil Biology* **32**: 35–9.

Kheirallah, A.M. (1990) Fragmentation of leaf litter by a natural population of the millipede *Julus scandinavius* (Latzel 1884). *Biology and Fertility of Soils* **10**: 202–6.

Kjøller, A.H., & Struwe, S. (2002) Fungal communities, succession, enzymes, and decomposition. In: R.G. Burns and R.P. Dick (eds.) *Enzymes in the Environment* pp. 267–84. Marcel Dekker, New York.

Lagerlöf, J., & Andrén, O. (1985) Succession and activity of microarthropods and enchytraeids during barley straw decomposition. *Pedobiologia* **28**: 343–57.

Lavelle, P., Lattaud, C., Trigo, D., & Barois, I. (1995) Mutualism and biodiversity in soils. *Plant and Soil* **170**: 23–33.

Lindo, Z., & Winchester, N.N. (2007) Oribatid mite communities and foliar litter decomposition in canopy suspended soils and forest floor habitats of western redcedar forests, Vancouver Island, Canada. *Soil Biology and Biochemistry* **39**: 2957–66.

Magill, A.H., & Aber, J.D. (2000) Variation in soil net mineralization rates with dissolved organic carbon additions. *Soil Biology and Biochemistry* **32**: 597–601.

Maraun, M., Martens, H., Migge, S., Theenhaus, A., & Scheu, S. (2003) Adding to "the enigma of soil animal diversity": fungal feeders and saprophagous soil invertebrates prefer similar food substrates. *European Journal of Soil Biology* **39**: 85–95.

Maraun, M., Schatz, H., & Scheu, S. (2007) Awesome or ordinary? Global diversity patterns of oribatid mites. *Ecography* **30**: 209–16.

Martin, A. (1991) Short- and long-term effects of the endogeic earthworm *Millsonia anomala* (Omodeo) (Megascolecidae, Oligochaeta) of tropical savannas, on soil organic matter. *Biology and Fertility of Soils* **11**: 234–8.

Menéndez, R., & Gutiérrez, D. (1999) Heterotrophic succession within dung-inhabiting beetle communities in northern Spain. *Acta Oecologia* **20**: 527–35.

Mikola, J., & Setälä, H. (1998) Productivity and trophic level biomasses in a microbial-based food web. *Oikos* **82**: 158–68.

Moore, J.C., Berlow, E.L., Coleman, D.C., *et al.* (2004) Detritus, trophic dynamics and biodiversity. *Ecology Letters* **7**: 584–600.

Neilson, R., & Boag, B. (2003) Feeding preferences of some earthworm species common to upland pastures in Scotland. *Pedobiologia* **47**: 1–8.

Niemelä, T., Renvall, P., & Penttilä, R. (1995) Interactions of fungi at late stages of wood decomposition. *Annales Botanici Fennici* **32**: 141–52.

Nieminen, J.K., & Setälä, H. (2001) Influence of carbon and nutrient additions on a decomposer food chain and the growth of pine seedlings in microcosms. *Applied Soil Ecology* **17**: 189–97.

Osler, G.H.R, Gauci, C.S., & Abbott, L.K. (2004) Limited evidence for short-term succession of microarthropods during early phases of surface litter decomposition. *Pedobiologia* **48**: 37–49.

Osono, T. (2006) Role of phyllosphere fungi of forest trees in the development of decomposer fungal communities and decomposition processes of leaf litter. *Canadian Journal of Microbiology* **52**: 701–16.

Osono, T., & Takeda H. (2001) Organic chemical and nutrient dynamics in decomposing beech leaf litter in relation to fungal ingrowth and succession during 3-year decomposition processes in a cool temperate deciduous forest in Japan. *Ecological Research* **16**: 649–70.

Paine, R.T. (1966) Food web complexity and species diversity. *American Naturalist* **100**: 65–74.

Paris, O.H., & Sikora, A. (1965) Radiotracer demonstration of isopod herbivory. *Ecology* **46**: 729–34.

Payne, J.A. (1965) A summer carrion study of the baby pig *Susa scrofa* Linnaeus. *Ecology* **46**: 592–602.

Peters, S., Koschinsky, S., Schwieger, F., & Tebbe, C.C. (2000) Succession of microbial communities during hot composting as detected by PCR-single strand-conformation polymorphism-based genetic profiles of small-subunit rRNA genes. *Applied and Environmental Microbiology* **66**: 930–6.

Pollierer, M.M., Langl, R., Körner, C., Maraun, M., & Scheu, S. (2007) The underestimated importance of belowground carbon input for forest soil animal food webs. *Ecology Letters* **10**: 729–36.

Pollierer, M.M., Langl, R., Scheu, S., & Maraun, M. (2009) Compartmentalization of the soil animal food web as indicated by dual analysis of stable isotope ratios ($^{15}N/^{14}N$ and $^{13}C/^{12}C$). *Soil Biology and Biochemistry* **41**: 1221–6.

Putnam, R.J. (1978) Patterns of carbon dioxide evolution from decaying carrion; decomposition of small mammal carrion in temperate systems 1. *Oikos* **31**: 47–57.

Renvall, P. (1995) Community structure and dynamics of wood-inhabiting Basidiomycetes on decomposing conifer trunks in northern Finland. *Karstenia* **35**: 1–51.

Ruess, L., Schmidt, I.K., Michelsen, A., & Jonasson, S. (2002) Responses of nematode species composition to factorial addition of carbon, fertiliser, bactericide and fungicide at two sub-arctic sites. *Nematology* **4**: 527–39.

Sabu, T.K., Vinod, K.V., & Vineesh, P.J. (2006) Guild structure, diversity and succession of dung beetles associated with Indian elephant dung in South Western Ghats forests. *Journal of Insect Science* **17**: 1–12.

Salmon, E.S., & Massee, G. (1901) Researches on coprophilous fungi. I. *Annals of Botany* **15**: 313–57.

Santos, P.F., & Whitford, W.G. (1981) The effects of microarthropods on litter decomposition in a Chihuahuan desert ecosystem. *Ecology* **62**: 654–63.

Scheu, S., & Schulz, E. (1996) Secondary succession, soil formation and development of a diverse community of oribatids and saprophagous soil macro-invertebrates. *Biodiversity and Conservation* **5**: 235–50.

Scheu, S., & Setälä, H. (2001) Multitrophic interactions in decomposer food-webs. In: T. Tscharntke and B.A. Hawkins (eds.) *Multitrophic Level Interactions*, pp. 223–64. Cambridge University Press, Cambridge.

Scheunemann, N., Scheu, S., & Butenschoen, O. (2010) Incorporation of decade old soil carbon into the soil animal food web of an arable system. *Applied Soil Ecology* **46**: 59–63.

Schneider, K., Migge, S., Norton, R.A., et al. (2004) Trophic niche differentiation in soil microarthropods (Oribatida, Acari): evidence from stable isotope ratios ($^{15}N/^{14}N$). *Soil Biology and Biochemistry* **36**: 1769–74.

Shimmel, S.M., & Darley, W.M. (1985) Productivity and density of soil algae in an agricultural system. *Ecology* **66**: 1439–47.

Stachowicz, J.J. (2001) Mutualism, facilitation, and the structure of ecological communities. *Bioscience* **51**: 235–46.

Swift, M.J., Heal, O.W., & Anderson, J.M. (1979). *Decomposition in Terrestrial Ecosystems*. University of California Press, Berkeley, CA.

Takeda, H. (1995) Changes in the collembolan community during the decomposition of needle litter in a coniferous forest. *Pedobiologia* **39**: 304–17.

Tedersoo, L., & Nara, K. (2010) General latitudinal gradient of biodiversity is reversed in ectomycorrhizal fungi. *New Phytologist* **185**: 351–4.

Tian, G., Brussard, L., & Kang, B.T. (1993) Biological effects of plant residues with contrasting chemical compositions under humid tropical conditions: effects on soil fauna. *Soil Biology and Biochemistry* **25**: 731–7.

Tiunov, A.V., & Scheu, S. (2005) Facilitative interactions rather than resource partitioning drive diversity-functioning relationships in laboratory fungal communities. *Ecology Letters* **8**: 618–25.

Torres, P.A., Abril, A.B., & Bucher, E.H. (2005) Microbial succession in litter decomposition in the semi-arid Chaco woodland. *Soil Biology and Biochemistry* **37**: 49–54.

Vreeken-Buijs, M.J., & Brussaard, L. (1996) Soil mesofauna dynamics, wheat residue decomposition and nitrogen mineralization in buried litterbags. *Biology and Fertility of Soils* **23**: 374–81.

Wall, D.H., & Moore, J.C. (1999) Interactions underground; soil biodiversity, mutualism, and ecosystem processes. *Bioscience* **49**: 109–17.

Wardle, D.A. (2002) *Communities and ecosystems, linking the aboveground and belowground components* pp. 205–9. Princeton University Press, Princeton, NJ.

Wardle, D.A., Yeates, G.W., Watson, R.N., & Nicholson, K.S. (1995) Development of the decomposer food-web, trophic relationships, and ecosystem properties during a three-year primary succession in sawdust. *Oikos* **73**: 155–66.

Watson, E.S., McClurkin, D.C., & Honeycutt, M.B. (1974) Fungal succession on loblolly pine and upland hardwood foliage and litter in north Mississippi. *Ecology* **55**: 1128–34.

Zimmer, M., Kautz, G., & Topp, W. (2005) Do woodlice and earthworms interact synergistically in leaf litter decomposition? *Functional Ecology* **19**: 7–16.

Zimmer, M., & Topp, W. (1997) Does leaf litter quality influence population parameters of the common woodlouse, *Porcellio scaber* (Crustacea: Isopoda)? *Biology and Fertility of Soils* **24**: 435–41.

CHAPTER 3.2

Patterns of Biodiversity at Fine and Small Spatial Scales

Matty P. Berg

3.2.1 The riddle of soil biodiversity

Life on earth is confined to very thin layers of soil and atmosphere. Wherever you take a handful of organically-rich soil, this small quantity of substrate contains billions of micro-organisms, kilometers of fungal hyphae, ten thousands of nematodes and other microfauna, and holds enchytraeids, mites, and springtails. The biodiversity in fertile soil is indeed mind-boggling (Bardgett 2005). It poses an interesting scientific problem to ecologists: how does this bewildering diversity reconcile with the classical niche exclusion principle (Kassen & Rainey 2004; Setälä et al. 2005)? This niche concept separates species based on their dissimilarity in habitat requirements, resource use, and environmental tolerances, and defines under what circumstances and where a particular species can exist (Hutchinson 1957). But how can a large collection of species that potentially competes for the same resources coexist on a small spatial scale? This question has occupied ecologists for many decades, and still does, and has been quoted as the "enigma of soil (animal) species diversity" (Anderson 1975).

Soil communities typically support a much greater biodiversity than do corresponding aboveground communities, while the factors that regulate their diversity are far less well understood (Wardle 2006). Part of this puzzle can be found in the nature of the soil environment. Soils as a habitat differ from other ecosystems. Soils are opaque, porous, and markedly heterogeneous in physical, chemical, and biological properties (Bardgett 2002; Setälä et al. 2005). Patterns in soil texture, water availability, pH, litter accumulation, nutrient availability, and plant cover change constantly across space and over time, due to factors such as disturbances, etc. The relationships between these soil properties result in a habitat that is both spatially and temporally very complex. The small-scale heterogeneity in resource availability and strong short-scale gradients in soil microclimate create high microhabitat diversity and unrivalled potential for resource utilization and niche partitioning, and hence species coexistence.

Although rich in species, it is intriguing to observe that diversity is not evenly distributed in soils. Not all species occur everywhere. High species richness is supported in small-scale patches of habitat that are often surrounded by areas of less or even infertile soil. Most species are, therefore, patchily distributed. These patterns of species diversity are the result of factors that drive species distribution and that operate at various spatial and temporal scales, from the global scale (biogeographical history and climate zones), to the landscape scale (soil pattern and topography), to the scale of the plot and the individual (soil structural complexity and resource chemical complexity) (Giller 1996; Ettema & Wardle 2002; Bardgett 2005; De Deyn & van der Putten 2005; Wardle 2006). The different spatiotemporal dimensions of these drivers make it difficult to explain diversity patterns at any scale, and ask for integration of knowledge obtained at different scales.

In this chapter the focus will be on spatial patterns of soil diversity at the small scale of the plot and at the fine scale of the individual. I first examine the spatial distribution of dominant and

Soil Ecology and Ecosystem Services. First Edition. Edited by Diana H. Wall *et al.*
© 2012 Oxford University Press. Published 2012 by Oxford University Press.

important soil organisms at these scales, and discuss the scale-relevant drivers that determine their patchy distributions. I then consider the spatial distribution of soil processes, such as nitrification and denitrification. Soil processes are critically dependent on the structure of soil communities and on functionally influential species. The patchy distribution of species will subsequently result in heterogeneity of soil functions. Finally, I will conclude with the observation that spatial scales are nested within each other and that the drivers that operate at the various scales interact and together determine species distribution. This observation calls for the integration of spatial scales to understand how heterogeneity and disturbances that operate at a hierarchy of scales affect species distributions and soil processes.

3.2.2 It is all a matter of scale

Soil organisms are patchily distributed at different spatial scales that range from the scale of continents and countries to a scale that can be as fine as a single sand grain (Fig. 3.2.1). Obviously, the determinants of diversity gradients over continents differ from the factors that drive the patchy distribution of organisms in a particular habitat or soil layer. For example, the distribution patterns of ground-beetles with a broad range in the western Palaearctic is strongly correlated with spatially structured variables related to current continental climate, such as ambient energy, precipitation, and actual evapotranspiration. Carabids, with a more restricted range to a single geographic region, are more strongly related to the geomorphology, especially to elevation (Schuldt & Assmann 2009). At the single country scale, spatial patterns in ground-beetle assemblages reflect the distribution of habitats, such as woodland, meadows, marshes, and heather fields (Scott & Anderson 2003). In contrast, species at the forest scale form aggregations due to the spatial distribution of microhabitat types, such as soil covered by shrubs, herbs, moss, or litter (Niemelä et al. 1992). Finally, at the finest scale, carabids differ in their vertical stratification in forest litter over a distance of only a few centimeters, which is explained by a variation in foraging and egg-laying behavior and larval survival (Loreau

1987). Similar scale-dependent effects on species distributions and important drivers have been observed for other groups, for example microbes (Horner-Devine et al. 2004; Green & Bohannan 2006; Nielsen et al. 2010), oribatid mites (Kallimanis et al. 2002; Nielsen et al. 2010), and macro-invertebrates (Dauber et al. 2005). These examples indicate that it is not easy to understand the spatial distribution of species because the various scale-dependent drivers operate at the same time.

Part of the difficulty to evaluate and explain species diversity and spatial distribution patterns in soil is to be found in the large range in body sizes and dispersal ability of taxa. Body size ranges from 1–2 µm in bacteria and fungi (body width <0.5 µm) to >10 cm in earthworms (body width >2 mm), a difference of more than five orders of magnitude (Swift et al. 1979). As a consequence, heterogeneity in resource availability on a scale of a few millimeters as perceived by microorganisms is not apparent for larger soil fauna. Species dissimilar in body size will, therefore, be aggregated on different spatial scales and differ markedly in dispersal ability. In small-sized taxa, passive dispersal by vectors, such as wind or larger animals, predominates over active dispersal. Conversely, in large-sized soil animals active dispersal prevails over passive dispersal. Dispersal ability determines to a great extent the reaction of soil organisms to shift in spatiotemporal dimensions of soil properties, for instance due to disturbances or the activity of larger soil vertebrates.

The huge range in spatial scales and the considerable variance in body size and dispersal ability of species call for a hierarchical approach to study spatial patterns of soil biodiversity. Determination of the proper spatial scale, the scale-dependent drivers, and interactions between subsequent scales is needed to help understand how ecological processes shape species distributions and community composition. In the following sections, I drill-down from the small scale of the plot to the fine scale of the organism and its behavior. At each scale, I discuss the spatial patterning, and the drivers that operate at these scales and drive the spatial distribution and patch sizes of soil organisms. I will not discuss larger spatial scales (i.e. the landscape and beyond). Although the importance of large-scale processes for local distribution patterns quickly

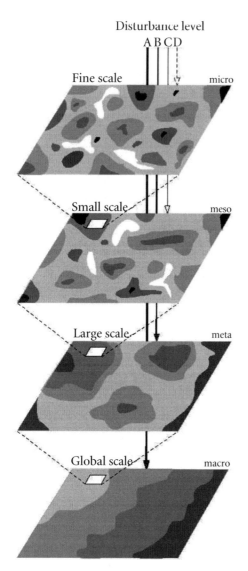

gains recognition, information for many groups is currently lacking. The focus, therefore, will be on the small and fine scale.

3.2.2.1 The small or mesoscale

A large part of the research on the spatial distribution of soil organisms has been concentrated on the small spatial scale, the scale of the plot, the soil community or the detrital food web (Fig. 3.2.1; Fig. 3.2.2 top panel). On this spatial scale, species share the same environment and interact with each other. The scale dimensions range from a few meters to over 100 meters. On the small spatial scale, horizontal heterogeneity in species distribution is very apparent and is often determined by the spatial distribution of soil properties.

3.2.2.1.1 Horizontal heterogeneity is the rule
Spatial scaling relationships can be used to invert the proper scale of horizontal heterogeneity in the distribution of organisms and soil properties. One important spatial scaling pattern is the distance–decay relationship, which describes how species abundance changes, or turns-over with geographic distance. When samples close to each other are more similar in abundance than samples taken further away (a case of heterogeneity), samples are so-called autocorrelated (Ettema & Wardle 2002). Using a geostatistical approach, various studies have shown that the range or scale of aggregation can vary from a few centimeters to several hundreds of

Figure 3.2.1 Spatial distribution of soil organisms occurs at nested spatial scales. At each spatial scale different environmental drivers shape the patterning in soil organism distributions. At the *fine scale or microscale* (sample or individual; mm to dm) soil aggregates, the root zone, and strong vertical stratification of soil climate and resources are the main drivers. At the *small scale or mesoscale* (plot or community; 1–100 m) individual plants, dung, and burrowing soil fauna are important factors. At the *large scale or metascale* (landscape; 100 m to 1 km) soil type, soil carbon, topography, and plant communities are major determinants. Finally, at the *global scale or macroscale* (country to continent; 1 km to >100 km) gradients in geomorphology, climate, energy input, and biomes are key factors. Similarly, a spatial hierarchy of life history traits of species occurs. From the fine to the large scale, the importance of functional trait shifts from traits that determine tolerance for stressful abiotic conditions (smaller scales), to traits that determine competitive ability and predator avoidance (intermediate scales), to traits that determine dispersal ability (larger scales). All patches will sooner or later be affected by some disturbance, which will create environmental heterogeneity. From A (a large, but infrequent disturbance, such as a hurricane) to D (a small, but frequent disturbance, such as a digging earthworm) the severity of the disturbance decreases, but the frequency increases from once a century (A) to monthly (D) (Bengtsson 2002). Hence, spatial scales interact with temporal scales. While large infrequent disturbances have the potential to affect all spatial scales, small frequent disturbances will only affect the smaller spatial scales. Disturbances also modify the feedbacks between spatial patterns of soil organisms and environmental heterogeneity. This interaction contributes to the enigma of soil biodiversity. (Hotspots of diversity or density: dark grey, cold spots: light grey or white). (After Ettema & Wardle 2002. Modified by permission of Elsevier Science Ltd.)

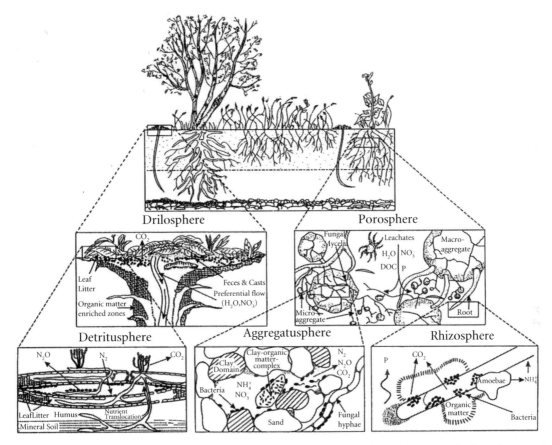

Figure 3.2.2 A hierarchical view of spheres or hotspots of activity, and drivers that determine the spatial patterning of soil organisms at small (three top panels) and fine (bottom panels) spatial scales. The top panel shows a cross section of the plot (small scale or mesoscale in Fig. 3.2.1), with mosaic of extensive root systems of trees and herbs. Most life occurs around roots and in the top few decimetres of the soil. Resources are transferred by plants to soils, which attracts, among others, borrowing macrofauna (*drilosphere*; local accumulation of organic matter in the form of litter, feces, and casts). Plants also create heterogeneity in soil physical properties (*porosphere*; patchiness in bulk density, water content, dissolved organic carbon and nutrient availability). At the fine scale or microscale, nested within the drilosphere and porosphere, patches of accumulated litter (*detritusphere*; well aerated areas of high quality organic matter, i.e. cellulose and aminoacids), root tips (*rhizosphere*; strong gradients in energy-rich products and a wide range of organic compounds), and soil aggregates (*aggregatusphere*; pores and voids surrounded by enclosed fine particulate organic matter bound with mineral particles) occur, that drive spatial patterns of soil biotic activity. (From Beare *et al.* 1995; drawing by Terry Moore. Reprinted by permission of Kluwer Academic Publishers.)

meters. For example, in sub-tropical forests ciliates have been shown to aggregate around 3–8 m (Acosta-Mercado & Lynn 2002), while others found no proof for spatial organization in protozoan communities in an upland grassland at scales above 10 cm (Griffiths 2002). Similarly, microbes in an organically managed alfalfa field showed a spatial heterogeneity around 28 m (Peigné 2009), while in a mixed spruce-birch stand spatial variability was observed at a scale of 4–8 m (Saetre 1999). Explanations for these apparent discrepancies in dimensions may be the different spatial scale of sampling, which is usually not smaller than one decimeter or meter, and vary considerably between studies.

The scale of aggregation also depends on species identity within taxa. For nematodes in a riparian wetland, patch sizes ranged from a minimum of 15 m for a *Chronogaster* sp. to a maximum of 66.5 m for an *Acrobeloides* sp. (Ettema *et al.* 1998, 2000; Fig. 3.2.3). Similarly, the spatial range of bacterivorous nematodes varied between 9.7–91 m in a corn-soybean rotation field, depending on the species (Liang et al. 2005). It is of interest to note that, in this field, dissimilar spatial distribution patterns were found between two adjacent soil layers. The spatial scale of heterogeneity for a *Cephalobus* sp. was larger at 0–10 cm than at 10–20 cm soil depth, a patch size of 13 m and 9.7 m, respectively. However, this soil layer effect was not observed for all species. Neotropical earthworms in a natural savannah and a grazed grass-legume pasture were aggregated at a scale of 20–40 m, although a limited number of species displayed only a very weak spatial pattern (Jiménez et al. 2006). Interestingly, in both ecosystems, two medium-sized species with similar patch sizes were spatially segregated. This alternation of patches was hypothesized to be the result of a high degree of niche overlap between the two earthworm species, a high potential for resource competition, followed by competitive exclusion. Moreover, earthworms in the pasture were spatially distributed at two nested spatial scales. This suggests that the earthworm distribution was influenced by two different drivers that operate at separate spatial scales.

Spatial scaling relationships can also be used to understand the spatial organization of community structure. For example, whole-bacterial community DNA in a rather homogenous wheat field was extracted from samples collected at a variety of separation distances ranging from 2.5 cm to 11 m (Franklin & Mills 2003). A remarkable degree of spatial autocorrelation was detected at multiple scales, ranging from 30 cm to more than 6 m. The presence of nested scales of variability in community composition is thought to be due to the variable response of different community subsets to the patchiness in soil properties. These properties are influenced by agricultural practices and management with fixed spatial intervals, such as associated with crop rows and aisles and soil compaction due to wheel traffic.

These are just a few of many examples indicating that the field observed differences in the spatial scale of aggregation depend on the organism (taxon, life stage, body size, mobility, competitive ability), the ecosystem (homogenous grassland, fields with evenly spaced crops, forests with unevenly spaced

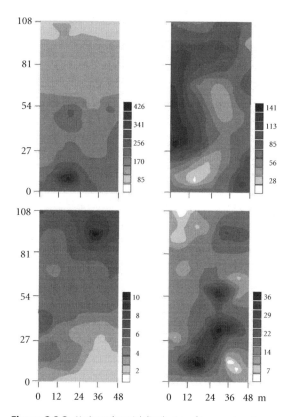

Figure 3.2.3 Horizontal spatial distribution of two species of bacterivorous nematodes and two environmental factors. The top panels show interpolation maps for *Acrobeloides* sp. (left; spatial range 15 m) and *Prismatolaimus* sp. (right: spatial range 52.2 m) across a wetland sampling area of 48 m by 108 m. The gray-scale units are 1000 ind. m^{-2}. Note that the spatial distribution of both bacterivorous nematodes do not overlap. The bottom panels show interpolation maps for microbial respiration (left; spatial range 21.3 m) and soil moisture (right: spatial range 15.3 m). The grey-scale units are g CO_2-C m^{-2} d^{-1} (microbial respiration), and mass/mass ratio (moisture, as percentage). *Chronogaster* is positively correlated with soil moisture (r = 0.40), while *Prismatolaimus* is negatively correlated with soil moisture (r = –0.57). No correlation is found between both species and microbial respiration. (For details see Ettema et al. 1998. Redrawn by permission of the Ecological Society of America.)

logs), the soil horizon (soil depth), and not unimportantly, the spatial scale of sampling. All these factors interact and contribute to the context-dependency of spatial heterogeneity in species distributions among sites. There is one intriguing study that mentions a lack of horizontal spatial patterning for all food web components in a coniferous forest plantation, from small-sized microbes to large-sized spiders (Berg & Bengtsson 2007). This unexpected result was ascribed to the absence of factors that may create heterogeneity. Large soil diggers, such as earthworms, were absent and there was no understory vegetation and woody debris. Moreover, trees were regularly spaced and a constant sample-tree trunk distance was maintained. The distance between samples, at least 2.5 m, might have been too coarse to detect heterogeneity; however, this study indirectly points to important drivers of horizontal spatial heterogeneity, which is discussed next.

3.2.2.1.2 Drivers of horizontal distribution
Soil communities are predominantly regulated by bottom-up forces, driven by increased energy and nutrient transfer via plant inputs to soil (Fig. 3.2.2). As plants differ in the amount of root exudation and litter production, their spatial distribution directly affects the patchiness of organisms, especially at the base of soil food webs. Heterogeneity in soil properties related to productivity can, therefore, drive the patchy distribution of soil organisms (Swift et al. 1979; Bardgett 2005; Berg & Bengtsson 2007). In addition, biotic interactions also have an impact on soil biodiversity and species distribution (Wardle 2006). Aboveground, herbivores can alter the structure of the vegetation, the distribution of plants, and their grazing activity affects the input of energy and nutrients into the soil. Belowground, digging macro-fauna modify the soil matrix and redistribute the soil properties that are related to soil fertility. Below, I explain the significance of plants, herbivores, and tunneling soil fauna for soil heterogeneity and species distributions, from highly patterned and sparsely vegetated soil in harsh environments to ecosystems with a more or less continuous vegetation cover under less extreme environmental conditions.

Dryland ecosystems with regular and severe cold and drought spells, such as arctic sites at higher latitudes, deserts or exposed parts of oceanic islands, are highly heterogeneous, over a scale of several meters up to more than 100 m. Life here is often restricted to vegetated patches, surrounded by more or less inhospitable dry rock or soil. For example, in the Antarctic, primary production is limited to algae and mosses that occur near ice-covered lakes and small transient streams. These discrete patches harbor a more diverse assemblage of rotifers, nematodes, and micro-arthropods than the surrounding dry soils (Simmons et al. 2009). Similarly, tussocks in arid ecosystems can be considered as biodiversity hotspots. In a dryland plant community dominated by the slow growing shrub *Coleogyne ramosissima* and the grass *Stipa hymenoides* that were growing at least 70 cm apart, microbes, protozoans, rotifers, tardigrades, and nematodes were consistently more abundant under plants than in interspaced soil (Housman et al. 2007). The spatial arrangement of tussocks influenced the distribution of soil organisms by modifying soil properties in their vicinity, not only fertility but also hydrology and temperature. In arid ecosystems, the presence of plants, and subsequently the coverage of soil by litter, also acts as a physical barrier. It reduces soil bulk density and temperature fluctuations, and prevents overland flow and erosion, causing higher soil water content under vegetation during dry periods (Sayer 2006). To a certain extent, this situation of vegetated soil separated by bare interspaced soil is equivalent to agricultural sites with rows of sown or planted crops (see for an example, Franklin & Mills (2003)).

Under less restrictive climate regimes and in ecosystems with a continuous vegetation cover, spatial patterning of soil communities is also plant driven. For example, more light and precipitation reaches the soil through forest canopy gaps. Here, different understory vegetations and soil environments are found with higher temperatures and stronger fluctuations in microclimate than in areas under a closed canopy. In a mixed beech forest, litter incubated under small canopy gaps showed a higher density of bacteria, fungi, protozoa, and nematodes in comparison to areas under a closed canopy. Interestingly, the effect of canopy openings on soil organism densities even overruled the effect of litter quality (Bjørnlund & Christensen 2005). Trees in forests with a closed canopy also affect soil organisms. The spatial patterning of microbes in a mixed spruce-birch stand, for instance, could be explained by the spatial arrangement of single trees (Saetre 1999). Spruce trees negatively affected photosynthetic active radiation, soil water content, the cover of mosses and associated herbs, and microbial biomass, but not microbial activity. Birch trees, in turn, stimulated both microbial biomass and activity.

Microbial biomass aggregated on a scale of 7–8 m, roughly the distance between spruce and birch trees, while microbial activity aggregated on a scale of 4–5 m, corresponding with the patch size of a single birch tree. Contrasts in the quality of litter and root exudates under coexisting tree species, in combination with dissimilar effects on soil climate create zones of influence that can explain the patchy distribution of many soil organisms in forests. Even in monocultures with equidistantly spaced trees and constant litter quality, the spatial variability in surface litter distribution due to interactions between the location of litter fall, gravity, wind, (micro)topography, and the distribution of efficient litter transformers can potentially affect the distribution of soil organisms.

To the human eye, homogeneous vegetations (e.g. meadows) are small-scale mosaics in plant community composition at close range. The contribution of these mosaics to the spatial distribution and diversity of soil organisms is explained by the often strong plant identity effects on soil organisms (Wardle 2006; van der Putten et al. 2009). Obviously, if soil organisms are tightly connected to the roots of a particular plant species (e.g. some mycorrhizal fungi, specialized root feeding nematodes, click-beetle, and snout-beetle larvae) their spatial distribution will closely follow and overlap that of their host. But, also if the relationship between soil organisms and plants is not as tight, plant attributes like productivity, litter quality and palatability, and rhizosphere effects will influence soil organisms. Although animals feeding on plant litter and microbes are seldom obligatory, they do show some sort of selectivity in their choice of litter type and fungal species. Therefore, the spatial configuration of different plant species may be a determining factor in the distribution of soil organisms. For example, within a relatively homogeneous *Saxifraga*-lichen heath in the Arctic, six dominant plant species were widely distributed, on a scale of 20 m^2 (Coulson et al. 2003). Distinct soil fauna communities (i.e. Collembola and oribatid mites) were identified between each plant species. As the six plant species differed in their spatial arrangement, so did the micro-arthropods that contribute to the community differentiation between these plants. The observed spatial distribution of micro-arthropods did not appear to be related to simple soil physical parameters, such as temperature, moisture, or soil depth. It was most likely explained by complex resource interactions between litter quality or quantity and associated microbes.

However, a wealth of studies do show that plant-induced heterogeneity in soil physical properties shape soil organism distributions. Plants then act as soil engineers. Soil engineers have the potential to modify soil structure, pore size, porosity and bulk density, water content, and also mix organic matter in the case of faunal engineers (Lavelle 2002). All these soil properties are important drivers of the spatial distribution of soil organisms (Fig. 3.2.3). Interestingly, the spatial arrangement of plants may also affect the spatial distribution of faunal engineers (i.e. earthworms, ants, and termites). In an Amazonian pasture, at a scale of <10 m^2, vegetation configuration explained 70% of the variation in the density and species richness of macrofauna engineers (Mathieu et al. 2009). The density of earthworms, ants, and termites decreased with increasing distances to the nearest grass tuft. Under tufts a lower soil temperature was measured, which is an important limiting factor in the local distribution of tropical macrofauna. In turn, soil fauna engineers modify the soil environment through their drilling and tunneling behavior and facilitate living conditions for other soil organisms (Lavelle 2002; Bronick & Lal 2005). They produce large amounts of solid organomineral structures, such as casts, pellets and walls of tunnels that last long and act as microhabitat or resource for smaller organisms. As a result, the patchy distribution of microbes and rotifers, nematodes, and other microfauna is nested within the sphere influenced by macrofauna (Beare et al. 1995; Pokarzhevskii et al. 2003). This functional domain in soil is referred to as the drilosphere (Fig. 3.2.2). Networks of earthworm tunnels and termite galleries and their direct sphere of influence have significant effects on spatial heterogeneity in organic matter turnover, porosity, bulk density, and water infiltration rate.

Given the impact of plants on the spatial arrangements of soil properties, one could hypothesize that aboveground herbivore–plant interactions may drive belowground patterns. Preferential grazing, but also trampling and dung deposition by large herbivores (vertebrates and invertebrates) are considered important factors driving plant species composition, biomass production, and vegetation dynamics in pastures. These herbivore-mediated effects alter the quantity and quality of resources, i.e. root exudates and litter, entering the soil through plants and creating small-scale heterogeneity in microclimate conditions (Bardgett & Wardle 2003). Large herbivores also exert direct effects on soil communities, for instance by trampling, which causes soil compaction, pore volume reduction, bulk density increment and water infiltration rate reduction. Trampling not only counteracts the drilosphere effects of soil engineers, but often strongly reduces the population densities of soil engineers and other soil organisms. In contrast, herbivores return large quantities of urine and dung to the soil. These waste products are highly spatially deposited and supply nutrients to the vegetation and offer undigested resources to many soil organisms. It is not easy to assess the relative contribution of these positive and negative herbivore activities to the patterning of plants and soil organisms. For example, large herbivores do not graze around dung. The enhancement in grazing intensity the further away from the dung pats causes a gradient in canopy height and trampling effects. Cattle dung pats have a small size, around 10 cm without any vegetation, while the patches surrounding the pats are larger, around 50 cm with a large canopy height (Gillet et al. 2010). Patches with a large canopy height also occur in areas without pats, where urine has been deposited and grazing pressure is low. These patches have a larger area, a dissimilar plant species composition, and a different spatial scaling. Therefore, the impact of herbivores on soil properties, either directly or indirectly via plants, occurs at various spatial scales. Interestingly, dung pats and other discrete resources, such as logs, large fungal fruiting bodies, fallen fruit, and carcasses all have a very specific community composition. These communities are composed of species that are highly dependent and adapted to a life in these discrete and often ephemeral habitats. Dung beetles, dung flies and other dung specialists in general have a good dispersal ability and are spatially structured at a the scale at which dung is deposited. The decay in community similarity with distance between discrete ephemeral habitats is, therefore, less steep than for non-specialist communities.

In summary, the evidence for the primacy of plants as structuring agents for soil organisms is strongest in extreme environments. However, in more homogeneous vegetation mosaics, plants also play a key role via plant-specific effects on soil properties, inputs that supply soil organisms with resources or due to their influence on the distribution of soil fauna engineers. It is important to note that the effects of specific plant species potentially can have long-lasting legacy effects belowground (van der Putten et al. 2009). These legacy effects of previous presence of particular plant species may remain for several months, such as for N-fixing legumes, to decades or centuries for trees. Similar legacy effects are contributed by soil fauna engineers. They create soil structures, such as the casts of earthworms that can last for months and mounds of termites that can last for centuries.

3.2.2.2 The fine or microscale

The fine spatial scale is defined as the scale of the sample, or the individual and its behavior (Fig. 3.2.1 top panel; Fig. 3.2.2 bottom row panels). Individuals share the same microhabitat and interact with each other on this scale. This scale ranges from a few millimeters or smaller for bacteria, to a few decimeters for micro- and mesofauna depending on the organism's body size. On this scale, vertical heterogeneity in species distribution with soil depth is often larger than horizontal heterogeneity within a soil horizon.

3.2.2.2.1 Horizontal heterogeneity

At the fine scale soil organisms generally have a patchy distribution as well. Geostatistical methods show clusters of bacteria in the top soil of arable fields with spatial ranges that vary between 250 μm–1600 μm. Below the top soil bacteria aggregate at spatial scales that vary between 0–990 μm. Sampling at an equivalent scale of resolution revealed that ammonium oxidizers occurred at randomly distributed patches of <180 μm, whereas nitrite oxidizers had patches of 250 μm. Patches with microbes were separated by 50–375 μm, depending on the group of oxidizer (Dechesne et al.

2007). The microscale distribution of ectomycorrhizae was also species-dependent in a red oak forest. Mycorrhizae of individual morphotypes showed a clear spatial patterning with the exception of one dominant species that was evenly distributed over soil monoliths. Approximately 70% of the morphotypes showed a scattered distribution pattern with clusters of 2–8 cm, and about 22% of the morphotypes showed clearly clustered distribution patterns, with clusters of 8–40 cm. Some morphotypes strongly overlapped in their distribution, while others were spatially separated (Gebhardt et al. 2009). Groups of Protozoa too were heterogeneously dispersed in a peat moss carpet. The biomass of testate amoebae was quite uniformly spread over a microtopographic surface, but species distribution maps suggested a spatial heterogeneity that varied with species. Spatial analyses revealed aggregations of testate amoebae of <10 cm, which were separated from each other by approximately 20 cm of *Sphagnum* tissue (Mitchell et al. 2000). These examples, among others, indicate that for surface area of a few decimeters the distribution of soil organisms is patchy. Likewise for the small spatial scale, variation in the spatial distribution observed in the field depends on the organism and its attributes (body size and width, mobility, the ability to release aggregation pheromones), the habitat (soil aggregate, pore size, root), the soil horizon (top soil, sub soil), and the spatial scale of sampling. What drives this patchiness in soil organism distribution?

3.2.2.2.2 Drivers of horizontal heterogeneity
An important driver that determines patchiness on the fine scale is soil structure, especially the spatial patterning and strength of soil aggregates. Soil aggregates are secondary particles formed through the combination of mineral particles with organic and inorganic substances. Soil organic matter acts as a binding agent and a nucleus in the formation of aggregates. Soil biota and their organic products, such as root exudates and excrements of soil fauna contribute to the development of soil structure and control the fine particle organic matter dynamics (Bronick & Lal 2005). The size, shape, and arrangement of aggregates determine the distribution of soil properties. The area of influence defined by the accumulation of soil aggregates and subsequently, the formation of habitable pore and void space, is often referred to as the aggregatusphere, while accumulations of macroaggregates dominated by faunal increments and fine particulate organic matter, define the detritusphere (Fig. 3.2.2). Studies that have adopted a scale of sampling that is fine enough to reveal the influence of soil aggregates, have observed a significant patterning on the millimeter scale. For example, in upland grasslands the variability in bacterial density at spatial scales of <1 mm could be linked to the occurrence of excrements of earthworms and enchytraeids (Bruneau et al. 2005). At closer inspection, soil thin slides revealed that bacteria cluster at a scale of 5–10 µm, the size of microaggregates. Here, bacteria are associated with polysaccharides and highly labile soil organic matter, and protected in micropores against predation by Protozoa and rotifers (Young & Ritz 2000; Dechesne et al. 2007). Fungal activity dominates in the less stable macroaggregates (>250 µm), where the fungi are associated with fresh organic matter. Factors that affect the relative distribution and spatial arrangement of micro- and macroaggregates, such as burrowing animals and roots, may explain differences in the spatial arrangement of bacteria and fungi. In this respect, as microaggregates are combined to form macroaggregates, the spatial arrangements of bacteria can be nested within the spatial scale of the fungi.

Roots not only influence the formation of soil particles. Their impact goes much further. They also directly affect the distribution of soil organisms in the surrounding of their roots, the rhizosphere (Fig. 3.2.2). Microbial growth in the rhizosphere is stimulated by the deposition of energy-rich products and a wide range of organic compounds. Rhizodeposition results in a 5–10 fold increase in microbial biomass compared to the bulk soil. The effects of root depositions, on top of the physical gradient in moisture and pH, are very local. The rhizosphere extends outwards for only a few tens of micrometers for most plants. Growth limiting substrates have the highest concentration on the root surface, the rhizoplane, which has the highest densities of microbes. The concentration differential between the root and the bulk soil results in fine scale diffusion gradients that shape the spatial distribution of microorganisms (Bazin et al. 1990; Fig. 3.2.4). The density of microbes, protozoa, and nematodes were highest at the root surface and between 0–1.8 mm from the root in an agricultural sandy loam planted with barley. Soil more than 1.8 mm from the root was hardly affected by these resources (Rønn et al. 1996). Besides this radial extension of the rhizosphere, roots can also be longitudinally characterized. Along the rhizoplane a striking gradient exists over the first 20 mm in the production of root depositions. At the tip (1–2 mm)

mainly mucilages are produced in large quantities. Here the microorganism densities are low. The root elongation zone occurs behind the root cap, and is rich in exudates and secretions. Here high densities of microorganisms occur. Some 10 cm behind the tip the first root hairs emerge. This is the zone where mycorrhizal fungi colonize the root. Above this zone organic matter mainly consists of lysates and sloughed cells from the root epidermis and cortex. This zone is also rich in microbes that can decompose more complex organic matter. This radial and longitudinal extend of the rhizosphere, together with spatial differences in root activity, creates a significant heterogeneity in organic matter quality and quantity, which strongly influence the fine-scale distribution of soil organisms.

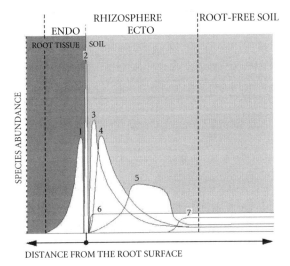

Figure 3.2.4 Micro-organisms may have different fine scale distributions in the rhizosphere, depending on their resource requirements and sensitivity for inhibitory products of other microorganisms. The area between the dashed lines, only a few millimeters wide, delineates the rhizopshere, and is divided into an area within the root (endo-rhizopshere) and outside the root (ecto-rhizopshere). The curves depict the densities of different species (A to G), or species groups. Species **1** occurs in the endo-rhizopshere, in intracellular space of root tissue. All other species live in the ecto-rhizopshere. Species **2** occurs only in the rhizoplane. Species **3** is physically excluded from the rhizoplane by species 2, but can utilize metabolites from species 2 and organic material from the root. Species **4** is physically excluded from the rhizoplane by species 2 and biochemically excluded by species 3, but can exploit organic material from the roots not used by species 2–3, as well as their metabolites. Species **5** utilized secondary products originating from the endo- and ecto-rhizopshere. Species **6** is physically excluded from the rhizoplane, but is not sensitive to the inhibitory metabolites of species 1–7. Species **7** is sensitive to the inhibitory metabolites of species 1–6, and is biochemically excluded from the rhizosphere. Note that the densities in the root-free soil are significantly lower than in the rhizosphere. (For details see Bazin et al. 1990. Redrawn by permission of John Wiley & Sons.)

3.2.2.2.3 Vertical heterogeneity rules

Roots not only occur in a horizontal plane, but also extend deeper into the soil. Therefore, the longitudinal characterization of the rhizosphere also has a vertical component. With depth, species composition along the rhizoplane will change, leading to vertical stratification of species and heterogeneous distribution patterns. Although often ignored, one of the most striking patterns in the spatial distribution of soil flora and fauna that can be observed is vertical with soil depth. Numerous studies have reported a significant vertical stratification of soil organisms, over a scale of sometimes not more than a few centimeters. For example, densities of bacterial cells differed significantly between horizons over a depth of 5 cm in upland grassland, which was linked to the occurrence of excrements of soil fauna (Bruneau et al. 2005). Morphotypes of ectomycorrhizal fungi showed a clear preference for a specific organic sublayer within the upper 3 cm of the organic horizon of a red oak forest (Gebhardt et al. 2009). Species of bacterivorous nematodes had a strong preference for a specific soil stratum in the upper 20 cm of an upland field (Liang et al. 2005). Species of micro-arthropods (Berg et al. 1998), and Carabid beetles (Loreau 1987) were clearly vertically stratified along the organic horizon of forests, over a range of 6 cm and 10 cm, respectively. Finally, macrofauna detritivores and predators had a distinct preference for the first 10 cm of the upper 30 cm of an Amazonian pasture (Mathieu et al.

2009). These studies, among others, show that groups of soil organisms, as well as many species, exhibit a distinct vertical stratification, leading to layer-specific species compositions. In one of the few studies that directly compared variability across space and time, vertical heterogeneity in food web composition at different levels of taxonomic resolution overruled horizontal and even temporal variability (Berg & Bengtsson 2007). Patterns in spatial distribution of soil organisms in this study of a pine plantation were stronger down the organic soil layers, over a range of only 5–6 cm, than they were across horizontal space, up to 50 m, or over time, over a period of almost 3 years.

This vertical stratification is extended to the aboveground compartment of ecosystems. For example, in an arctic dryland with a thin crust of moss and algae, various groups of microfauna and micro-arthropods were more abundant in the aboveground material compared to the soil beneath (Simmons et al. 2009). More distant from the soil surface are suspended soils, i.e. leaves, seeds, flower heads, and other organic debris, moss and lichens that accumulate in tree forks in the forest canopy. In these spatially discrete soil islands in the forest canopy the composition of oribatid mite communities were significantly different from forest floor assemblages (Lindo & Winchester 2008). Mite communities of forest soil and canopy suspended soil had only 40% similarity in species composition, which was independent of season and year of sampling.

3.2.2.2.4 Drivers of vertical heterogeneity
The main drivers of the vertical stratification of soil organisms are soil physical properties, microclimate, and especially organic matter quality. In highly stratified soils, such as forests on acid sand, a short-scale gradient exists in soil organic matter quality. Litter that enters the soil is decomposed by a large array of soil organisms. The resultant of the activity of the decomposer community is an increase in litter complexity and a reduction in litter quality with soil depth (Hansen & Coleman 1998). Fresh litter is degraded to smaller fragments mixed with excrements by the feeding activity of detritivores, but is also reduced in quality. Fresh litter contains high amounts of easily accessible substrates, such as sugars and cellulose, while older and partly decomposed organic matter consists mainly of recalcitrant material due to decomposition by microorganisms. Each year a new layer of litter tops-up the older organic matter, and in the absence of bioturbation by earthworms a strong gradient in litter quality, fragment size, and pore volume exists that has a strong impact on the vertical distribution of soil organisms. Indeed, assemblages of soil micro-arthropods in a pine stand showed a significant vertical microstratification, with a species-specific composition for each organic layer (Berg et al. 1998). Furthermore, it was shown that variability in spatial distribution, both horizontally and vertically, of food web constituents was significantly correlated with the variability in organic matter decomposability, a proxy for litter quality (Berg & Bengtsson 2007). Although organic matter properties drive the patchy distributions of soil biota, gradients in microclimatic conditions in the organic horizon will also play a role (Krab et al. 2010). Gradual changes in litter quality and microclimate down the organic profile occur in concert. Superficial living species are exposed to fluctuating temperatures and relative humidity and, in more open ecosystems, to greater ultraviolet radiation (Verhoef et al. 2000). They must be able to tolerate large oscillations in microclimatic conditions. Deep-living species are more buffered against strong variation in microclimatic conditions that occur at the surface. The increase in dissimilarity in tolerance levels down the soil profile adds to the maintenance of a strong vertical microdistribution of many soil fauna species.

In soils that are bioturbated by soil fauna engineers or in agricultural fields that are tilled, organic matter is mixed through the soil. These activities disturb the fine-scale stratification of organic matter and homogenize gradients in organic matter quantity and quality (reviewed in Young & Ritz 2000). In these less stratified soils the spatial scale of gradients in organic matter availability, soil properties, and microclimate will be much larger than in stratified soils due to the absence of a thick litter layer. Nevertheless, soil properties, including texture, pH, and availability of water and nutrients vary with soil depth, and a gradient in microclimatic conditions exists. For instance, microbial communities varied greatly along a soil profile on limestone, reflecting gradients in carbon availability, pH, and water content (Hansel *et al.* 2008). Some bacteria were largely unique to each of the distinguished horizons. It is interesting to observe that a vertical gradient in soil properties is frequently followed by a distinct vertical microstratification of organisms, often with species-specific community compositions in adjacent soil layers. Vertical heterogeneity, therefore, contributes to the spatial complexity in soils, and highlights the existence of high species diversity. In the next section I will discuss the consequences of vertical and horizontal spatial patchiness of soil organisms for the functioning of the soil.

3.2.3 Spatial distribution of soil functions

Resource availability, soil properties, and microclimatic conditions all determine the composition and distribution of soil biota. In turn, the structure of soil communities is known to affect the location and rate of biochemical transformations in soils. Soil processes, such as mineralization, ammonification, nitrification, denitrification, and nitrogen fixation are all regulated by soil organisms. These soil functions underlay the important ecosystem services nutrient cycling, soil fertility and productivity. As soil processes are depending on the activity of soil biota, it is to be expected that the spatial patterning of soil communities, governed by the many abiotic and biotic drivers, leads to patchiness in soil functions and, subsequently, ecosystem services. Indeed, for many soil functions spatial heterogeneity in soil process rates have been reported, especially for processes that rely on microorganisms, for example denitrification. Denitrification is the conversion of nitrate to gaseous N by heterotrophic bacteria. It is a process that can only occur at anaerobic microsites, for instance at the centre of soil aggregates, in earthworm casts or in water films surrounding soil particles. At a fine scale, aerobic and anaerobic microsites (0.5–3 mm) exist in close proximity to each other and anaerobic patches are separated by only a few millimeters, resulting in very different denitrification rates between neighboring microsites. The controlling factors for the spatial arrangement of anaerobic patches are the aggregate, particle size, and the substrate composition distributions

(Young & Ritz 2000). The nitrogen (N) gases produced by heterotrophs diffuse through soil pores, voids, and small cracks to the atmosphere. Diffusion may result in patterns of gaseous N fluxes that occur at larger spatial scales than the scale of the microsites where the producers of these gases are active. This scaling difference between producers and soil processes has, for instance, been shown for nitrifying bacteria and nitrate leaching. In this example, a high horizontal spatial variation in both nitrate and ammonium concentrations was observed in leachates from intact soil cores taken from a Scots pine forest soil, at a scale of a few meters to <1 m (Laverman *et al.* 2002). The presence of hotspots of nitrate leaching was explained by low levels of denitrification at these microsites. Subsequently, these intact soil cores were incubated to quantify nitrate production by nitrifying bacteria. Although no significant differences were observed between cores with a high nitrate leaching before and after incubation, the spatial scale at which nitrate was produced by the bacteria was much finer than the scale at which nitrate was leached (Fig. 3.2.5). However, contrary to the processes of nitrification and denitrification, most soil functions do not solely depend on the presence of a few particular species, but on an array of species that often belong to different taxa. The sum of the activities of multiple species with different reactions to environmental variability and disturbances will result in patterning of soil processes on a small rather than a fine spatial scale.

Notwithstanding the often conspicuous vertical stratification of decomposer organisms, not much is known on heterogeneity in soil functions down the soil profile, especially on a fine scale. It is assumed, and partly shown, that in a stratified habitat the impact of soil fauna on soil processes is through their microstratification. For example, in temperate forest litter micro-arthropods interact with fungi and the net resultant for nutrient fluxes depends on the organic horizon where the interaction occurs. In the top layer consisting of freshly fallen litter, grazing of soil fauna on fungi stimulates the growth of fungi, which results in immobilization of nutrients in the microbial biomass (Faber 1991). Just under the top organic layer the feeding activity of soil fauna will increase net mineralization and nutrient mobilization rates. Here nutrients are released from

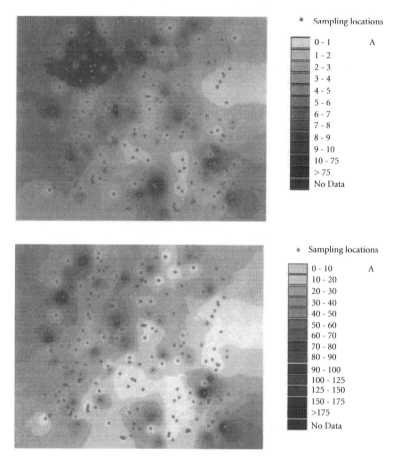

Figure 3.2.5 Horizontal spatial distribution of nitrate leached from intact soil column (top panel) and nitrate produced after incubation of the same intact soil columns for 3 weeks at 18°C (bottom panel) across a 50-year-old coniferous forest stand sampling area of 20 by 20 m. The gray-scale units are µg N g^{-1} dry soil. Note that although the values of leached nitrate from the field and after incubation are significantly correlated, the spatial range of nitrate production is smaller than that of nitrate leaching. (For details see Laverman *et al.* 2002. Reprinted by permission of Elsevier Science Ltd.)

the microbial biomass and may affect plant growth by nutrient uptake by the roots or interference with mycorrhizal colonization of the roots. A study in an acid forest soil indeed revealed a strong fine-scale distribution of soil functions (Clays-Josserand *et al.* 1988). In 1-cm thick slices of the organic layer, significant differences in nitrification potential, hence nutrient leaching, were observed, often with the highest potential in the layer with fragmented litter under the top layer. Similarly, in stratified forest soils a strong decrease in decomposition rate was observed between adjacent versus non-adjacent organic layers, which was attributed to the spatial patterning of the decomposer community (Berg & Bengtsson 2007). Based on these studies it is anticipated that a more fine-scale resolution of sampling will reveal that many soil functions and ecosystem services will have a significant vertical patterning.

These results, among others, indicate that the spatial patterning of soil organisms and the functions they perform largely overlap, but that the scale of patchiness of biota and soil processes can differ. Moreover, not all soil organisms do function at similar spatial scales, and soil processes that rely on an array of organisms, such as decomposition, probably will have a different spatial patterning compared to soil processes that depend on a limited set of organisms, such as nitrification and denitrification. Unfortunately, not many studies have simultaneously measured patchiness in soil processes and the distribution of the organisms that regulate these functions. Additionally, in many studies the focus has been on quantifying variability in fluxes on scales relevant to management or global cycles and not on scales necessary to understand the mechanisms that underlay variability in fluxes, which requires a finer scale of study.

3.2.4 Spatial scales are nested

In many studies spatial variability in distribution of soil organisms is usually viewed as random noise. Contrary to this general view, I have shown that species are not randomly distributed but often exhibit a relatively predictable spatial structure, at scales that range from a few mm to >100m. Spatial explicit studies have provided insights into the factors that drive and explain the observed spatial distribution of populations and activity of soil organisms and mediate soil functions, at a hierarchy of spatial scales. At each spatial scale abiotic and biotic factors have been identified that explain aggregated patterns in species distribution. At the small-scale plant growth is primarily structuring the distribution of soil organisms, while at the fine-scale soil structure and physicochemical properties explain patterns in species distribution.

Environmental controls on the spatial patterning of soil biota operate at various spatial scales due to the substantial variance in body size of soil organisms (Ettema & Wardle 2002; Fig. 3.2.1). Heterogeneity in soil properties at the finest spatial scale as perceived by small-bodied soil organism is as such not apparent for large-bodied soil fauna. Soil with fine-scale heterogeneity in soil texture is for an earthworm rather homogeneous. They in turn react to environmental heterogeneity that occurs at a larger spatial scale, for instance in vegetation composition. Conversely, heterogeneity at the plot scale as perceived by larger-bodied organisms is as such not apparent for small-bodied organisms. A patchy vegetation composition at the plot scale is not noticeable by enchytraeids with a limited home-range and dispersal ability. Furthermore, the activity of soil organisms at one hierarchical level can influence spatial heterogeneity in resources and biotic interactions at other hierarchical levels (Beare et al. 1995; Wardle 2006). Large-bodied soil engineers create habitat for small-bodied organisms, can locally modify the physical and chemical properties of the soil habitat, and promote the dispersal of smaller organisms. These biologically relevant spheres of influence operate at different spatial scales (Beare et al. 1995; Fig. 3.2.2), and result in a spatial distribution of small-bodied organisms that is nested within the larger-scale patchy distribution of large-bodied species.

Notwithstanding the evidence I have presented on the importance of fine- and small-scale environmental heterogeneity for species distribution and soil functions, data is accumulating that heterogeneity at the large- and even global-scale also can affect soil organism distributions. Most microorganisms and soil fauna can disperse and are not restricted in their occurrence to a single soil sample or community but can be found over a country, continent or beyond (Wolters 2001; Ettema & Wardle 2002; Bardgett 2005; De Deyn & van der Putten 2005). Large-scale gradients in soil type, soil carbon, and topography, and changes in plant communities along chronosequences all affect the distribution of soil organisms across landscapes (Fierer et al. 2009; Nielsen et al. 2010). Even on a global-scale, latitudinal patterns are observed in the abundance and diversity of species, such as the gradual decrease in ant diversity from the tropics to the poles in the northern and southern hemisphere (Kusnezov 1957) or global patterns in microbial and faunal communities which are explained by broad-scale patterns in soil pH and soil C:N ratios (Fierer et al. 2009; Chu et al. 2010). These studies show that spatial heterogeneity in soil organism distribution occurs on nested scales that exceed the plot scale (Fig. 3.2.1), and that we cannot ignore these larger spatial scales.

Crossing the "spatial barrier" between the hierarchical levels seems to be the biggest challenge for future investigations in soil biodiversity and soil functions (Wolters 2001). The importance of large-scale processes for local distribution patterns of soil organisms quickly gains recognition, and has several implications for future studies. First, the traditional niche-based approaches in community ecology alone will not be sufficient to improve our understanding of what drives the diversity of life in soil, and of the structure and functioning of terrestrial communities. Spatial aggregation of soil organisms is not only influenced by processes that operate on a local scale, such as tolerance for extreme environmental conditions, reproduction and competition, but also by processes that work on a more global scale, i.e. the ability to arrive at a certain place by dispersal, for instance after an large infrequent disturbance (Fig. 3.2.1). Community assembly approaches, with a focus on the dispersal ability of organisms, could be included in our studies on soil biodiversity. Second, soil organisms are not only distributed unevenly across space but also in time. Temporal variation in the life span of high quality patches, frequency of disturbances, and seasonal effects all contribute to the spatial heterogeneity of species distributions. Space and time therefore interact, and the outcome is scale dependent. At the fine-scale temporal variation can change over time on a scale from hours, in the case of soil moisture levels to days when digging soil macro invertebrates disturb the structure of the soil (Bengtsson 2002; see also Fig. 3.2.1). At this short time scale small-scale patches are relatively stable, while fine-scale patches are not. Similarly, at the small-scale, dung patches are slowly decomposed, leading to succession of species and shifts in community composition on a scale of months to a year. At this time scale temporal variability in dung deposition and quality will not affect the patchiness of the landscape. To influence the spatial heterogeneity of landscapes a large disturbance is necessary, such as a flooding or a severe storm. While large infrequent disturbances have the potential to affect all spatial scales, small frequent disturbances will only affect the smaller spatial scales. Disturbances create and maintain environmental heterogeneity at hierarchical spatial scales and are therefore an important component of soil biodiversity.

Finally, the enigma of soil species biodiversity is not really an enigma at all. The bewildering diversity in soil is largely attributable to the nested set of ecological worlds in soil. Patterns in the spatial distribution of abiotic and biotic drivers of soil organisms at a hierarchy of scales provide important clues about the underlying mechanisms that structure ecological communities and determine soil functions. Soil ecological studies should make us aware of this insight to be able to move forward from merely showing how diverse soils are to identifying the key factors that drive the diversity of soil organism. Only if we know the rules that govern soil biodiversity can we expect to evaluate the consequences of soil biodiversity changes for ecosystem functioning.

References

Acosta-Mercado, D., & Lynn, D.H. (2002) A preliminary assessment of spatial patterns of soil ciliate diversity in two subtropical forests in Puerto Rico and its implications for designing an appropriate sampling approach. *Soil Biology Biochemistry* **34**: 1517–20.

Anderson, J.M. (1975) The enigma of soil animal species diversity. In: Vanek J. (eds.) *Progress in soil zoology*, pp. 51–8. Academia, Prague.

Bardgett, R.D. (2002) Causes and consequences of biological diversity in soil. *Zoology* **105**: 367–74.

Bardgett, R.D. (2005) *The biology of soil: a community and ecosystem approach*. Oxford University Press, Oxford.

Bardgett, R.D., & Wardle, D.A. (2003) Herbivore-mediated linkages between aboveground and belowground communities. *Ecology* **84**: 2258–68.

Bazin, M.J., Markham, P., Scott, E., & Lynch, J.M. (1990) Population dynamics and rhizosphere interactions. In: Lynch, J.M. (ed.) *The rhizosphere*, pp. 99–127. John Wiley & Sons, Chichester.

Beare, M.H., Coleman D.C., Crossley D.A., Hendrix P.F., & Odum E.P. (1995) A hierarchical approach to evaluating the significance of soil biodiversity to biogeochemical cycling. *Plant and Soil* **170**: 5–22.

Bengtsson, J. (2002) Disturbance and resilience in soil animal communities. *European Journal of Soil Biology* **38**: 119–25.

Berg, M.P., & Bengtsson J. (2007) Spatial and temporal variation in food web composition. *Oikos* **116**: 1789–804.

Berg, M.P., Kniese, J.P. Bedaux, J.J.M., & Verhoef, H.A. (1998) Dynamics and stratification of functional groups of micro- and mesoarthropods in the organic layer of a Scots pine forest. *Biology and Fertility of Soils* **26**: 268–84.

Bjørnlund, L., & Christensen, S. (2005) How does litter quality and site heterogeneity interact on decomposer food webs of a semi-natural forest? *Soil Biology Biochemistry* **37**: 203–13.

Bronick, C.J., & Lal, R. (2005) Soil structure and management: a review. *Geoderma* **124**: 3–22.

Bruneau, P.M.C., Davidson, D.A., Grieve, I.C., Young, I.M., & Nunan, N. (2005) The effects of soil horizons and faunal excrements on bacterial distribution in an upland grassland soil. *FEMS Microbiology Ecology* **52**: 139–44.

Chu, H., Fierer, N., Lauber, C.L., Caporaso, J.G., Knight, R. & Grogan, P. (2010) Soil bacterial diversity in the Arctic is not fundamentally different from that found in other biomes. *Environmental Microbiology* **12**: 2998–3006.

Clays-Josserand, A., Lensi, R., & Gourbiere, F. (1988) Vertical distribution of nitrification potential in an acid forest soil. *Soil Biology Biochemistry* **20**: 405–6.

Coulson, S.J., Hodkinson, I.D., & Webb, N.R. (2003) Microscale distribution patterns in high arctic soil arthropods communities: the influence of plant species within the vegetation mosaic. *Ecography* **26**: 801–9.

Dauber, J., Purtauf, T., Allspach, A., Frisch, J., Voigtländer, K., & Wolters, V. (2005) Local vs. landscape controls on diversity: a test using surface-dwelling soil macroinvertebrates of differing mobility. *Global Ecology and Biogeography* **14**: 213–21.

De Deyn, G.B., & van der Putten, W.H. (2005) Linking aboveground and belowground diversity. *Trends in Ecology and Evolution* **20**: 625–33.

Dechesne, A., Pallud, C., & Grundmann, G.L. (2007) Spatial distribution of bacteria at the microscale in soil. In: Franklin, R., Mills, A. (eds.), *The spatial distribution of microbes in the environment*, pp. 87–107. Springer, Berlin.

Ettema, C.H., Coleman, D.C., Vellidis, G., Lowrance, R., & Rathbun, S.L. (1998) Spatiotemporal distributions of bacterivorous nematodes and soil resources in a restored riparian wetland. *Ecology* **79**: 2721–34.

Ettema, C.H., Rathbun, S.L., & Coleman, D.C. (2000) On spatiotemporal patchiness and the coexistence of five species of *Chronogaster* (Nematoda: Chronogasteridae) in a riparian wetland. *Oecologia* **125**: 444–52.

Ettema, C.H., & Wardle, D.A. (2002) Spatial soil ecology. *Trends in Ecology & Evolution* **17**: 177–83.

Faber, J.H. (1991) Functional classification of soil fauna. A new approach. *Oikos* **62**: 110–17.

Fierer, N., Strickland, M.S., Liptzin, D., Bradford, M.A. & Cleveland, C.C. (2009) Global patterns in belowground communities. *Ecology Letters* **12**: 1238–49.

Franklin, R.B., & Mills, A.L. (2003) Multi-scale variation in spatial heterogeneity for microbial community structure in an eastern Virginia agricultural field. *FEMS Microbial Ecology* **44**: 335–46.

Gebhardt, S., Wöllecke, J., Münzenberger, B., & Hüttl, R.F. (2009) Microscale spatial distribution patterns of red oak (*Quercus rubra* L.) ectomycorrhizae. *Mycological Progress* **8**: 245–57.

Giller, P.S. (1996) The diversity of soil communities, the poor man's tropical rainforest. *Biodiversity and Conservation* **5**: 135–68.

Gillet, F., Kohler, F., Vandenberghe, C., & Buttler, A. (2010) Effect of dung deposition on small-scale patch structure and seasonal vegetation dynamics in mountain pastures. *Agriculture, Ecosystems & Environment* **135**: 34–41.

Green, J., & Bohannan, B.J.M. (2006) Spatial scaling of microbial biodiversity. *Trends in Ecology and Evolution* **21**: 501–7.

Griffiths, B.S. (2002) Spatial distribution of soil protozoa in an upland grassland. *European Journal of Protistology* **37**: 371–3.

Hansel, C.M., Fendrof, S., Jardine, P.M., & Francis, C.A. (2008) Changes in bacterial and archaeal community structure and functional diversity along a geochemically variable soil profile. *Applied and Environmental Microbiology* **74**: 1620–33.

Hansen, R.A., & Coleman, D.C. (1998) Litter complexity and composition are determinants of the diversity and species composition of oribatid mites (Acari: Oribatida) in litterbags. *Applied Soil Ecology* **9**: 17–23.

Horner-Devine, M.C., Carney, C.M., & Bohannan, B.J.M. (2004) An ecological perspective on bacterial biodiversity. *Proceedings of the Royal Society of London, Series B* **271**: 113–22.

Housman, D.C., Yeager, C.M., Darby, B.J., Sanford, R.L., Kuske, C.R., Neher, D.A., & Belnap, J. (2007) Heterogeneity of soil nutrient and subsurface biota in a dryland ecosystem. *Soil Biology and Biochemistry* **39**: 2138–49.

Hutchinson, G.E. (1957) Concluding remarks. *Cold Spring Harbour Symposium on Quantitative Biology* **22**: 415–27.

Jiménez, J., Decaëns, T., Rossi, J. (2006) Stability of spatio-temporal distribution and niche overlap in neotropical earthworm assemblages. *Acta Oecologica* **30**: 299–311.

Kallimanis, A.S., Argyropoulou, M.D., & Sgardelis, S.P. (2002) Two scale patterns of spatial distribution of oribatid mites (Acari, Cryptostigmata) in a Greek mountain. *Pedobiologia* **46**: 513–25.

Kassen, R., & Rainey, P.B. (2004) The ecology and genetics of microbial diversity. *Annual Review in Microbiology* **58**: 207–31.

Krab, E.J., Oorsprong, H., Berg, M.P., & Cornelissen, J.H.C. (2010) Turning northern peatlands upside down: disentangling microclimate and substrate quality effects on vertical distribution of Collembola. *Functional Ecology* **24**: 1362–9.

Kusnezov, N. (1957) Numbers of species of ants in faunae of different latitudes. *Evolution* **11**: 298–9.

Lavelle, P. (2002) Functional domains in soils. *Ecological Research* **17**: 441–50.

Laverman, A.M., Borgers, P., & Verhoef, H.A. (2002) Spatial variation in net nitrate production in a N-saturated coniferous forest soil. *Forest Ecology and Management* **161**: 123–32.

Liang, W., Jiang, Y., Liu, Q.L.Y., & Wen, D. (2005) Spatial distribution of bacterivorous nematodes in a Chinese Ecosystem Research Network (CERN) site. *Ecological Research* **20**: 481–6.

Lindo, Z., & Winchester, N.N. (2008) Scale dependent diversity patterns in arboreal and terrestrial oribatid mite (Acari: Oribatida) communities. *Ecography* **31**: 53–60.

Loreau, M. (1987) Vertical distribution of activity of carabid beetles in a beech forest floor. *Pedobiologia* **30**: 173–8.

Mathieu, J., Grimaldi, M., Jouquet, P., Rouland, C., Lavelle, P., Desjardins, T., & Rossi, J.P. (2009) Spatial patterns of grasses influence soil macrofauna biodiversity in Amazonian pastures. *Soil Biology Biochemistry* **41**: 586–93.

Mitchell, E.A.D., Borcard, D., Buttler, A.J., Grosvernier, Ph., Gilbert, D., & Gobat, J.M. (2000) Horizontal distribution patterns of testate amoeba (Protozoa) in a *Sphagnum magellanicum* carpet. *Microbial Ecology* **39**: 290–300.

Nielsen, U.N., Osler, G.H.R., Campbell, C.D., Burslem, D.F.R.P. & van der Wal, R. (2010) The influence of vegetation type, soil properties and precipitation on the composition of soil mite and microbial communities at the landscape scale. *Journal of Biogeography* **37**: 1317–28.

Niemelä, J., Haila, Y., Halme, E., Pajunen, T., & Punttila, P. (1992) Small-scale heterogeneity in the spatial distribution of carabid beetles in the southern Finnish taiga. *Journal of Biogeography* **19**: 173–81.

Peigné, J., Vian, J.F., Cannavacciuolo, M., Bottollier, B., & Chaussod, R. (2009) Soil sampling based on field spatial variability of soil microbial indicators. *European Journal of Soil Biology* **45**: 488–95.

Pokarzhevskii, A.D., van Straalen, N.M., Zaboev, D.P., & Zaitsev, A.S. (2003) Microbial links and element flows in nested detrital food webs. *Pedobiologia* **47**: 213–24.

Rønn, R., Griffiths, B.S., Ekelund, F., & Christensen, S. (1996) Spatial distribution and successional pattern of microbial activity and micro-faunal populations on decomposing barley roots. *Journal of Applied Ecology* **33**: 662–72.

Saetre, P. (1999) Spatial patterns of ground vegetation, soil microbial biomass and activity in a mixed spruce-birch stand. *Ecography* **22**: 183–92.

Sayer, E.J. (2006) Using experimental manipulation to assess the roles of leaf litter in the functioning of forest ecosystems. *Biological Reviews* **81**: 1–31.

Schuldt, A., & Assmann, T. (2009) Environmental and historical effects on richness and endemism patterns of carabid beetles in the western Palaearctic. *Ecography* **32**: 705–14.

Scott, W.A., & Anderson, R. (2003) Temporal and spatial variation in carabid assemblages from the United Kingdom Environmental Change Network. *Biological Conservation* **110**: 197–210.

Setälä, H., Berg, M.P., & Jones, T.H. (2005) Trophic structure and functional redundancy in soil communities. In: Bardgett, R.D., Usher, M.B., Hopkins, D.W. (eds.) *Biological diversity and function in soils*, pp. 236–49. Cambridge University Press, Oxford.

Simmons, B.L., Wall, D.H., Adams, B.J., Ayres, E., Barrett, J.E., & Virginia, R.A. (2009) Terrestrial mesofauna in above- and below-ground habitats: Taylor Valley, Antarctica. *Polar Biology* **32**: 1549–58.

Swift, M.J., Heal, O.W., & Anderson, J.M. (1979) *Decomposition in terrestrial ecosystems*. University of California Press, Berkeley, CA.

Van der Putten, W.H., Bardgett, R.D., de Ruiter, P.C., et al. (2009) Empirical and theoretical challenges in aboveground-belowground ecology. *Oecologia* **161**: 1–14.

Verhoef, H.A., Verspagen, J.M.H., & Zoomer, H.R. (2000) Direct and indirect effects of ultraviolet-B radiation on soil biota, decomposition and nutrient fluxes in dune grassland soil systems. *Biology and Fertility of Soils* **31**: 366–71.

Wardle, D.A. (2006) The influence of biotic interactions on soil biodiversity. *Ecology Letters* **9**: 870–86.

Wolters, V. (2001) Biodiversity of soil animals and its function. *European Journal of Soil Biology* **37**: 221–7.

Young, I.M., & Ritz, K. (2000) Tillage, habitat space and function of soil microbes. *Soil & Tillage Research* **53**: 201–13.

CHAPTER 3.3

Linking Soil Biodiversity and Human Health: Do Arbuscular Mycorrhizal Fungi Contribute to Food Nutrition?

Pedro M. Antunes, Philipp Franken, Dietmar Schwarz, Matthias C. Rillig, Marco Cosme, Martha Scott, and Miranda M. Hart

3.3.1 Soil health is linked to human health and global food security

Human health is fundamentally dependent upon plants, as they provide most of our nutrients through the food chain. Because plants acquire elements from the soil, it follows that human health is ultimately linked to soil health. Unfortunately, data suggest that up to 60% of the world's arable land is deficient in some nutrients (Fageria *et al.* 2008), and micronutrient deficiencies are very common in plants growing in alkaline soils, which represent roughly 30% of agricultural soils around the world (White & Broadley 2009). In areas where soils lack certain necessary elements, people who survive on a local, plant-based diet could suffer malnutrition associated with certain mineral imbalances (White & Zasoski 1999). Up to 75% of the world's poor live in rural areas and practice subsistence agriculture (Anriquez & Stlovkal 2008). Many of these people survive on a diet that is entirely local and plant based, making them especially vulnerable to micronutrient malnutrition.

3.3.1.1 Low nutrient crops

Regardless of soil nutrient levels, people subsisting on grains may lack adequate levels for optimum, or even basic, nutrition (Mayer *et al.* 2008). Exclusively plant-based diets in general can lead to malnutrition, particularly if seeds are the primary source of calories because plants have tightly regulated controls governing the type and amount of nutrients allocated to seeds (Cordain 1999). This is particularly true for cultures that process their grains, thereby removing even more limiting nutrients. In the developed world, most malnutrition is due to diets that are based on highly processed grain products (Welch & Graham 1999). In addition, the nutrient content of agricultural products has significantly decreased in the past 50 years in Europe and North America most likely due to dilution effects caused by intensive agricultural management and/or plant breeding practices, which have focused almost exclusively on yield (White & Broadley 2005, 2009; Davis 2009). While this "dilution effect" has resulted in higher nutrient use efficiency (NUE), the resulting crops may not be as nutritious simply because they provide a less concentrated supply of nutrients.

3.3.1.2 Linking soils to human health

Recently, the soil and its biota have been identified as key players in what is being referred to as the "second green revolution," that is, one that sustains not only the world's food supply, but the environment, as well (Royal Society 2009). While rhizosphere microbes are well known to improve plant nutrient status, people are only just beginning to directly link soil health to human health (Morrissey *et al.* 2004; He & Nara 2007). This is a very complex issue that spans agricultural practices, soil microbiology, food culture, and global food security.

In the following sections, we highlight some of the first studies linking rhizosphere microbes and food quality. We propose that manipulation or management of soil microbes may be an accessible and sustainable approach to minimize malnutrition in many systems. Finally, to illustrate how microbes may be key mediators of human nutrition, we use the arbuscular mycorrhizal (AM) symbiosis as a model, as AM fungi form an important mutualism with almost every food crop (He & Nara 2007). However, we consider that many rhizosphere microbes may have the potential to influence the nutrient quality of crops.

3.3.2 Traditional ways of boosting crop nutrients

Humans have been concerned with increasing the nutritional value of food since essential nutrients were first described (Kaluski *et al.* 2003). With the industrialization of food production in the last century, nutrients can be augmented at various steps along the production line. However, the following approaches, while successful in many cases, are often beyond the reach for most of the world's population.

3.3.2.1 Diversified diet

The simplest solution to nutrient malnutrition is to eat a more varied diet. There is huge opportunity and need to introduce new plant species in human diets. Only about 15 of the 250,000 plant species known to science provide 90% of the word's human food consumption (Pimentel & Pimentel 1996). Plants differ greatly in which nutrients they are able to store in edible tissues (Marschner 1996; Grusak & DellaPenna 1999). Even within varieties of the same species, there are genetically imposed upper limits for nutrients (Zhao *et al.* 2009). A varied diet is straightforward for developed countries, which can import produce to ensure year-round access. However, people living at subsistence levels generally depend on locally produced plant staples (White & Broadley 2005). If the soils where these are grown are nutrient poor, it is almost impossible for them to receive adequate nutrition. So while a varied diet may be a good approach for relatively few, it is an unreasonable goal for much of the world's population.

3.3.2.2 Fertilize

Perhaps the most obvious solution to reduce human malnutrition is to fertilize soils directly with the deficient minerals. This has been done with some success in Finland (Ekholm *et al.* 2007), Malawi (Denning *et al.* 2009), and in Turkey (Cakmak *et al.* 1999). However, to be done effectively, this requires detailed knowledge about soil nutrient status. It also requires farmers having access to fertilizers, and guidance to manage their soils.

Physiological constraints on nutrient content of seeds mean that it is not a feasible approach for all crops. Fertilizer application may result in a yield increase without a concomitant increase in nutrient levels (Grusak *et al.* 1999; Rengel *et al.* 1999; Liu *et al.* 2007), effectively treating hunger without treating malnutrition. Similarly, fertilization does not always result in increased nutrition in the edible portion of the plants (e.g. calcium is not found in seeds of grasses in significant amounts, so fertilizing calcium deficient soils would not alleviate a calcium deficiency; Rengel *et al.* 1999). Finally, not all nutrient deficiency is due to a lack of soil nutrients. Often nutrients are present, but inaccessible to plants, or bound by plants and not translocated to edible organs.

Logistical difficulties aside, there are more pernicious problems with the use of fertilizer to boost food nutrient levels, as much of the applied nitrogen fertilizer does not end up in our food supply. Rather, it can be converted into nitrous oxide, a significant greenhouse gas, or it can be washed away as leachate or surface run off, where it collects in the water table or ends up in lakes and oceans causing massive algal "blooms" (Anderson *et al.* 2002; Nosengo 2003; Beman *et al.* 2005).

3.3.2.3 Fortify

Post-harvest fortification involves adding deficient nutrients (or nutrients removed during processing) back into foods during processing. In many countries, for example, flours contain additives such as essential B-vitamins. While this is undoubtedly effective for many nutrition related-diseases, it assumes that people will have access to processed foods, which is not economically possible for most of the world's malnourished people. Further, post-harvest fortification makes the food chain from producer to consumer even longer, and violates many of the values held by consumers and producers who are searching for foods derived from local and sustainable agricultural practices.

3.3.2.4 Biofortification

Biofortification of staple crops has become an increasingly popular solution to malnutrition (see reviews Nestel *et al.* 2006; Mayer *et al.* 2008; White & Broadley 2009). Biofortified crops have enhanced, or concentrated levels of nutrients either through selection of highly functioning genotypes, or through plant breeding programs that select for genotypes with higher nutrient status via higher tissue concentrations of essential minerals (Butelli *et al.* 2008), increased synthesis of factors which enhance nutrient uptake (George *et al.* 2005), decreased levels of antinutrients (such as oxalates or phytic acid; Zhu *et al.* 2007), or genotypes modified to produce essential compounds *de novo* (i.e. golden rice; Enserink 2008).

While biofortified crops may provide essential nutrients, there are conflicting issues surrounding their wide scale adoption. Most significantly, the creation of specialized lines is expensive. Modified crops, whether they are selected varieties, hybrids, or genetically modified organisms (GMOs) require farmers to have access to novel seed sources, which may not be affordable or desirable. While there are few reports that some GMO crops or changes associated with their management alter microbial communities (Dunfield & Germida 2003; Hart *et al.* 2009; Powell *et al.* 2009), if future varieties manipulate compounds important for microbial signaling (which is possible since many of these signals are important human nutrients), such plants could greatly alter soil microbial communities.

In summary, current approaches to boosting crop nutrient levels rely heavily on industrialized food production systems. While this is an effective way of producing nutritious food for countries that can pay for it, there remain economic and social barriers for many countries.

3.3.3 A critical role for soil microbes

An alternative to traditional ways of boosting crop nutrients may reside in the soil. Interactions between beneficial microorganisms and plants have been well studied. These include root-dwelling microbes as well as other endophytic and epiphytic microbes found in roots and shoots (Rengel & Marschner 2005). These symbiotic associations are well known to influence the nutrient status of plants based on their ability to access minerals, most importantly in nutritionally stressed environments (Jumpponen 2001; Jeffries *et al.* 2003; Pawlowska & Charvat 2004; Cardoso & Kuyper 2006; Yang *et al.* 2009; Johnson *et al.* 2010).

Many organisms are already used as "biofertilizers" (e.g. *Rhizobia*, *Azospirillum*, AM fungi). However, these organisms are typically used to boost yield, and have not been used with the intent to modify the nutrient profile of food crops. The following sections are examples of major groups of rhizosphere microbes that are known to benefit plants, and may potentially be used to manipulate the nutrient value of food crops.

3.3.3.1 Rhizobia

Perhaps the most well known "biofertilizers" are bacteria belonging to a large group that form symbioses with legumes, and fix atmospheric nitrogen within root nodules. They have been developed commercially since the 1890s and are widely accepted as a yield-enhancing biofertilizer (e.g. Peoples & Craswell 1992; Rai 2006).

3.3.3.2 Cyanobacteria

Until recently, cyanobacteria were the major source of nitrogen (N) for rice cultivation in Asia (Vaishampayan et al. 2001). The use of inorganic fertilizers, however, is leading to serious environmental problems by interfering with the *Azolla-Anabaena* symbiosis (Bouman *et al.* 2002). Globally, despite its potential, the use of cyanobacteria in agriculture is almost non-existent (Hashem 2001).

3.3.3.3 Plant growth promoting rhizobacteria (PGPR) and free-living N_2 fixing bacteria

Plant growth promoting rhizobacteria (PGPR) represent a diverse assemblage of soil bacteria that stimulate plant growth through various, often unknown, modes of action (Vessey 2003). They can promote plant growth through phosphorus (P) solubilization (Richardson 2001; Gyaneshwar *et al.* 2002), enhanced nutrient uptake, synthesis of plant hormones (Rodriguez & Fraga 1999; Persello-Cartieaux *et al.* 2003) and systemically induce resistance against pathogens (Beckers & Conrath 2007).

3.3.3.4 Mycorrhizal fungi

Ectomycorrhizal fungi are a taxonomically diverse group of fungi that associate with the roots of many tree species. These fungi facilitate seedling establishment by enhancing the uptake of essential nutrients in exchange for tree-supplied photosynthate. For many years, the forest industry has used ectomycorrhizal fungi to inoculate new seedlings (e.g. Trappe 1977; Molina 1982). As many large, dry, oily seed or fruit producing trees (i.e. tree nuts) such as chestnuts, hazelnuts, or walnuts are ectomycorrhizal (Smith & Read 2008), there is promise that these fungi might also be important for food production.

AM fungi are the most widespread plant-microbe symbiosis. These root-associated symbionts demonstrate strong growth promoting effects for most plants, including almost all edible crops (Smith & Read 2008). Commercial inoculants are now being marketed for use in agriculture, horticulture, and restoration (Schwartz *et al.* 2006).

Moreover, recent studies have shown that AM fungal species contain high genetic diversity (Stockinger *et al.* 2009; Croll *et al.* 2009; Borstler *et al.* 2010), which can be manipulated and "bred" for specific uses (see Sanders 2010)

3.3.3.5 Non-mycorrhizal fungal endophytes

There is a wide variety of fungi that colonize both above- and belowground plant tissue, conferring pathogen resistance, among other benefits (Petrini 1996; Saikkonen *et al.* 1998). The use of such fungi to improve crop nutrition has received little attention (Varma *et al.* 1999; Weiss *et al.* 2004; Mandyam & Jumpponen 2005). Although there is evidence that they increase yield, and induce tolerance to abiotic and resistance to biotic stress (Waller *et al.* 2005), there are no commercial inocula available.

However, even if microbial associations in the soil are able to improve the nutrient quality of food crops and enhance human nutrition, these effects may be destroyed by current agricultural practices (Brito *et al.* 2008). An important aspect of modern cropping systems is the use of high-performing cultivars, hybrids, or genetically modified lines. Historically, the nutrient content of the crop has not been a priority for plant breeders (Wissuwa *et al.* 2009) and most new lines are developed especially for increased yield and/or long-term storage and transportation (Araus *et al.* 2008). An unforeseen byproduct of this selection is the loss of mycorrhizal effect seen in most crops (Sawers *et al.* 2006). As crop lines have been selected for performance in high nutrient environments, their reliance on AM fungi to scavenge nutrients has decreased (Hetrick *et al.* 1992, 1993; Zhu *et al.* 2001). This effect has also been observed for *Rhizobia* (Kiers *et al.* 2007). In addition, breeding plants for pathogen resistance is thought to have produced less mycorrhizal dependent lines likely due to unintended alterations of plant-microbial communication pathways necessary for symbiotic establishment (Toth *et al.* 1990).

Crops are typically grown in monoculture, with periods of fallow, despite evidence that intercropping or arboriculture increase soil biodiversity and overall system productivity (Gavito & Miller 1998;

Szumigalski & van Acker 2006; Chifflot *et al.* 2009). Many studies have shown that fallow (or intercropping with non-AM plants) significantly reduces inoculum potential of agricultural soil (Harinikumar & Bagyaraj 1988; Drijber *et al.* 2000). The reverse is also true; a recent review showed that intercropping dicots and graminaceous species can effectively "biofortify" crops with zinc (Zn) and iron (Fe) (Zuo & Zhang 2009).

The use of inorganic fertilizer has been shown to decrease soil biodiversity (Kleijn *et al.* 2009). The mechanisms driving this relationship are not clear. It has been suggested that excess nutrients may disrupt other geochemical cycles, or may favor more aggressive, fertilizer tolerant microbes that are able to dominate in high-input agrosystems (Oehl *et al.* 2004). There is considerable evidence showing reduced biodiversity of AM fungi in agricultural fields as a result of continued use of inorganic fertilizers (Helgason *et al.* 1998). There is also indication that high-input agricultural systems select for less effective AM fungal communities (Johnson 1993; Corkidi *et al.* 2002; Ryan & Graham 2002; Anderson & Cairney 2004).

Tillage is a widely practiced method of aerating the top layers of soil to prevent soil compaction and reduce weed growth. However, there are many data demonstrating a negative effect of tillage on soil microbial communities (Simmons & Coleman 2008). This is particularly true for mycorrhizal fungi, since tillage destroys mycelia and severs connections between plants and the common mycelial network that greatly expands the ability of the plant root system to scavenge for nutrients (Miller 2000; Jansa *et al.* 2003; Antunes *et al.* 2006a; Mathimaran *et al.* 2007).

Herbicides and pesticides may directly reduce soil biodiversity by eliminating competing weeds, herbivorous insects, or crop pathogens. Pesticides, while controlling pests and pathogens, can also affect non-target species including microbes that form an important component of the rhizosphere (Engelen *et al.* 1998; El Fantroussi *et al.* 1999)

3.3.4 Using rhizosphere microbes to create healthier food

Despite the considerable body of evidence linking microbes to plant nutrient uptake, there are very few studies that have focused on how they interact with nutrients important for human health, and fewer still that focus on the edible portions of crops (Table 3.3.1).

In general, there are two primary mechanisms to improve plant nutrient content via microbial symbionts such as AM fungi: they can directly increase the net uptake from soil to plant through various channels, or they can create changes in the plant and/or soil, which allow the plant to take up or utilize more nutrients. Via the mechanisms outlined in the following sections, it is possible that improved plant nutrition due to microbes may translate into improved human health. Numbers in parentheses refer to mechanisms outlined in Fig. 3.3.1.

3.3.4.1 Direct effects

The role of rhizosphere microbes on plant nutrition can be broadly characterized as either direct or indirect. Direct mechanisms enhance a plant's ability to *access* nutrients directly from the environment. These mechanisms could be important for human nutrition under low nutrient conditions, where fertilizer applications are not feasible due to either economics, cropping practices, or environmental reasons. Under such conditions, resident (or applied) microbes could enhance the nutrient content of AM crops over non-AM crops.

3.3.4.1.1 Increased nutrient uptake
The most studied benefit from AM fungi is enhanced P acquisition of AM plants. This is largely due to the increased surface area of AM plants, whereby the fungal mycelium acts as an extension of the host root system, either directly acquiring nutrients, or acting in tandem with the host root system to absorb nutrients (1–3) (Smith *et al.* 2009). P ions have poor mobility in soil (Tinker & Nye 2000) and without AM fungi plants can quickly deplete adjacent soil of all available P. AM hyphal networks have been shown to increase that distance to >11 cm from the rhizoplane in manipulative experiments (Li *et al.* 1991). These hyphal networks serve also to increase absorption of other slowly mobile nutri-

Table 3.3.1 Studies showing the effect of rhizosphere microbes on nutrient levels in food crops

Crop	Rhizosphere microbe	Plant tissue analyzed	Impacts on nutritional status of crop	Reference
Pomegranate	PGPR (*Azotobacter chroococcum* plus *Glomus mosseae*)	Leaf	K, Ca, Mg, Cu, Zn, Mn, and Fe concentration in leaf tissue was significantly higher than in controls	Aseri et al. 2008
Banana	PGPR (phosphate solubilizing bacteria (PSB)–*Pseudomonas fluorescent PS* and *Bacillus megaterium* BM)	Fruit	Plants fertilized with both microbes had higher N, P, and K in the fruit pulp	Attia et al. 2009
Ancho Chile	AM fungi (*Glomus fasciculatum*, plus two mixed AM fungi (local soils))	Fruit	Inoculation increased fruit chlorophyll, carotenes, and xanthophylls in drought stressed plants	Mena-Violante, et al. 2006
Onions	AM fungi (mixed *Glomus* spp.)	Bulb	AM fungi increased some flavonol glycosides, but had no effect on organosulfur compounds	Perner et al. 2008
Tomato	PGPR (*Bacillus amyloiquefaciens*, *Bacillus pumilus*) plus AM fungi (*Glomus intraradices*)	Shoot	Mixed inoculation prevented loss of shoot N and P despite reduction of applied fertilizer	Adesemoye et al. 2009
Pepper	AM fungi (*G. clarum*)	Fruit	Inoculation in salt stressed plants increased N, P, and K concentration in fruit to non-stressed levels	Kaya et al. 2009
Corn	AM fungi (*G. intraradices, G. mosseae, G. etunicatum*)	Leaf	Inoculation with AM fungi increased P, Fe, Mn, and Zn in leaf tissue subjected to soil compaction	Miransari et al. 2008
Wheat	AM fungi (*Glomus* spp.).	Shoot	Inoculation lead to increases in P, K, Fe, Mn, Zn, and Cu	Miransari et al. 2009
Sage (*Salvia* sp.)	AM fungi (*G. mosseae, G. intraradices*)	Leaf and Root	AM fungi increased P above controls, but not as much as P fertilization. There was no effect of AM fungi on phenolics, rosmarinic acid, and essential oils.	Nell et al. 2009
Wheat and field pea	AM fungi (natural field inoculum)	Shoot	Wheat from fields with high AM fungi colonization levels had increased Zn concentration.	Ryan and Angus 2003
Wheat and field pea	AM fungi (*G. intraradices, Sc. calosporum*)	Shoot	Inoculated wheat had higher Zn and P levels, only in non-fertilized treatments	Ryan and Angus 2003
Leek	AM fungi (*G. intraradices, G. claroideum, G. mosseae*)	Shoot and root	Pre-inoculated field plants had higher P, Zn, Cu, and N	Sorensen et al. 2008

Plant	Inoculant	Tissue	Result	Reference
Basil	AM fungi (G. mosseae)	Shoot	AM fungi had no effect on rosmarinic acid, caffeic acid, essential oils, or phenolics	Toussaint et al. 2008
Tomato	AM fungi (Glomus sp.)	Fruit	AM fungi increased both lycopene and β-carotene levels in fruit	Ulrichs et al. 2008
Lettuce	AM fungi (G. mosseae)	Leaf	P, K, Ca, MG, Cu, and Fe increased in inoculated plants in all treatments. Zn and S increased only under low levels of fertilization	Azcón et al. 2003
Basil	PGPR (Baccilus subtilis)	Leaf	Inoculation increased terpenes (α-terpinol, euglenol)	Banchio et al. 2009
Artemisia annua	AM fungi (G. macrocarpus, G. fasiculatum)	Leaf	Inoculation increased P, Zn, and Fe, essential oils and artemisin levels were also higher.	Chaudhary et al. 2008
Basil	AM fungi (G. mosseae, Gigaspora margarita, Gigaspora rosea)	Leaf	Only Gi. rosea increased α-terpinol, but all AM fungi altered the ratio of different essential oils.	Copetta et al. 2006
Grapes	AM fungi (G. mosseae)	Leaf	Inoculation increased leaf P, K, and B	Karagiannidis et al. 2007
Pepper	PGPR (Methlobacterium oryzae) plus AM fungi (Acaullospora longula, G. clarum, and G. intraradiaces)	Shoot and root	Combined inoculation increased N, P, K, Ca, Zn, Cu, Fe, and Mn. Chlorophyll concentration also increased.	Kim et al. 2010
Artichoke	AM fungi (G. intraradices, G. mosseae)	Leaf, flower	Inoculation increased total phenolic content in field and greenhouse	Ceccarelli et al. 2010
Strawberry	AM fungi (G. intraradices)	Fruit	Inoculation resulted in changes of berry color, increased K and Cu, and increases in certain phenols, anthocyanins, quercetin, and catechin	Castellanos-Morales et al. 2010
Tea	AM fungi (mixed inoulum)	Leaf	Inoculation resulted in increased amino acid and protein concentration, increased polyphenols, caffeine and sugar	Singh et al. 2010

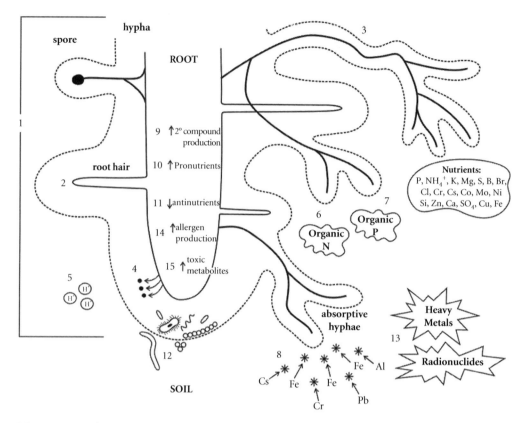

Figure 3.3.1 A summary of potential mechanisms and pathways through which AM fungi influence the nutritional value of plants. Numbers refer to points addressed within the text.

ents, especially Zn and copper (Cu) (Lambert et al. 1979). In clover, 79%, 50%, and 60% of shoot P, Zn, and Cu, respectively, was shown to be taken up via hyphal pathways (Marschner 1995). There is increasing evidence that AM fungi are important for plant N nutrition, with AM plants showing increased uptake for both organic and inorganic N sources (Leigh et al. 2009), particularly in N-limiting conditions (Schimel & Bennett 2004; Richardson et al. 2009; Johnson 2010).

3.3.4.1.2 Changes in plant/soil chemistry
It is well known that AM plants have altered exudate chemistry (4) (Jones et al. 2004; Rengel & Marschner 2005), which in combination with local soil conditions, can improve the uptake of certain nutrients. These exudates can also alter the pH of the soil (5) (Li et al. 1991), making certain ions more mobile. While changes in pH are often concurrent with nutrient uptake (Bago et al. 1996), there is no evidence that hyphae themselves induce changes in pH (Antunes et al. 2007).

3.3.4.1.3 Nutrient solubilization and transport
Microbial symbionts can significantly enhance plant access to nutrients that are usually unavailable due to intrinsic enzymatic capacity. There is growing evidence that AM fungi can hydrolyze organic N (Leigh et al. 2009; Whiteside et al. 2009; Hodge & Fitter 2010) and P directly (6, 7) (Marschner & Dell 1994; Koide & Kabir 2000; Feng et al. 2003), greatly increasing the number of nutrient pools available to plants. Whether this is true for other nutrients is not known.

The presence of AM fungal transporters may further enhance a plant's ability to take up limiting nutrients. AM fungal transporters have been identified for P (two transporters; Harrison & Vanbuuren 1995; Benedetto et al. 2005), and NH_4 (Lopez-Pedrosa et al. 2006; Guether et al. 2009). Recent studies show enhanced supfur (S) uptake and fungal transcripts for S-assimilatory genes in a Daucus carrota and Glomus intraradices symbiosis (Allen & Shachar-Hill 2009). However, it is possible that more transporters, in particular for nutrients important for human nutrition, may be identified with the completion of the genome of the AM fungus Glomus intraradices (Martin et al. 2008).

3.3.4.1.4 Iron chelation
Studies have shown leaf Fe concentration to increase in the presence of AM fungi (reviewed in Haselwandter & Winkelmann 2007), though there is no conclusive evidence that AM fungi produce iron-chelating siderophores. It is possible that AM fungi are able to use siderophores produced from other fungi or hyphal-associated bacteria, or possess alternate mechanisms of reducing iron (Johnson 2008). It is not known whether plants can use these putative siderophores and transporters themselves (Winkelmann 2007). If they can, association with AM fungi would enhance a plant's ability to capture Fe even further[R1]. It is also possible that siderophores, if present, can also bind to other metals, including lead (Pb), chromium (Cr), aluminum (Al), cesium (Cs), and actinide ions (Renshaw et al. 2002) which may or may not enhance plant nutrient status, depending on the metal (8). Of course, this will only translate into improved food quality if improved uptake occurs in the edible plant portion.

3.3.4.2 Indirect effects

Indirect mechanisms evoke *changes in the host plant itself*, which result in improved nutrient status, either through increased metabolite production or by decreasing inhibitors. Colonization by AM fungi creates a cascade of effects in host plants. In different non-targeted approaches, numerous plant genes were identified being modulated during root colonization by an AM fungus (for review see Arlt et al. 2009). A large percentage of these genes were not affected by P fertilization of the plants, which indicates that many physiological changes occur independently of improved P uptake.

Many of the compounds produced by plants in response to AM fungal colonization function as antioxidants in the plant and also in our diet (9) (Toussaint et al. 2007). There is mounting evidence that these compounds are critical for optimal human health and disease prevention (Halliwell 1996). There are an estimated 10,000 secondary metabolites in plants; also referred to as phytonutrients. These are included in four general groups; flavonoids, terpenes, alkaloids, and sulphur containing compounds (Kochian & Garvin 1999). Because of the enormous number of metabolites involved, and their large potential role in human health, the effect of microbes on secondary plant metabolites and human nutrition represents an exciting new field where studies are urgently needed.

Phenols, in particular flavonoids, are important for human health, as they form a large group of important antioxidants found in fruits, vegetables, tea and wine (Yao et al. 2004). There is considerable evidence that root colonization by microbes, including rhizobia and AM fungi, stimulate phenol and flavonoid production in plant roots (Peipp et al. 1997; Larose et al. 2002; Devi & Reddy 2002; Zhu & Yao 2004; Antunes et al. 2006b; Carlsen et al. 2008; Araim et al. 2009; Dardanelli et al. 2008; Perner et al. 2008; Ceccarelli et al. 2010; Zubek 2010). More recent studies have begun to look at flavonoids in other plant tissues. For example, Khaosaad et al. (2008) found a larger concentration of the isoflavone biochanin A in the root and shoot of AM red clover but reduced shoot levels of genistein. *Glomus mosseae* enhanced the concentration of different secondary metabolites in a plant of medicinal interest (*Begonia malabarica* Lam. (Begoniaceae)), including total flavonoid content; effects were further enhanced by co-inoculation with PGPB (Selvaraj et al. 2008). (see Table 3.3.1 for further examples).

Colonization of plant roots by AM fungi has been shown to induce the accumulation of secondary metabolites arising from terpenoid metabolism (Krishna et al. 2005; Rapparini et al. 2008). Related studies also show that AM fungi induce higher levels of carotenoids in roots (Klingner et al. 1995; Akiyama & Hayashi 2002; Fester et al. 2005; Nell et al. 2010; Walter et al. 2000) leaves (Manoharan et al. 2008; Rapparini et al. 2008,) and fruits (Ulrichs et al. 2008). Carotenoids are precursors of vitamin A and can also function as antioxidants.

Alkaloids are produced by many plants (Levin 1976) and there is some indication that AM fungi may influence the content of certain alkaloids and alexins in seeds (Yao et al. 2003; Rojas-Andrade et al. 2003; Abu-Zeyad et al. 1999). Considering that these compounds have important pharmacological effects (e.g. caffeine, nicotine, morphine, or the antimalarial drug quinine) it is surprising that very little is known about the relationship between the AM symbiosis and their synthesis.

Sulphur containing compounds include the glucosinolates and other compounds found in plants of the Brassicaceae and related families (Fahey et al. 2001). Glucosinolates can also be found in plant families that form mycorrhizal associations such as Capparaceae, Caricaceae, Resedaceae, and Tropaeolaceae (Vierheilig et al. 2000; Wang & Qiu 2006). Another sulfur containing compound with strong antioxidant properties, allicin (see Vaidya et al. 2009) can be found in garlic and onions and other species in the family Alliaceae (Block 1985). There is increasing evidence that sulfur-containing compounds are produced in certain plants in response to AM fungal colonization (Vierheilig et al. 2000; Pongrac et al. 2008), and evidence that these compounds can improve human health is accumulating (Vekerk et al. 2009).

Mycorrhiza can also affect pro-nutrients or anti-nutrients, that is, compounds that enhance/reduce plant production or uptake of minerals/vitamins without themselves being nutrients. Ascorbic acid (vitamin C) is a common example of a pro-nutrient. Colonization by AM fungi enhances plant ascorbic acid production (Subramanian et al. 2006) which in turn enhances our ability to access plant-derived iron (10).

AM fungi may also inhibit anti-nutrients (11). One example is phytic acid, a compound produced by plants that binds available metal ions, making them unavailable to humans. Plants with high levels of phytic acid are not good food sources for humans, despite their nutrient content. AM fungi are known to hydrolyze phytic acid in the soil (Feng et al. 2003) and there are some indications that AM fungi can affect phytic acid levels within plant tissues. Recently, Ryan et al. (2008) showed that the bioavailability of Zinc (given as the ratio of phytic acid levels to Zn) in grains of several common crops was enhanced by both AM colonization and a reduction of P-fertilizer.

AM fungi create what is referred to as the "hyphosphere": or the community of microorganisms, dependent on, and living around the mycorrhizal hyphae (12) (Marschner 1995). This microcosm facilitates the colonization of many PGPB (Andrade et al. 1998; Rengel & Marschner 2005; Artursson et al. 2006; Gentili & Jumpponen 2006; Toljander et al. 2007), which can influence plant nutrient content in many more ways. For example, interactions between AM fungi and PGPB have been consistently reported, most notably in the tripartite symbiosis between AM fungi, rhizobacteria and legume crops (Toro et al. 1998; Barea et al. 2002; Caravaca et al. 2004; Cabello et al. 2005). Interactions between different microbial groups may have synergistic effects on plant nutrients that we have yet to discover.

3.3.5 Negative effects of microbes on food quality

Under specific conditions, it is possible that microbial effects on crops could also produce food that is less nutritious, even toxic. While nutrient uptake and secondary metabolite production are generally regarded as beneficial to both plants and their consumers, there may be situations where these functions produce crops that are toxic to consumers.

Concentration of metals in soil may lead to excessive uptake by crops (13) (Gerritse et al. 1983; Nriagu & Pacyna 1988; McLaughlin et al. 1999; Agbenin et al. 2009), consequently posing a threat for humans (White & Zasoski 1999). It has been found that heavy metals accumulate less in tissues of AM plants (Chen et al. 2005), which could be due to the high heavy metal absorption capacity of AM fungal hyphae compared to other organisms (Joner et al. 2000). Conversely, it has also been reported that AM plants accumulate relatively high metal concentrations in their shoots (Rabie 2005) but it remains unclear what conditions are responsible for increased/decreased metal uptake.

AM fungi might stimulate the synthesis of harmful, rather than beneficial, secondary metabolites, yet there are very few data addressing this possibility (15). For example, while many secondary metabolites are toxic to humans in very high concentrations, a few, such as linamarin in cassava and solanin in potatoes (i.e. glycoalkaloids) are potentially harmful at regular occurring concentrations (Friedman 1997). Other compounds that may be influenced by AM fungi are pro-oxidant (i.e. agents capable of generating toxic oxygen species) or interfere with the absorption of other nutrients (e.g. phytic acid, phenolic compounds) (Frossard et al. 2000). As is the case for metal uptake, there are examples of AM fungi increasing and decreasing secondary metabolites (i.e. Ryan et al. 2008). Clearly more research is needed to resolve this issue.

Soil contamination by radioactive isotopes (13) (e.g. cesium, ^{134}Cs or ^{137}Cs, uranium ^{233}U), results from mining, nuclear weapon testing, or accidents at nuclear power plants. These contaminants represent a serious risk upon entry in the food chain. Although there is evidence that AM fungi have the capacity to accumulate and translocate radionuclides to plant roots, evidence suggests that they may reduce root-to-shoot translocation (Rufyikiri et al. 2004; Dupr de Boulois et al. 2008a,b). Despite the evidence of mycorrhizal mediated mechanisms of enhanced metal tolerance, there is a need to study the movement of radioactive isotopes in AM fungi and their movement into plants in contaminated sites.

Numerous genes expressed in roots colonized by AM fungi encode proteins involved in plant defense responses usually produced after infection with pathogenic microorganisms (14) (Gianinazzi-Pearson et al. 1996; Garcia-Garrido & Ocampo 2002). First analyses suggested that the expression of the corresponding genes is only transient during early infection stages of AM fungal colonization and repressed during late stages, but subsequent research determined some are also expressed in arbusculated cells when the symbiosis is fully established (Grunwald et al. 2004; Hohnjec et al. 2006). A recent study showed that fruits of tomato plants inoculated with *G. mosseae* showed increased expression for 6/8 genes encoding for putative allergens (Schwarz et al. 2011). Although differences between AM plants and controls were not detectable in skin-prick tests, it highlights the potential for microbes to affect the allergenicity of food crops.

3.3.6 The full potential of soil microbes to improve human health

While we have some knowledge about how soil microbes influence plant health (Table 3.3.1); we know little about how they influence our own health. To fully exploit the capacity of soil organisms to improve human health, many basic questions about plant/soil microbial community structure and functioning must be addressed, including the following:

1) *Nutrient loading in edible plant tissues* It is unlikely that all crops will respond equally to rhizosphere microbes because of physiological homeostatic mechanisms within the plant (i.e. tighter controls on nutrient accumulation in seeds versus leaves and stems) (Grusak et al. 1999; Rengel et al. 1999). It is possible that leaf, stem, and root based crops will respond most positively, because nutrient allocation to seeds may be less flexible, but this remains to be tested. While there exists considerable data about the role of AM fungi on plant root/shoot tissue nutrient concentrations, studies that target the edible portions (grains, fruits, tubers, etc.) are few. Such information is needed to determine the true value of rhizosphere microbes in human nutrition.

2) *Differential effects of microbes* It has been well established that AM fungi, like other rhizosphere microbes, are multifunctional (Newsham et al. 1995) and display a certain degree of host preference (Lerat et al. 2003; Bever et al. 2009). Thus it is unlikely that one microbial "inoculant" will serve to enhance all nutrients, in all cropping systems. On the host plant side, AM fungal responsiveness should also be considered as a new trait in breeding programs (Khalil et al. 1994; Gahoonia & Nielsen 2004; Galvan et al. 2008).

3) *"Culturing" healthy soils* What are the best practices for incorporating microbes that enhance food quality into agricultural systems? Are beneficial microbes already present in the soil but limited by management practices? Does microbial diversity matter? It has been shown that more diverse assemblages (i.e. rhizosphere communities from organically versus conventionally managed soils) perform better for a variety of functions (van der Heijden *et al.* 1998; Verbruggen *et al.* 2010). It may be that desired effects in the plant may result from synergistic interactions of different microbial populations. While there is great potential for rhizosphere microbes to mediate crop nutrient levels, much work remains to be done to realize these effects. Whether or not specific inocula need be developed or whether sufficient effects can be realized managing endemic microbes are pressing topics for research.

3.3.7 Conclusion

There is a strong argument for the importance of soil health in the nutritional value of crops. Increases in yields, while important, should not be the only goal driving crop development. Crop quality, in terms of nutrient value, and soil sustainability are equally important. Empirical evidence to support the positive impact of soil microbes on nutrient level of plants and contribution to human nutrition is scant. However, we consider that some well-designed experiments may reveal a novel opportunity to improve human health through sustainable farming practices using soil microbes.

References

Abu-Zeyad, R., Khan, A.G., & Khoo, C. (1999) Occurrence of arbuscular mycorrhiza in *Castanospermum australe* A. Cunn. & C. Fraser and effects on growth and production of castanospermine. *Mycorrhiza* **9**: 111–17.

Adesemoye, A.O., Torbert, H.A., & Kloepper, J.W. (2009) Plant growth-promoting rhizobacteria allow reduced application rates of chemical fertilizers. *Microbial Ecology* **58**: 921–9.

Agbenin, J.O., Danko, M., & Welp, G. (2009) Soil and vegetable compositional relationships of eight potentially toxic metals in urban garden fields from northern Nigeria. *Journal of the Science of Food and Agriculture* **89**: 49–54.

Akiyama, K., & Hayashi, H. (2002) Arbuscular mycorrhizal fungus-promoted accumulation of two new triterpenoids in cucumber roots. *Bioscience, Biotechnology and Biochemistry* **66**: 762–9.

Allen, J.W., & Shachar-Hill, Y. (2009) Sulfur transfer through an arbuscular mycorrhiza. *Plant Physiology* **149**: 549–60.

Anderson, D., Glibert, P., & Burkholder, J. (2002) Harmful algal blooms and eutrophication: Nutrient sources, composition, and consequences. *Estuaries Coasts* **25**: 704–26.

Anderson, I.C., & Cairney, J.W.G. (2004) Diversity and ecology of soil fungal communities: increased understanding through the application of molecular techniques. *Environmental Microbiology* **6**: 769–79.

Andrade, G., Linderman, R.G., & Bethlenfalvay, G.J. (1998) Bacterial associations with the mycorrhizosphere and hyphosphere of the arbuscular mycorrhizal fungus *Glomus mosseae*. *Plant and Soil* **202**: 79–87.

Anriquez, G., & Stlovkal, L. (2008) *Rural population changes in developing countries: lessons for policy making*. ESA Working Paper No. 09-09. The Food and Agriculture Organization of the United Nations, Rome.

Antunes, P.M., de Varennes, A., Zhang, T., & Goss, M.J. (2006a) The tripartite symbiosis formed by indigenous arbuscular mycorrhizal fungi, *Bradyrhizobium japonicum* and soya bean under field conditions. *Journal of Agronomy and Crop Science* **192**: 373–8.

Antunes, P.M., de Varennes, A., Rajcan, I., & Goss, M.J. (2006b) Accumulation of specific flavonoids in soybean (*Glycine max* (L.) Merr.) as a function of the early tripartite symbiosis with arbuscular mycorrhizal fungi and *Bradyrhizobium japonicum* (Kirchner) Jordan. *Soil Biology and Biochemistry* **38**: 1234–42.

Antunes, P.M., Schneider, K., Hillis, D., & Klironomos, J.N. (2007) Can the arbuscular mycorrhizal fungus *Glomus intraradices* actively mobilize P from rock phosphates? *Pedobiologia* **51**: 281–6.

Araim, G., Saleem, A., Arnason, J.T., & Charest, C. (2009) Root colonization by an arbuscular mycorrhizal (AM) fungus increases growth and secondary metabolism of purple coneflower, *Echinacea purpurea* (L.) Moench. *Journal of Agriculture and Food Chemistry* **57**: 2255–8.

Araus, J.L., Slafer, G.A., Royo, C., & Serret, M.D. (2008) Breeding for yield potential and stress adaptation in cereals. *Critical Reviews in Plant Science* **27**: 377–412.

Arlt, M., Schwarz, D., & Franken, P. (2009) Analysis of mycorrhizal functioning using transcriptomics. In: Az

Azcón-Aguilar, C., Barea, J.M., Gianinazzi, S., Gianinazzi-Pearson, V. (eds.) *MycorrhizasÑfunctional processes and ecological impact*, pp. 1–14. Springer, Berlin.

Artursson, V., Finlay, R.D., & Jansson, J.K. (2006) Interactions between arbuscular mycorrhizal fungi and bacteria and their potential for stimulating plant growth. *Environmental Microbiology* **8**: 1–10.

Aseri, G.K., Jain, N., Panwar, J., Rao, A.V., & Meghwal, P.R. (2008) Biofertilizers improve plant growth, fruit yield, nutrition, metabolism and rhizosphere enzyme activities of Pomegranate (*Prunica granatum* L.) in Indian Thar Desert. *Scientia Horticulturae* **117**: 130–5.

Attia, M., Ahmed, M.A., & El-Sonbaty, M.R. (2009) Use of biotechnologies to increase growth, productivity and fruit quality of Maghrabi banana under different rates of phosphorus. *World Journal of Agricultural Sciences* **5**: 211–20.

Azcón, R., Ambrosano, E., & Charest, C. (2003) Nutrient acquisition in mycorrhizal lettuce plants under different phosphorus and nitrogen concentration. *Plant Science* **165**: 1137–45.

Bago, B., Vierheilig, H., Piche, Y., & Azcón-Aguilar, C. (1996) Nitrate depletion and pH changes induced by the extraradical mycelium of the arbuscular mycorrhizal fungus *Glomus intraradices* grown in monoxenic culture. *New Phytologist* **133**: 273–80.

Banchio, E., Xie, X., Zhang, H., & Par, P.W. (2009) Soil bacteria elevate essential oil accumulation and emissions in sweet basil. *Journal of Agriculture and Food Chemistry* **57**: 653–7.

Barea, J.M., Toro, M., Orozco, M.O., Campos, E., & Azcón, R. (2002) The application of isotopic (P-32 and N-15) dilution techniques to evaluate the interactive effect of phosphate-solubilizing rhizobacteria, mycorrhizal fungi and Rhizobium to improve the agronomic efficiency of rock phosphate for legume crops. *Nutrient Cycling in Agroecosystems* **63**: 35–42.

Beckers, G.J.M., & Conrath, U. (2007) Priming for stress resistance: from the lab to the field. *Current Opinion in Plant Biology* **10**: 425–31.

Beman, J.M., Arrigo, K.R., & Matson, P.A. (2005) Agricultural runoff fuels large phytoplankton blooms in vulnerable areas of the ocean. *Nature* **434**: 211–14.

Benedetto, A., Magurno, F., Bonfante, P., & Lanfranco, L. (2005) Expression profiles of a phosphate transporter gene (GmosPT) from the endomycorrhizal fungus *Glomus mosseae*. *Mycorrhiza* **15**: 620–7.

Bever, J.D., Richardson, S.C., Lawrence, B.M., Holmes, J., & Watson, M. (2009) Preferential allocation to beneficial symbiont with spatial structure maintains mycorrhizal mutualism. *Ecology Letters* **12**: 13–21.

Block, E. (1985) The chemistry of garlic and onions. *Scientific American* **252**: 114–19.

Borstler, B., Thiery, O., Sykorova, Z., Berner, A., & Redecker, D. (2010) Diversity of mitochondrial large subunit rDNA haplotypes of *Glomus intraradices* in two agricultural field experiments and two semi-natural grasslands. *Molecular Ecology* **19**: 1497–511.

Bouman, B.A.M., Castaeda, A.R., & Bhuiyan, S.I. (2002) Nitrate and pesticide contamination of groundwater under rice-based cropping systems: past and current evidence from the Philippines. *Agriculture, Ecosystems and the Environment* **92**: 185–99.

Brito, I., Goss, M.J., Carvalho, M., Tuinen, D., & Antunes, P.M. (2008) Agronomic management of indigenous mycorrhizas. In: Varma, A. (eds.) *Mycorrhiza*, pp. 375–402. Springer, Berlin.

Butelli, E., Titta, L., Giorgio, M., Mock, H.P., Matros, A., Peterek, S., Schijlen, E., Hall, R.D., Bovy, A.G., Luo, J., & Martin, C. (2008) Enrichment of tomato fruit with health-promoting anthocyanins by expression of select transcription factors. *Nature Biotechnology* **26**: 1301–8.

Cabello, M., Irrazabal, G., Bucsinszky, A.M., Saparrat, M., & Schalamuk, S. (2005) Effect of an arbuscular mycorrhizal fungus, *Glomus mosseae*, and a rock-phosphate-solubilizing fungus, *Penicillium thomii*, on *Mentha piperita* growth in a soilless medium. *Journal of Basic Microbiology* **45**: 182–9.

Cakmak, I., Kalayci, M., Ekiz, H., Braun, H.J., Kilinc, Y., & Yilmaz, A. (1999) Zinc deficiency as a practical problem in plant and human nutrition in Turkey: A NATO-science for stability project. *Field Crops Research* **60**: 175–88.

Caravaca, F., Alguacil, M.M., Azcón, R., Diaz, G., & Roldan, A. (2004) Comparing the effectiveness of mycorrhizal inoculation and amendment with sugar beet, rock phosphate and *Aspergillus niger* to enhance field performance of the leguminous shrub *Dorycnium pentaphyllum* L. *Applied Soil Ecology* **25**: 169–80.

Cardoso, I.M., & Kuyper, T.W. (2006) Mycorrhizas and tropical soil fertility. *Agriculture, Ecosystems and the Environment* **116**: 72–84.

Carlsen, S.C.K., Understrup, A., Fomsgaard, I.S., Mortensen, A.G., & Ravnskov, S. (2008) Flavonoids in roots of white clover: interaction of arbuscular mycorrhizal fungi and a pathogenic fungus. *Plant and Soil* **302**: 33–43.

Castellanos-Morales, V., Villegas, J., Wendelin, S., Vierheilig, H., Ederc, R., & Cardenas-Navarroa, R. (2010) Root colonisation by the arbuscular mycorrhizal fungus *Glomus intraradices* alters the quality of strawberry fruits (*Fragaria ananassa* Duch.) at different nitrogen levels. *Journal of the Science of Food and Agriculture* **90**: 1774–82.

Ceccarelli, N., Curadi, M., Martelloni, L., Sbrana, C., Picciarelli, P., & Giovannetti, M. (2010) Mycorrhizal colonization impacts on phenolic content and antioxidant properties of artichoke leaves and flower heads two years after field transplant. *Plant and Soil* **335**: 311–23.

Chaudhary, V., Kapoor, R., & Bhatnagar, A.K. (2008) Effectiveness of two arbuscular mycorrhizal fungi on concentrations of essential oil and artemisinin in three accessions of *Artemisia annua* L. *Applied Soil Ecology* **40**: 174–81.

Chen, B.P.R., Borggaard, O.K., Zhu, Y., & Jakobsen, I. (2005) Mycorrhiza and root hairs in barley enhance acquisition of phosphorus and uranium from phosphate rock but mycorrhiza decreases root to shoot uranium transfer. *New Phytologist* **165**: 591–8.

Chifflot, V., Rivest, D., Olivier, A., Cogliastro, A., & Khasa, D. (2009) Molecular analysis of arbuscular mycorrhizal community structure and spores distribution in tree-based intercropping and forest systems. *Agriculture, Ecosystems and the Environment* **131**: 32–9.

Copetta, A., Lingua, G., & Berta, G. (2006) Effects of three AM fungi on growth, distribution of glandular hairs, and essential oil production in *Ocimum basilicum* L. var. Genovese. *Mycorrhiza* **16**: 485–94.

Cordain, L. (1999) Cereal grains: Humanity's double-edged sword. In: A.P. Simopoulos (ed.) Evolutionary aspects of nutrition and health. diet, exercise, genetics and chronic disease. *World Review of Nutrition and Dietetics Basel, Karger* **84**: 19–73.

Corkidi, L., Rowland, D.L., Johnson, N.C., & Allen, E.B. (2002) Nitrogen fertilization alters the functioning of arbuscular mycorrhizas at two semiarid grasslands. *Plant and Soil* **240**: 299–310.

Croll, D., Giovannetti, M., Koch, A.M., Sbrana, C., Ehinger, M., Lammers, P.J., & Sanders, I.R. (2009) Nonself vegetative fusion and genetic exchange in the arbuscular mycorrhizal fungus *Glomus intraradices*. *New Phytologist* **181**: 924–37.

Dardanelli, M.S., Fernandez de Curdoba, F.J., *et al.* (2008) Effect of *Azospirillum brasilense* coinoculated with *Rhizobium* on *Phaseolus vulgaris* flavonoids and Nod factor production under salt stress. *Soil Biology and Biochemistry* **40**: 2713–21.

Davis, D.R. (2009) Declining fruit and vegetable nutrient composition: What is the evidence? *HortScience* **44**: 15–19.

Denning, G., Kabambe, P., Sanchez, P., *et al.* (2009) Input subsidies to improve smallholder maize productivity in Malawi: Toward an African green revolution. *PLoS Biology* **7**: 2–10.

Devi, M.C., & Reddy, M.N. (2002) Phenolic acid metabolism of groundnut (*Arachis hypogaea* L.) plants inoculated with VAM fungus and *Rhizobium*. *Plant Growth Regulators* **37**: 151–6.

Drijber, R.A., Doran, J.W., Parkhurst, A.M., & Lyon, D.J. (2000) Changes in soil microbial community structure with tillage under long-term wheat-fallow management. *Soil Biology and Biochemistry* **32**: 1419–30.

Dunfield, K.E., & Germida, J.J. (2003) Seasonal changes in the rhizosphere microbial communities associated with field-grown genetically modified canola (*Brassica napus*). *Applied and Environmental Microbiology* **69**: 7310–18.

Dupr de Boulois, H., Joner, E.J., Leyval, C., *et al.* (2008a) Impact of arbuscular mycorrhizal fungi on uranium accumulation by plants. *Journal of Environmental Radioactivity* **99**: 775–84.

Dupr de Boulois, H., Joner, E.J., Leyval, C., *et al.* (2008b) Role and influence of mycorrhizal fungi on radiocesium accumulation by plants. *Journal of Environmental Radioactivity* **99**: 785–800.

Ekholm, P., Reinivuo, H., Mattila, P., *et al.* (2007) Changes in the mineral and trace element contents of cereals, fruits and vegetables in Finland. *Journal of Food Composition and Analysis* **20**: 487–95.

El Fantroussi, S., Verschuere, L., Vweaxhuweew, L., Verstraete, W.l., & Top, E.M. (1999) Effect of phenylurea herbicides on soil microbial communities estimated by analysis of 16S rRNA gene fingerprints and community-level physiological profiles. *Applied and Environmental Microbiology* **65**: 982–8.

Engelen, B., Meinken, K., von Wintzintgerode, F., Heure, H., Malkomes, H.P., & Bakhaus, H. (1998) Monitoring impact of a pesticide treatment on bacterial soil communities by metabolic and genetic fingerprinting in addition to conventional testing procedures. *Applied and Environmental Microbiology* **64**: 2814–21.

Enserink, M. (2008) Tough lessons from golden rice. *Science* **320**: 468–71.

Fageria, N.K., Baligar, V.C., & Li, Y.C. (2008) The role of nutrient efficient plants in improving crop yields in the twenty first century. *Journal of Plant Nutrition* **31**: 1121–57.

Fahey, J.W., Zalcmann, A.T., & Talalay, P. (2001) The chemical diversity and distribution of glucosinolates and isothiocyanates among plants. *Phytochemistry* **56**: 5–51.

Feng, G., Song, Y.C., Li, X.L., & Christie, P. (2003) Contribution of arbuscular mycorrhizal fungi to utilization of organic sources of phosphorus by red clover in a calcareous soil. *Applied Soil Ecology* **22**: 139–48.

Fester, T., Wray, V., Nimtz, M., & Strack, D. (2005) Is stimulation of carotenoid biosynthesis in arbuscular mycorrhizal roots a general phenomenon? *Phytochemistry* **66**: 1781–6.

Friedman, M. (1997) Chemistry, biochemistry, and dietary role of potato polyphenols. A Review. *Journal of Agronomy and Food Chemistry* **45**: 1523–40.

Frossard, E., Bucher, M., MŠchler, F., Mozafar, A., & Hurrell, R. (2000) Potential for increasing the content and bioavailability of Fe, Zn and Ca in plants for human nutrition. *Journal of the Science of Food and Agriculture* **80**: 861–79.

Gahoonia, T.S., & Nielsen, N.E. (2004) Root traits as tools for creating phosphorus efficient crop varieties. *Plant and Soil* **260**: 47–57.

Galvan, G.A., Burger, K., Kuiper, T.W., Kik, C., & Scholten, O.E. (2008) *Breeding for improved responsiveness to arbuscular mycorrhizal fungi in onion.* Plant Research International, Wageningen UR publication, Wageningen.

Garcia-Garrido, J.M., & Ocampo, J.A. (2002) Regulation of the plant defence response in arbuscular mycorrhizal symbiosis. *Journal of Experimental Botany* **53**: 1377–86.

Gavito, M.E., & Miller, M.H. (1998) Changes in mycorrhiza development in maize induced by crop management practices. *Plant and Soil* **198**: 185–92.

Gentili, F., & Jumpponen, A. (2006) Potential and possible uses of bacterial and fungal biofertilizers. In: Rai, M.K. (ed.) *Handbook of Microbial Biofertlizers,* pp. 1–28. Food Products Press, Binghampton, NY.

George, T.S., Simpson, R.J., Hadobas, P.A., & Richardson, A.E. (2005) Expression of a fungal phytase gene in *Nicotiana tabacum* improves phosphorus nutrition of plants grown in amended soils. *Plant Biotechnology Journal* **3**: 129–40.

Gerritse, R., Van Driel, W., Smilde, K., & Van Luit, B. (1983) Uptake of heavy metals by crops in relation to their concentration in the soil solution. *Plant and Soil* **75**: 393–404.

Gianinazzi-Pearson, V., Dumas-Gaudot, E., Gollotte, A., Tahiri-Alaoui, A., & Gianinazzi, S. (1996) Cellular and molecular defence-related root responses to invasion by arbuscular mycorrhizal fungi. *New Phytologist* **133**: 45–57.

Grunwald, U., Nyamsuren, O., Tarnasloukht, M., et al. (2004) Identification of mycorrhiza-regulated genes with arbuscule development-related expression profile. *Plant Molecular Biology* **55**: 553–66.

Grusak, M.A., & DellaPenna, D. (1999) Improving the nutrient composition of plants to enhance human nutrition and health. *Annual Review of Plant Physiology and Plant Molecular Biology* **50**: 133–61.

Grusak, M.A., Pearson, J.N., & Marentes, E. (1999) The physiology of micronutrient homeostasis in field crops. *Field Crops Research* **60**: 41–56.

Guether, M., Neuhauser, B., Balestrini, R., Dynowski, M., Ludewig, U., & Bonfante, P. (2009) A mycorrhizal-specific ammonium transporter from Lotus japonicus acquires nitrogen released by arbuscular mycorrhizal fungi. *Plant Physiology* **150**: 73–83.

Gyaneshwar, P., Kumar, G.N., Parekh, L.J., & Poole, P.S. (2002) Role of soil microorganisms in improving P nutrition of plants. *Plant and Soil* **245**: 83–93.

Halliwell, B. (1996) Antioxidants in human health and disease. *Annual Review of Nutrition* **16**: 33–50.

Harinikumar, K.M., & Bagyaraj, D.J. (1988) Effect of crop rotation on native vesicular arbuscular mycorrhizal propagules in the soil. *Plant and Soil* **110**: 77–80.

Harrison, M.J., & Vanbuuren, M.L. (1995) A phosphate transporter from the mycorrhizal fungus *Glomus versiforme. Nature* **378**: 626–9.

Hart, M.M., Powell, J.R., Gulden, R.H., et al. (2009) Separating the effect of crop from herbicide on soil microbial communities in glyphosate-resistant corn. *Pedobiologia* **52**: 253–62.

Haselwandter, K., & Winkelmann, G. (2007) Siderophores of symbiotic fungi. In: Chincholkar, S.B., Varma, A., (eds.), *Microbial Siderophores. Soil Biology Series, Vol. 12,* pp. 91–103. Springer, Berlin.

Hashem, M.A. (2001) Problems and prospects of cyanobacterial biofertilizer for rice cultivation. *Functional Plant Biology* **28**: 881–8.

He, X., & Nara, K. (2007) Element biofortification: can mycorrhizas potentially offer a more effective and sustainable pathway to curb human malnutrition? *Trends in Plant Science* **12**: 331–3.

Helgason, T., Daniell, T.J., Husband, R., Fitter, A.H., & Young, J.P.W. (1998) Ploughing up the wood-wide web? *Nature* **394**: 431–431.

Hetrick, B.A.D., Wilson, G.W.T., & Cox, T.S. (1992) Mycorrhizal dependence of modern wheat-varieties, landraces, and ancestors. *Canadian Journal of Botany* **70**: 2032–40.

Hetrick, B.A.D., Wilson, G.W.T., & Cox, T.S. (1993) Mycorrhizal dependence of modern wheat cultivars and ancestorsÑa synthesis. *Canadian Journal of Botany* **71**: 512–18.

Hodge, A., & Fitter, A. (2010) Substantial nitrogen acquisition by arbuscular mycorrhizal fungi from organic material has implications for N cycling. *Proceedings of the National Academy of Sciences of the United States of America* **107**: 13754–9.

Hohnjec, N., Henckel, K., Bekel, T., et al. (2006) Transcriptional snapshots provide insights into the molecular basis of arbuscular mycorrhiza in the model legume *Medicago truncatula. Functional Plant Biology* **33**: 737–48.

Jansa, J., Mozafar, A., Kuhn, G., et al. (2003) Soil tillage affects the community structure of mycorrhizal fungi in maize roots. *Ecological Applications* **13**: 1164–76.

Jeffries, P., Gianinazzi, S., Perotto, S., Turnau, K., & Barea, J.M. (2003) The contribution of arbuscular mycorrhizal fungi in sustainable maintenance of plant health and soil fertility. *Biology and Fertility of Soils* **37**: 1–16.

Johnson, L. (2008) Iron and siderophores in fungal-host interactions. *Mycological Research* **112**: 170–83.

Johnson, N.C. (1993) Can fertilization of soil select less mutualistic mycorrhizae? *Ecological Applications* **3**: 749–57.

Johnson, N.C. (2010) Resource stoichiometry elucidates the structure and function of arbuscular mycorrhizas across scales. *New Phytologist* **185**: 631–47.

Johnson, N.C., Wilson, G.W.T., Bowker, M.A., Wilson, J.A., & Miller, R.M. (2010) Resource limitation is a driver of local adaptation in mycorrhizal symbioses. *Proceedings of the National Academy of Sciences of the United States of America* **107**: 2093–8.

Joner, E., Briones, R., & Leyval, C. (2000) Metal-binding capacity of arbuscular mycorrhizal mycelium. *Plant and Soil* **226**: 227–34.

Jones, D.L., Hodge, A., & Kuzyakov, Y. (2004) Plant and mycorrhizal regulation of rhizodeposition. *New Phytologist* **163**: 459–80.

Jumpponen, A. (2001) Dark septate endophytesÑare they mycorrhizal? *Mycorrhiza* **11**: 207–11.

Kaluski, D.N., Tulchinsky, T.H., Haviv, A., et al. (2003) Addition of essential micronutrients to foodsÑimplication for public health policy in Israel. *IMAJ* **5**: 277–80.

Karagiannidis, N., Nikolaou, N., Ipsilantis, I., & Zioziou, E. (2007) Effects of different N fertilizers on the activity of Glomus mosseae and on grapevine nutrition and berry composition. *Mycorrhiza* **18**: 43–50.

Kaya, C., Ashraf, M., Sonmez, O., et al. (2009) The influence of arbuscular mycorrhizal colonisation on key growth parameters and fruit yield of pepper plants grown at high salinity. *Scientia Horticulturae* **12**: 1–6.

Khalil, S., Loynachan, T.E., & Tabatabai, M.A. (1994) Mycorrhizal dependency and nutrient-uptake by improved and unimproved corn and soybean cultivars. *Agronomy Journal* **86**: 949–58.

Khaosaad, T., Krenn, L., Medjakovic, S., et al. (2008) Effect of mycorrhization on the isoflavone content and the phytoestrogen activity of red clover. *Journal of Plant Physiology* **165**: 1161–7.

Kiers, E.T., Hutton, M.G., & Dennison, R.F. (2007) Human selection and the relaxation of legume defences against ineffective rhizobia. *Proceedings of the Royal Society B* **274**: 3119–26.

Kim, K., Yim, W., Trivedi, P., et al. (2010) Synergistic effects of inoculating arbuscular mycorrhizal fungi and *Methylobacterium oryzae* strains on growth and nutrient uptake of red pepper (*Capsicum annum* L.) *Plant and Soil* **327**: 429–40.

Kleijn, D., Kohler, F., Baldi, A., et al.. (2009) On the relationship between farmland biodiversity and land-use intensity in Europe. *Proceedings of the Royal Society B* **276**: 903–9.

Klingner, A., Bothe, H., Wray, V., & Marner, F.J. (1995) Identification of a yellow pigment formed in maize roots upon mycorrhizal colonization. *Phytochemistry* **38**: 53–5.

Kochian, L.V., & Garvin, D.F. (1999) Agricultural approaches to improving phytonutrient content in plants: An overview. *Nutrition Reviews* **57**: S13–S18.

Koide, R.T., & Kabir, Z. (2000) Extraradical hyphae of the mycorrhizal fungus *Glomus intraradices* can hydrolyse organic phosphate. *New Phytologist* **148**: 511–17.

Krishna, H., Singh, S.K., Sharma, R.R., Khawale, R.N., Grover, M., & Patel, V.B. (2005) Biochemical changes in micropropagated grape (*Vitis vinifera* L.) plantlets due to arbuscular-mycorrhizal fungi (AMF) inoculation during ex vitro acclimatization. *Scientia Horticulturae* **106**: 554–67.

Lambert, D.H, Baker, D.E., & Cole, H. (1979) The role of mycorrhizae in the interactions of P with Zn, Cu and other elements. *Soil Science Society of America Journal* **43**: 976–908.

Larose, G., Chenevert, R., Moutoglis, P., Gagne, S., Piche, Y., & Vierheilig, H. (2002) Flavonoid levels in roots of *Medicago sativa* are modulated by the developmental stage of the symbiosis and the root colonizing arbuscular mycorrhizal fungus. *Journal of Plant Physiology* **159**: 1329–39.

Leigh, J., Hodge, A., & Fitter, A.H. (2009) Arbuscular mycorrhizal fungi can transfer substantial amounts of nitrogen to their host plant from organic material. *New Phytologist* **181**: 199–207.

Lerat, S., Lapointe, L., Gutjahr, S., Piche, Y., & Vierheilig, H. (2003) Carbon partitioning in a split-root system of arbuscular mycorrhizal plants is fungal and plant species dependent. *New Phytologist* **157**: 589–95.

Levin, D.A. (1976) Alkaloid-bearing plants: An ecogeographic perspective. *American Naturalist* **110**: 261–84.

Li, X.L., George, E., & Marschner, H. (1991) Extension of the phosphorus depletion zone in VA-mycorrhizal white clover in a calcareous soil. *Plant and Soil* **136**: 41–8.

Liu, J.Y., Maldonado-Mendoza, I., Lopez-Meyer, M., Cheung, F., Town, C.D., & Harrison, M.J. (2007) Arbuscular mycorrhizal symbiosis is accompanied by local and systemic alterations in gene expression and an increase in disease resistance in the shoots. *Plant Journal* **50**: 529–44.

Lopez-Pedrosa, A., Gonzalez-Guerrero, M., Valderas, A., Azcón-Aguilar, C., & Ferrol, N. (2006) GintAMT1 encodes a functional high-affinity ammonium transporter that is expressed in the extraradical mycelium of *Glomus intraradices*. *Fungal Genetics and Biology* **43**: 102–10.

Mandyam, K., & Jumpponen, A. (2005) Seeking the elusive function of the root-colonizing dark septate endophytic fungus. *Studies in Mycology* **53**: 173–89.

Manoharan, P.T., Pandi, M., Shanmugaiah, V., Gomathinayayagam, S., & Balasubramanian, N. (2008) Effect of vesicular arbuscular mycorrhizal fungus on the physiological and biochemical changes of five different tree seedlings grown under nursery conditions. *African Journal of Biotechnology* **7**: 3431–6.

Marschner, H. (1995) *Mineral nutrition of higher plants*. Academic Press, San Diego, CA.

Marschner, H. (1996) Mineral nutrient acquisition in nonmycorrhizal and mycorrhizal plants. *Phyton-Annales Rei Botanicae* **36**: 61–8.

Marschner, H., & Dell, B. (1994) Nutrient-uptake in mycorrhizal symbiosis. *Plant and Soil* **159**: 89–102.

Martin, F., Gianinazzi-Pearson, V., Hijri, M., et al. (2008) The long hard road to a completed *Glomus intraradices* genome. *New Phytologist* **180**: 747–50.

Mathimaran, N., Ruh, R., Jama, B., Verchot, L., Frossard, E., & Jansa, J. (2007) Impact of agricultural management on arbuscular mycorrhizal fungal communities in Kenyan ferra sol. *Agriculture, Ecosystems and the Environment* **119**: 22–32.

Mayer, J.E., Pfeiffer, W.H., & Beyer, P. (2008) Biofortified crops to alleviate micronutrient malnutrition. *Current Opinion in Plant Biology* **11**: 166–70.

McLaughlin, M.J., Parker, D.R., & Clarke, J.M. (1999) Metals and micronutrientsÑfood safety issues. *Field Crops Research* **60**: 143–63.

Mena-Violante, H., Ocampo-JimŽnez, O., Dendooven, L., et al. (2006) Arbuscular mycorrhizal fungi enhance fruit growth and quality of chile ancho (*Capsicum annuum* L. cv San Luis) plants exposed to drought. *Mycorrhiza* **16**: 261–7

Miller, M.H. (2000) Arbuscular mycorrhizae and the phosphorus nutrition of maize: A review of Guelph studies. *Canadian Journal of Plant Science* **80**: 47–52.

Miransari, M., Bahrami, H.A., Rejali, F., & Malakouti, M.J. (2008) Using arbuscular mycorrhiza to reduce the stressful effects of soil compaction on wheat (*Triticum aestivum* L.) growth. *Soil Biology and Biochemistry* **40**: 1197–206.

Miransari M., Bahrami, H.A, Rejali, F., & Malakouti, M.J. (2009) Effects of arbuscular mycorrhiza, soil sterilization, and soil compaction on wheat (*Triticum aestivum* L.) nutrients uptake. *Soil Tillage Research* **104**: 48–55.

Molina, R. (1982) Use of the ectomycorrhizal fungus Laccarialaccata in forestry. I. Consistency between isolates in effective colonization of containerized conifer seedlings. *Canadian Journal of Forest Research* **12**: 469–73.

Morrissey, J.P., Maxwell J., Dow, G., Mark, L., & O'Gara, F. (2004) Are microbes at the root of a solution to world food production? *EMBO Reports* **5**: 922–6.

Nell, M., Votsch, M., Vierheilig, H.M., et al. (2009) Effect of phosphorus uptake on growth and secondary metabolites of garden sage (*Salvia officinalis* L.). *Journal of Science Food and Agriculture* **89**: 1090–6.

Nell, M., Wawrosch, C., Steinkellner, S., Vierheilig, H., Kopp, B., & Lossi, A. (2010) Root colonization by symbiotic arbuscular mycorrhizal fungi increases sesquiterpenic acid concentrations in *Valeriana officinalis* L. *Planta Medica* **76**: 393–8.

Nestel, P., Bouis, H.E., Meenakshi, J.V., & Pfeiffer, W. (2006) Biofortification of staple food crops. *The Journal of Nutrition* **136**: 1064–7.

Newsham, K.K., Fitter, A.H., & Watkinson, A.R. (1995) Multifunctionality and biodiversity in arbuscular mychorrhizas. *Trends in Ecology & Evolution* **10**: 407–11.

Nosengo, N. (2003) Fertilized to death. *Nature* **425**: 894–5.

Nriagu, J.O., & Pacyna, J.M. (1988) Quantitative assessment of worldwide contamination of air, water and soils by trace metals. *Nature* **333**: 134–9.

Oehl, F., Sieverding, E., Mader, P., et al. (2004) Impact of long-term conventional and organic farming on the diversity of arbuscular mycorrhizal fungi. *Oecologia* **138**: 574–83.

Pawlowska, T.E., & Charvat, I. (2004) Heavy-metal stress and developmental patterns of arbuscular mycorrhizal fungi. *Applied and Environmental Microbiology* **70**: 6643–9.

Peipp, H., Maier, W., Schmidt, J., Wray, V., & Strack, D. (1997) Arbuscular mycorrhizal fungus-induced changes in the accumulation of secondary compounds in barley roots. *Phytochemistry* **44**: 581–7.

Peoples, M.B., & Craswell, E.T. (1992) Biological nitrogen fixation: Investments, expectations and actual contributions to agriculture. *Plant and Soil* **141**: 13–39.

Perner, H., Rohn, S., Driemel, G., et al. (2008) Effect of nitrogen species supply and mycorrhizal colonization on organosulfur and phenolic compounds in onions. *Journal of Agricultural and Food Chemistry* **56**: 3538–45.

Persello-Cartieaux, F., Nussaume, L., & Robaglia, C. (2003) Tales from the underground: molecular plant-rhizobacteria interactions. *Plant, Cell Environment* **26**: 189–99.

Petrini, O. (1996) Ecological and physiological aspects of host specificity in endophytic fungi In: Redlin, S.C., Carris, L.M. (eds.) *Endophytic Fungi in Grasses and Woody Plants*, pp. 87–100. APS Press, St Paul, MN.

Pimentel, D., & Pimentel, M. (1996) *Food, Energy and Society*. University Press of Colorado, Boulder, CO.

Pongrac, P., Vogel-Mikus, K., Regvar, M., Tolra, R., Poschenrieder, C., & Barcelo, J. (2008) Glucosinolate profiles change during the life cycle and mycorrhizal colonization in a Cd/Zn hyperaccumulator *Thlaspi praecox* (Brassicaceae). *Journal of Chemical Ecology* **34**: 1038–44.

Powell, J.R., Campbell, R.G., Dunfield, K.E., et al. (2009) Effect of glyphosate on the tripartite symbiosis formed by *Glomus intraradices, Bradyrhizobium japonicum*, and genetically modified soybean. *Appied Soil Ecology* **41**: 128–36.

Rabie, G.H. (2005) Contribution of arbuscular mycorrhizal fungus to red kidney and wheat plants tolerance grown in heavy metal-polluted soil. *African Journal of Biotechnology* **4**: 332–45.

Rai, M.K. (2006) *Handbook of microbial biofertilizers*. The Haworth Press, Inc., Binghamton, NY.

Rapparini, F., Liusia, J., & Penuelas, J. (2008) Effect of arbuscular mycorrhizal (AM) colonization on terpene emission and content of *Artemisia annua* L. *Plant Biolology* **10**: 108–22.

Rengel, Z., Batten, G.D., & Crowley, D.E. (1999) Agronomic approaches for improving the micronutrient density in edible portions of field crops. *Field Crops Research* **60**: 27–40.

Rengel, Z., & Marschner, P. (2005) Nutrient availability and management in the rhizosphere: exploiting genotypic differences. *New Phytologist* **168**: 305–12.

Renshaw, J.C., Robson, G.D., Trinci, A.P.J., et al. (2002) Fungal siderophores: structures, functions and applications. *Mycological Research* **106**: 1123–42.

Richardson, A.E. (2001) Prospects for using soil microorganisms to improve the acquisition of phosphorus by plants. *Australian Journal of Plant Physiology* **28**: 897–906.

Richardson, A.E., Barea, J.M., McNeill, A.M., & Prigent-Combaret, C. (2009) Acquisition of phosphorus and nitrogen in the rhizosphere and plant growth promotion by microorganisms. *Plant and Soil* **321**: 305–39.

Rodriguez, H., & Fraga, R. (1999) Phosphate solubilizing bacteria and their role in plant growth promotion. *Biotechnology Advances* **17**: 319–39.

Rojas-Andrade, R., Cerda-Garcia-Rojas, C.M., et al. (2003) Changes in the concentration of trigonelline in a semiarid leguminous plant (*Prosopis laevigata*) induced by an arbuscular mycorrhizal fungus during the presymbiotic phase. *Mycorrhiza* **13**: 49–52.

Royal Society (2009) *Reaping the Benefits: Science and the sustainable intensification of global agriculture*. Royal Society Policy document 11/09, London.

Rufyikiri, G., Declerck, S., & Thiry, Y. (2004) Comparison of 233U and 33P uptake and translocation by the arbuscular mycorrhizal fungus *Glomus intraradices* in root organ culture conditions. *Mycorrhiza* **14**: 203–7.

Ryan, M.H., & Angus, A.J. (2003) Abuscular mycorrhizae in wheat and field pea crops on a low P soil: increased Zn-uptake but no increase in P-uptake or yield. *Plant and Soil* **250**: 225–39.

Ryan, M.H., & Graham, J.H. (2002) Is there a role for arbuscular mycorrhizal fungi in production agriculture? *Plant and Soil* **244**: 263–71.

Ryan, M.H., McInerney, J.K., Record, I.R., & Angus, J.F. (2008) Zinc bioavailability in wheat grain in relation to phosphorus fertiliser, crop sequence and mycorrhizal fungi. *Journal of the Science of Food and Agriculture* **88**: 1208–16.

Saikkonen, K., Faeth, S.H., Helander, M., & Sullivan, T.J. (1998) Fungal edophytes: A Continuum of interactions with host plants. *Annual Review of Ecology and Systematics* **29**: 319–43.

Sanders, I.R. (2010) "Designer" mycorrhizas: Using natural genetic variation in AM fungi to increase plant growth. *ISME Journal* **4**: 1081–3.

Sawers, R.J., Gutjahr, C., & Paszkowski, U. (2006) Cereal mycorrhiza: an ancient symbiosis in modern agriculture. *Trends in Plant Science* **13**: 93–7.

Schimel, J.P., & Bennett, J. (2004) Nitrogen mineralization: challenges of a changing paradigm. *Ecology* **85**: 591–602.

Schwartz, M.W., Hoeksema, J.D., Gehring, C.A., et al. (2006) The promise and the potential consequences of the global transport of mycorrhizal fungal inoculum. *Ecology Letters* **9**: 501–15.

Schwarz, D., Welter, S., George, E., et al. (2011) Impact of arbuscular mycorrhizal fungi on the allergenic potential of tomato. *Mycorrhiza* **21**: 341–9.

Selvaraj, T., Rajeshkumar, S., Nisha, M.C., Wondimu, L., & Tesso, M. (2008) Effect of Glomus mosseae and plant growth promoting rhizomicroorganisms (PGPR's) on growth, nutrients and content of secondary metabolites in *Begonia malabarica* Lam. *Maejo International Journal of Science and Technology* **2**: 516–25.

Simmons, B.L., & Coleman, D.C. (2008) Microbial community response to transition from conventional to conservation tillage in cotton fields. *Applied Soil Ecology* **40**: 518–28.

Singh, S., Pandey, A., Kumar, B., & Palni, L.M.S. (2010) Enhancement in growth and quality parameters of tea *Camellia sinensis* (L.) O. Kuntze through inoculation with arbuscular mycorrhizal fungi in an acid soil. *Biology and Fertility of Soils* **46**: 427–33.

Smith, F.A., Grace, E.J., & Smith, S.E. (2009) More than a carbon economy: nutrient trade and ecological sustainability in facultative arbuscular mycorrhizal symbioses. *New Phytologist* **182**: 347–58.

Smith, S.E., & Read, D.J. (2008) *Mycorrhizal symbioses*. Academic Press, London.

Sorensen, J.N., Larsen, J. & Jakobsen, I. (2008) Pre-inoculation with arbuscular mycorrhizal fungi increases early nutrient concentration and growth of field-grown leeks under high productivity conditions. *Plant and Soil* **307**: 135–47.

Stockinger, H., Walker, C., & Schussler, A. (2009) *"Glomus intraradices* DAOM197198," a model fungus in arbuscular mycorrhiza research, is not Glomus intraradices. *New Phytologist* **183**: 1176–87.

Subramanian, K.S., Santhanakrishnan, P., & Balasubramanian, P. (2006) Responses of field grown tomato plants to arbuscular mycorrhizal fungal colonization under varying intensities of drought stress. *Scientia Horticulturae* **107**: 245–53.

Szumigalski, A.R., & van Acker, R.C. (2006) The agronomic value of annual plant diversity in crop-weed systems. *Canadian Journal of Plant Science* **86**: 865–74.

Tinker, P.B.N., & Nye, P.H. (2000) *Solute movement in the rhizosphere*. Oxford University Press, Oxford.

Toljander, J.F., Lindahl, B.D., Paul, L.R., Elfstrand, M., & Finlay, R.D. (2007) Influence of arbuscular mycorrhizal mycelial exudates on soil bacterial growth and community structure. *FEMS Microbiology Ecology* **61**: 295–304.

Toro, M., Azcón, R., & Barea, J.M. (1998) The use of isotopic dilution techniques to evaluate the interactive effects of Rhizobium genotype, mycorrhizal fungi, phosphate-solubilizing rhizobacteria and rock phosphate on nitrogen and phosphorus acquisition by *Medicago sativa*. *New Phytologist* **138**: 265–73.

Toth, R., Toth, D., Starke, D., & Smith, D.R. (1990) Vesicular-arbuscular mycorrhizal colonization in *Zea mays* affected by breeding for resistance to fungal pathogens. *Canadian Journal of Botany* **68**: 1039–44.

Toussaint, J., Smith, F., & Smith, S. (2007) Arbuscular mycorrhizal fungi can induce the production of phytochemicals in sweet basil irrespective of phosphorus nutrition. *Mycorrhiza* **17**: 291–7.

Toussaint, J.P., Kraml, M., Nell, M., *et al.* (2008) Effect of *Glomus mosseae* on concentrations of rosmarinic and caffeic acids and essential oil compounds in basil inoculated with *Fusarium oxysporum* f. sp. *Plant Pathology* **57**: 1109–16.

Trappe, J.M. (1977) Selection of fungi for ectomycorrhizal inoculation in nurseries. *Annual Reviews in Phytopathology* **15**: 203–22.

Ulrichs, C., Fischer G., BŸttner, C., & Mewis, I. (2008) Comparison of lycopene, β-carotene and phenolic contents of tomato using conventional and ecological horticultural practices, and arbuscular mycorrhizal fungi (AMF). *Agronomia Colombiana* **26**: 40–6.

Vaidya, V., Ingold, K.U., & Pratt, D.A. (2009) Garlic: source of the ultimate antioxidants-sulfenic acids. *Angewande Chemie* **48**: 157–60.

Vaishampayan, A., Sinha, R.P., Hader, D.P., *et al.* (2001) Cyanobacterial biofertilizers in rice agriculture. *Botanical Review* **6**: 453–516.

van der Heijden, M.G.A., Klironomos, J.N., Ursic, M., *et al.* (1998) Mycorrhizal fungal diversity determines plant biodiversity, ecosystem variability and productivity. *Nature* **396**: 69–72.

Varma, A., Savita, V., Sudha, Sahay, N., Butehorn, B., & Franken, P. (1999) Piriformospora indica, a cultivable plant-growth-promoting root endophyte. *Applied and Environmental Microbiology* **65**: 2741–4.

Vekerk, K., Schriner, M, Krumbein, A., *et al.* (2009) Glucosinolates in Brassica vegetables: the influence of the food supply chain on intake, bioavailability and human health. *Molecular Nutrition & Foods Research* **53**: S219–65.

Verbruggen, E., Rsling, W.F.M., Gamper, H.A., *et al.* (2010) Positive effects of organic farming on below-ground mutualists: large-scale comparison of mycorrhizal fungal communities in agricultural soils. *New Phytologist* **186**: 968–79.

Vessey, J.K. (2003) Plant growth promoting rhizobacteria as biofertilizers. *Plant and Soil* **255**: 571–86.

Vierheilig, H., Bennett, R., Kiddle, G., Kaldorf, M., & Ludwig-Muller, J. (2000) Differences in glucosinolate patterns and arbuscular mycorrhizal status of glucosinolate-containing plant species. *New Phytologist* **146**: 343–52.

Waller, F., Achatz, B., Baltruschat, H., *et al* (2005) The endophytic fungus *Piriformospora indica* reprograms barley to salt-stress tolerance, disease resistance, and higher yield. *Proceedings of the National Academy of Science of the United States of America* **102**: 13386–91.

Walter, M.H., Fester, T., & Strack, D. (2000) Arbuscular mycorrhizal fungi induce the non-mevalonate methylerythritol phosphate pathway of isoprenoid biosynthesis correlated with accumulation of the "yellow pigment" and other apocarotenoids. *Plant Journal* **21**: 571–8.

Wang, B., & Qiu, Y.L. (2006) Phylogenetic distribution and evolution of mycorrhizas in land plants. *Mycorrhiza* **16**: 299–363.

Weiss, M., Selosse, M.A., Rexer, K.H., Urban, A., & Oberwinkler, F. (2004) Sebacinales: a hitherto overlooked cosm of heterobasidiomycetes with a broad mycorrhizal potential. *Mycological Research* **108**: 1003–10.

Welch, R.M., & Graham, R.D. (1999) A new paradigm for world agriculture: meeting human needsÑProductive, sustainable, nutritious. *Field Crops Research* **60**: 1–10.

White, J.G., & Zasoski, R.J. (1999) Mapping soil micronutrients. *Field Crops Research* **60**: 11–26.

White, P.J., & Broadley, M.R. (2005) Biofortifying crops with essential mineral elements. *Trends in Plant Science* **10**: 586–93.

White, P.J., & Broadley, M.R. (2009) Biofortification of crops with seven mineral elements often lacking in human dietsÑiron, zinc, copper, calcium, magnesium, selenium and iodine. *New Phytologist* **182**: 49–84.

Whiteside, M.D., Treseder, K.K., & Atsatt, P.R. (2009) The brighter side of soils: Quantum dots track organic nitrogen through fungi and plants. *Ecology* **90**: 100–8.

Winkelmann, G. (2007) Ecology of siderophores with special reference to the fungi. *Biometals* **20**: 379–92.

Wissuwa, M., Mazzola, M., & Picard, C. (2009) Novel approaches in plant breeding for rhizosphere-related traits. *Plant and Soil* **321**: 409–30.

Yang, J., Kloepper, J.W., & Ryu, C.M. (2009) Rhizosphere bacteria help plants tolerate abiotic stress. *Trends in Plant Science* **14**: 1–4.

Yao, L.H., Jiang, Y.M., Shi, J., Tomas-Barberan, F.A., Datta, N., Singanusong, R., & Chen, S.S. (2004) Flavonoids in food and their health benefits. *Plant Foods for Human Nutrition* **59**: 113–22.

Yao, M.K., Desilets, H., Charles, M.T., Boulanger, R., & Tweddell, R.J. (2003) Effect of mycorrhization on the accumulation of rishitin and solavetivone in potato plantlets challenged with *Rhizoctonia solani*. *Mycorrhiza* **13**: 333–6.

Zhao, F.J., Su, Y.H., Dunham, S.J., et al. (2009) Variation in mineral micronutrient concentrations in grain of wheat lines of diverse origin. *Journal of Cereal Science* **49**: 290–5.

Zhu, C., Naqvi, S., Gomez-Galera, S., Pelacho, A.M., Capell, T., & Christou, P. (2007) Transgenic strategies for the nutritional enhancement of plants. *Trends in Plant Science* **12**: 548–55.

Zhu, H.H., & Yao, Q. (2004) Localized and systemic increase of phenols in tomato roots induced by *Glomus versiforme* inhibits *Ralstonia solanacearum*. *Journal of Phytopathology* **152**: 537–42

Zhu, Y.G., Smith, S.E., Barritt, A.R., & Smith, F.A. (2001) Phosphorus (P) efficiencies and mycorrhizal responsiveness of old and modern wheat cultivars. *Plant and Soil* **237**: 249–55.

Zubek, S., Stojakowska, A., Anielska, T., & Turnau, K. (2010) Arbuscular mycorrhizal fungi alter thymol derivative contents of *Inula ensifolia* L. *Mycorrhiza* **20**: 497–504.

Zuo, Y., & Zhang, F. (2009) Iron and zinc biofortification strategies in dicot plants by intercropping with gramineous species. A review. *Agronomy for Sustainable Development* **29**: 63–71.

CHAPTER 3.4

Ecosystem Influences of Fungus-Growing Termites in the Dry Paleotropics

Gregor W. Schuurman

3.4.1 Introduction

Termites are a diverse group of >2600 social insect species with major ecosystem influences (particularly on soils) around the world (Kambhampati & Eggleton 2000; Thorne *et al.* 2000). Termites evolved from a late-Jurassic, cockroach-like ancestor into three primitive, Cretaceous period families (i.e. Hodotermitidae, Mastotermitidae, and Termopsidae) and dispersed across Pangaea's tropical and temperate regions. Their historic existence is now preserved in the amber-based fossil record (Thorne *et al.* 2000). The Cretaceous–Tertiary (K–T) transition was a turbulent time during which formerly significant groups (e.g. dinosaurs) declined dramatically, while others (e.g. mammals) radiated and expanded their ecological influence. Rapid termite evolution in the early-Tertiary produced the four modern termite families (i.e. Termitidae, Kalotermitidae, Serritermitidae, and Rhinotermitidae), which together account for >95% of today's genera and species (Kambhampati & Eggleton 2000; Thorne *et al.* 2000). Termites have evolved important physical and chemical defenses since the Jurassic, but as individual organisms they remain comparatively delicate members of the soil fauna. They owe their ecological success and global influence less to individual fortitude than to cooperation—both in their sophisticated social systems and in their diverse and efficient digestive symbioses (Bignell 2000). Workers in a soil-dwelling termite colony collaboratively move mountains, relatively speaking, to excavate vast nest cavities and subterranean tunnel networks that protect a colony's reproductive pair, offspring, and foraging parties from aboveground predators and climate. Nest cavities can occupy thousands of liters, and a single colony's tunnel network can extend 50 m from the nest and cover thousands of square meters with a total length exceeding 7 km (Darlington 1982; Traniello & Leuthold 2000). Many soil-dwelling termites extend their use of soil shielding above the soil realm itself, and painstakingly build a layer of protective soil sheeting over their aboveground runways and the detritus they consume. Large mounds, extensive subterranean tunnel networks, and vast, continually shifting networks of soil sheeting have long suggested that these creatures occupy an engineering role in soil ecosystems (Lee & Wood 1971; Lobry de Bruyn & Conacher 1990).

The trophic effects of termites also depend on close cooperation. Termites use symbiotic relationships with a wide range of primary decomposers to consume a very broad array of litter types with high assimilation. Digestive symbionts of termites include protists, methanogenic Archaea, bacteria, and fungi (Bignell 2000). The influence of termites on plant-litter decomposition has been clearly demonstrated through exclusion experiments in North America, Asia, and Africa (see references in Schuurman (2005)). Termites are nearly globally distributed in tropical and sub-tropical regions, with some outliers in warm temperate environments. They dominate soil arthropod assemblages across much of the dry tropics, and even dry temperate ecosystems such as the Chihua-

Soil Ecology and Ecosystem Services. First Edition. Edited by Diana H. Wall *et al.*
© 2012 Oxford University Press. Published 2012 by Oxford University Press.

Figure 3.4.1 A *Macrotermes* mound under an Acacia tree (*Acacia* cf. *erioloba*). Deception Valley, central Kalahari, Botswana. (Photo credit: S. Crausbay.)

huan and Sonoran deserts of the southwestern USA (Aanen & Eggleton 2005; Whitford *et al*. 1981, 1982). This dominance often translates into strong influences, not only on decomposition, but also on additional ecosystem functions and attributes, including carbon and nutrient flows, soil properties, hydrology, and ultimately landscape structure. Nowhere is this influence stronger than in tropical dry systems dominated by the Macrotermitinae—a subfamily within the Termitidae that cultivates a fungal exosymbiont. Although macrotermitines' generic radiation occurred in the early-Tertiary African rainforest (Aanen & Eggleton 2005), several genera emigrated out of their natal biome to colonize Africa's expanding Miocene Epoch semiarid savannas and woodlands. Expansion into this drier habitat was so successful that today, fungus-growers are considered keystones and ecosystem-engineers across much of the dry paleotropics (Fig. 3.4.1).

3.4.2 Fungus-growers

The most spectacular and successful termite strategy for arid-region living is the digestive mutualism of fungus-growing termites with a monophyletic group of basidiomycete or white-rot fungi (*Termitomyces* spp.). All of the approximately 330 members of the subfamily cultivate the fungus in subterranean chambers on a medium of masticated plant detritus known as fungus comb (Fig. 3.4.2). Both termites and fungi are obligate mutualists—fungus-growers obtain energy and nutrients from degraded fungus combs or *Termitomyces* tissue itself and *Termitomyces* species occur only in living macrotermitine colonies (Rouland-Lefèvre 2000; Aanen *et al*. 2002; Rouland-Lefèvre *et al*. 2002).

3.4.2.1 Evolution and biogeography

Macrotermitines are a monophyletic subfamily, and a sister clade to the rest of the large and diverse family Termitidae, which accounts for 70% of all termite species (Inward *et al*. 2007). All fungus-growing termites and *Termitomyces* strains originated from a single transition to fungiculture (Aanen *et al*. 2002) in the African rainforest (Aanen & Eggleton 2005) about 60 million years ago (Brandl *et al*. 2007). Early-Tertiary macrotermitine radiation

Figure 3.4.2 A close-up view of a *Macrotermes michaelseni* fungus-chamber shelf, showing fungus comb with spherical white asexual conidial structures of the *Termitomyces* fungus. (Photo credit: G. Schuurman.)

in the African rainforest subsequently produced ten currently recognized genera, five of which colonized Africa's expanding Miocene Epoch semiarid savannas and woodlands and are now more species-rich outside of their natal biome. These five include three small-bodied subterranean-nesters (i.e. *Allodontermes*, *Ancistrotermes*, and *Microtermes*), the largest-bodied macrotermitine (i.e. *Macrotermes*) which is mostly a mound-builder, and morphologically variable and likely polyphyletic genus *Odontotermes* (Aanen et al. 2002; Darlington et al. 2008), which has a more even mixture of subterranean and mound-building species.

Termites are poor dispersers compared with other social insects, because colonies require bi-parental founding and because the winged reproductives (alates) are weak fliers with a very brief nuptial flight. Macrotermitine dispersal is even further constrained because in most species the fungal symbiont must disperse independently (i.e. horizontal transmission; exceptions are *Microtermes* species and *Macrotermes bellicosus*; Nobre et al. 2010). Macrotermitine distribution is limited to Africa and parts of Asia. Fungus-growers are absent from Papua and Australia, and the group's longest known trans-oceanic dispersal was the single-event colonization by *Microtermes* of Madagascar 13 million years ago (Nobre et al. 2010). *Microtermes*, *Ancistrotermes*, *Macrotermes*, and *Odontotermes* dispersed into tropical Asia during the Middle/Upper Miocene (Engel & Krishna 2004; Aanen & Eggleton 2005; Darlington et al. 2008) when the African-Arabian and Asian plates joined and closed off the Tethys seaway. Within Asia, *Odontotermes* and *Macrotermes* are both broadly distributed from India to Indonesia and are in fact more species-rich in Asia than in their natal continent (Kambhampati & Eggleton 2000). *Microtermes* occurs across the southern latitudes of the continent, but does not extend to Southeast Asia's islands, and *Ancistrotermes* is limited to the forests of Southeast Asia (Aanen & Eggleton 2005). The other major dry-system genus—*Allodontermes*—remains a southern African endemic.

3.4.2.2 Fungus and fungiculture

The macrotermitine exosymbionts, white-rot fungi, are powerful aerobic decomposers that can break down lignin and therefore make cellulose more accessible and digestible (Hyodo et al. 2000). *Termitomyces* nutritionally enriches the fungus comb by reducing its C:N ratio from 200–1200:1 (Cornwell *et al.* 2009) to values much closer to those of termites (10:1–5:1; Traniello & Leuthold 2000). Workers consume plant litter on the soil surface and quickly transport it to the fungus chamber where they excrete this minimally degraded material in the form of "primary feces." They join these feces together into complex three-dimensional shapes to expose maximum fungus-comb surface area to the warm, damp, aerobic conditions that white-rot fungi require. Fungus comb structure varies among macrotermitine taxa from sponge shapes to elaborate folded lamina (e.g. Fig. 3.4.2; Darlington *et al.* 2008), but in all cases *Termitomyces* mycelia grow through the fungus comb and over a period of weeks to months (Wood 1978) convert the undigested plant tissue into nutritious vegetative fungal nodules on the surface of the comb. Termites feed on the nodule-rich lower portions of the comb and generally add new primary feces to the top of the comb (Hyodo *et al.* 2000; Rouland-Lefèvre 2000; Mueller *et al.* 2005). As a result, nutrients derived from plant litter are ultimately incorporated into termite tissue. Further, nodules contain asexual spores (conidia) that persist in workers' guts and automatically inoculate primary feces with *Termitomyces* during gut passage to effectively perpetuate this fungicultural system (Mueller *et al.* 2005).

Mueller and colleagues (2005) argue that insect fungiculture is agriculture because, in a review of nine independent transitions to fungiculture in three insect orders (macrotermitines, leaf-cutting ants, and ambrosia beetles), the following criteria were fulfilled: 1) habitual planting; 2) cultivation aimed at the improvement of growth conditions for the crop; 3) harvesting of the cultivar for food; and 4) obligate nutritional dependency on the crop. In the case of fungus-growing termites, and unlike leaf-cutting ants, the crop is fully domesticated and has never reverted to the wild state. *Termitomyces* depends entirely on the host colony for food, shelter from competition, and a stable, rainforest-like microclimate of well-oxygenated air, near-saturation humidity, and a stable, warm (~27–30°C) temperature (Aanen & Eggleton 2005). Macrotermitine species create this fungus-garden microhabitat within the thermally stable subterranean environment either by expanding the nest chamber itself, creating a diffuse network of small fungus chambers linked with tunnels, or both. Large, centralized fungus gardens (e.g. Fig. 3.4.3) generate sufficient CO_2 to require ventilation, and species in several genera construct sophisticated surface mounds that carry out gas exchange with the external atmosphere while maintaining internal microclimatic homeostasis (Turner 1994, 2000). Studies of fungus garden homeostasis and mound function show that fungus-garden temperatures in several mound-building species maintain these stable conditions, even where external climate varies greatly (Aanen & Eggleton 2005).

White-rot fungi are efficient, aerobic, generalist decomposers, and this "external rumen" lends fungus-growing termites great versatility and resource-use efficiency. Observations and stable

Figure 3.4.3 A 4.6 m high *Macrotermes michaelseni* mound excavated to reveal the fungus chamber. The king and queen are housed in a specially constructed cell at the center of the fungus chamber. (Photo credit: G. Schuurman.)

isotope studies confirm that individual macrotermitine species' diets are broad, and include wood, grass, leaves, herbivore dung, and even ungulate hooves and horns (Wood 1978; Schuurman 2006b; pers. obs.). Additionally, the fungus comb can act as a food store to meet seasonal peaks in colony nutrient and energy needs (e.g. during annual alate production), or be drawn down to sustain a colony through episodes when flooding, drought, or other conditions preclude foraging (Bignell & Eggleton 2000; Schuurman 2006a). Given this fungal digestive breadth and efficiency, and the fact that a colony's fungus comb typically greatly exceeds colony biomass and respiration, macrotermitines not surprisingly harvest litter at rates up to six times that of other termites—per unit biomass or area—and generally dominate assemblages where they occur (Bignell & Eggleton 2000; Rouland-Lefèvre 2000).

3.4.3 Fungus-grower influences on ecosystem processes

"Where termites are dominant soil animals, as in much of the warm temperate and tropical regions of the world, the organic cycle as it is generally presumed to function is drastically modified. Large quantities of plant tissue are taken from fallen litter and sometimes from living plants, often moved to a central storage point in a nest or mound, and intensely degraded so that the residue may contain little that is useful to other organisms." (Lee & Wood 1971)

3.4.3.1 Decomposition

Fungus-growing termites create a distinct, "self-contained" decomposition pathway marked by thoroughness, efficiency, versatility, and substantial climatic independence. Nutrients and carbon brought into the macrotermitine-*Termitomyces* decomposition system are essentially completely mineralized via a digestive pathway that can dramatically accelerate decomposition rates relative to background free-living microbial decomposition. Ironically, the termite-fungus decomposition system is most competitive against *in situ* microbial decomposition not in the rainforest where it arose, but instead in the arid tropical biomes where free-living microbial decomposition is episodic and frequently suppressed by climate (Aanen & Eggleton 2005). In general, termite influence as a whole increases along gradients of increasing aridity, and in Africa and Asia fungus-growers in particular respond to an aridity gradient with increased dominance (Deshmukh 1989; Bignell & Eggleton 2000; and see references in Schuurman (2006a)). Where macrotermitines dominate soil arthropod assemblages, they can match mammalian herbivore biomasses and plant-matter consumption rates, even in the African savanna where megaherbivores are abundant (Bignell & Eggleton 2000; Deshmukh 1989).

Fungus-grower dominated assemblages mediate about 1–2% of all C-mineralization in rainforests, but in dry Africa they can mediate up to 20% of C-mineralization, and consume as much as 1500 kg ha^{-1} of dry litter annually including almost all wood litter, up to 50% of all grass litter, and a major share of mammalian dung, particularly in the dry season (Wood & Sands 1978; and see references in Bignell & Eggleton (2000) and Schuurman (2005)). When fungus-growers are present in a termite assemblage, they can increase the mass loss rate of woody litter, relative to that of an assemblage free of fungus-growers, by 17-fold (Schuurman 2005). Macrotermitines are also abundant in Asia's highly seasonal dry forests, where they mineralize ~10% of annual aboveground litterfall in forests receiving < 2000 mm annual rainfall (Yamada *et al.* 2005), and can consume > 30% of leaf litterfall (Matsumoto & Abe 1979). Field-based, stable isotope studies show that many macrotermitine species are flexible, generalist feeders (Lepage *et al.* 1993; Schuurman 2006b) that can therefore affect decomposition rates of all plant litter types. However, their effects on woody-litter pool size are particularly strong because they rapidly and dramatically expand the surface area of even freshly fallen litter, and because they transport the material to a homeostatic rainforest-like microclimate where *Termitomyces* so effectively decomposes lignocellulose. Cornwell *et al.* (2009) argue that termites stand apart from all other insects in their impact on global wood decomposition, and should be recognized and included alongside microbial decomposition and fire generally the greatest carbon mineralizer in dry tropical systems—as the primary agents of wood turnover in global carbon models. Essentially, the high abundance of macrotermitines in dry systems substantially reduces the pool of surface litter of all kinds and of soil carbon, and rapidly redirects nutrients and energy in litterfall away from the episodically active soil food web and into the termite colony and its predator food chain (Jones 1990; Sugimoto et al. 2000; Traniello & Leuthold 2000).

3.4.3.2 Nutrient and energy flows

Macrotermitines direct a powerful nutrient and energy stream from the soil surface down into the colony. This stream bypasses the soil almost entirely. Fungus-growers concentrate and retain nutrients within termite tissue, rather than within the mound and surrounding soil, because they re-incorporate their feces into the fungus comb (Wood 1978; Noirot & Darlington 2000). Through predator–prey relationships, these nutrients ultimately flow out of the mound to higher trophic levels. These nutrient and energy flows to the larger food web are likely substantial because macrotermitines attract a host of specialized and generalist predators. Generalist predators consume termites from nests and foraging parties, and include a broad array of reptiles, numerous paleotropical ground-feeding birds (jungle fowl, quail, partridge, hornbills, babblers, etc.), mammals such as the Himalayan black bear and Malayan sun bear, and ants (Mathur 1962; Lee & Wood 1971; Lepage & Darlington 2000; Traniello & Leuthold 2000; Wong et al. 2002; Measey et al. 2009). Specialist predators are equally diverse. Doryline army ants in Africa, for example, are a family that invades macrotermitine nests and can destroy as many as a third of the nests in an area in a year (Lepage & Darlington 2000). Much larger mammals such as pangolins (scaly anteaters), aardvarks, and sloth bears specialize in breaking open mounds and consuming termite colonies (Joshi et al. 1997; Mathur 1962), and chimpanzees across equatorial Africa create tools and sometimes tool kits to "fish" termites from within termite mounds (Sanz et al. 2004; Bogart & Pruetz 2008). Chimpanzees in semiarid Fongoli, Senegal eat termites—primarily *Macrotermes*—year-round as a significant component of their diet, and Bogart and Pruetz (2008) cite evidence that *Australopithecus robustus* used a bone tool to fish for termites in a similar ecological context 1–1.8 million years ago to suggest that insects may be an underappreciated component of the early hominid diet.

A termite colony's annual alate brood represents a significant seasonal nutrient pulse from the colony into the environment (Fig. 3.4.4). The synchronous nuptial flight is clearly a predator-satiation strategy, as these poorly defended and extremely fat- and protein-rich animals are broadcast thickly into the air and onto the ground. Their brief flight attracts a suite of aerial predators that includes dragonflies and other aerial insects, insectivorous bats, and a wide range of birds from crows and bulbuls to swallows, owls, and kites (Mathur 1962). When alates shed their wings and begin to search for a mate and excavate a nest, they become prey to terrestrial predators including ants, cockroaches, frogs, toads, lizards, and a broad group of terrestrial mammalian opportunists from mice to foxes and bears (Mathur 1962; Wong et al. 2002).

Humans across Africa and Asia consume macrotermitine alates, and preserve (desiccate) and trade them, and they also eat foraging soldiers and excavate termite mounds to eat the termite queen as a delicacy (Mathur 1962; Van Huis 2003; Sileshi et al. 2009). The nutritional significance of macrotermitines in the southern African savanna has long been appreciated by one of the world's oldest cultures—the indigenous hunter-gatherer San—who equate fat with spiritual potency and believe that termites were the first meat that their god gave to humans (Mguni 2006). San maximize the alate harvest by plugging flight holes to channel the emergence, and in times when large mammals are scarce they excavate large mounds and winnow larvae from soil nest structures using reed mats (Mguni 2006). In rock art features known as formlings dating back thousands of years, the San depict macrotermitines and their fungal symbionts alongside another well-known fat-rich spiritual icon, the eland antelope (Mguni 2006). This elevation of termites into San cosmology reflects extremely long-term human recognition of the significance of macrotermitine energy and nutrient flows in the dry paleotropics.

Figure 3.4.4 Alates—winged reproductive termites—emerge from holes created by the colony's workers. Alates take flight and seek mates in a synchronous seasonal dispersal event, the nuptial flight. Few succeed; instead, the vast majority immediately enters the predatory food web in an ecologically significant nutrient pulse. (Photo credit: P. & B. Pickford.)

3.4.4 Fungus-growers as ecosystem engineers

3.4.4.1 Soils

Macrotermitines move, sort, and aggregate soil in remarkable feats of collaborative engineering that create vast subterranean tunnel networks, layers of aboveground soil sheeting (e.g. Fig. 3.4.5) that protect foraging parties, and exquisitely engineered homeostatic nests and fungus chambers (Fig. 3.4.3). Termites are soil ecosystem engineers (Jouquet et al. 2006) whose influence on soil properties has long been appreciated (Lee & Wood 1971; Lobry de Bruyn & Conacher 1990). Termites are further sub-classified as "extended phenotype engineers" because they build structures to maintain and improve colony growth and fitness, and these biogenic structures tend to create environmental heterogeneity. A fungus-growing termite assemblage creates resource heterogeneity at a wide range of scales, from the dense tunnel networks that course through every cubic meter of the macrotermitine soil landscape, to mounds that represent distinct and important landscape-level habitat "islands" for a range of plants and animals (e.g. Fig. 3.4.6).

Fungus-growers continually engineer the subsoil and the soil surface, as each colony develops and maintains its subterranean tunnel network and forages aboveground. Termite species diversity can be high at small spatial scales (e.g. Schuurman 2006a), and therefore an abundant and diverse assemblage can riddle the soil landscape with spatially overlapping but mutually exclusive tunnel networks that vary in size, depth, and other attributes. Fungus-growing termites differ from many other termites in that they do not use their feces to construct their nests and other structures (Noirot & Darlington 2000). Instead, they use only soil to engineer tunnels, galleries, and even the large nest and fungus-garden cavities (Fig. 3.4.3) that must remain structurally stable through the wet season. Macrotermitines require high-clay soil for nests and fungus chambers, which they obtain either by physically sorting the soil to adjust its clay content (Jouquet et al. 2002a) or by seeking out and transporting clay-rich soil from layers several meters below (Lee & Wood 1971; Noirot & Darlington 2000). Investment varies according to function—nests and fungus chamber walls and shelves have high proportions of clay for both structural stability and to retain moisture and maintain a

Figure 3.4.5 Soil sheeting covers dry grass stalks to protect foraging termites from predators and desiccation. A hollow shell of soil sheeting often persists long after termites have completely removed the detritus, and its shape indicates whether the object was a leaf, a twig, or a pile of herbivore dung. Texture of the sheeting (soil aggregate size) correlates with termite body size—here intermediate-sized aggregates indicate that an *Odontotermes* foraging party is at work. (Photo credit: G. Schuurman.)

Figure 3.4.6 A lone fan palm (*Hyphaene petersiana*) occupies scarce high ground on an Okavango Delta floodplain in northern Botswana. Long-term hydrological cycles, biotic influences, and tectonic activity collectively generate highly variable flooding regimes—an annually inundated floodplain may go dry for years or even decades—and termites including the mound-building *Macrotermes michaelseni* quickly colonize the newly available dry land. The returning flood will kill these pioneers because their nests lie at and below ground level (see Fig. 3.4.3), but if the dry episode has lasted for a few years or more then the colony will leave behind a mound of earth that extends above the flood line. The top of this "island" constitutes flood-proof habitat for fortunate plants and animals, including subsequent termite colonists. (Photo credit: D. Hamman.)

maximally humid microclimate (Lee & Wood 1971), whereas temporary aboveground runways and sheeting (Fig. 3.4.5) used to protect soldiers and mature workers match the local surface soil (Villenave et al. 2009). Although in most systems this behavior enhances clay content relative to background surface-soil abundances, on the flat Laikipia Plateau in Kenya the clay content in the "black cotton" vertisols is so high that soils oscillate seasonally between being waterlogged and drying and cracking. Here, mound-building O. montanus species-groups K and E (Darlington et al. 2008) have successfully colonized the landscape by sorting soil in the opposite direction to create raised mounds of comparatively sandy content (Fox-Dobbs et al. 2010).

Macrotermitine tunnel-building and foraging activity creates small-scale resource heterogeneity throughout the upper few meters of soil, and the long-term consequence of this activity across the landscape is a substantially modified soil environment (Lee & Wood 1971). The influence of termite tunnels on soil water infiltration has been widely observed and experimentally demonstrated (Lee & Wood 1971; Holt & Lepage 2000), and traditional agriculture manages macrotermitine foraging and tunnel-building to enhance soil porosity and water storage capacity. Termite-mediated vertical soil movement is estimated in the tons ha^{-1} year^{-1} range, and can result in an accumulation of several centimeters per millennium (Lee & Wood 1971). Further, all fungus-grower architecture is based on a building block of soil particles aggregated together with a worker's highly adhesive saliva, and workers may actually chemically transform (e.g. weather) the soil as they handle it (Jouquet et al. 2002b). Buried stone lines, patterns of laterite formation, and other evidence suggests that over millennia termites may be a dominant agent of tropical soil profile development (Lobry de Bruyn & Conacher 1990; Holt & Lepage 2000).

3.4.4.2 Mound-builder impacts

Mounds, particularly those created by *Macrotermes* and *Odontotermes*, are profound examples of soil engineering not just among termites but among all living creatures (Turner 2000), and they generate distinct and powerful ecosystem effects. A mound-building colony does not create a mound until the colony and its fungus garden are large enough to require ventilation for gas exchange (Noirot & Darlington 2000). Mound design is intricate and species-specific. Mounds in the open-chimney style and those with continuous but porous mound walls will both harness wind power to drive internal air currents that facilitate gas exchange, and may at times be used for evaporative cooling (Lee & Wood 1971; Noirot & Darlington 2000; Turner 2000). Over time, mounds can grow to be as tall as 9 m (Lee & Wood 1971; pers. obs.), and the mound itself essentially becomes a long-term landscape feature that can consist of over ~1000 m^3 of mineral soil that has been vertically transported, sorted, and worked by the colony (Noirot & Darlington 2000).

Mounds are adaptive structures that must maintain the nest and fungus chamber's high humidity and thermal homeostasis as the colony grows, and also resist erosion and intrusion. Therefore, macrotermitines continually repair and modify both the interior and exterior in response to changing colony needs and environmental conditions, or damage. Ultimately, mounds will become damaged or eroded, and as they do so, they often create a sloped outwash pediment that is impermeable to water and extends up to 15 m outward from the edge of the mound itself (Lee & Wood 1971). This pediment can shed rainfall as surface runoff and shield underlying soil layers from leaching (Lee & Wood 1971), and can also protect the nest from temporary flooding and burrowing nest predators (Turner 2006).

The mounds that termites mold out of subterranean mineral soil and thrust up into the aboveground realm often constitute discrete islands of distinct resource concentration and microtopography that support species or communities not found elsewhere. In the African savanna, macrotermitine mound abundance generally ranges from 1–4 ha^{-1}, but can be higher and can represent as much as 30% of the soil surface (Lee & Wood 1971). This discrete habitat can therefore be ecologically significant in its own right. Such significance, however, should not be assumed, because it depends on the degree to which the mound differs from the surrounding matrix. Mounds in some systems with plentiful clays in surface soil layers, for instance, can be similar in this respect to surrounding surface soil (Holt

& Lepage 2000), but in other systems can be profoundly different (Lee & Wood 1971). Macrotermitine mounds often have lower organic-matter content than surrounding surface soils because termites construct them from mineral subsurface soil and add only saliva (Lee & Wood 1971). They also tend to have a higher pH than surrounding soils, and to be enriched in calcium and exchangeable cations (Lee & Wood 1971; Lobry de Bruyn & Conacher 1990; Jouquet et al. 2006). Nitrogen is more abundant in mounds than in adjacent soils in some systems (e.g. Laikipia Plateau *O. montanus* mounds), but exceptions are common (Lee & Wood 1971; Lobry de Bruyn & Conacher 1990). As a result of these impacts on hydrology, water holding capacity, chemistry, nutrient-content, and soil texture, mounds—particularly mounds on nutrient-poor soils—can support very distinct and important flora, including mound-pediment specialists (Lee & Wood 1971; Dangerfield et al. 1998; Bloesch 2008; Brody et al. 2010).

3.4.4.3 Emergent landscape effects

The macrotermitine colony's strong trophic links with the detritus pool and predators are obvious (Fig. 3.4.4), but the ecosystem influences of the physical mound are not uniformly apparent. Mounds reach far beyond their pediments and into the surrounding matrix to exert powerful cryptic influences on critical ecosystem processes and long-term system structure and development. A clear example comes from the well-studied Laikipia Plateau. Here, slightly elevated (0.5 m high) *O. montanus* mounds are enriched in nitrogen because they are protected from waterlogging and leaching, unlike the surrounding soils, and they encourage faster growth and higher reproduction in the dominant tree (*Acacia drepanolobium*) by reducing the need for N_2 fixation (Brody et al. 2010). Significantly, reduced N_2 fixation extends beyond the edge of the mound, and this cryptic influence ultimately reaches >50% of the trees in the system despite the fact that mound structures cover only 20% of the soil surface (Fox-Dobbs et al. 2010). An experimental herbivore-exclosure component to the study shows that this effect arises directly from the mound itself—rather than as an indirect consequence of attracting mammalian browsers—and demonstrates that termite effects on this critical ecological process dwarf herbivore effects (Fox-Dobbs et al. 2010).

Mound-dwelling vegetation communities can differ in many ways from the surrounding ecosystem, and therefore they frequently represent resource hotspots for animals (see references in Jouquet et al. (2006), Levick et al. (2010), and Pringle et al. (2010)). Exclosure experimentation and careful observation show that mounds can exert powerful indirect effects—through herbivores—by concentrating megaherbivore foraging on and near the mound (Levick et al. 2010). A remote-sensing study in low-fertility granitic savanna in Kruger National Park in South Africa compared vegetation in a long-term herbivore exclosure with adjacent habitat to show that termite mounds differentially attract megaherbivores (elephants and giraffes)—apparently because mounds are enriched in nutrients. This effect significantly influences vegetation structure as far as 20 m from mound centers. Although the mounds only cover 5% of the landscape, mound influence on vegetation structure ultimately extends to 20% of the landscape (Levick et al. 2010). As in the Laikipia savanna where termites' direct effects on nitrogen fixation are significantly stronger than those of herbivores, a herbivore-exclosure component in this Southern African savanna reveals how termites' indirect effects on vegetation, mediated through their direct effects on herbivore behavior, dwarf the background herbivore impact on vegetation structure.

A mound's impact on its landscape is explicitly spatial and essentially radial. Collective and emergent mound impacts on a landscape depend not only on mound density but also on mound placement. Termite distribution is often heterogeneous because of differential soil composition, subterranean water-holding capacity, inundation frequency, predation, and vegetation, as well as interspecific resource competition and aggression (Darlington 1982; Bignell & Eggleton 2000). Nevertheless, intraspecific aggression is common (Traniello & Leuthold 2000) and therefore conspecific mounds are rarely found next to each other. As a result, mound distribution at some local and even landscape scales in homogenous landscapes like the Laikipia

Plateau—where water-logging likely simplifies the macrotermitine assemblage by excluding subterranean-nesting species—can be highly regular and substantially magnify mounds' trophic influence (Pringle *et al.* 2010)

Macrotermitine engineering may also exert powerful system-level hydrological effects. The excavations necessary to create and maintain large epigeal mounds with their tall spires, expansive internal cavities, ventilation shafts, and subterranean tunnels radiating outward from the nest create a region of modified and hollowed-out soil extending 10–12 m down below the colony (Turner 2006). Turner argues that rainfall washes off the outwash pediment and penetrates the foraging tunnel-rich soil region immediately beyond the pediment, likely creating a local influence on vegetation, and then percolates downward and inward towards this modified and porous soil region beneath the mound where it becomes suspended in impermeable layers several meters below the mound. Trees in dry savannas on or near termite mounds retain their leaves much longer into the dry season than do conspecifics further away (Turner 2006 and references therein), and widely accepted estimates of daily water loss from large *Macrotermes* mounds can only be offset by termites actively transporting water into the mound to maintain a humid microclimate. This transport would be energetically cheap, and whether through passive percolation or active transportation, the mound can essentially be thought of as a dry-savanna "water gatherer" (Turner 2006) whose soil engineering makes a substantial contribution to hydrologic balance of its ecosystem, and has significant influences on dry-savanna phenology. Such phenological effects likely entrain additional herbivore-mediated indirect effects on the surrounding aboveground landscape.

Termite mounds may also exert powerful and cryptic influence on their ecosystems on long timescales and even larger spatial scales. Mounds can persist for decades or centuries, because queens can live as long as two decades and can sometimes be succeeded by a reproductive offspring, and because mounds can also be re-colonized some time after the preceding colony expires (pers. obs.). Mound-builder influence can be particularly influential on long-term landscape development and structure in flat, periodically flooded landscapes where occasional dry episodes are of sufficient length to allow mound-builders to colonize and live long enough to create raised microhabitat for woody-species (e.g. Fig. 3.4.6; Dangerfield *et al.* 1998; Bloesch 2008; Brody *et al.* 2010). In the Okavango Delta on the Kalahari sands of northern Botswana, for example, *Macrotermes michaelseni* engineers this habitat, and woody species in turn colonize and vastly expand these island nuclei through transpiration that precipitates silica and calcite (Fig. 3.4.7; Dangerfield *et al.* 1998). In this way, the interaction between termites, flooding regimes, and vegetation maintains an array of terrestrial islands within an otherwise episodically aquatic flatland matrix.

3.4.5 Synthesis

Macrotermitines are a profoundly influential soil-fauna group. However, their full range and depth of ecosystem influences is still emerging and being recognized (e.g. Cornwell *et al.* 2009) and incorporated into conservation and management as science develops new tools and as paradigms shift (Levick *et al.* 2010; Pringle *et al.* 2010). Where macrotermitines are dominant, they clearly rival or exceed herbivores in mineralizing primary productivity, governing major ecosystem processes, and ultimately engineering the critical and familiar patterns and structures of dry paleotropical ecosystems—even in dry savannas known for abundant and influential ungulate populations.

Macrotermitine distribution is highly variable across dry tropical Africa and Asia, and macrotermitine abundance as a whole can vary dramatically across an ecological landscape, with powerful effects on decomposition rates (e.g. Schuurman 2005). Fungus-growing genera and species can display great dietary breadth and flexibility, and therefore considerable overlap. However, the lineages (genera) that independently migrated into the dry tropical biomes millions of years ago also differ in important ways. Macrotermitine abundance, distribution, and assemblage composition can vary according to geography, elevation, soil composition, predation type and intensity, flooding frequency, aboveground habitat effects on wind-driven mound function, and other disturbance regimes,

Figure 3.4.7 A constellation of termite-mound islands creates landscape structure deep in the Okavango Delta, Botswana. (Photo credit: K. Collins.)

and these effects often influence termite genera differentially (Bignell & Eggleton 2000; Sugimoto et al. 2000; Inoue et al. 2006; Schuurman 2006a).

Consequently, macrotermitine assemblages in the dry paleotropics can differ dramatically not only in their overall, collective abundance and impact on decomposition (e.g. Schuurman 2005), but also in the representation and relative abundances of the constituent *Macrotermes*, *Odontotermes*, *Ancistrotermes*, *Allodontermes*, and *Microtermes* species. Genera and even congeners differ in whether or not they build mounds, and how they are distributed across the landscape, and species differ in their nest types, colony and body sizes, seasonality of feeding, degree to which they sort or transport soil, and therefore in many of their significant ecological impacts (Lee & Wood 1971; Bignell & Eggleton 2000; Schuurman 2006a). The most significant distinction within fungus-growers is between mound-builders and soil-dwellers, because mounds generate their own emergent landscape-level effects on the surrounding ecosystem. Mound-builders tend to be larger colonies with larger territories and they tend to exert enough microclimatic control of their nests and fungus chambers to allow their seasonal foraging patterns to be dictated by internal colony reproductive needs rather than by climatic constraints that govern subterranean-nesters (Schuurman 2006a).

Land-use change and intensification can cause dramatic and progressive loss of overall termite species richness (Bignell & Eggleton 2000; reviewed in Jones et al. (2003)). Although such changes also have a simplifying effect within the macrotermitinae, some groups such as *Ancistrotermes* and *Microtermes*—a soil-dweller that can nest deep underground and consume crop residues—may persist or even flourish as other groups decline (Bignell & Eggleton 2000; and see references in Schuurman (2006a)). Rainfall governs assemblage composition and dominance among termite feeding groups at large spatial scales—soil-feeding groups dominate wet systems, fungus-growers dominate semi-arid and arid systems, and finally non-macrotermitines persist at generally low abundances in the most extremely arid systems (Bignell & Eggleton 2000) such as the Sahara. As a consequence, major changes in precipitation in the fungus-grower range could dramatically alter ecosystem process and structure through indirect effects on the decomposer community.

3.4.6 Take-home messages

- Fungus-growing termites are a key group within the soil fauna, and their social behavior and powerful digestive exosymbiosis with an aerobic white-rot fungus allows them to accelerate decomposition and reduce litter and soil carbon pools.
- Macrotermitines transfer nutrients and energy to the predatory food web via both generalist and specialist predators.
- Macrotermitines engineer the soil environment in ways that allow them to dominate arid systems and dramatically transform the soil itself on long time scales.
- Macrotermitine mounds are long-term landscape features that can function as resource hotspots, and can also exert powerful direct and indirect effects that reach far beyond the mound itself.
- Humans have long recognized, used, managed, and even revered macrotermitine ecological services.
- Macrotermitine abundance and assemblage composition are influenced by a wide variety of environmental influences including climate and land use, and macrotermitine distributions and therefore ecological effects are likely to change in response to global change.

3.4.7 Future directions

Macrotermitine landscapes of the dry paleotropics are the cradle of humanity (Darlington 2005), and still support traditional societies that incorporate macrotermitine ecosystem services into their agriculture, land management, and diet. New research including remote sensing demonstrates that these ecosystem impacts include powerful direct and indirect effects extending far beyond the mound itself, and that mound distribution patterns within the landscape can generate additional emergent ecosystem impacts. Pringle *et al.* (2010) and Levick *et al.* (2010) argue that pattern-forming organisms that influence ecosystems at the landscape level should be the focus of conservation and retention under human management and modification of the landscape because they are critically important to maintaining ecosystem structure, creating heterogeneity, and maximizing delivery of ecosystem services.

To maintain these critical ecosystem components in natural and semi-natural systems in a changing world, and to harness and enhance their ecosystem services in human landscapes, we must better understand what governs macrotermitine abundance and distribution, how this variation influences fungus-grower impact across the landscape, and how each fungus-growing taxon will respond to global change. This synthesis requires that we more strongly integrate emerging landscape-level studies with the larger body of community- and species-level termite research to provide critical ecological and taxonomic context. Traditional human societies across the macrotermitine world are also a critical information resource. Indigenous knowledge and traditions can provide important insights into ecology and management of this important ecosystem engineer and nutritional resource, which can in turn be shared broadly to maximize delivery of macrotermitine ecosystem services.

References

Aanen, D.K., & Eggleton, P. (2005) Fungus-growing termites originated in African rain forests. *Current Biology* **15**: 651–5.

Aanen, D.K., Eggleton, P., Rouland-Lefèvre, C., Guldberg-Frøslev, T., Rosendahl, S., & Boomsma, J.J. (2002) The evolution of fungus-growing termites and their mutualistic fungal symbionts. *Proceedings of the National Academy of Sciences of the United States of America* **99**: 14887–92.

Bignell, D.E. (2000) Introduction to symbiosis. In: T. Abe, D.E. Bignell & M. Higashi (eds.) *Termites: Evolution, Sociality, Symbioses, Ecology*, pp. 189–208. Kluwer Academic, Dordrecht.

Bignell, D.E., & Eggleton, P. (2000) Termites in ecosystems. In: T. Abe, D.E. Bignell, & M. Higashi (eds.) *Termites:*

Evolution, Sociality, Symbioses, Ecology, pp. 363–87. Kluwer Academic, Dordrecht.

Bloesch, U. (2008) Thicket clumps: A characteristic feature of the Kagera savanna landscape, East Africa. *Journal of Vegetation Science* **19**: 31–44.

Bogart, S.T., & Pruetz, J.D. (2008) Ecological context of savanna chimpanzee (*Pan troglodytes verus*) termite fishing at Fongoli, Senegal. *American Journal of Primatology* **70**: 605–12.

Brandl, R., Hyodo, F., von Korff-Schmising, M., et al. (2007) Divergence times in the termite genus *Macrotermes* (Isoptera: Termitidae). *Molecular Phylogenetics and Evolution* **45**: 239–50.

Brody, A.K., Palmer, T.M., Fox-Dobbs, K., & Doak, D.F. (2010) Termites, vertebrate herbivores, and the fruiting success of *Acacia drepanolobium*. *Ecology* **91**: 399–407.

Cornwell, W.K., Cornelissen, J.H.C., Allison, S.D., et al. (2009) Plant traits and wood fates across the globe: rotted, burned, or consumed? *Global Change Biology* **15**: 2431–49.

Dangerfield, J.M., McCarthy, T.S., & Ellery, W.N. (1998) The mound-building termite *Macrotermes michaelseni* as an ecosystem engineer. *Journal of Tropical Ecology* **14**: 507–20.

Darlington, J.P.E.C. (1982) The underground passages and storage pits used in foraging by a nest of the termite *Macrotermes michaelseni* in Kajiado, Kenya. *Journal of Zoology, London* **198**: 237–47.

Darlington, J.P.E.C. (2005) Distinctive fossilised termite nests at Laetoli, Tanzania. *Insectes Sociaux* **52**: 408–9.

Darlington, J.P.E.C., Benson, R.B., Cook, C.E., & Walker, G. (2008) Resolving relationships in some African fungus-growing termites (Termitidae, Macrotermitinae) using molecular phylogeny, morphology, and field parameters. *Insectes Sociaux* **55**: 256–65.

Deshmukh, I. (1989) How important are termites in the production ecology of African savannas? *Sociobiology* **15**: 155–68.

Engel, M.S., & Krishna, K. (2004) Family-group names for termites (Isoptera). *American Museum Novitates* **3432**: 1–9.

Fox-Dobbs, K., Doak, D.F., Brody, A.K., & Palmer, T.M. (2010) Termites create spatial structure and govern ecosystem function in an East African savanna. *Ecology* **91**: 1296–307.

Holt, J.A., & Lepage, M. (2000) Termites and soil properties. In: T. Abe, D.E. Bignell and M. Higashi (eds.) *Termites: Evolution, Sociality, Symbioses, Ecology*, pp. 389–407. Kluwer Academic, Dordrecht.

Hyodo, F., Inoue, T., Azuma, J.I., Tayasu, I., & Abe, T. (2000) Role of the mutualistic fungus in lignin degradation in the fungus-growing termite *Macrotermes gilvus* (Isoptera; Macrotermitinae). *Soil Biology and Biochemistry* **32**: 653–8.

Inoue, T., Takematsu, Y., Yamada, A., et al. (2006) Diversity and abundance of termites along an altitudinal gradient in Khao Kitchagoot National Park, Thailand. *Journal of Tropical Ecology* **22**: 1–4.

Inward, D.J.G., Vogler, A.P., & Eggleton, P. (2007) A comprehensive phylogenetic analysis of termites (Isoptera) illustrates key aspects of their evolutionary biology. *Molecular Phylogenetics and Evolution* **44**: 953–67.

Jones, D.T., Susilo, F.X., Bignell, D.E., Hardiwinoto, S., Gillison, A.N., & Eggleton, P. (2003) Termite assemblage collapse along a land-use intensification gradient in lowland central Sumatra, Indonesia. *Journal of Applied Ecology* **40**: 380–91.

Jones, J.A. (1990) Termites, soil fertility and carbon cycling in dry tropical Africa, a hypothesis. *Journal of Tropical Ecology* **6**: 291–305.

Joshi, A.R., Garshelis, D.L., & Smith, J.L.D. (1997) Seasonal and habitat-related diets of sloth bears in Nepal. *Journal of Mammalogy* **78**: 584–97.

Jouquet, P., Dauber, J., Lagerlof, J., Lavelle, P., & Lepage, M. (2006) Soil invertebrates as ecosystem engineers: intended and accidental effects on soil and feedback loops. *Applied Soil Ecology* **32**: 153–64.

Jouquet, P., Lepage, M., & Velde, B. (2002a) Termite soil preferences and particle selections: strategies related to ecological requirements. *Insectes Sociaux* **49**: 1–7.

Jouquet, P., Mamou, L., Lepage, M., & Velde, B. (2002b) Effect of termites on clay minerals in tropical soils: fungus-growing termites as weathering agents. *European Journal of Soil Science* **53**: 521–7.

Kambhampati, S., & Eggleton, P. (2000) Taxonomy and phylogeny of termites. In: T. Abe, D.E. Bignell, & M. Higashi (eds.) *Termites: Evolution, Sociality, Symbioses, Ecology*, pp. 1–24. Kluwer Academic, Dordrecht.

Lee, K.E., & Wood, T.G. (1971) *Termites and Soils*. Academic Press, London.

Lepage, M., Abbadie, L., & Mariotti, A. (1993) Food habits of sympatric termite species (Isoptera, Macrotermitinae) as determined by stable carbon isotope analysis in a Guinean savanna (Lamto, Cote d'Ivoire). *Journal of Tropical Ecology* **9**: 303–11.

Lepage, M., & Darlington, J.P.E.C. (2000) Population dynamics of termites. In: T. Abe, D.E. Bignell, & M. Higashi (eds.) *Termites: Evolution, Sociality, Symbioses, Ecology*, pp. 333–61. Kluwer Academic, Dordrecht.

Levick, S.R., Asner, G.P., Kennedy-Bowdoin, T., & Knapp, D.E. (2010) The spatial extent of termite influences on herbivore browsing in an African savanna. *Biological Conservation* **143**: 2462–7.

Lobry de Bruyn, L.A., & Conacher, A.J. (1990) The role of termites and ants in soil modification: a review. *Australian Journal of Soil Research* **28**: 55–93.

Mathur, R.N. (1962) Enemies of termites (white ants). In: *Termites in the Humid Tropics, Proceedings of the New Delhi Symposium* pp. 137–40. UNESCO, New Delhi.

Matsumoto, T., & Abe, T. (1979) The role of termites in an equatorial rain forest ecosystem of West Malaysia. II. Leaf litter consumption of the forest floor. *Oecologia* **38**: 261–74.

Measey, G.J., Armstrong, A.J., & Hanekom, C. (2009) Subterranean herpetofauna show a decline after 34 years in Ndumu Game Reserve, South Africa. *Oryx* **43**: 284–7.

Mguni, S. (2006) Iconography of termites' nests and termites: symbolic nuances of formlings in Southern African San rock art. *Cambridge Archaeological Journal* **16**: 53–71.

Mueller, U.G., Gerardo, N.M., Aanen, D.K., Six, D.L., & Schultz, T.R. (2005) The evolution of agriculture in insects. *Annual Review of Ecology, Evolution, and Systematics* **36**: 563–95.

Nobre, T., Eggleton, P., & Aanen, D.K. (2010) Vertical transmission as the key to the colonization of Madagascar by fungus-growing termites? *Proceedings of the Royal Society of London B* **277**: 359–65.

Noirot, C., & Darlington, J.P.E.C., (2000) Termite nests: architecture, regulation and defence. In: T. Abe, D.E. Bignell and M. Higashi (eds.) *Termites: Evolution, Sociality, Symbioses, Ecology* pp. 121–39. Kluwer Academic, Dordrecht.

Pringle, R.M., D.F., Doak, A.K., Brody, R., Jocqué, & Palmer, T.M. (2010) Spatial pattern enhances ecosystem functioning in an African savanna. *PLoS Biol* **8**: e1000377.

Rouland-Lefèvre, C. (2000) Symbiosis with fungi. In: T. Abe, D.E. Bignell, & M. Higashi (eds.) *Termites: Evolution, Sociality, Symbioses, Ecology*, pp. 189–306. Kluwer Academic, Dordrecht.

Rouland-Lefèvre, C., Diouf, M.N., Brauman, A., & Neyra, M. (2002) Phylogenetic relationships in Termitomyces (Family Agaricaceae) based on the nucleotide sequence of ITS: a first approach to elucidate the evolutionary history of the symbiosis between fungus-growing termites and their fungi. *Molecular Phylogenetics and Evolution* **22**: 423–9.

Sanz, C., Morgan, D., & Gulick, S. (2004) New insights into chimpanzees, tools, and termites from the Congo Basin. *The American Naturalist* **164**: 567–81.

Schuurman, G. (2005) Decomposition rates and termite assemblage composition in semiarid Africa. *Ecology* **86**: 1236–49.

Schuurman, G. (2006a) Foraging and distribution patterns in a termite assemblage dominated by fungus-growing species in semi-arid northern Botswana. *Journal of Tropical Ecology* **22**: 277–87.

Schuurman, G. (2006b) Termite diets in dry habitats of the Okavango Delta region of northern Botswana: A stable isotope analysis. *Sociobiology* **47**: 373–90.

Sileshi, G.W., Nyeko, P., Nkunika, P.O.Y., Sekematte, B.M., Akinnifesi, F.K., & Ajayi, O.C. (2009) Integrating ethnoecological and scientific knowledge of termites for sustainable termite management and human welfare in Africa. *Ecology and Society* **14**: 48.

Sugimoto, A., Bignell, D.E., & MacDonald, J.A. (2000) Global impact of termites on the carbon cycle and atmospheric trace gases. In: T. Abe, D.E. Bignell, & M. Higashi (eds.) *Termites: Evolution, Sociality, Symbioses, Ecology*, pp. 409–36. Kluwer Academic, Dordrecht.

Thorne, B.L., Grimaldi, D.A., & Krishna, K. (2000) Early fossil history of the termites. In: T. Abe, D.E. Bignell, & M. Higashi (eds.) *Termites: Evolution, Sociality, Symbioses, Ecology*, pp. 77–94. Kluwer Academic, Dordrecht.

Traniello, J.F.A., & Leuthold, R.H. (2000) Behavior and ecology of foraging in termites. In: T. Abe, D.E. Bignell, & M. Higashi (eds.) *Termites: Evolution, Sociality, Symbioses, Ecology*, pp. 409–36. Kluwer Academic, Dordrecht.

Turner, J.S. (1994) Ventilation and thermal constancy of a colony of a southern African termite (*Odontotermes transvaalensis*, Macrotermitinae). *Journal of Arid Environments* **28**: 231–48.

Turner, J.S. (2000) *The extended organism: the physiology of animal-built structures*. Harvard University Press, Cambridge, MA.

Turner, J.S. (2006) Termites as mediators of the water economy of arid savanna ecosystems. In: P. D'Odorico & A. Porporato (eds.) *Dryland Ecohydrology*, pp. 303–13. Springer, Netherlands.

van Huis, A. (2003) Insects as food in sub-Saharan Africa. *Insect Science and its Application* **23**: 163–85.

Villenave, C., Djigal, D., Brauman, A., & Rouland-Lefèvre, C. (2009) Nematodes, indicators of the origin of the soil used by termites to construct biostructures. *Pedobiologia* **52**: 301–7.

Whitford, W.G., Meentemeyer, V., Seastedt, T.R., et al. (1981) Exceptions to the AET model: deserts and clearcut forests. *Ecology* **62**: 275–7.

Whitford, W.G., Steinberger, Y., & Ettershank, G. (1982) Contributions of subterranean termites to the "economy" of Chihuahuan desert ecosystems. *Oecologia* **55**: 298–302.

Wong, S.T., Servheen, C., & Ambu, L. (2002) Food habits of Malayan Sun Bears in lowland tropical forests of Borneo. *Ursus* **13**: 127–36.

Wood, T.G. (1978) Food and feeding habits of termites. In: M.V. Brian (ed.) *Production Ecology of Ants and Termites*, pp. 55–80. Cambridge University Press, Cambridge.

Wood, T.G., & Sands, W.A. (1978) The role of termites in ecosystems. In: M.V. Brian (ed.) *Production Ecology of Ants and Termites*, pp. 245–92. Cambridge University Press, Cambridge.

Yamada, A., Inoue, T., Wiwatwitaya, D., *et al.* (2005) Carbon mineralization by termites in tropical forests, with emphasis on fungus combs. *Ecological Research* **20**: 453–60.

CHAPTER 3.5

The Biogeography of Microbial Communities and Ecosystem Processes: Implications for Soil and Ecosystem Models

Mark A. Bradford and Noah Fierer

3.5.1 Predicting environmental responses of soil processes

We are entering a period of rapid and pronounced environmental change. Understanding how this change will influence soil communities is essential for accurate prediction of future climate, biogeochemical cycles, and human well-being. Such reliable prediction is the fundamental test of scientific understanding (i.e. can we use knowledge of underlying mechanisms to predict accurately how phenomena change across space and time). Predicting how soils respond to environmental change will be central to adaptive management of ecosystems, given our reliance on soils for sustained food production, water purification, carbon storage, and nutrient retention. This ability to predict with certainty how environmental change will influence soil processes requires mechanistic understanding of how soils work. Developing this mechanistic understanding is perhaps the "Grand Challenge" for soil ecologists if we are to advance our basic understanding of soils and apply this knowledge to effective environmental management.

Mechanistic understanding improves our confidence in predictions of the future that extrapolate observed relationships between regulatory variables (e.g. temperature) and process rates (e.g. soil respiration) (Reynolds et al. 2001). Perhaps paradoxically, there may be much less agreement in predicted outcomes between ecosystem models when more mechanistic understanding is included.

Although we might have "faith" that the true ecosystem response lies somewhere within the predictions from the models, predictions from mathematical models that rely on empirical (or statistical) relationships may be much more similar. Empirical models comprise the majority of ecosystem models used to predict soil biogeochemical processes (see Box 3.5.1). Our confidence in the predictions by empirical models of future ecosystem response relies on the assumption that observations of regulatory variable-process rate relationships hold across space and time (Reynolds et al. 2001). So, we are faced with a dilemma. Do we develop and employ mechanistic models that may be more accurate but also more uncertain or do we follow the more traditional approach, relying on empirical models that may predict ecosystem processes with more certainty but less accuracy? There is likely no single right answer to this question when developing models to predict how ecosystems will respond to environmental change. A family of models that covers this continuum of approach—between empiricism and mechanism—will permit us to determine where model predictions agree. Where they disagree likely indicates the greatest uncertainty in model predictions, and it is in addressing these disagreements that future research might yield most insight.

The soil ecology community predominately relies on more empirical models and this reliance has repercussions beyond the discipline and can influence global policy. For example, the soil submodels

Soil Ecology and Ecosystem Services. First Edition. Edited by Diana H. Wall *et al.*
© 2012 Oxford University Press. Published 2012 by Oxford University Press.

> **Box 3.5.1 Comparison of mechanistic and empirical models**
>
> Most commonly it is the perspective of an individual, not model structure, which determines whether a model is classed as mechanistic or empirical. In reality, these two terms distinguish opposite ends of a continuum. Mechanistic models are designed to explain outcomes with formulations that represent actual physical, chemical, and biological processes, such as increased catalytic rates of respiratory enzymes as temperature increases. Empirical models are often referred to as "correlative or statistical models" and are designed to predict—as opposed to explain—outcomes. To enable this prediction, formulations are used that best describe relationships between two measured variables, such as temperature and soil respiration rates, with no necessary regard for the underlying mechanism. Empirical models are typically parameterized—at least in the ecological and biogeochemical sciences—using regression relationships derived from data. When we extrapolate these relationships we are assuming they are robust across space and time, even if the physical, chemical, and biological conditions under which the data were collected differ markedly to those in the new location or time for which we make the prediction. With mechanistic models the argument is that—if we understand and represent the underlying mechanisms accurately in the model formulations—then we can make robust predictions for other locations and times. The challenges with mechanistic models are thus twofold: we have to both understand the mechanisms and then be able to represent them accurately when formulating models. These challenges are not trivial, and may delimit the very edges of our understanding of soils.

embedded within the coupled atmosphere–biosphere carbon cycle models used by the Intergovernmental Panel on Climate Change (IPCC) to predict future climate are largely empirical. The IPCC has high uncertainty in some of the assumptions inherent to the embedded soil models, and this leads to uncertainty in climate prediction. For example, there is marked uncertainty as to whether warmer temperatures will lead to the net loss of soil carbon that then generates a positive feedback to global warming (Denman *et al.* 2007). Given the vast published literature on this topic (e.g. Davidson & Janssens 2006; Bradford *et al.* 2008; Allison *et al.* 2010), it is unsettling that we—the soil ecology community—cannot reach agreement on the question of how soil carbon stocks will respond to temperature change, especially given the societal need for resolution. One plausible reason for this disagreement is that biological and ecological mechanisms are typically represented implicitly—not explicitly—in soil models (Bahn *et al.* 2010).

3.5.2 Misplaced physics envy in soil models

Soils are not merely physical and chemical entities, yet we generally model soil processes by assuming they can be represented simply as a product of physics and chemistry. This assumption is not actually at odds with the fact the models are usually also implicitly rooted in fundamental biological understanding. But what it does mean is that we assume we can omit biology (and within "biology" we include ecology) from the models as a force that could shape responses of soil processes to perturbation. That is, we assume physicochemical controls on soil processes are unaffected by phenotypic or genotypic changes in biota, and biotic interactions, across time and space. Yet at the same time we're aware that soil organisms—at least in part—mediate many soil processes (e.g. respiration and nitrification). The omission of biology from soil models is perhaps best illustrated through example. Take the approach of using a single, first-order decay equation to estimate how the rate of soil carbon decomposition varies with temperature (e.g. Kirschbaum 2004; Knorr *et al.* 2005). The equations are typically parameterized by data collected across a large range in one or more controlling factors, in this case temperature. Variation in temperature is generated by use of broad spatial gradients (e.g. elevation, latitude) or short temporal changes (e.g. incubators, daily variation). Using these statistical relationships to project how soil carbon decomposition responds to environmental change assumes that the biology is invariant across space and time. In other words, understanding of biological mechanism is unnecessary for accurately predicting how soil processes will respond to environmental change.

Social scientists are well aware of the shortcomings of using principles from classical physics to predict response of human (and hence biological) systems to perturbation (for an excellent commentary see Bernstein et al. 2000). Application of principles such as 1) invariance, 2) probability, and 3) simplicity "promise" a level of certainty in prediction that will not be realized where biological processes influence outcomes. With invariance (1) we assume that history does not matter, with responses a fixed function of controlling variables. In human systems, we can understand that "experience" influences decision-making, so outcomes of perturbations differ if a person is naïve or familiar with the event. Soil organisms may not exhibit higher learning but directional selection in gene frequencies provides a mechanism whereby history and past conditions can influence outcomes of perturbations. For example, the development of resistance of microbial populations chronically exposed to antibiotics influences whether populations are extirpated when antibiotics are applied in acute doses. Likewise, microbial communities that have developed under conditions of moisture stress are likely to have accumulated osmoregulants and other physiological strategies to cope better with future drought events.

With probability (2) the assumption is that all individuals are equally likely to respond identically—hence we can predict rates of radioactive decay even if we can't predict which atom actually undergoes decay. Yet we know that soil organisms sort out along an r to K life-history continuum (Fierer et al. 2007)—or at least exhibit metabolically alert strategies—so whereas r-selected individuals might rapidly increase their intrinsic population growth rates if resources are made available, K-selected organisms will display a lagged response. In practice this could mean that, for example, the decomposition of labile versus more recalcitrant carbon pools could exhibit different temperature responses if microbes with r versus K strategies, respectively, decompose these different carbon pools. Admittedly, there is burgeoning debate about whether we can assume—at least when explaining species abundance distributions—that species exhibit identical life-history strategies (e.g. Clark 2009; Rosindell et al. 2010). Yet recent work highlights that those patterns attributed to this assumption of identical strategies can equally well be explained by statistical artefact introduced when dealing with large data sets, and that the same data sets used to make this "probability argument" can actually only have arisen through complex (i.e. ecological niche) processes (Warren et al. 2011). In other words, the assumption of probability might accurately predict patterns (or process rates) that we observe, but for the wrong mechanistic reasons.

Finally, with simplicity (3) the assumption is that only a few variables influence outcomes and that these variables can be measured with accuracy. Yet even if simplicity holds then phenomena that should be predictable from deterministic parameters can be inherently unpredictable because of spatial complexity. For example, Bernstein et al. (2000) observe that the arrangement of balls on a pool table, following their break, should be predictable from a few physical laws. Yet, when we introduce spatial complexity across different pool tables, such as in the exact lay of the table, the nap of the felt, the curvature of each ball, and so on, we introduce fine-scale complexity that means the resulting distribution of the balls is unpredictable (approximating one that is random). The development of the subfield of "spatial soil ecology" is testament to the fact that rates of biogeochemical processes in soils are influenced—at least to some degree—by spatial complexity (Ettema & Wardle 2002). For example, the presence of anaerobic microsites may not be captured in models with insufficient spatial complexity, meaning that denitrification might be observed but not predicted for soils that are, on average, aerobic.

No soil ecologist would argue that soils follow the rules of invariance, probability, or simplicity. When we assume these rules apply (i.e. through our modeling efforts that "black box" the biology) we recognize that we are abstracting reality. There are two primary reasons for such an abstraction. The first is pragmatic; you have to start somewhere and the best place to start is with simplifying assumptions. Mathematical models force us to formalize our understanding, highlighting to us the assumptions we make, the evidence supporting these

> **Box 3.5.2 Messages soil ecologists can communicate when we "black box" soil communities**
>
> When we, as a community of soil ecologists, submit mathematical models of soil biogeochemical processes to the wider scientific and policy community we communicate important information about the state-of-the-art of our understanding of how soil communities influence ecosystem function. The majority of our models "black box" soil communities and, in doing so, they do not account for how changes in soil communities across time or space might influence biogeochemical pools and fluxes. These models are embedded in management and policy efforts of global significance, such as projections of future climate. It is therefore imperative that we make a statement concerning how the wider scientific community should interpret the black boxing of soil communities. We consider we have three main options and that numbers two and three likely best represent the state of our science:
>
> 1. Black boxing is valid because the effect of soil communities on ecosystem function aggregates to a common, invariant response across environmental variation (e.g. temperature) in time and space. You can have high certainty in the projections of our models.
> 2. We black box because the complexity of how soil communities influence function is beyond our current modeling capabilities. This is the best we have, so be skeptical of the certainty of our model projections.
> 3. We black box because it is the simplest assumption we can make, and we recognize the need for research to test this assumption robustly because it might be wrong. This is the best we have, so be skeptical of the certainty of our model projections.

assumptions, and the need for empirical research to test them. In this regard they provide the most effective tool for integration of theory and empirical research to advance basic understanding. They also permit us to make predictions that can inform policy and adaptive management. Even though we are aware of limitations in our knowledge and measurements, such modeling efforts must be favored over inaction. What we have to be careful with is the message we present to those outside of the soil ecology community—i.e. do we wish to communicate that we recognize our abstraction is likely wrong, or that we think the abstraction is a fair generalization (see Box 3.5.2)? If the latter then this gives us the second reason for abstracting the biology in soil models: we believe the abstraction is valid. That is, we can reliably predict how ecosystem processes will respond to a perturbation because biological phenomena aggregate to a common, invariant response (e.g. Rosindell *et al.* 2010). This is the rationale often invoked for black boxing the biology in soil and ecosystem models, and the rationale is presented under the hypotheses of functional redundancy, similarity, and equivalence. Although these hypotheses have been applied to soil communities in an inclusive framework encompassing both animals and microorganisms (e.g.

Andrén & Balandreau 1999), we restrict our discussion to bacteria and fungi (collectively "microbes"). This is not to minimize the potential role of animals in shaping ecosystem processes (e.g. Schmitz 2008) but because in modeling soil and ecosystem processes, we typically discuss the microbes as the primary agents of biogeochemical transformations.

3.5.3 Functional redundancy, similarity, equivalence, and biogeography

Terms in ecology are often used to convey multiple meaning (e.g. consider "niche" when used by Grinnel vs. Hutchinson), which creates confusion. The terms functional redundancy, similarity, and equivalence are applied interchangeably in ecology to comparisons between species, between communities, and as properties of a community (Resetarits & Chalcraft 2007; Allison & Martiny 2008; Strickland *et al.* 2009). Following Allison and Martiny (2008), "functional redundancy" is the ability of one microbial taxon to carry out a process at the same rate as another under the same environmental conditions. Species loss can occur without change in function if a community has many taxa that are functionally redundant (Fig. 3.5.1). The same can even be true where taxa are not functionally redundant but their

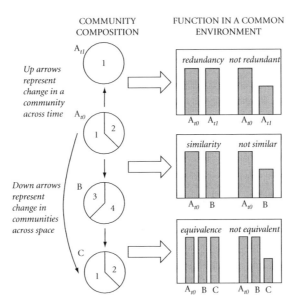

Figure 3.5.1 Organization across space and time of the terms functional redundancy, similarity, and equivalence when applied to the biogeochemical processes performed by soil microbial communities. Circles represent communities. Shown within each community is the abundance of each taxon (denoted by the numbers 1–4). For community A, a perturbation that shifts the community composition from time zero to time one (through the loss of taxon 2) but does not alter function—in a common environment—supports the hypothesis of redundancy. In comparing different communities (i.e. across space), the hypothesis of similarity is supported when a change in community composition does not alter function; again assuming the comparison is made in a common environment. The hypothesis of equivalence is less restrictive: there is no burden on the investigator to demonstrate a difference in richness, relative abundance, or phylogenetic structure of taxa in an assemblage (collectively referred to as "composition") to falsify the hypothesis of equivalence, as there is with redundancy and similarity. Instead, a change in for example phenotypic abundance or physiology within a community—without a necessary change in composition—that alters function in a common environment disproves the hypothesis of functional equivalence. Note that in the figure community C can be classed as functionally non-equivalent to both communities A_{t0} and B, despite having the same composition as A_{t0}.

collective positive and negative responses to disturbance sum to zero net change in aggregate community function: a so-called "portfolio effect." Allison and Martiny (2008) define "functional similarity" as the ability of two microbial communities to carry out a function at a similar rate under the same environmental conditions, regardless of differences in community composition. The same criterion of common environmental conditions applies to the definition of functional equivalence (sensu Strickland et al. 2009), which is the ability of two microbial communities to carry out a functional process at a similar rate, regardless of differences in community composition, physiology, and/or interaction strengths (see Fig. 3.5.1). In this regard, mechanisms underlying functional similarity are a subset of the mechanisms that might drive functional equivalence. That is, similarity refers to the functional effects of shifts in the taxonomic or phylogenetic composition of communities, whereas equivalence includes the effects of these shifts and/or 1) changes in the phenotypes (including physiology) of the members of a community, and 2) interactions between individuals. Indeed, Allison and Martiny (2008) restrict use of the terms redundancy and similarity to differences in microbial community composition—where they define "composition" as the richness, relative abundance, and phylogenetic structure of taxa in an assemblage. The definition of functional equivalence is much less restrictive.

Here we focus on functional equivalence because we argue "equivalence" (as opposed to redundancy or similarity per se) is the assumption that is made when using soil models, that black box microbial communities, to predict ecosystem process rates. This is best illustrated by example. Imagine two identical communities: one maintained at a constant temperature and one to which we apply a step increase in temperature. With temperature increase we would expect enhanced physiological rates (e.g. respiration) because of the positive effect of temperature on the rate of enzyme-mediated reactions. Although physiological rates increase, there is not necessarily a change in organismal physiology and if temperature is returned to control conditions then both communities function equivalently. If the temperature treatment alters community composition, then functional redundancy and similarity are only falsified if the compositional change elicits functional differences when control conditions are restored. If functional differences are observed, say through altered physiology (e.g. shifts in enzyme expression), but there is no measurable change in composition, then there is no evidence to reject the redundancy and similarity hypotheses. However, functional equivalence is falsified. Allison *et al.* (2010) incorporate microbial physiology in a soil model and show that by doing so respiration rates in warmed soils, that are returned to ambient conditions on reaching steady-state, are lower than in control soils given the role of physiological response in reducing microbial biomass and extracellular enzyme abundance.

In subsuming the hypotheses of redundancy and similarity, the hypothesis of functional equivalence has broad application to the question over whether the biogeography of microorganisms can be causally linked to variation in ecosystem function. The two overarching factors that shape biogeographic patterns are environment and history. The role of environment is nicely summarized by Baas Becking's evocative paradigm that "everything is everywhere, *but*, the environment selects" (de Wit & Bouvier 2006). The mechanism underlying this hypothesis is global dispersal where propagules of all microbial taxa are ubiquitously dispersed across the globe (Martiny *et al.* 2006). Growth and survival is then determined (non-randomly) by environmental conditions and there is at least some evidence that this factor structures microbial assemblages at coarse levels of phylogenetic resolution (e.g. Fierer & Jackson 2006; Lauber *et al.* 2009). In soil models, we can reliably black box microbial communities if Baas Becking's hypothesis can be extrapolated to function and if we can identify the key environmental factors driving the microbial process of interest. Under such a scenario, we can still recognize that microbial communities exert proximate control on ecosystem process rates. Yet, because the environment ultimately shapes the community, its role is implicit in the model when we regulate the function of the black box with environmental variables such as temperature and moisture.

We only require the second part of Baas Becking's hypothesis to hold (i.e. "the environment selects [function]") for microbial communities to exert only proximate control on ecosystem processes. This is pertinent to the "black-box debate" because evidence is accumulating that microbes are not all globally dispersed—at least not at rates that obviate the role of historical contingencies in shaping communities—even at spatial scales of only a few meters (Ramette & Tiedje 2007). Even with limited dispersal we can, however, still invoke a number of properties of microbial communities that might serve to homogenize microbial functional potentials across communities; namely high taxon abundance and diversity, rapid evolutionary adaptation, and prolific growth rates (Allison & Martiny 2008; Green *et al.* 2008). We can apply the same arguments to justify disbelief in the functional equivalence hypothesis. Under limited dispersal, these same properties of microbial communities might facilitate rapid genetic differentiation. When populations are isolated geographically, they can solve the same survival problem different ways, which might have different functional implications. Such phenomena introduce contingencies in biological systems that make them "unique" (e.g. Jacob 1977; Levin 1998). In this regard they differ fundamentally from physical and chemical entities in that the rules change (evolution) across time, making replication difficult and history fundamental to understanding their current structure and function. Arguments that we can ignore the identity of microbial taxa (e.g. Falkowski *et al.* 2008) and instead focus on the 500 or so enzyme systems underpinning biogeochemical cycling—have not considered that many enzyme systems (e.g. aerobic respiration) are conserved across the domains of life (Hochachka & Somero 2002) but yield markedly different process rates under the same environmental conditions. Indeed it is now well supported that rates of biogeochemical processes such as litter decomposition, measured in the field, are not only a product of the environment but also adaptation/specialization of the soil community (Ayres *et al.* 2009).

The question that remains is how to test the hypothesis of functional equivalence if we are to justify a microbial-explicit modeling approach. Before reviewing approaches in the next section, we justify the methodological advantage of testing for functional equivalence, as opposed to redundancy or similarity. The advantage is that to test for equivalence there is no requirement to demonstrate differences in microbial community composition. When attempting to falsify functional redundancy or similarity there is the requirement to prove shifts in the structure of microbial communities are associated with shifts in function. This is not trivial—the structure of microbial communities can be analyzed at many different levels of resolution (from the phylum to the strain level), and we do not know which level of phylogenetic resolution is most closely related to function. And with macro-organisms, such as angiosperms, we now know that different genotypes within a single population can markedly influence ecosystem process rates (Whitham *et al.* 2008). Such a genetic shift in the composition of soil communities is unlikely to be detected even with the most advanced molecular techniques available, given that the researcher—before quantifying the genotypic diversity of a population—must first identify the taxon within the many thousands present. There are numerous other issues that complicate the association of community shifts with functional shifts, including microbial dormancy (who is active and when), horizontal gene transfer, and the fact that we often lack a fundamental understanding of which taxa are likely responsible for a specific process. So, if we follow philosophical arguments that a scientific hypothesis must be falsifiable, the methodological issue of being able to demonstrate difference in composition could be used to question the validity of even posing the hypotheses of functional redundancy and similarity for soil communities.

3.5.4 Experimental tests of functional equivalence

Reed and Martiny (2007) comprehensively evaluate three approaches used to test whether there is a necessary relationship between microbial communities and ecosystem function. They focus on approaches that consider "whole communities" (i.e. those that we find extant at field sites). The three approaches have increasing power to tease out causal relationships between microbial communities and function. The first approach is long-term environmental manipulation. These types of studies identify correlative relationships between microbial communities and function, such as the predominance of more *r*-selected phylotypes under N fertilization (e.g. Ramirez *et al.* 2010). Common garden experiments, where a microbial community is manipulated through environmental treatment at a common location, permits short-term investigation of links between microbial communities and function. In the longer-term the change in the environment is considered itself to play a role in regulating biogeochemical rates (i.e. the idea of the environment as the ultimate control) meaning that as with long-term environmental manipulations, correlative and not causative relationships are evaluated. The third approach—and that approach with the most power to identify causation—is the use of reciprocal transplants where the transplant is the community (e.g. a soil monolith) and the environment is considered common to the location where the different communities are brought together. Reed and Martiny (2007) highlight that even with this approach caveats include the fact that the environment of the transplanted units may not fully equilibrate with the new environment. We might ask how long it would take for a soil core transplanted from a pine to an oak forest to assume the environment of the oak forest. For example, total soil carbon and soil texture are parameters that are unlikely to change in such an instance, and even if controlled for there is likely to be immigration from the surrounding community which obscures clear relationships between microbial communities and function that are separate from the environmental conditions.

All three of the approaches evaluated by Reed and Martiny (2007) reduced—to at least some extent—the issue we have in field observation where it is not possible to move beyond correlation (to causation) between microbial communities and function, given the confounding issue of environmental variation. Admittedly, observations are the classical starting point for scientific investigation—and such approaches can identify potentially important environmental and microbial factors that might regulate biogeochemistry (e.g. Strickland *et al.* 2010). Yet to link microbial community composition unquestionably to function requires tightly controlled experiments. These are provided in the form of experimental assemblies of known isolates. Strict control of environmental conditions (e.g. in a bioreactor) permits identification of ultimate causation through the microbial community in terms of regulating process rates. Such studies unambiguously show that microbial composition and diversity influence productivity and nutrient cycling (Bell *et al.* 2005). Where the environment is slightly more realistic (e.g. wood disks), studies with natural isolates have even shown that the assembly history of the community influences decomposition and carbon release rates (Fukami *et al.* 2010). The major limitation of these approaches is, of course, the difficulties associated with assembling the complex communities found in the field. If we assume that a gram of soil may contain as many as 10,000 taxa, the drastic simplification of the experimental assemblages (~10 taxa) is certainly not representative of the enormous taxon diversity in natural microbial communities. This means that mechanisms—such as the portfolio effect and functional redundancy—are strongly selected against in isolate experiments as agents creating functional equivalence. In addition, if ~1% of taxa are culturable and these taxa likely represent more r-selected organisms, then experiments with isolates select for a narrow slice of the ecological strategies observed in natural communities.

Strickland *et al.* (2009) present an experimental approach for testing functional equivalence that is a compromise between whole community and cultured isolate approaches. Recognizing the joint needs of establishing a highly diverse microbial community and to have a common but realistic environment, they established experimental microcosms with milled and sterilized leaf litter (the environment), and inoculated these with whole communities through introduction of a small mass of soil. They combined common garden (a single leaf litter) with reciprocal transplant (litters and soils crossed from multiple sites) approaches and measured carbon mineralization rates from the microcosms across 300 days. The microbial community inocula explained as much as 86% of the variation in mineralization rates, providing strong support for the hypothesis that functional dissimilarity, not functional equivalence, can be important and biogeochemically-relevant in soil communities.

The common caveat to tests of functional equivalence is that we expect the community to modify the environment as the experiment progresses which, in turn, modifies the community. The crux here is modification of the environment. We only falsify equivalence if microbial community effects are compared "under common environmental conditions." This need to hold the environment constant explicitly recognizes that functional rates vary directly as environmental factors change. Indeed, covariation between functional rates and controlling factors is central to all soil models, whether they treat the microbial community as a black box or not. To identify microbial communities as ultimate controllers of biogeochemical process rates requires differences under constant environmental conditions. In Strickland *et al.* (2009) there was high certainty that the initial environments were essentially the same, but with time the environments diverged as decomposition proceeded and the communities shaped the environmental conditions (e.g. the carbon pools available, nutrient levels, pH, etc.). It then leaves us with the conclusion that initial functional differences in the community led to overall differences in function that themselves might have been a product of the community and/or altered environment. Separation of the effects of community composition and the environment is challenging, and clearly we need the suite of experimental approaches outlined in this section to robustly challenge the hypothesis of functional equivalence. Criticism of one approach to support another fails to recognize the advantages each brings to the discourse (e.g. Fukami *et al.* 2010) and will only reinforce the hypothesis in the absence of appropriate falsification. Given the common assumption of the functional equivalence hypothesis in ecosystem modeling there is a scientific and societal need to test it appropriately. If we do so we will advance basic understanding of soil ecology and permit reliable evaluation of modeled ecosystem responses and feedbacks to environmental change.

3.5.5 Putting ecology into soil models

The soil (and ecosystem) models applied widely to address questions related to feedbacks to global warming, carbon stocks across regional gradients, and nitrogen dynamics across agricultural management regimes, all "black box" the microbial community. They assume functional equivalence in their parameterization, validation, and prediction. Indeed, we often parameterize ecosystem models using data collected across space, and then apply these parameterizations to make predictions across time for a system. This approach assumes microbial communities do not ultimately influence biogeochemical cycling either in time or space. So what happens when we construct models that open up the black box and permit the identity of taxa, their physiology, and biomass to influence ecosystem process rates? An in-depth model evaluation and review is beyond the scope of this chapter, but below we present a brief discussion to identify future modeling needs by highlighting the implications of relaxing the "black box" assumption.

Using multi-pool, soil organic carbon (SOC) models—which black box soil communities—Kirschbaum (2004) and Knorr et al. (2005) showed that depletion of labile SOC pools could explain the ephemeral augmentation of soil respiration under simulated warming. Once the warmed systems reached a steady-state—i.e. the labile pools had been depleted to a constant value—respiration rates matched carbon input rates, which were unchanged from pre-warming conditions. That inputs equal outputs is expected for any steady-state system. Yet the studies demonstrated that the conventional way we model soil carbon can predict observed respiration responses to soil warming. Indeed, there is empirical support for this substrate-depletion mechanism (Bradford et al. 2008). However, at the same site, there is also empirical support that the microbial communities adjust to the thermal regime in a manner that influences respiration rates (Bradford et al. 2008). So where the conclusions of Kirschbaum (2004) and Knorr et al. (2005) went too far was in arguing that by finding evidence for the substrate-depletion hypothesis they had falsified alternate hypotheses that microbial community responses explained observed patterns of respiration to warming. Demonstrating that one mechanism can explain an observed pattern does not falsify alternate mechanisms that might equally recreate the same pattern (i.e. the absence of evidence is not evidence of absence). To evaluate the competing hypotheses of functional equivalence and redundancy, we require soil models that open-up the black box by explicitly modeling microbial dynamics to evaluate the role of microbes in driving biogeochemical processes.

Allison et al. (2010) present one of an emerging family of microbe-explicit models (e.g. Lawrence et al. 2009). They compare conventional multi-pool, SOC models to an enzyme-based approach that represents solubilization of SOC by extracellular enzymes, microbial assimilation of dissolved organic carbon compounds, and the expected negative relationship between temperature and microbial growth efficiencies. Model predictions were most sensitive to this latter parameter. In response to sustained warming, microbial biomass was reduced because less of the carbon assimilated by the microbes was allocated to growth. This served to decrease the abundance of microbial extracellular enzymes that solubilize SOC, creating a negative feedback to warming-induced losses of SOC. By explicitly modeling the microbial dynamics they found no evidence for positive feedback to climate warming through loss of SOC to the atmosphere, a finding contrary to most black box soil models. The finding is significant because when we relax the assumption of functional equivalence, model predictions that influence policy may well differ from those derived from more conventional modeling approaches.

Soil models that relax the assumption of functional equivalence are not a recent phenomenon (e.g. Hunt et al. 1987) but, like their earlier counterparts, they have not yet been incorporated into modeling efforts that might influence policy and practice on environmental issues such as global climate change and carbon emissions (e.g. those models used in Denman et al. 2007). They are also deterministic, and so do not permit context-dependent histories that are likely essential to generating functional dissimilarity to shape functional outcomes that are not fully reversible. For example, in the Allison *et al.* (2010) model returning the system to a pre-warming state will eventually permit microbial biomass, extracellular enzyme abundance, and carbon stocks to recover to pre-warming conditions. Given context-dependent histories, we rarely expect changes in ecological systems to be fully reversible upon restoration of original environmental conditions (Levin 1998). Maybe such context-dependency is too difficult to incorporate into current modeling efforts. However, the tractability of including microbial dynamics in deterministic soil models has been demonstrated (e.g. Lawrence *et al.* 2009; Allison *et al.* 2010) and provides a likely productive direction for exploring the implications of assuming functional equivalence vs. dissimilarity when projecting biogeochemical response and feedbacks to environmental change.

3.5.6 Revisiting the functional paradigm in soil ecology

Soil models—including those used in the coupled carbon cycle models to project climate change—typically assume that soil communities are functionally equivalent. To put this in colloquial terms, "it doesn't matter who is there, nor in what form, number, or location," because every soil community is essentially a black box that functions the same way under the same environmental conditions when we look across space or time (see Schimel 2001). Application of this hypothesis assumes that functional responses to a disturbance (e.g. temperature change) can be described with a single, mathematical equation. In making this assumption "history"—in its broadest sense—is disregarded as a force that influences the functioning of microbial communities through changes in biomass, composition, or the physiology of soil taxa. For example, even the decrease in total microbial biomass observed by Allison *et al.* (2010)—when they warmed a system and assumed a negative effect of temperature on microbial growth efficiencies—would fail to elicit an initial difference in respiration rates between a system pre and post a warming disturbance if we assume functional equivalence. Few ecosystem ecologists would argue—if we held composition and the environment constant—that microbial biomass was irrelevant to ecosystem process rates in soils (nor would any plant ecologist dare argue that plant biomass and community type are irrelevant to predicting photosynthetic rates). Yet we have to accept that as soil ecologists the dominant paradigm we espouse through our ecosystem modeling is that soil microbial communities are homogenously functioning units across space and time, which exhibit invariant functional responses to changes in controlling factors such as temperature, even where microbial biomass differs.

Recent models (e.g. Lawrence *et al.* 2009; Allison *et al.* 2010) challenging the functional equivalence paradigm have not yet been coupled with efforts that provide the scientific basis for policy to mitigate and adapt to environmental problems (e.g. Denman *et al.* 2007). Even within academic circles, soil biology appears to have had little influence in shaping general ecological knowledge (Barot *et al.* 2007). If our field is to advance knowledge and application outside of its own perimeters, then we must take a fresh look at the paradigms of functional redundancy, similarity, and equivalence and—and if we find them lacking—challenge application of these paradigms where soil ecological knowledge is applied outside of our field. This includes application in areas of high societal importance, such as the coupled atmosphere-biosphere carbon cycle models for projecting feedbacks to climate change (Denman *et al.* 2007). Plant ecologists have engaged with atmospheric modelers in these realms, and soil ecologists must now do the same if we wish to make a robust claim that soil biology need be considered when managing ecosystems and climate in the face of environmental change.

References

Allison, S.D., & Martiny, J.B.H. (2008) Resistance, resilience, and redundancy in microbial communities. *Proceedings of the National Academy of Sciences of the United States of America* **105**: 11512–19.

Allison, S.D., Wallenstein, M.D., & Bradford, M.A. (2010) Soil-carbon response to warming dependent on microbial physiology. *Nature Geoscience* **3**: 336–40.

Andrén, O., & Balandreau, J. (1999) Biodiversity and soil functioning—from black box to can of worms? *Applied Soil Ecology* **13**: 105–8.

Ayres, E., Steltzer, H., Simmons, B.L., et al. (2009) Home-field advantage accelerates leaf litter decomposition in forests. *Soil Biology & Biochemistry* **41**: 606–10.

Bahn, M., Janssens, I.A., Reichstein, M., Smith, P., & Trumbore, S. (2010) Soil respiration across scales: towards and integration of patterns and processes. *New Phytologist* **186**: 292.

Barot, S., Blouin, M., Fontaine, S., Jouquet, P., Lata, J.C., & Mathieu, J. (2007) A tale of four stories: soil ecology, theory, evolution and the publication system. *PLoS One* **2**: e1248.

Bell, T., Newman, J.A., Silverman, B.W., Turner, S.L., & Lilley, A.K. (2005) The contribution of species richness and composition to bacterial services. *Nature* **436**: 1157–60.

Bernstein, S., Lebow, R.N., Stein, J.G., & Weber, S. (2000) God gave physics the easy problems: adapting social science to an unpredictable world. *European Journal of International Relations* **6**: 43–76.

Bradford, M.A., Davies, C.A., Frey, S.D., et al. (2008) Thermal adaptation of soil microbial respiration to elevated temperature. *Ecology Letters* **11**: 1316–27.

Clark, J.S. (2009) Beyond natural science. *Trends in Ecology & Evolution* **24**: 8–15.

Davidson, E.A., & Janssens, I.A. (2006) Temperature sensitivity of soil carbon decomposition and feedbacks to climate change. *Nature* **440**: 165–73.

de Wit, R., & Bouvier, T. (2006) "Everything is everywhere, but, *the environment selects*"; what did Baas Becking and Beijerinck really say? *Environmental Microbiology* **8**: 755–8.

Denman, K.L., Brasseur, G., Chidthaisong, A., et al. (2007) Couplings between changes in the climate system and biogeochemistry. In: Solomon, S., Qin, D., Manning, M. et al. (eds.) *Climate change 2007: The physical science basis. Contribution of working group I to the fourth assessment report of the Intergovernmental Panel on Climate Change.* Cambridge University Press, Cambridge.

Ettema, C.H., & Wardle, D.A. (2002) Spatial soil ecology. *Trends in Ecology & Evolution* **17**: 177–83.

Falkowski, P.G., Fenchel, T., & Delong, E.F. (2008) The microbial engines that drive Earth's biogeochemical cycles. *Science* **320**: 1034–9.

Fierer, N., Bradford, M.A., & Jackson, R.B. (2007) Toward an ecological classification of soil bacteria. *Ecology* **88**: 1354–64.

Fierer, N., & Jackson, R.B. (2006) The diversity and biogeography of soil bacterial communities. *Proceedings of the National Academy of Sciences of the United States of America* **103**: 626–31.

Fukami, T., Dickie, I.A., Wilkie, J.P., et al. (2010) Assembly history dictates ecosystem functioning: evidence from wood decomposer communities. *Ecology Letters* **13**: 675–84.

Green, J.L., Bohannan, B.J.M., & Whitaker, R.J. (2008) Microbial biogeography: from taxonomy to traits. *Science* **320**: 1039.

Hochachka, P.W., & Somero, G.N. (2002) *Biochemical adaptation: Mechanism and process in physiological evolution.* Oxford University Press, New York.

Hunt, H.W., Coleman, D.C., Ingham, E.R., et al. (1987) The detrital food web in a shortgrass prairie. *Biology and Fertility of Soils* **3**: 57–68.

Jacob, F. (1977) Evolution and tinkering. *Science* **196**: 1161–6.

Kirschbaum, M.U.F. (2004) Soil respiration under prolonged soil warming: are rate reductions caused by acclimation or substrate loss? *Global Change Biology* **10**: 1870–7.

Knorr, W., Prentice, I.C., House, J.I., & Holland, E.A. (2005) Long-term sensitivity of soil carbon turnover to warming. *Nature* **433**: 298–301.

Lauber, C., Knight, R., Hamady, M., & Fierer, N. (2009) Soil pH as a predictor of soil bacterial community structure at the continental scale: a pyrosequencing-based assessment. *Applied and Environmental Microbiology* **75**: 5111–20.

Lawrence, C.L., Neff, J.C., & Schimel, J.S. (2009) Does adding microbial mechanisms of decomposition improve soil organic matter models? A comparison of four models using data from a pulsed rewetting experiment. *Soil Biology & Biochemistry* **41**: 1923–34.

Levin, S.A. (1998) Ecosystems and the biosphere as complex adaptive systems. *Ecosystems* **1**: 431–6.

Martiny, J.B.H., Bohannan, B.J.M., Brown, J.H., et al. (2006) Microbial biogeography: putting microorganisms on the map. *Nature Reviews Microbiology* **4**: 102–12.

Ramette, A., & Tiedje, J.M. (2007) Multiscale responses of microbial life to spatial distance and environmental heterogeneity in a patchy ecosystem. *Proceedings of the National Academy of Sciences of the United States of America* **104**: 2761–6.

Ramirez, K.S., Lauber, C.L., Knight, R., Bradford, M.A., & Fierer, N. (2010) Consistent effects of nitrogen fertilization on soil bacterial communities in contrasting systems. *Ecology* **91**: 3463–70.

Reed, H.E., & Martiny, J.B.H. (2007) Testing the functional significance of microbial composition in natural communities. *FEMS Microbiology Ecology* **62**: 161–70.

Resetarits, W.J., & Chalcraft, D.R. (2007) Functional diversity within a morphologically conservative genus of predators: implications for functional equivalence and redundancy in ecological communities. *Functional Ecology* **21**: 793–804.

Reynolds, J.F., Bugmann, H., & Pitelka, L. (2001) How much physiology is needed in forest gap models for simulating long-term vegetation response to global change? Limitations, potentials, and recommendations. *Climatic Change* **51**: 541–57.

Rosindell, J., Hubbell, S.P., & Etienne, R.S. (2010) Protracted speciation revitalizes the neutral theory of biodiversity. *Ecology Letters* **13**: 716–27.

Schimel, J.P. (2001) Biogeochemical models: implicit vs. explicit microbiology. In: Schulze, E.D., Harrison, S.P., Heimann, M., *et al.* (eds.) *Global biogeochemical cycles in the climate system.* Academic Press, San Diego, CA.

Schmitz, O.J. (2008) Effects of predator hunting mode on grassland ecosystem function. *Science* **319**: 952–4.

Strickland, M.S., Callaham, M.A., Jr, Davies, C.A., *et al.* (2010) Rates of *in situ* carbon mineralization in relation to land-use, microbial community and edaphic characteristics. *Soil Biology & Biochemistry* **42**: 260–9.

Strickland, M.S., Lauber, C., Fierer, N., & Bradford, M.A. (2009) Testing the functional significance of microbial community composition. *Ecology* **90**: 441–51.

Warren II, R.J., Skelly, D.K., Schmitz, O.J., & Bradford, M.A. (2011) Universal ecological patterns in college basketball communities. *PLoS One* **6**: e17342.

Whitham, T.G., Difazio, S.P., Schweitzer, J.A., *et al.* (2008) Extending genomics to natural communities and ecosystems. *Science* **320**: 492–5.

CHAPTER 3.6

Biogeography and Phylogenetic Community Structure of Soil Invertebrate Ecosystem Engineers: Global to Local Patterns, Implications for Ecosystem Functioning and Services and Global Environmental Change Impacts

Lijbert Brussaard, Duur K. Aanen, Maria J.I. Briones, Thibaud Decaëns, Gerlinde B. De Deyn, Tom M. Fayle, Samuel W. James, and Tânia Nobre

3.6.1 Introduction

Biodiversity is the variety of nature in terms of the abundance and distributions of, and interactions between genotypes, species, communities, and ecosystems (Ash et al. 2009). Ecosystem services are the benefits people derive from nature. If we want to establish the links between the two (if any, and causal or not), we have to recognize that biodiversity differs, both in absolute terms and in terms of what we know of it, between different global locations, continentally, regionally, or locally. The same holds true for the ecosystem services required by people from nature and the existing or desired governance structures to manage biodiversity and ecosystem services. Hence, the objective of this chapter is to assess global-to-local scale geographical patterns of soil biodiversity and to relate those to ecological processes as a prerequisite for understanding and managing the provision of ecosystem services and the responses to drivers of global environmental change.

This chapter deals specifically with the soil invertebrate ecosystem engineers *sensu* Hastings et al. (2007), notably the termites, ants, earthworms, and enchytraeids, which leave physical traces on or in the soil that can outlive the engineers: excrements, mounds, burrows, moved particles, and soil aggregates.

The geography of the soil ecosystem engineers cannot be understood without considering the geography of soils. Following Jenny (1941), organisms, including the soil biota, are one of the soil-forming factors, in addition to (and interrelated with) parent material, climate, topography and time. Our entry point for the connection between the geography of soil and soil ecosystem engineers is that any location on earth has experienced considerable change over time due to the drift of the continents, changes in topography, climate changes, and evolutionary changes of the biota, all imprinting a legacy on the thin skin of the earth that currently constitutes the living soil. The current variation in solar energy interacts across the globe with that legacy and with the current complex patterns of global water distribution to form the major biomes that are the playgrounds of today's evolutionary and ecological processes (Fig. 3.6.1).

Soil Ecology and Ecosystem Services. First Edition. Edited by Diana H. Wall *et al.*
© 2012 Oxford University Press. Published 2012 by Oxford University Press.

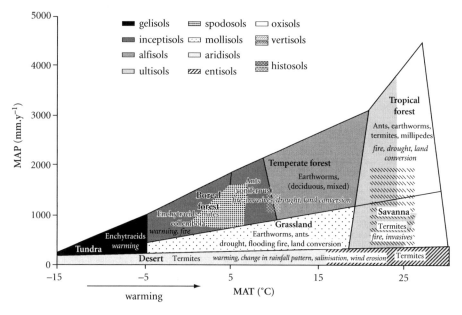

Figure 3.6.1 Dominant soil types, typical soil forming invertebrates and major global change threats across biomes with characteristic mean annual temperature (MAT) and precipitation (MAP). Warming causes increased nutrient cycling rates, extreme precipitation patterns, increasing catastrophic disturbances (floods, droughts, erosion, fires). Land conversion is an important non-climatic disturbance factor. Mean Annual Temperature was calculated from Climatic Research Unit (CRU) monthly temperature data (2000–2006), http://www.cru.uea.ac.uk/cru/data/temperature/. Mean Annual Precipitation was calculated from Global Precipitation Climatology Centre (GPCC) monthly precipitation data (2000–2006), http://www.esrl.noaa.gov/psd/data/gridded/data.gpcc.html. Global soil order map for 2006 was obtained from the United States Department of Agriculture (USDA). Biomes were derived from the International Geosphere-Biosphere Program (IGBP) Land Cover Characterization (LCC) database. (Recalculated after De Deyn et al. 2008 by Rogier de Jong.)

A comprehensive understanding of how these evolutionary and ecological factors have shaped present-day communities is aimed at in the analysis of phylogenetic community structure and in trait-based ecology. The conceptual foundation of trait-based ecology consists of observed trait distributions, initially derived from the pool of possible traits of individual organisms and selected by performance filters, i.e. environmental (biotic and abiotic) filters selecting against traits with inadequate local fitness. The filtering results in a certain community composition, associated with a certain level of ecosystem functioning (Fig. 3.6.2).[1]

[1] The hypothesis of environmental filtering and subsequent density-dependent processes has to be tested against the null hypothesis of random species assembly (Hubbell 2001). Testing this hypothesis is not the subject of this chapter, but we note that in one area in which much research has been done, i.e. plant–herbivore ecology, random patterns of plant community phylogenetic structure are less frequently observed than non-randomness, resulting in clustering of trait distributions in species assemblages (Cavander-Bares et al. 2009).

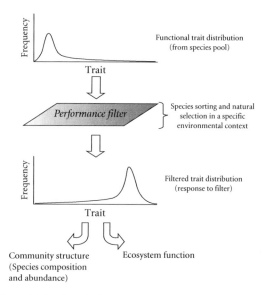

Figure 3.6.2 Relationship between trait distribution and performance filter. The functional trait distribution is filtered by the environment based on performance (the match between the trait and the environment) via natural selection and/or ecological sorting at a particular space/time location. (From Webb et al. 2010, with permission from John Wiley & Sons.)

In trait-based ecology, a trait is a well-defined property of organisms, usually measured at the individual level and used comparatively across species (McGill *et al.* 2006). In general, organisms are characterized in terms of their multiple biological attributes such as physiological, morphological, behavioral, or life-history traits. They can henceforth be encompassed in *functional trait groups*, depending on the ecosystem function under study, making the functional group concept less static. For further details see Brussaard (Chapter 1.3, this volume).

Phylogenetic community structure is the pattern of phylogenetic relatedness of species distributions within and among communities. The concept goes beyond a mere phylogenetic approach to biogeography at species level in that phylogenetic attraction of related species, driven by environmental filtering, and phylogenetic repulsion, possibly caused by competition, simultaneously occur and can be made explicit (Helmus *et al.* 2007). Hence, phylogenetic community structure and trait-based ecology help to reveal the historical and contemporary processes driving the assembly of biological communities and how they determine ecosystem functioning.

The processes that drive the organization of species in a focal area operate over varying temporal scales and depend fundamentally on the spatial scale of analysis (Fig. 3.6.3). At the broadest, (supra-) continental spatial scale, species distributions are determined largely by biogeographical processes that involve speciation, extinction, and dispersal. These processes occur over long temporal scales. At intermediate, regional/biome scales, as depicted in Fig. 3.6.3, dispersal varies with the dispersal ability of the organisms. Dispersal can alter patterns of species distributions as they become established through ecological sorting processes which occur when species are filtered out or added, as related to their physiological tolerances. The environment can include both abiotic factors (temperature, soil texture, soil moisture, light availability, pH) and biotic factors (symbionts, pollinators, hosts, prey). It follows that similar habitats in different regions may have different numbers of species, because differing histories of the areas have led to occupancy by different clades of the major taxa (in our case: of the soil invertebrate ecosystem engineers). In addition, different clades have different potentials for diversification, while dif-

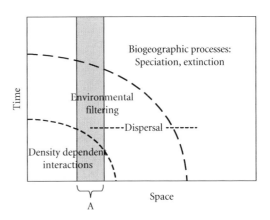

Figure 3.6.3 At a given spatial scale A, species distributions depend on multiple factors. (From Cavander-Bares *et al.* 2009, with permission from John Wiley & Sons.)

ferent lengths of time of occupancy also play a role in determining the numbers of extant species (Webb *et al.* 2002). At the smallest, local, i.e. neighborhood scales, density-dependent processes are likely to operate most intensively. These processes may include competition, disease, herbivory, interspecific gene flow, facilitation and mutualism, and may interact with the abiotic environment to reinforce or diminish environmental filtering (Cavander-Bares *et al.* 2009).

Following a section with an account of macroecological patterns in soil invertebrate community distributions, we give phylogeny-based accounts of biogeographic patterns of the major soil invertebrate engineer groups (termites, ants, earthworms, and enchytraeids). We also provide examples of phylogenetic community structure at continental-to-local levels and explore the possible effects of drivers of global environmental change on community composition and associated ecosystem functioning and services. Following these group-specific sections, we will discuss trait-based ecology of soil invertebrate ecosystem engineers more generally, with a view to the possible effects of global environmental change on ecosystem functioning and services.

3.6.2 Macroecological patterns in soil invertebrate communities

3.6.2.1 Area–diversity relationships

Area–species richness relationships have been documented for a number of aboveground organisms, but have seldom been studied for soil organisms (Decaëns 2010). The theory of island biogeography predicts that species richness of a given group increases with island area and decreases with island distance from the mainland (Gaston 2000). In mainlands, it is also predicted that species richness increases with spatial observation scale. The few studies that describe such patterns for soil invertebrates include community assessment of ants of the Melanesian archipelago (Wilson 1974), of mites in islands and mainlands (Maraun et al. 2007), of springtails (Ulrich & Fiera 2009), and of European earthworms (Judas 1988). In most of these cases, species richness of ants and micro-arthropods also tends to be lower in islands than in mainlands for a given area (Wilson 1974; Stanton & Tepedino 1977; Maraun et al. 2007; Ulrich & Fiera 2009), which highlights a negative impact of habitat isolation on community species richness (Blondel 1995). On the other hand, Jonsson et al. (2009), while studying the community structure of six groups of soil invertebrates (i.e. ground-dwelling spiders, web-building spiders, beetles, collembolans, mites, and nematodes) in boreal lake islands of northern Sweden, found that taxonomic richness was either neutrally or negatively related to island size, and either neutrally or positively related to island isolation.

Patterns supporting the predictions of island biogeography theory have been explained by a number of factors (Gaston 2000; Decaëns 2010). For instance, larger observation scales may encompass higher habitat diversity and greater environmental stability, thus leading to a higher ratio between immigration/speciation and extinction processes. Increasing observation scales also encompass a higher number of nested levels of organization (from habitat patches to successional stages and landscape units), which results in a larger size of species pools through higher potential spatial and ecological segregation. Additionally, species richness calculation at increasing spatial scales requires the compilation of increasingly larger data sets, which may result in an enhanced probability of detecting rare species. Patterns that depart from island biogeography predictions such as those described by Jonsson et al. (2009) have been attributed to the combined and confounding effects of several factors, but not necessarily those predicted as important by island biogeography theory, in determining invertebrate species richness in some island systems. For instance, they propose that an increase in species richness with island isolation may result from increased habitat heterogeneity due to greater disturbance from climatic events, and lower levels of predation from birds.

3.6.2.2 Latitudinal gradients

A remarkably low number of studies have explored latitudinal variations in soil animal communities (Bardgett et al. 2005; De Deyn & van der Putten 2005; Decaëns 2010). A few studies have described such variation for oribatid mites (Maraun et al. 2007), ants (Kusnezov 1957), springtails (Ulrich & Fiera 2009), and termites (Eggleton 1994; Lavelle & Spain 2001). Some authors suggested that this pattern may not apply for all soil organisms due to the cosmopolitan nature of many edaphic species or to a relatively low latitudinal variability of soil conditions (Wardle 2002; De Deyn & van der Putten 2005; Maraun et al. 2007). In many cases, however, the absence of a recognized latitudinal pattern for soil animals can easily be ascribed to a deficit of sampling in intertropical regions. For instance, Maraun et al. (2007) only found an increase in diversity going from boreal to temperate latitudes, after which diversity leveled off, but Decaëns (2010) suggested that this was mainly due to an unbalanced amount of data in favor of temperate countries. Lavelle (1983, 1986), using data on earthworm communities, suggested that enhanced efficiency of mutualism under tropical climates may be one of the causes of latitudinal gradients in soil animal communities. Other factors classically proposed, to explain the increase in species richness from high to low latitudes, include the increase in geographic areas towards the equator or the increase of the heterogeneity, productivity or environmental stability (both past and actual) of habitats (Huston 1994; Brown & Lomolino 1998; Gaston & Spicer 1998; Gaston 2000).

A frequent feature of latitudinal patterns is that the shape of the variation in biodiversity is often asymmetric (Gaston 2000). This was, for example, described by Dunn et al. (2009) for ants and by Eggleton (1994) for termites, who found that southern hemisphere sites were more diverse than northern hemisphere sites. Dunn *et al.* (2009) emphasized that most of this asymmetry could be explained statistically by differences in contemporary climate, local species richness of ants being positively correlated with temperature, and negatively associated with temperature range, and precipitation. Another part of the asymmetry in the pattern was explained by the greater climate stability of the southern hemisphere during the Eocene epoch.

3.6.2.3 Altitudinal gradients

The few studies that explored altitudinal variation in soil biodiversity have focused on invertebrates. Some have described a continuous decrease in the total number of species by altitudinal stratum in a given geographical area, like earthworms in France (Bouché 1972; Dahmouche 2007), ants in the Smoky Mountains (Cole 1940), termites in Sarawak (Collins 1980), and dung beetles in Spain (Romero-Alcaraz & Ávila 2000). These patterns may be explained by different factors including the increasingly harsh abiotic conditions (in particular temperature), the reduced levels of primary productivity and ecosystem carrying capacity, and the smaller habitat areas at higher elevation.

On the other hand, as reported for many aboveground organisms, altitudinal variations in soil biodiversity may present strong local and/or taxonomic specificity (Brown & Lomolino 1998; Decaëns 2010). Many taxonomic groups show a humpback response to elevation gradients. For example, Collins (1980) reported a peak of taxonomic diversity at 500–1,200 m for beetle communities, and at 1,300–1,700 m for dipteran assemblages in Sarawak. Loranger *et al.* (2001) reported a humpback distribution of springtail species richness across an altitudinal transect from 950–2,150 m in the French Alps. These results may partly be explained by altitudinal variation in environmental factors (e.g. rainfall, temperature, pH or organic matter quality) and an increased influence of dispersal barriers (Gaston 2000).

3.6.2.4 Landscape modification gradients

A related pattern to the area–species diversity relationships is the response of communities to habitat fragmentation gradients (Gaston & Spicer 1998). This has been directly addressed in a few studies focusing on micro-arthropods (Rantalainen *et al.* 2005), termites (Fonseca De Souza & Brown 1994), ants (Suarez *et al.* 1998; Carvalho & Vasconcelos 1999; Vasconcelos *et al.* 2006), and ground beetles (Barbosa & Marquet 2002; Driscoll & Weir 2005).

Some recent studies have underpinned the complexity and the specificity of the responses within or between broad taxonomic groups. For example, Davies (2002) illustrated opposite responses of two termite functional groups to the fragmentation of Amazonian rain forest (positive response for litter and wood feeders, negative for geophagous species). Sousa *et al.* (2006) studied Collembolan community patterns in comparable gradients from forested to agricultural dominated landscapes in eight European countries. Although species richness patterns were not fully concordant among the different countries, they found that high species richness was associated with high landscape heterogeneity (i.e. high number of land use units). They also report that the average local richness of forest patches decreased along the gradient, whereas the opposite pattern occurred for open habitat assemblages, showing that both ecological groups were sensitive to the fragmentation of their preferred habitat at the landscape scale. Assuming that different soil organisms have different dispersal and colonization capacities, Hedlund *et al.* (2004) predicted a relative resistance of bacterial-based communities, and a higher vulnerability of fungal-based organisms.

3.6.2.5 Concluding remarks

Understanding the driving factors of soil invertebrate communities across different spatial and temporal scales is of primary importance if we wish to predict soil responses to global changes and the impact these changes will have on the delivery of ecosystem services. We are, however, far from reaching a clear picture of these patterns, in part because of the still unsatisfactory knowledge of the taxonomy and phylogeny of most groups of soil animals (Decaëns *et al.* 2006, 2008a). It is thus of prime importance to address the strong taxonomic deficit in most groups of soil organisms. This will likely be achieved through implementing new genomic taxonomical methods such as DNA barcoding (Rougerie *et al.* 2009; Richard *et al.* 2010). As suggested by Decaëns *et al.* (2008a), this approach has great potential for species identification, for stimulating accurate soil biodiversity surveys or any ecological or biogeographical research based on species lists, and for helping soil taxonomists to solve taxonomic and phylogenetic problems. In light of these exciting new tools, the efforts in the next sections to connect biogeography, taxonomy, and phylogeny within ants, termites, earthworms, and enchytraeids, have to be seen as preliminary rather than conclusive for gaining insight in the effects of global environmental change drivers on ecosystem functions and services, associated with the community structure and abundance of soil invertebrate ecosystem engineers.

3.6.3 Termite biogeography and phylogenetic community structure

3.6.3.1 Introduction

Termites are eusocial insects, phylogenetically nested within cockroaches (Inward *et al.* 2007). Colonies show complex division of labor, with reproductives, soldier and worker castes. Workers forage, and dead plant material is the main food source of almost all species. However, the plant material used is diverse, ranging from wood to organic material present in soil. Termites are renowned soil ecosystem engineers (Lavelle *et al.* 1997) as they have a highly significant impact on pedogenesis, soil properties, and soil functions. Hence, they play an important role as mediators of soil ecological processes (e.g. Lee & Wood 1971; Pearce 1997; Bignell & Eggleton 2000). They can translocate large quantities of soil, promote soil stability and water permeability, and change soil chemistry (Bignell 2006), and they can degrade and utilize even the most recalcitrant residues of dead plant material such as lignin, cellulose, and humus (Rouland-Lefèvre & Bignell 2002). Termites owe a large part of this efficiency to their gut symbionts, which include microorganisms of all major taxa: Archaea, Bacteria, and Eucarya (Bignell 2000). Any extant termite is associated with representatives of at least two of these groups. The "higher termites" (family Termitidae) have retained their bacterial symbionts, but typically lack the protozoan gut symbionts that most other termite families have. A single subfamily in this terminal clade, the Macrotermitinae has evolved a unique "agricultural" ectosymbiosis with basidiomycete fungi of the genus *Termitomyces*. This niche differentiation has allowed termites to gain an immense impact on the global terrestrial carbon cycle, exceeding the cumulative decomposition roles of other arthropods and only being surpassed by the mammalian herbivores (Bignell *et al.* 1997). It is believed that termites mediate between 2–5% of the CO_2 flux from all terrestrial sources (Sanderson 1996; Bignell *et al.* 1997; Sugimoto *et al.* 2000). For example, in moist savannah systems, estimations of population respiration rates indicated that roughly 20% of carbon mineralization could be attributed to, mostly fungus-growing, termites (Wood & Sands 1978). In contrast, in the African rainforests, their relative contribution to decomposition is relatively low—about 1–2% of all C mineralization—in spite of their much higher taxonomic diversity in this habitat (Bignell & Eggleton 2000). In forests, tree metabolism dominates the carbon fluxes, making the relative contribution of the termites small, although the absolute abundance and biomass may be higher than in the savannahs (Bignell *et al.* 1997). Fungus-growing termites are only found in the Old World. In the New World, their ecological role may be covered to some degree by other fungus-growing social insects, the attine ants. However, attine ant food consists mainly of fresh rather than dry plant material (Mueller *et al.* 2005).

3.6.3.2 Continental scale

Termites have been divided into seven families (Fig. 3.6.4), 281 genera, and about 2,600 species (Eggleton 2000). The highest diversity is found in the "higher termites," comprising ca. 84% of the world termite species. Termites show rather extreme differences in their distributional range, so that at the same latitude, and under approximately similar environmental conditions, there are huge differences

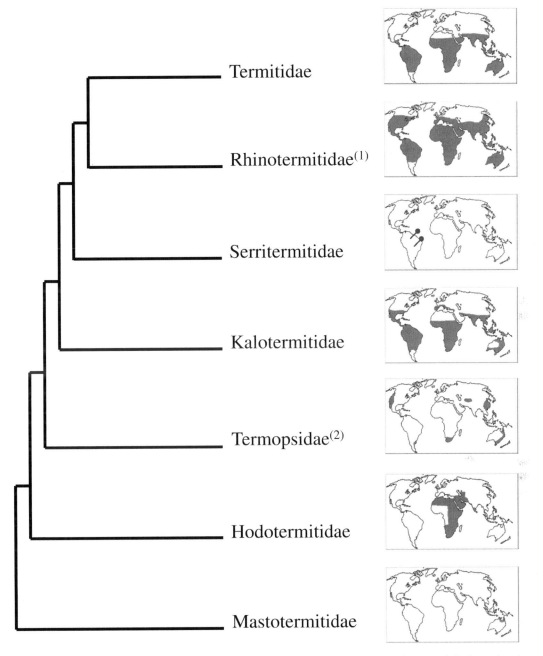

Figure 3.6.4 Schematic representation of termite family relationships (Inward et al. 2007) and their general patterns of distribution (based on Eggleton 2000). Rhinotermitidae is shown as a monophyletic group whereas it appears paraphyletic because the Brazilian Serritermitidae nest within this family. However, the status of the monotypic Serritermitidae as a separated family is debatable (see Kambhampati & Eggleton 2000). Termopsidae is likewise a paraphyletic group with the Hodotermitidae clade nesting within it and not being basal to it as represented here. On the distribution patterns: Mastotermitidae has only one extant member restricted to the non-rainforest parts of northern Australia; Hodotermitidae are restricted to the Old World and can be found in habitats ranging from dry savannah to arid grassland; Termopsidae are the most widely scattered of all the major termite groups and nest and feed on rotting logs; Kalotermitidae are widely distributed and live in water-stressed habitats; Serritermitidae consist of only two known species present at Santarém and Coxipó da Ponte in Brazil (Emerson & Krishna 1975) and likely inhabit savannah-like habitats (cerrado); Rhinotermitidae are widely distributed in habitats ranging from wet to dry; Termitidae comprise, amongst others, the fungus growers restricted to the Old World, and soil feeders present in the Neotropical and Oriental regions.

across regions and continents which can be explained by historical factors but also, to some degree, by habitat differences. Overall, in temperate and cold regions, they are virtually absent, whereas they are very abundant in tropical regions. Within their distributional range, termite species richness drops very fast from roughly 10° north and south of the equator (Eggleton 2000). This drop is much stronger north than south, which has been attributed to the relative protection of the south from the effects of glaciation (Eggleton 2000). The African fauna has a much higher genus richness than both neotropical and oriental faunas.

Some authors (e.g. Pearce 1997) suggest that termites might have been present since the end of the Permian period (Paleozoic era), but there are no fossil records to indicate that they have a pre-Cretaceous history (Thorne et al. 2000). Recently however, Engel et al. (2009) suggested that termites have diverged from cryptocercid roaches in the Late Jurassic, making termites the oldest group of eusocial animals, pre-dating the origin of ants by some 35 million years. A strong historical effect is apparent: phenetic patterns of the genus structure composition are closely explained by the paleogeographic history of the continents and it seems that the divergence from the basal groups happened already on the separating landmasses (Eggleton 2000). In this way, vicariance has played an important role for basal groups. The other (more diverged) groups, however, have recently dispersed over long distances (Emerson 1955).

It is unknown which factors have been causal in determining termite global distribution, but it has been suggested that differences in the histories and relative ages of the various forest blocks of the continents are important (Eggleton 2000). Additionally, an interesting correlation has been found between nesting habit and genus range size and species richness within each genus (Eggleton 2000): genera with at least one species nesting in wood have wider overall ranges than subterranean or mound-building genera. This has been interpreted as evidence for the hypothesis that nesting in wood facilitates dispersal by rafting, or, more recently, by humans via timber transport (Bess 1970).

A case in point is the fungus-growing termites (subfamily Macrotermitinae), which have evolved an ectosymbiosis with basidiomycete fungi of the genus *Termitomyces* (Aanen et al. 2002). Their ancestral habitat has been identified through phylogenetic reconstruction (Aanen & Eggleton 2005) as African rainforests, where their taxonomic diversity is highest. The main radiation leading to the extant taxa probably took place in this habitat just before the expansion of the savannah, at least 30 million years ago (mya; Brandl et al. 2007; Nobre et al. 2011). Subsequently, four out-of-Africa migrations have occurred into Asia and a single one into Madagascar (Aanen & Eggleton 2005; Nobre et al. 2010), so that historical reasons have constrained this group to the Old World only, whereas it seems that their preferred habitat is not restricted to this geographic region.

Several functional classification schemes have been proposed, based on nesting type and primary food (Fig. 3.6.5; Abe 1987; Donovan et al. 2001; Eggleton & Tayasu 2001). Abe (1987) was the first to separate termite species according to the degree to which their nesting and feeding substrates overlap, temporally and spatially. In Abe's classification, single-piece nesters corresponds to termites that feed and nest in the same discrete substrate and intermediate nesters are the ones that, besides nesting in their feeding substrate, also forage outside their colony centre. In these two categories, we can only find wood feeders. The separate-piece nesters actively forage for their feeding substrate away from the nest and these termites have a wide range of feeding substrates. The soil feeders were given a separate status because this term is used for termites that nest and feed in soil, a substrate that is not discrete but continuous. Later, Donovan et al. (2001) presented a new quantitative functional classification of four feeding groups. Their classification was based on the consistent match between morphological character states of the workers and their gut contents across the humification gradient between living plant tissue and soil organic matter. Based on the consideration that the substrate's positional relation to the nest centre and the humification state of the substrate consumed were of more ecological importance than the substrate itself, Eggleton and Tayasu (2001) proposed a "lifeway" classification from a combination of the Abe and Donovan classifications.

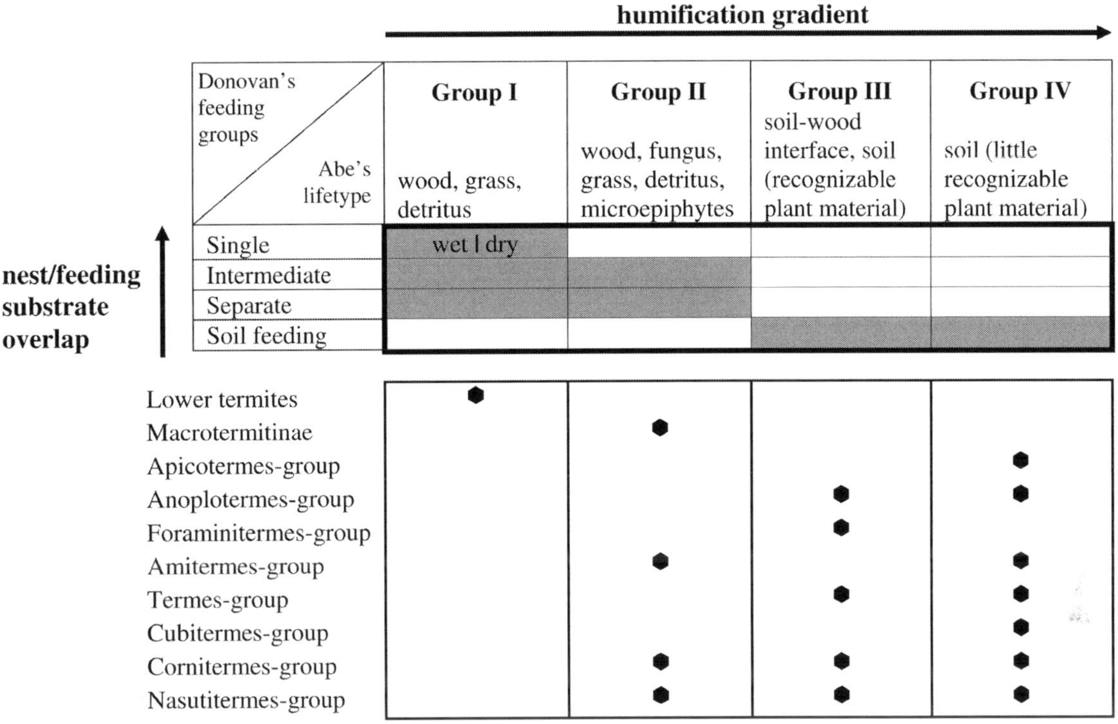

BIOGEOGRAPHY AND PHYLOGENETIC COMMUNITY STRUCTURE OF SOIL INVERTEBRATE ECOSYSTEM ENGINEERS

Donovan's feeding groups / Abe's lifetype	Group I wood, grass, detritus	Group II wood, fungus, grass, detritus, microepiphytes	Group III soil-wood interface, soil (recognizable plant material)	Group IV soil (little recognizable plant material)
Single	wet | dry			
Intermediate				
Separate				
Soil feeding				

nest/feeding substrate overlap ↑

humification gradient →

	Group I	Group II	Group III	Group IV
Lower termites	●			
Macrotermitinae		●		
Apicotermes-group				●
Anoplotermes-group			●	●
Foraminitermes-group			●	
Amitermes-group		●		●
Termes-group			●	●
Cubitermes-group				●
Cornitermes-group		●	●	●
Nasutitermes-group		●	●	●

Figure 3.6.5 Diagram showing the eight lifeway categories (grey boxes). Distinction between Donovan's feeding groups III and IV is based on worker morphology, and especially intestinal anatomy. The soil feeders are not classified by lifetype because the distinction between single-piece and separate-piece breaks down for many species. (Adapted from Donovan et al. 2001 and Eggleton & Tayasu 2001.)

This classification provides a useful basis for understanding termite phylogenetic community structure and how it relates to their functioning in terms of carbon substrate use. For example, both wood feeders and soil feeders are biogeographically widely distributed. However, whereas wood feeders inhabit a very wide range of habitats, from wet to dry—with even one genus being able to survive in true desert (*Psammotermes*)—soil feeders are more restricted to savannahs and especially tropical forests (Eggleton & Tayasu 2001). Their contribution to carbon and nitrogen mineralization in relation to feeding habits is summarized in Eggleton and Tayasu (2001), where they show that $\delta^{13}C$ values are determined mostly by the vegetation in which the termites live, whereas $\delta^{15}N$ correlates more tightly with the feeding groups. Soil feeders—(III) and (IV)—thus face low carbon availability so that their nutrient strategy (C and N) delivers a high throughput as both C and N are limiting.

The ecological consequences of the much higher proportion of soil-feeding termites in Africa, compared to both Oriental and Neotropical faunas, have not been studied but may be important for the turnover rate of organic material in the soil (Eggleton 2000). The separate nest builders—Sep(I) and Sep(II)—form underground or aboveground nests, made externally of soil, and thus resulting in an even higher impact on soil turnover. This effect is even more pronounced in mound builders, as they bring soil from deep horizons to the surface, affecting soil characteristics, especially physicochemical properties, as well as infiltration, water status, and rainfall use efficiency, leading to nutrient-enriched microhabitats. Termite mounds, therefore, trap water and nutrients when compared to the surrounding soil, and in this way provide for increased primary production, which in turn fosters the build-up of food chains (herbivore and predator secondary production), making mounds hotspots of ecosystem productivity (reviewed in Schmitz 2010).

3.6.3.3 Regional scale

The known habitats of termites range from dry deserts, steppes, prairies, and Mediterranean shrublands to the wettest tropical forests, with different assemblages occurring in different habitats (Bignell & Eggleton 2000). Not only phylogenetic community structure is strongly influenced by historical processes, but as a consequence, also the functional diversity. The majority of studies on the ecological importance of termites are focused on tropical forests and savannahs, where functional diversity patterns and the influence of environmental factors on assemblages have been analyzed (e.g. Bignell *et al.* 1997; Eggleton & Tayasu 2001; Eggleton *et al.* 2002; Davies *et al.* 2003, Jones *et al.* 2003), and where the impact of termites in the ecosystem (on the decomposition process, on soil physical and chemical properties, on carbon and nitrogen mineralization, on vegetation composition and thus in shaping higher trophic levels) is acknowledged as substantial (Wood & Sands 1978; Bignell 2006). In warm temperate and subtropical biomes, termite diversity is much lower, but their impact on forest systems is still apparent (Nobre *et al.* 2009).

3.6.3.4 Local scale

Local species richness strongly depends on environmental factors such as rainfall, vegetation type, temperature, and altitude (Williams 1966; Eggleton 2000). In general, wood and grass feeding termites are favored by decreasing rainfall, in contrast to soil feeding termites (Bignell 2006). Because most termites are soil-dwelling, creating extensive gallery systems in the soil and excavating mineral material to build mounds and runways, soil type is crucial for them. Some soils do not support termites at all, like excessively sandy soils, semi-permanently waterlogged soils, and severely cracking vertisols, (Holt *et al.* 1980; Wood 1988; Bignell & Eggleton 2000).

Over the entire group, the food sources of termites are (in different taxa) extremely diverse, and include both live and dead vegetation, wood, humus, dung, fungi, and even lichens. However, they all depend on cellulosic material to some extent. Different termite species are adapted to handle different stages in the decomposition process. This ranges from hardly humified plant material (wood, litter, and grass), via more degraded plant material to the most humified substrate present, on which the soil-feeding termites feed. Even within the wood feeders some groups feed preferentially on wood previously decomposed by microorganisms (e.g. Lenz *et al.* 1991; Rouland-Lefèvre 2000), probably owing to an enhanced nutrient content, decreased wood density, and a probable reduction of toxic allelochemicals (e.g. Waller *et al.* 1990). In parallel, also soil feeders can use a range of soils, from humus layers on organic rich soil with high levels of recognizable plant material to soil with high mineral content. Generally, wood feeders feed on a higher energy resource that is patchily distributed, and soil feeders make use of low energy but universally abundant substrate.

3.6.3.5 Global change drivers, termite community structure, and ecosystem functioning

Global change may impact termites in three ways. First, increasing temperatures are expected to extend distributional ranges. The distributional range of termites is concentrated in tropical and subtropical areas and quickly drops at higher latitudes. Within Europe, there is evidence that species of the genus *Reticulitermes* are moving north (Clément *et al.* 2001). However, some of this movement has been linked mainly to human interference as some species seem to be restricted to cities (Vieau 1993; Kutnik *et al.* 2004). However, though global warming is a factor, the cities' central heating systems seem to be more important, as illustrated by the now well-established *Reticulitermes* colonies near the harbor of Hamburg, inhabiting the pipe work of the district heating system. Colonies of this subterranean termite could be found in locations as far apart as Sauton, north Devon (UK), and Santiago and Valparaiso (Chile) and in numerous French and

Italian cities north of the natural distribution area of the genus.

Second, global change results in conversion and fragmentation of habitats. For example, human impact has led to increased fragmentation of tropical forests. The effects of forest fragmentation on the species composition of tropical termite assemblages have been addressed in several studies (Fonseca de Souza & Brown 1994; Eggleton et al. 1996, 1997; Davies 2001; Davies et al. 2003). These studies have generally shown that termite assemblages are very sensitive to habitat disturbance in the medium-to-long term, but rather insensitive in the short term (<5 years). The long-term effect of fragmentation is thus a change in species composition and decrease in species diversity.

Third, deforestation and conversion to agriculture can result in very considerable declines in species richness and changes in the composition of termite communities (Eggleton & Bignell 1995; Lavelle & Pashanasi 1989). The drop in species richness seems to be accompanied by the appearance of r-strategy species (Collins 1980; Johns 1992), i.e. species that are favored by disturbance and are good colonizers. Amongst the favored species one may find those with the potential to become pests. A further consequence of global change may therefore be the increased likelihood of the spread of invasive species (Leniaud et al. 2009).

3.6.4 Ant biogeography and phylogenetic community structure

3.6.4.1 Introduction

The common ancestor of all living ants probably lived in the early Cretaceous period, although there is not a consensus on this (range between 115–169 mya), (Brady et al. 2006; Moreau et al. 2006). Since this time ants have radiated to produce >20,000 species (Fig. 3.6.6; Bolton 2010), with 12,606 described (Agosti & Johnson 2010). They have come to abound around the world, only being absent from the very coldest areas. Although most species are predatory to a greater or lesser degree, the group performs a wide variety of ecosystem functions. Many species fill primary consumer roles either by feeding on honeydew (Blüthgen et al. 2003; Davidson et al. 2003) or by cultivating fungus on harvested leaves (Wilson 1974). The seeds of 35% of all herbaceous plants may be distributed by ants in some habitats (Beattie 1985) and ground nesting species can turn over large volumes of soil (Lyford 1963; Whitford 2000). It is not our intention to provide an exhaustive review of studies detailing the contributions of ants to ecosystem functioning. For this, see reviews by Folgarait (1998) and Crist (2009). Rather, this section uses examples from what is known about the broad patterns of this radiation and to discuss how ant biogeography and phylogeny might impact the contribution of this group to ecosystem functioning.

3.6.4.2 Continental scale

The main large-scale biogeographic division in ant communities is that between the Old and New World faunas. There are numerous genera, and even some subfamilies, that exist only on either one side of the Atlantic or the other. The impact that this division has on the ways that ants contribute to ecosystem functioning depends to a great extent on the time at which particular traits evolved. For example, army ants (the clade comprising the Aenictinae, Dorylinae, and Ecitoninae, nested within the "dorylomorphs," Fig. 3.6.6) are abundant predators in most tropical forests and affect not only their prey but also the many other species that live within or around their colonies. Army ants evolved once in the mid-Cretaceous (94–116 mya) Period, and the current pan-tropical distribution of this group seems to be due to their Gondwanan origin (Brady 2003). On the other hand, the existence of traits that only evolved relatively recently may mean that certain contributions to ecosystem functioning are restricted to one side or the other of this inter-continental biogeographic divide. For example, leaf cutter ants (tribe Attini in the Myrmicinae), which harvest living plant material and use it to cultivate the fungus on which they feed, can be dominant herbivores in tropical forests (Rao et al. 2002). However, because the trait of leaf harvesting and fungal cultivation evolved only 8–12 mya in the Neotropics (Schultz & Brady 2008), the ants carrying out this function are presently only found in that region. The cultivation of fungus for food from live plant material might in turn directly

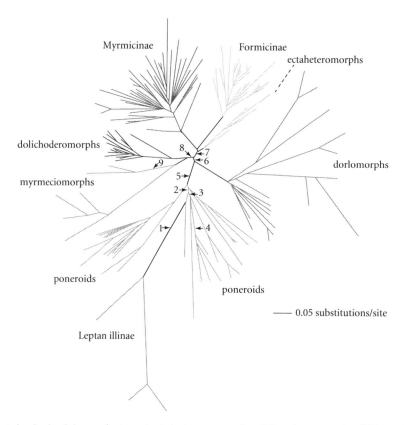

Figure 3.6.6 Unrooted molecular phylogeny of ants constructed using sequences from 162 species representing all 20 ant subfamilies. Numbers denote various possible rootings. Rooting 1 is most likely, although rootings 2–5 are also possible. Rootings 6–9 give significantly worse fits to the data. The "ectaheteromorphs" comprise the subfamilies Heteroponerinae and Ectatomminae; the "dolichoderomorphs" the Aneuretinae and the Dolichoderinae; the "dorylomorphs" the Leptanilloidinae, Cerapachyinae, Aenictogitoninae, Dorylinae, Aenictinae, and the Ecitoninae; the "myrmeciomorphs" the Myrmeciinae and the Pseudomyrmecinae; the "poneroids" the Agroecomyrmecinae, Amblyoponinae, Paraponerinae, Ponerinae, and Proceratiinae. (After Brady et al. 2006, with permission from National Academy of Sciences, U.S.A.)

affect the distribution and longevity of the clade. Obligate mutualisms such as this one mean that both partners need to arrive simultaneously for colonization to be successful, and extinction of one partner inevitably leads to the extinction of both (coextinction; Koh et al. 2004), potentially making the clade both less dispersive and less long-lived.

3.6.4.3 Regional scale

Large-scale biogeographical considerations such as the ones described earlier apply mainly to functionally relevant traits that evolve infrequently. Many of the traits that determine the contributions of ants to ecosystem functioning are highly labile, i.e. easy to evolve and lose. For example, the trait of excavating a nest in the ground, and consequently contributing to soil turnover, can be found in many ant genera. In these cases, it is the overall density of colonies and their rates of activity that are important, although it should be noted that there will be some variation between species in terms of their contribution to this ecosystem process. Other such labile functional traits include seed dispersal (Garrido et al. 2002), predation (Jeanne 1979), and foraging for protein and other baits (Fayle et al. 2011; Gove & Majer 2006). For these traits, broader, non-phylogenetic patterns of ant presence and activity relating to environmental filters on community assembly become more important; for example, the declining gradient of ant diversity and abundance from the equator to the poles (Dunn *et al.* 2009). We would expect the rates of processes in these cases to track these gradients, which have been shown for predation (Jeanne 1979), although not, to our knowledge, for any other process.

3.6.4.4 Local scale

Phylogenetic structuring of ant communities at local scales is likely to reflect interactions between species rather than the biogeographical history of the group. One of the most extreme examples of this is the case of invasive ant species that often displace native ants and other invertebrates, and can have profound impacts on entire ecosystems (e.g. O'Dowd *et al.* 2003). Not only do these invaders reduce the diversity of native species, they can also alter the phylogenetic community structure of the remaining species. In fact, the structure of the resident community is likely to have an impact on the functioning of the ecosystem in terms of resilience against invasion by non-native species. Lessard *et al.* (2009) conducted a meta-analysis of 12 published studies and found that the arrival of invasive ants changed the phylogenetic structure of ant communities, with invaded sites supporting a more phylogenetically clumped community than uninvaded ones. That is, species loss was non-random with respect to phylogeny. Interestingly, the native species that were lost to produce this clumped distribution were not always closely related to the invader.

There remains much we do not know about the way ant phylogeny and biogeography at a range of scales affect the contribution of this group to ecosystem functioning in natural habitats. Besides the continuation of taxonomic work to further reveal the phylogeny of ants, particularly at the species level, a critical area of future research is to examine the traits that determine the degree to which each species contributes to particular ecosystem processes. The creation of global databases for such traits, building on those providing taxonomic and distributional information (Dunn *et al.* 2007; Agosti & Johnson 2010; Fisher 2010), is the first step towards a more comprehensive understanding of ant biodiversity and ecosystem functioning.

3.6.4.5 Global change drivers, ant community structure, and ecosystem functioning

Since ants contribute substantially to the functioning of soil ecosystems, it is of vital importance to understand how global environmental change affects ant communities, and how this in turn alters the way that ecosystems function. There are three aspects of global environmental change that are particularly relevant for ants: habitat conversion, the spread of invasive ant species and climate change. Habitat conversion and invasion by non-natives usually result in loss of species richness (Dunn 2004) and a change in species composition (e.g. Fayle *et al.* 2010). In some cases the nature of interaction networks within communities is also altered, with an increase in the strength of negative interspecific interactions during habitat degradation (Floren *et al.* 2001), but a decrease in interaction strengths when the community is invaded by a non-native species (Sanders *et al.* 2003). The impacts of climate change are less clear, although the strong response of ant communities to latitudinal (Dunn *et al.* 2009) and altitudinal gradients (Brühl *et al.* 1999) in climate strongly indicate that this group will be affected.

While we have some idea about the ways that global changes affect ant communities, we know much less about how these changes will affect the contribution of ants to ecosystem functioning. Habitat fragmentation does not seem to substantially alter the mutualistic relationships between ants and plants (Bruna *et al.* 2005), nor is there any difference in foraging rate for ants beneath trees in plantations and those in the plantation matrix (Gove & Majer 2006). Ant species richness is related to the rate at which nutrients are redistributed in rain forest, but only at very small spatial scales, at which there is no response of species richness to even quite drastic habitat conversion (Fayle *et al.* 2011). It is not known how species richness and the functional composition of ant communities are tied to nutrient redistribution at larger spatial scales. Even less is known about the ways that climate change will affect ant-mediated processes contributing to ecosystem functioning. The gradient of decreasing ant species richness and predatory function away from the equator found by Jeanne (1979) indicates that increasing temperature may increase ant species richness locally and also increase predation rates. Interestingly, this pattern only holds true in forests and not in cleared areas, indicating that the responses of ant diversity and related ecosystem functions to changes in climate potentially differ between habitat types. It is clear that large-scale manipulative experiments are required in order to fully investigate potential changes in ant contribution to ecosystem functioning. These are currently ongoing for both habitat fragmentation and conversion (Ewers *et al.* 2011) and for climate change (Dunn *et al.* 2010).

3.6.5 Earthworms

3.6.5.1 Introduction

Phylogenetic community structure and trait-based ecology of earthworms are not well-explored topics. We approach the coupling of both topics in a slightly backwards fashion, by first explaining the basics of earthworm functional groups, and then considering the phylogeography of earthworms. This order is dictated by the need to use the functional group terminology in the discussion of phylogenetic community structure. Eventually we return to the consideration of traits, and finally the responses of earthworms and associated ecosystem services to global change drivers.

Earthworms, like termites (Eggleton & Tayasu 2001), have been assigned a categorization of species into functional groups, known by the names given to them by Bouché (1977): epigeic, endogeic, and anecic. They are based on morphological and behavioral traits (Lee 1959; Bouché 1977). Epigeic earthworms live in and feed on nearly pure organic matter substrates, such as thick forest floor leaf packs, fallen tree trunks, epiphyte mats, suspended arboreal soils, and leaf axils of palms, bromeliads, Pandanaceae, and some ferns. They are usually small-bodied, darkly pigmented, have the ability to move rapidly, have minimal development of intestinal surface area, and have high reproductive rates and short life spans. Anecics form a relatively permanent burrow system and emerge to feed on surface organic matter such as dead leaves. A typical anecic worm has a large body with a pigmented anterior end, and a minimal degree of intestinal surface development. Endogeics live in mineral soil and consume soil organic matter. Lavelle (1983) divided the endogeic category into three, based on the degree of decomposition or humification of the organic matter consumed, from less humified to more: polyhumic, mesohumic and oligohumic. The first can resemble epigeics in some characteristics of coloration, but they are usually less deeply colored, larger, and form burrows. Oligohumic endogeics are unpigmented, large bodied, with high degree of intestinal surface area augmentation, and have no escape behavior. Mesohumics are also unpigmented and otherwise are intermediate. These functional groups were analyzed across multiple biomes by Lavelle (1983; Fig. 3.6.7). On a biomass basis, tropical soils contained the greatest diversity and evenness of functional groups, dominated by endogeics and having a uniquely tropical group, the oligohumic endogeics feeding on low quality soil organic matter. Temperate zones (mostly European data) had roughly 50% of community earthworm biomass in the anecic category, bearing in mind that anecics are typically large-bodied earthworms. Cold climates trended towards dominance by epigeics and polyhumic endogeics.

Boreal forests generally lack earthworms for historical reasons (past permafrost, glaciations), but a few epigeic acid-tolerant species like *Dendrobaena octaedra* can live along the southern margins of the boreal forest biome. Low quality of soil organic matter (SOM) and plant litter in this biome (De Deyn *et al.* 2008) are significant obstacles to most earthworms, yet the epigeic gut morphology is quite different from that of the oligohumic endogeics also existing on low quality SOM. The difference is that the epigeics have access to large volumes of SOM and litter, and ingest nearly 100% organic matter. Temperate forests support the full range of earthworm functional groups, exploiting a range of resource types corresponding to the mix of plant tissue traits generated by the forest vegetation (De Deyn *et al.* 2008). Tropical wet forests are similar, except that rapid decomposition drastically reduces the resource base for epigeics, who may be found more reliably in suspended organic soils created within or by epiphytes. In seasonal tropical forests, though the range of plant resources may be broad

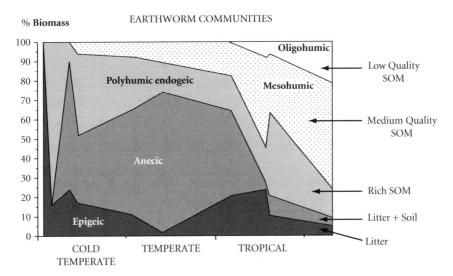

Figure 3.6.7 Percentages of biomass in different ecological categories of earthworm communities along a thermo-latitudinal gradient. Arrows to the right indicate the main food source used. (Modified from Lavelle & Spain 2001, with permission from Springer Science + Business Media B.V.)

and litter layers better-developed, the environmental stresses of the dry season eliminate most epigeic niches and promote endogeic or anecic earthworm activity. A similar result is found in all grassland/savannah biomes, with emphasis on endogeic functional groups and near absence of epigeics.

3.6.5.2 Continental scale

At the largest time/space scale, earthworm community composition has been driven by evolution *in situ* on stable landmasses. Changing land area relationships have divided supracontinental faunas into the present continental faunas and, in a few cases, have delivered new taxonomic groups to a continent. Many paleogeographic interpretations of earthworm biogeography have been made (Bouché 1983; Omodeo 2000; James 2004; Blakemore 2006) in combination with phylogenetic hypotheses. We now have a more robust estimate of earthworm phylogeny at the family level (James & Davidson, unpublished), a simplified version of which is in Fig. 3.6.8. In this phylogeny there are many branching points which do not lend themselves to straightforward vicariance interpretations at the time scale of the separation of Gondwana from Laurasia, nor in relation to the breakup of those two supercontinents. For example, Glossoscolecidae (s.s.; South America) appears as the sister taxon to Eudrilidae (Africa), a simple transatlantic relationship which would in isolation be consistent with the opening of the South Atlantic. However, the Ocnerodrilidae are found on both sides of the divide, and share a more recent common ancestor with Eudrilidae than with the Glossoscolecidae. The same is true of the Benhamiinae, with genera on both sides of the South Atlantic, and an even more recent common ancestor than the Ocnerodrilidae–Eudrilidae node. If any of the above bifurcations was approximately contemporaneous with the opening of the South Atlantic, it was the division of the Benhamiinae. All others are of greater age. Pre-pangean continental earthworm faunas can only be speculated, but the existence of various apparent relicts far from their nearest relatives (e.g. Biwadrilidae, Kynotidae in Fig. 3.6.8; James & Davidson, unpublished data) suggests that there were such faunas, and that earthworms are an old element of terrestrial communities.

3.6.5.3 Regional scale

With most earthworm families distributed at continental or subcontinental scales, phylogenetic community structure is independent across the scale

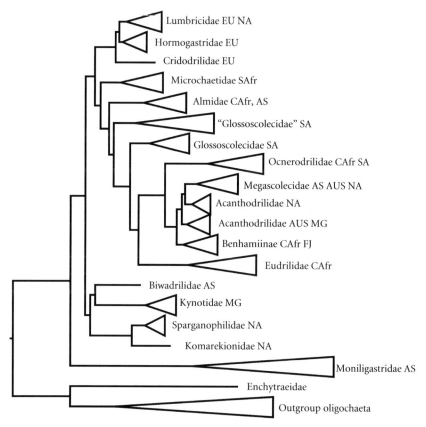

Figure 3.6.8 Phylogeny of earthworm families rooted with aquatic oligochaete outgroups (Bayesian analysis; 18s, 28s, and 16s genes), collapsed to family level. "Glossoscolecidae" indicates a section of that family which falls outside a more strictly defined Glossoscolecidae. Geographical locations of the sampled taxa: AS Asia, AUS Australia, CAfr Central Africa, EU Europe, FJ Fiji, MG Madagascar, NA North America, SA South America, SAfr South Africa.

units. Within continents, those families with diverse genera probably had a long history of evolution and occupation of the land mass. By comparison across higher taxonomic groups it is evident that the range of functional group characteristics has been repeated many times, though not in precisely the same way or to the same degree. Several of these, such as the number and locations of gizzards, pigmentation, pattern of musculature, the setal arrangement, and the degree of intestinal surface area enhancement, have clearly converged in many different earthworm lineages. In some faunas the anecic functional group appears to be missing, in spite of a long history. In one case the evolution of one functional group of species from another appears to have been rapid (Perel-Vsevolodova 1987).

Therefore, it is better to look within these units of scale at the diversity within families or other supra-specific taxa. In some earthworm faunas, earthworm functional group traits can be strongly linked to phylogeny. Forest epigeic worms in the eastern USA are exclusively *Bimastos*, while the endogeics are one or more of *Eisenoides*, *Diplocardia*, and *Komarekiona*. Southern Brazilian *Glossoscolex* contains separate wetland and mesic soil clades, and epigeics, if present at all, are typically *Urobenus* (James & Brown 2006). Southern France earthworm communities usually have *Scherotheca* anecic species, *Prosellodrilus* mesohumic endogeics, other endogeics such as *Gatesona*, *Aporrectodea*, and infrequently some *Dendrobaena* epigeic species (Bouché 1972). Details of distributions often show little range overlap between congeners, exclusive of invasive species. For example, in most of Southern France one can only find a single species of *Scherotheca* on a given site, (Bouché 1972). This pattern is suggestive of phylogenetic repulsion driven by competition among similar taxa (Helmus *et al.* 2007). Considering that most earthworm species are narrow endemics, apparent phylogenetic repulsion could be quite common. Species pairs rarely or never coexisting in northern France included intrageneric pairs and intergeneric pairs, but all such infrequent pairings were between earthworms of the same ecological type (Decaëns *et al.* 2008b). This suggests that niche-based prediction of competitive exclusion is more likely to be correct than taxon-based prediction.

The other process mentioned by Helmus *et al.* (2007), the coexistence of related taxa under a phylogenetic attraction model driven by environmental filtering, is also possible. There is a clade of epigeic Lumbricidae including *Eisenia*, *Allolobophoridella*, *Dendrodrilus*, and *Bimastos* (North American) (James & Davidson, unpublished data). Forested habitats favorable to such worms should have a strong phylogenetic community structure and clustering of traits linked to the epigeic function. This is partially confirmed in Decaëns *et al.* (2008b), but there were also epigeics from other lumbricid clades. If phylogenetic attraction is to be invoked in this case, it implies that the most recent common ancestor of these epigeics was also epigeic, and other lifestyles evolved from that ancestral state. This hypothesis competes with the alternative that environmental filtering operated on independently-derived epigeic groups. It would be instructive to investigate the assembly rules of natural earthworm communities, rather than those generated by chance introductions influenced by human activity. Other functional groups are less strongly connected to Lumbricidae phylogeny, with anecics spread across at least five genera apparently by convergent evolution. However, the scale-dependence is important, because a given genus may provide the majority of anecic species functions within a given area.

One set of traits known for many decades, but only recently explored in detail, is the presence of nephridial symbiotic bacteria in many earthworm lineages (Knop 1926; Schramm *et al.* 2003; Davidson & Stahl 2006, 2008; Pinel *et al.* 2008; Lund *et al.* 2009; Davidson & James, unpublished data). Best known in the Lumbricidae, where such bacteria are ubiquitous, the diverse lineages of symbiotic bacteria include obligate symbionts. One lineage, *Verminephrobacter* (Pinel *et al.* 2008) has many genomic rearrangements typical of symbiotic bacteria and has a large number of genes involved in nitrogenous compound transport and transformation. It is transmitted vertically by maternal means (the egg parent is the source of infection to the embryos in the egg capsules) and is only acquired during embryonic development (Davidson & Stahl 2006, 2008; Davidson *et al.* 2010). The phylogenetic distribution of symbiotic bacteria shows a potential ancestral symbiosis retained in the Lumbricidae, multiple independent acquisitions of other symbionts, one or more losses in the divergences leading to other earthworm families, and further independent acquisitions in some of those families (Davidson & James, unpublished data). Nephridial morphology and earthworm ecological type appear to have some relationship to the nephridial symbiont status of earthworms. Ecological significance, contribution to ecosystem processes, and impact on individual fitness remain to be investigated.

3.6.5.4 Local scale

Worms and many other soil biota are low-dispersal, and so are expected to have high inherent geographical structure. This is most relevant to the smallest spatial scales of phylogenetic community structure. In topographically complex regions, of which New Zealand and mountainous central Europe are the best known for earthworms, closely related species of very similar ecology occupy small ranges. Phylogenetic community structure and trait-based earthworm ecology should be meaningful and predictive. These could be combined with environmental data for niche-based range prediction, and further explored in global change models. Cryptic species diversity is a potential complication in this connection, and also in the examination of the ecology and distributions of common anthropochorous species (Novo *et al.* 2009, 2010; James *et al.* 2010). Earthworms have very deep mitochondrial COI divergence (10–20%; Pérez-Losada *et al.* 2005; King *et al.*

2008; Rougerie et al. 2009; Chang et al. 2009) in the DNA barcode region (Hebert et al. 2003a,b), even between some morphologically identical lineages. At this time there has been no investigation of the functional equivalence of cryptic species, but morphological evidence would indicate that any differentiation is probably small. Abiotic conditions and litter/SOM quality are important influences on earthworm community structure at local to regional scales (Fig. 3.6.7; De Deyn et al. 2008).

3.6.5.5 Global change drivers, earthworm community structure, and ecosystem functioning

Anthropogenic changes relevant to earthworm communities include habitat alteration, invasive species, and climate change. Assuming an indigenous set of earthworm species, intense disturbances such as deforestation typically result in the loss of a large percentage of those species. Disturbance intensity and frequency are important variables for both extinction of endemics and colonization by invasive species (Hendrix et al. 2008). Limited recolonization post-disturbance can occur depending on the extent and severity. Sanchez de Leon et al. (2003) observed that native Puerto Rican earthworms repopulated abandoned agricultural land, but that the soils were still dominated by an invasive species. The outcomes of habitat alteration and succession are variable across earthworm communities. The majority of earthworm ecology has been done on species tolerant of habitat alterations, most of which are now considered invasive earthworms.

Invasive earthworm success in population establishment varies with habitat disturbance factors and the presence or absence of a resident community of earthworms. Phylogenetic content replacement is the rule, and traits may be lost, gained, or redistributed as a result of invasions. In a typical tropical scenario, the invasive endogeic *Pontoscolex corethrurus* may be the only species present after disturbance and propagule arrival. Temperate zones globally typically acquire invasive European Lumbricidae or some of the East Asian *Amynthas*. In North America, where anecic species are unusual, *Lumbricus terrestris* introduces this trait set into forests and grasslands. However, the majority of the distribution of the invasive Lumbricidae in Europe and North America is in regions previously earthworm-free in the wake of glaciation. In such locations earthworm invasions may be significant agents of ecosystem change (Hale et al. 2005; Hendrix et al. 2008).

Climate change effects on earthworm communities have not been closely explored with experimental procedures. Distributional data in temperate zones indicate a strong influence of glaciation history on modern distributions, and a very slow rebound from refugia to deglaciated land, where invasive species now dominate (e.g. Frelich et al. 2006; Bouché 1972). The slowness indicates that range shifts of continental scale may take between 0.5 and 1 million years, with regional shifts an order of magnitude less. Here we assume a spread rate of 10 m year^{-1}, typical of invasive species in unpopulated soils (van Rhee 1969; Stockdill 1982) and no significant geographical barriers such as rivers, inhospitable soils, mountains, etc. Northern plant communities are out of equilibrium with modern climate (Davis 1989), and yet move faster than earthworms (~200 m year^{-1}). Rapid climate changes make it unlikely that earthworm communities and their ecosystem services will survive in current form, because the environmental filters may change faster than worms can respond. Ecological sorting depends on the mobility of the species pool as well as the rate of environmental driver change. Here we are considering a pool that is essentially immobile relative to the rates of environmental change anticipated. Under the conservative estimates of climate change, the future of earthworm ecosystem services may depend on invasive species in large, topographically simple regions. Topographic complexity provides steeper environmental gradients over short distances, which lend themselves better to climate driven range shifts by non-vagile soil organisms. Otherwise, medium-term evolutionary change may be the response mechanism available to earthworms.

3.6.6 Enchytraeids

3.6.6.1 Introduction

Enchytraeidae are small oligochaete worms (6–50 mm in length), adapted to semiterrestrial and terrestrial environments (Christensen & Glenner 2010), including intertidal sands and ice sheets (Erséus & Rota 2003; Hartzell et al. 2005). They show a global distribution from the Arctic to the tropics (Nurminen 1965; Petersen & Luxton 1982; Standen 1988; Didden 1993). Up to 700 species have been described, of which 650 are considered valid (Erséus 2005). Traditionally, they are assumed to be mainly microbivores (Didden 1993). Recent research sustains that they are also saprovores, consuming organic matter that is on average 5–10 years old (Briones & Ineson 2002), but with great feeding flexibility, allowing them to access the more recalcitrant C sources when competitive pressures force them to switch to a different diet in response to warming (Briones et al. 2007a, 2010).

Most research on Enchytraeidae has focused on Europe and is based on a low number of samples from a few localities, taken over a short period of time (Briones et al. 2007b). However, from these studies it is possible to conclude that climate exerts profound effects in determining their geographical patterns. They are more abundant in temperate rainy climates with moderate or cold summers than in areas with dry, warm summers such as alpine meadows, tropical grasslands and tropical rainforests or in snowy areas and tundras (Briones et al. 2007b). In addition to climatic factors, soil pH, soil type, and vegetation cover also appear to play important roles in determining their presence and abundance. Accordingly, moorland soils, which are characterized as cold and extremely wet and having low-growing vegetation (shrubs, grasses, and mosses), high organic matter content and low pH (<4), sustain the highest numbers (Peachey 1959; Springett 1967; Briones et al. 2007b; see also review by Briones (2009)).

3.6.6.2 Continental scale

Detailed descriptions of biogeographical patterns are still scarce. Only recently the phylogenetic relationships of these small sized oligochaetes have started to be unveiled. The first studies using molecular data, which included not only enchytraeids but also other members of the Crassiclitellata (such as earthworms) suggested that they are a monophyletic group (Siddall et al. 2001; Erséus & Källersjö 2004; Rousset et al. 2008). More recently, Christensen and Glenner (2010) produced the first phylogenetic analysis for 14 enchytraeid species representative of nine genera, which rendered two main clades: the early segregated paraphyletic group comprising *Lumbricillus* and *Enchytraeus*, exploiting sea shore habitats (r-strategists) and a later segregated larger monophyletic group including species living on decaying organic matter in more stable terrestrial ecosystems (K-strategists) (Christensen & Glenner 2010). According to this study, the sea shore could have then represented the first semi-terrestrial habitat which allowed early colonization until a well-established land vegetation provided an appropriate environment for later diversification. This is in agreement with the results from a previous phylogenetic analysis concluding that enchytraeids originated from an aquatic environment (Rousset et al. 2008), but contradicts early studies suggesting that the family has a terrestrial origin, either from South America or a contiguous southern land mass (Coates 1989) or from the Arctic (Dash 1990). This lack of agreement on a terrestrial or aquatic origin is a reflection of the inclusion of taxonomically problematic genera such as *Fridericia* and *Marionina* (e.g. Schmelz 2003; Schmelz & Collado 2008; Dózsa-Farkas 2009) in the phylogenetic trees, biased sampling towards a high proportion of non-terrestrial species (Rousset et al. 2008) and extensive radiation processes occurring in some genera (Erséus et al. 2010).

Interestingly, the most recent and more complete analysis of a multigene data set combining mitochondrial and nuclear genes (Erséus et al. 2010) and including 86 species (belonging to 14 genera) corroborated the existence of a basal dichotomy of the Enchytraeidae family but with a different evolutionary history. Accordingly, the Enchytraeidae root would be represented by those enchytraeid species with a tropical distribution, while the sister group would comprise the remaining enchytraeid taxa which can be further subdivided into two clades, each with both terrestrial and marine species (and including the previously considered ancestral genera, *Lumbricillus* and *Enchytraeus*) (Fig. 3.6.9). This different tree topology is the result of a much greater number of outgroups (17 in Erséus et al. 2010, and two in Christensen and Glenner 2010) and thus, a wider variety of Oligochaeta families and provides stronger evidence that tropical soils were perhaps the first inland habitat successfully colonized by enchytraeids from where major genera diversification took place. It is also likely that the enchytraeid invasion of a greater range of habitats took place on several occasions (Coates 1989). The majority of the species included in the potentially most primitive clade (*Hemienchytraeus, Achaeta, Guaranidrilus,* and *Tupidrilus*) are only found in the southern hemisphere, with many only known from South America (Fig. 3.6.9). Although detailed descriptions of biogeographical patterns are still scarce, it has been speculated that climatic factors (extreme temperatures, long periods of frost/snow cover) as well as soil factors (e.g. organic matter content and pH) exert control over their geographi-

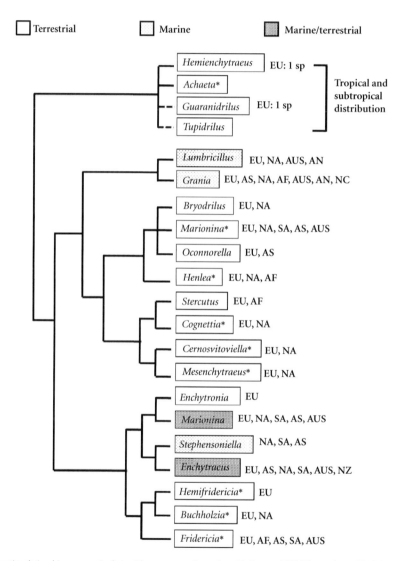

Figure 3.6.9 Schematic relationships among Enchytraeidae genera, redrawn from Erséus *et al.* (2010) together with their recorded geographical distributions. *Terrestrial genera with marine species (http://www.inhs.uiuc.edu/~mjwetzel/ENCH/MarEnchList.html). Distributions from Rota (1995), Rota and Erséus (2003), Schmelz and Collado (2010) and the GBIF Data Portal, data.gbif.org (accessed on 12 December 2010). Geographical locations of the sampled taxa: AS Asia, AUS Australia, AF Africa, EU Europe, NA North America, SA South America, NZ New Zealand, AN Antarctica, NC New Caledonia.

cal distribution (Rota *et al.* 1998). Thus, species such as *Enchytronia parva* and *Enchytraeus lacteus* are mainly restricted to the southern boreal zone (Nurminen 1970; Kairesalo 1978) and *Buchholzia fallax* does not even reach this latitude (Nurminen 1967). Similarly, a mean annual temperature threshold of 16 °C has been found for *Cognettia sphagnetorum* (Briones *et al.* 2007b), above which it has never been recorded. Both ecological (habitat preferences, physiological tolerances, dispersal abilities, etc.) and historical data (geological history, vegetation patterns and geomorphology of the area sampled) are needed to elaborate a more complete picture of enchytraeid dispersal from the southern land mass.

3.6.6.3 Regional scale

Under a constant climate environmental factors such as soil pH, soil type, and organic matter quality and quantity also appear to have an important influence in controlling enchytraeid community structure. In non-wooded habitats, low organic matter content, together with higher pH values, usually result in a more diverse but less numerous enchytraeid community, whereas the acidic conditions of the organic-rich soils lead to high numbers of enchytraeids but dominated by two to three species (Standen 1984) and with *Cognettia. sphagnetorum* usually representing 95% of the total community. In forested ecosystems, the quality of litter materials and the concentration of microbial biomass therein appear to be most important for enchytraeid populations (Scheu *et al.* 2003). Indeed, changes in food quality as well as in abiotic conditions seem to explain the observed transient increases in enchytraeid abundance after clear-felling in the short term (Huhta 1976; Lundkvist 1983; Siira-Pietikäinen *et al.* 2001; Nieminen 2009). In contrast, the long term (>10 years) responses varied and numbers could either be lower than the control forests (Lundkvist 1983; Bengtsson *et al.* 1997) or leveled out (Siira-Pietikäinen & Haimi 2009). However, higher mortality rates of enchytraeids have been recorded a few years after tree-felling during the summer months and very few recovered in the following month, suggesting that microclimate conditions exert a much stronger influence and could compromise their future existence (Uhia & Briones 2002).

3.6.6.4 Local scale

Although vegetation cover and diversity do not seem to affect their abundance, some studies have associated enchytraeids to certain plant microhabitats. For example, in grasslands significantly higher densities of enchytraeids have been found under *Dactylis glomerata* and *Trifolium subterraneum* than under *Rumex obtusifolius* and *T. pratense* (Wardle *et al.* 2003). Similarly, *Juncus squarrosus* moors usually sustain relatively high population numbers (Peachey 1959; Springett 1967; Whittaker 1974; Briones *et al.* 1997, 2007b) and generally higher densities have been recorded in coniferous forest compared to deciduous woodlands (Kasprzak 1986 in Kapusta *et al.* 2003). This could be possibly related to changes in soil acidity, with numbers declining as pH increases (Standen 1984) and has led to the conclusion that enchytraeid communities are more sensitive to soil variations within the same plant communities than to successional vegetation changes (Nowak 2001). Furthermore, acidic conditions in soils usually prevent the colonization by other ecosystem engineers such as earthworms, with only a few exceptions of some acid-tolerant species; consequently, enchytraeids become the dominant faunal group in terms of live biomass in these ecosystems (Cragg 1961; Coulson & Whittaker 1978).

High population density does not often go with high species richness. For example, in peatland soils enchytraeid communities are not usually very diverse with typically only five to six species (Springett 1967, 1970; Standen & Latter 1977; Briones *et al.* 1997). This contrasts with the species-rich wet habitats such as river or lake banks (Rota *et al.* 1998), limestone areas (Springett 1967, 1970; Standen & Latter 1977; Rota *et al.* 1998) and taiga and tundra habitats (Piper *et al.* 1982; Christensen & Dózsa-Farkas 1999), where up to 18 species have been recorded but with lower overall enchytraeid densities. Again, changes in soil pH and organic matter content appear to be the main explaining factors and as a result, species richness has been reported to be positively correlated to soil pH (Standen 1984; Nowak 2001) and negatively with organic matter content in the topsoil layers (Kapusta et al. 2003). However, life-history traits have also played an important role in shaping enchytraeid communities. For example, the more diverse enchytraeid communities in the Arctic tundras are characterized by a high degree of endemism which indicates an undisturbed long history (Christensen & Dózsa-Farkas 2006). This has led to the suggestion that enchytraeids are slow dispersers (Hartzell et al. 2005) and that the most likely sources of postglacial colonization were glacial refugia (Christensen & Dózsa-Farkas 2006).

In functional terms, enchytraeids have long been considered "litter transformers," together with microarthropods (Lavelle 1996). Accordingly, their main role in ecosystems would be depositing casts (excrements) which are hotspots for microbial activity and, hence, mineralization. Their fecal pellets can constitute a large proportion of the soil horizons in upland grasslands and evergreen forest floors (Davidson et al. 2002; Bruneau et al. 2004; Tagger et al. 2008). Because their assimilation efficiency is very low, they have to ingest large quantities of organic matter (up to 0.75 Mg ha^{-1} year^{-1} of soil; Didden 1990), and considerable amounts of undigested material are deposited (Martin & Marinissen 1993).

Regarding their role as bioturbators, limited research has been done (Van Vliet et al. 1995, 1998; Tyler et al. 2001; Roithmeier & Pieper 2009), but from these studies it is obvious that enchytraeids reduce soil compaction and, consequently, increase hydraulic conductivity (Van Vliet et al. 1998) and, hence, nutrient leaching potential of water-extractable compounds such as dissolved organic carbon, nitrate, calcium and magnesium (Roithmeier & Pieper 2009). Furthermore, in moorlands and wetlands, bioturbation can increase oxygen concentration and modify the chemical gradients in soil profiles by increasing the fluxes of nutrients (Mermillod-Blondin & Lemoine 2010). In those soil systems where they coexist with earthworms, enchytraeids seem to have a more important role in soil structure dynamics because they comminute the feces of other animals and as a result they increase the pore volume of the soil (Topoliantz et al. 2000). In contrast, in soils without earthworms, soil homogenization is much less complete, but by ingesting and excreting clay and organic matter, enchytraeids appear to be the dominant bioturbators within the subsurface horizons of such soils (Tyler et al. 2001). Therefore, enchytraeids could be considered to be the ecosystem engineers of the organic layers, whereas earthworms play an essential role in structuring the organo-mineral layers (Tagger et al. 2008).

3.6.6.5 Global change drivers, enchytraeid community structure, and ecosystem functioning

Among global change drivers, climatic factors, namely temperature and rainfall regimes, seem to shape global enchytraeid distribution (Briones et al. 2007b). And so, accordingly, cold and wet environments with mild summers are consistently linked to greater densities of enchytraeids, whereas summer droughts drastically reduce total numbers (Briones et al. 1997; Beylich & Achazi 1999). Because their response to abiotic factors is species-dependent (Briones et al. 1997), behavioral adaptations exhibited by certain species (namely C. sphagnetorum) could allow them to survive in the deeper layers, which would result in accelerated turnover of the less labile C substrates occurring there, and with implications for water quality, greenhouse gas emissions, and the fate of carbon stocks in general (Briones et al. 2007b, 2010). Therefore, any alterations in enchytraeid communities may have important implications for ecosystem services in those biomes where they are dominant.

3.6.7 Trait-based ecology of soil invertebrate ecosystem engineers with a view to the possible effects on global environmental change and ecosystem functioning and services

The community assembly of soil invertebrate ecosystem engineers cannot be understood without considering their food resources. With the exception of predatory ants and ants tending sap-sucking insects, their food is dead organic matter or fungi cultivated on it. The physical environment, resource quality and organisms control organic matter transfers in terrestrial ecosystems (Swift *et al.* 1979) in a hierarchical fashion whereby the physical environment is of overall, and resource quality is of specific importance in structuring soil organism communities (Lavelle *et al.* 1993). This assumption also underlies Fig. 3.6.1 in that different major groups of soil invertebrate ecosystem engineers are globally associated dominantly with biomes that are characterized by combinations of climate and soil type. Yet, at regional to local scales a clear hierarchy of environment and resources is not straightforward in soil invertebrate ecosystem engineers, because, by definition, they modify their habitats to suit their niche requirements. Consequently, resource quality will be an important habitat-selecting factor, influencing ecosystem functioning and services (Tian *et al.* 1997; Wolters 2000; Fox *et al.* 2006).

The filtered trait distribution of Fig. 3.6.2 can be used to predict changes in either species composition and abundance or ecological processes (Fig. 3.6.10), such as decomposition, nutrient cycling and carbon sequestration, and responses to drivers of global environmental change.

We suggest that the distinct patterns across biomes in belowground community biomass and composition (Fierer *et al.* 2009) and in soil mesofauna-mediated litter decomposition rates, which is quantitatively important for the C balance, especially in temperate and wet tropical biomes (Wall *et al.* 2008), can be explained by trait associations of soil invertebrates and plants. Through plant tissue/litter quality, plants will elicit response traits in soil-dwelling litter transformers, including soil invertebrate ecosystem engineers that are detritivorous or humivorous, and which result in effect traits on ecosystem functions and associated services. Trait associations are also to be expected between plants and leaf-cutter ants, but seem less probable between plants and the majority of predatory ants.

Plant associations with soil organisms are crucial for supporting soil carbon input from primary production as well as for carbon output and carbon stabilization in soil (Brussaard & Juma 1996; Wolters 2000; De Deyn *et al.* 2008). The carbon assimilation efficiency of decomposers (usually < 20%, but >50% in termites), the efficiency with which plants use mineralized nutrients, and the level of interactions between organic and mineral soil fractions throughout the soil profile (via rooting and soil invertebrate ecosystem engineers), determine the balance between soil carbon input and output (De Deyn *et al.* 2008).

These considerations are important, because, as traits arise as innovations along the tree of life, often reflecting their biogeographical origins, and tending to be shared by species that have common ancestry (phylogenetic history), it can been argued that phylogenetic diversity is better correlated with ecological processes than species richness (Cavander-Bares *et al.* 2009; Donoghue *et al.* 2009). Hence, the analysis of phylogenetic community structure should be helpful when using a trait-based, i.e. *functional trait-group* approach to understanding (environmental impacts on) ecosystem functioning and associated ecosystem services (Webb *et al.* 2010).

Kozak *et al.* (2008) suggest that the use of GIS data on environmental variation can be relevant in studying processes of speciation, genetic divergence among populations and evolution of traits. One application is in ecological niche modeling, which allows one to predict the range of a species from its known distribution, but evolutionary processes and species interactions also have to be taken into account (Fig. 3.6.11). By extension to the multiple species at a single site, it will eventually become possible to predict community composition, and therefore: 1) estimate the importance of phylogenetic community structure across landscape or regional-scale patches of habi-

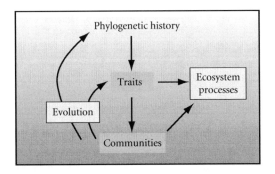

Figure 3.6.10 The central role of traits in understanding evolutionary and ecological processes. (From Cavander-Bares *et al.* 2009, with permission from John Wiley & Sons.)

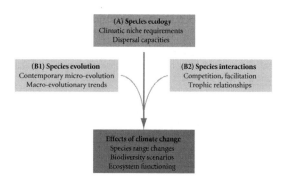

Figure 3.6.11 Mechanisms that should be considered in order to adequately project the effect of climate change on species ranges and species assemblages (*bottom box*). The top box (*A*) and main arrow represent the most widely used approaches of species distribution models. Boxes (*B1* and *B2*) and side arrows represent the mechanisms that have been little envisaged to forecast climate changes effects, so far, namely that species evolve and interact. (From Lavergne et al. 2010, with permission from Annual Reviews.)

tat; 2) determine the degree to which species or phylogenetically-structured sets of species co-occur and are therefore non-random samples of a species pool; and 3) provide an analogous approach to the distribution of organismal traits, ecosystem processes and ecosystem services across spatial scales.

This approach will be a major challenge, not only because it will require the set-up of an appropriate trait database, but also because only a simultaneous approach of trait-based ecology of all major soil invertebrate ecosystem engineer groups will yield insight in their evolutionary and ecological relationships with ecosystem functioning and services and the impacts of global environmental change drivers. To take this approach is also urgent, because most soil invertebrate ecosystem engineers are K-strategists. Whereas their legacies in soil will outlive their presence for mostly unknown periods of time, they will be difficult to recover, once they are gone.

Acknowledgments

We thank Rogier de Jong, Laboratory of Geo-Information Science, Wageningen University, Wageningen, The Netherlands, for making the calculations underlying Fig. 3.6.1, based on primary data. We are also very grateful to an anonymous reviewer for diligent comments on an earlier version of the manuscript and to Caroline Wiltink for editing the references.

References

Aanen, D.K., & Eggleton, P. (2005) Fungus-growing termites originated in African rain forest. *Current Biology* **15**: 851–5.

Aanen, D.K., Eggleton, P., Rouland-Lefevre, C., Guldberg-Froslev, T., Rosendahl, S., & Boomsma, J.J. (2002) The evolution of fungus-growing termites and their mutualistic fungal symbionts. *Proceedings of the National Academy of Sciences of the United States of America* **99**: 14887–92.

Abe, T. (1987) Evolution of life types in termites. In: S. Kawano, J.H. Connell & T. Hidaka (eds.) *Evolution and coadaptation in biotic communities*, pp. 125–48. University of Tokyo Press, Tokyo.

Agosti, D., & Johnson, N.F. (eds.) (2010) Antbase. World Wide Web electronic publication. [online] http://www.antbase.org (accessed 23 September 2010).

Ash, N., Jürgens, N., Leadley, P., et al. (2009) *Assessing, Monitoring and Predicting Biodiversity Change—bioDiscovery Science Plan and Implementation Strategy*. Report No. 7. DIVERSITAS, Paris.

Barbosa, O., & Marquet, P.A. (2002) Effects of forest fragmentation on the beetle assemblage at the relict forest of Fray Jorge, Chile. *Oecologia* **132**: 296–306.

Bardgett, R.D., Yeates, G.W., & Anderson, J.M. (2005) Patterns and determinants of soil biological diversity. In: R.D. Bardgett, M.B. Usher & D.W. Hopkins (eds.) *Biological diversity and function in soils*, pp. 100–18. Cambridge University Press, New York.

Beattie, A.J. (1985) *The Evolutionary Ecology of Ant-Plant Mutualisms*. Cambridge University Press, New York.

Bengtsson, J., Persson, T., & Lundkvist, H. (1997) Long-term effects of logging residue addition and removal on macroarthropods and enchytraeids. *Journal of Applied Ecology* **34**: 1014–22.

Bess, H.A. (1970) Termites of Hawaii and the oceanic islands. In: K. Krishna & F.M. Weesner (eds.) *Biology of termites*, pp. 449–76. Academic Press, New York.

Beylich, A., & Achazi, R.K. (1999) Influence of low soil moisture on enchytraeids. *Newsletter on Enchytraeidae* **6**: 49–58.

Bignell, D.E. (2000) Introduction to symbiosis. In: T. Abe, D.E. Bignell, & M. Higashi (eds.) *Termites: evolution, sociality, symbiosis, ecology*, pp. 189–208. Kluwer, Dordrecht.

Bignell, D.E. (2006) Termites as soil engineers and soil processors. In: H. König & A. Varma (eds.) *Soil biology*, Vol. 6, pp. 183–220. Springer-Verlag, Berlin.

Bignell, D.E., & Eggleton, P. (2000) Termites in ecosystems. In: T. Abe, D.E. Bignell, & M. Higashi (eds.) *Termites: evolution, sociality, symbiosis, and ecology*, pp. 263–388. Kluwer Academic Publishers, Dordrecht.

Bignell, D.E., Eggleton, P., Nunes, L., & Thomas, K.L. (1997) Termites as mediators of carbon fluxes in tropical forests: budgets for carbon dioxide and methane emissions. In: A.D. Watt, N.E. Stork, & M.D. Hunter (eds.) *Forests and insects*, pp. 109–34. Chapman and Hall, London.

Blakemore, R.J. (2006) A series of searchable texts on earthworm biodiversity, ecology and systematics from various regions of the world. In: M.T. Ito & N. Kaneko (eds.) CD-ROM. Yokohama National University, Japan.

Blondel, J. (1995) *Biogéographie. Approche écologique et évolutive*. Masson, Paris.

Blüthgen, N., Gebauer, G., & Fiedler, K. (2003) Disentangling a rainforest food web using stable isotopes: dietary diversity in a species-rich ant community. *Oecologia* **137**: 426–35.

Bolton, B. (2010) *The World Ants*. [online] http://www.antweb.org/world.jsp (accessed 23 September 2010).

Bouché, M.B. (1972) *Lombriciens de France. Ecologie et Systematique*. INRA publication 72–2. Institut National de la Recherche Agronomique, Paris.

Bouché, M.B. (1977) Stratégies lombriciennes. In: U. Lohm & T. Persson (eds.) *Soil organisms as components of ecosystems*, pp. 122–32. Swedish Natural Research Council, Stockholm.

Bouché, M.B. (1983) The establishment of earthworm communities. In: J.E. Satchell (eds.) *Earthworm ecology from Darwin to vermiculture*, pp. 431–48. Chapman and Hall, London.

Brady, S.G. (2003) Evolution of the army ant syndrome: the origin and long-term evolutionary stasis of a complex of behavioral and reproductive adaptations. *Proceeding of the National Academy of Sciences* **100**: 6575–9.

Brady, S.G., Schultz, T.R., Fisher, B.L., & Ward, P.S. (2006) Evaluating alternative hypotheses for the early evolution and diversification of ants. *Proceeding of the National Academy of Sciences* **103**: 18172–7.

Brandl, R., Hyodo, F., Korff-Schmising, M., et al. (2007) Divergence times in the termite genus *Macrotermes* (Isoptera: Termitidae). *Molecular Phylogenetics and Evolution* **45**: 239–50.

Briones, M.J.I. (2009) Uncertainties related to the temperature sensitivity of soil carbon decomposition. In: P. Baveye, J. Mysiak & M. Laba (eds.) *Uncertainties in environmental modelling and consequences for policy making*, pp. 317–35. Springer, New York.

Briones, M.J.I., Garnett, M.H., & Ineson, P. (2010) Soil biology and warming play a key role in the release of "old C" from organic soils. *Soil Biology and Biochemistry* **42**: 960–7.

Briones, M.J.I., & Ineson, P. (2002) Use of ^{14}C carbon dating to determine feeding behaviour of enchytraeids. *Soil Biology and Biochemistry* **34**: 881–4.

Briones, M.J.I., Ineson, P., & Heinemeyer, A. (2007b) Predicting potential impacts of climate change on the geographical distribution of enchytraeids: a meta-analysis approach. *Global Change Biology* **13**: 2252–69.

Briones, M.J.I., Ineson, P., & Piearce, T.G. (1997) Effects of climate change on soil fauna; responses of enchytraeids, Diptera larvae and tardigrades in a transplant experiment. *Applied Soil Ecology* **6**: 117–34.

Briones, M.J.I., Ostle, N., & Garnett, M.H. (2007a) Invertebrates increase the sensitivity of non-labile carbon to climate change. *Soil Biology and Biochemistry* **39**: 816–18.

Brown, J.H., & Lomolino, M.V. (1998) *Biogeography*. Sinauer Associates, Sunderland, MA.

Brühl, C.A., Mohamed, M., & Linsenmair, K.E. (1999) Altitudinal distribution of leaf litter ants along a transect in primary forests on Mount Kinabalu, Sabah, Malaysia. *Journal of Tropical Ecology* **15**: 265–77.

Bruna, E.M., Vasconcelos, H.L., & Heredia, S. (2005) The effect of habitat fragmentation on communities of mutualists: Amazonian ants and their host plants. *Biological Conservation* **124**: 209–16.

Bruneau, P.M.C., Davidson, D.A., & Grieve, I.C. (2004) An evaluation of image analysis for measuring changes in void space and excremental features on soil thin sections in an upland grassland soil. *Geoderma* **120**: 165–75.

Brussaard, L., & Juma, N.G. (1996) Organisms and humus in soils. In: A. Piccolo (ed.) *Humic substances in terrestrial ecosystems*, pp. 329–59. Elsevier, Amsterdam.

Carvalho, K.S., & Vasconcelos, H.L. (1999) Forest fragmentation in central Amazonia and its effects on litter-dwelling ants. *Biological Conservation* **91**: 151–7.

Cavander-Bares, J., Kozak, K.H., Fine, P.V.A., & Kembel, S.W. (2009) The merging of community ecology and phylogenetic biology. *Ecology Letters* **12**: 693–715.

Chang, C.H., Rougerie, R., & Chen, J.H. (2009) Identifying earthworms through DNA barcodes: pitfalls and promise. *Pedobiologia* **52**: 171–80.

Christensen, B., & Dózsa-Farkas, K. (1999) The enchytraeid fauna of the Palearctic tundra (Oligochaeta, Enchytraeidae). *Biologiske Skrifter Dan Vid Selsk* **52**: 1–37.

Christensen, B., & Dózsa-Farkas, K. (2006) Invasion of terrestrial enchytraeids into two postglacial tundras: North-eastern Greenland and the Arctic archipelago of

Canada (Enchytraeidae, Oligochaeta). *Polar Biology* **29**: 454–66.

Christensen, B., & Glenner, H. (2010) Molecular phylogeny of Enchytraeidae (Oligochaeta) indicates separate invasions of the terrestrial environment. *Journal of Zoological Systematics and Evolutionary Research* **48**: 208–12.

Clément, J.L., Bagnères, A.G., Uva, P., *et al.* (2001) Biosystematics of *Reticulitermes* termites in Europe: morphological, chemical and molecular data. *Insectes Sociaux* **48**: 202–15.

Coates, K.A. (1989) Phylogeny and origins of Enchytraeidae. *Hydrobiologia* **180**: 17–33.

Cole, A.C., Jr. (1940) A guide to the ants of the Great Smoky Mountains National Park, Tennessee. *American Midland Naturalist* **24**: 1–88.

Collins, N.M. (1980) The distribution of soil macrofauna on the West Ridge of Gunung (Mount) Mulu, Sarawak. *Oecologia* **44**: 263–75.

Coulson, J.C., & Whittaker, J.B. (1978) The ecology of moorland animals. In: O.W. Heal & D.F. Perkins (eds.) *Production ecology of British moors and montane grasslands*, pp. 52–93. Springer, Berlin.

Cragg, J.B. (1961) Some aspects of the ecology of moorland animals. *Journal of Ecology* **49**: 477–506.

Crist, T.O. (2009) Biodiversity, species interactions, and functional roles of ants (Hymenoptera: Formicidae) in fragmented landscapes: a review. *Myrmecological News* **12**: 3–13.

Dahmouche, S. (2007) *Début de biogéographie des vers de terre de France*, Master thesis.

Dash, M.C. (1990) Oligochaeta: Enchytraeidae. In: D.L. Dindal (eds.) *Soil biology guide*, pp. 311–40. Wiley & Sons, New York.

Davidson, D.A., Bruneau, P.M.C., Grieve, I.C., & Young, I.M. (2002) Impacts of fauna on an upland grassland soil as determined by micromorphological analysis. *Applied Soil Ecology* **20**: 133–43.

Davidson, D.W., Cook, S.C., Snelling, R.R., & Chua, T.H. (2003) Explaining the abundance of ants in lowland tropical rainforest canopies. *Science* **300**: 969–72.

Davidson, S.K., Powell, R.J., & Stahl, D.A. (2010) Transmission of a bacterial consortium in *Eisenia fetida* egg capsules. *Environmental Microbiology* **12**: 2277–88.

Davidson, S.K., & Stahl, D.A. (2006) Transmission of nephridial bacteria of the earthworm, *Eisenia foetida*. *Applied and Environmental Microbiology* **72**(1): 769–75.

Davidson, S.K., & Stahl, D.A. (2008) Selective recruitment of bacteria during embryogenesis of an earthworm. *International Society for Microbial Ecology Journal* **2**: 510–18.

Davies, R.G. (2001) Patterns of Termite Functional Diversity: Discussions and Comments on the Manuscript. From Local Ecology to Continental History. Ph.D Thesis.

Davies, R.G. (2002) Feeding group responses of a Neotropical termite assemblage to rain forest fragmentation. *Oecologia* **133**: 233–42.

Davies, R.G., Eggleton, P., Jones, D.T., Gathorne-Hardy, F.J., & Hernandez, L.M. (2003) Evolution of termite functional diversity: analysis and synthesis of local ecological and regional influences on local species richness. *Journal of Biogeography* **30**: 847–77.

Davis, M.B. (1989) Lags in vegetation response to greenhouse warming. *Climate Change* **15**: 75–82.

De Deyn, G.B., Cornelissen, J.H.C., & Bardgett, R.D. (2008) Plant functional traits and soil carbon sequestration in contrasting biomes. *Ecology Letters* **11**: 516–31.

De Deyn, G.B., & van der Putten, W.H. (2005) Linking aboveground and belowground diversity. *Trends in Ecology & Evolution* **20**: 625–33.

Decaëns, T. (2010) Macroecological patterns in soil communities. *Global Ecology and Biogeography* **19**: 287–302.

Decaëns, T., Jiménez, J.J., Gioia, C., Measey, J., & Lavelle, P. (2006) The values of soil animals for conservation biology. *European Journal of Soil Biology* **42**: S23–S38.

Decaëns, T., Lavelle, P., & Jiménez, J. (2008a) Priorities for conservation of soil animals. *CAB reviews: perspectives in agriculture, veterinary science, nutrition and natural resources* **3**: S23–S38.

Decaëns, T., Margerie, P., Aubert, M., Hedde, M., & Bureau, F. (2008b) Assembly rules within earthworm communities in North-Western France-a regional analysis. *Applied Soil Ecology* **30**: 321–35.

Didden, W.A.M. (1990) Involvement of Enchytraeidae (Oligochaeta) in soil structure evolution in agricultural fields. *Biology and Fertility of Soils* **9**: 152–8.

Didden, W.A.M. (1993) Ecology of terrestrial Enchytraeidae. *Pedobiologia* **37**: 2–29.

Donoghue, M.J., Yahara, T., Conti, E., *et al.* (2009) *Providing an Evolutionary Framework for Biodiversity Science—bioGenesis Science Plan and Implementation Strategy*. Report No. 6. Diversitas, Paris.

Donovan, S.E., Eggleton, P., & Bignell, D.E. (2001) Gut content analysis and a new feeding group classification of termites. *Ecological Entomology* **26**: 356–66.

Dózsa-Farkas, K. (2009) Review of the *Fridericia* species (Oligochaeta: Enchytraeidae) possessing two spermathecal diverticula and description of a new species. *Journal of Natural History* **43**: 1043–65.

Driscoll, D.A., & Weir, T. (2005) Beetle responses to habitat fragmentation depend on ecological traits, habitat condition, and remnant size. *Conservation Biology* **19**: 182–94.

Dunn, R., Sanders, N., Ellinson, A., & Gotelli, N. (2010) Climate change and ants: an experimental approach. [Online] http://cl3-dunnlabg5.zo.ncsu.edu/~dunn_lab/duke_forest_site/index.html.

Dunn, R.R. (2004) Managing the tropical landscape: a comparison of the effects of logging and forest conversion to agriculture on ants, birds, and lepidoptera. *Forest Ecology and Management* **191**: 215–24.

Dunn, R.R., Agosti, D., Andersen, A.N., et al. (2009) Climatic drivers of hemispheric asymmetry in global patterns of ant species richness. *Ecology Letters* **12**: 324–33.

Dunn, R.R., Sanders, N.J., Fitzpatrick, M.C., et al. (2007) Global ant (Hymenoptera: Formicidae) biodiversity and biogeography—a new database and its possibilities. *Myrmecological News* **10**: 77–83.

Eggleton, P. (1994) Termites live in a pear-shaped world: a response to Platnick. *Journal of Natural History* **28**: 1209–12.

Eggleton, P. (2000) Global patterns of termite distribution. In: T. Abe, D.E. Bignell, & M. Higashi (eds.) *Termites: evolution, sociality, symbioses, ecology*, pp. 25–51. Kluwer Academic Publishers, Dordrecht.

Eggleton, P., & Bignell, D.E. (1995) Monitoring the response of tropical insects to changes in the environment: troubles with termites. In: R. Harringon & N. E. Stork (eds.) *Insects in a changing environment*, pp. 473–97. Academic Press, London.

Eggleton, P., Bignell, D.E., Hauser, S., Dibog, L., Norgrove, L., & Madong, B. (2002) Termite diversity across an anthropogenic disturbance gradient in the humid forest zone of West Africa. *Agriculture Ecosystems and Environment* **90**: 189–202.

Eggleton, P., Bignell, D.E., Sands, W.A., Mawdsley, N.A., Lawton, J.H., Wood, T.G., & Bignell, N.C. (1996) The diversity, abundance and biomass of termites under differing levels of disturbance in the Mbalmayo Forest Reserve, Southern Cameroon. *Philosophical Transactions of the Royal Society of London B* **351**: 51–68.

Eggleton, P., Homathevi, R., Jeeva, D., Jones, D.T., Davies, R.G., & Maryati, M. (1997) The species richness and composition of termites (Isoptera) in primary and regenerating lowland dipterocarp forest in Sabah, east Malaysia. *Ecotropica* **3**: 119–28.

Eggleton, P., & Tayasu, I. (2001) Feeding groups, lifetypes and the global ecology of termites. *Ecological Research* **16**: 941–60.

Emerson, A.E., & Krishna, K. (1975) The termite family Serritermitidae (Isoptera). *American Museum Novitates* **2570**: 1–31.

Emerson, W.V. (1955) Geographical origins and dispersions of termite genera. *Chicago Natural History Museum* **37**: 465–521.

Engel, M.S., Grimaldi, D.A., & Krishna, K. (2009) Termites (Isoptera): their phylogeny, classification, and rise to ecological dominance. *American Museum Novitates* **3650**: 1–27.

Erséus, C. (2005) Phylogeny of oligochaetous Clitellata. *Hydrobiologia* **535**: 357–72.

Erséus, C., & Källersjö, M. (2004) 18S rDNA phylogeny of Clitellata (Annelida). *Zoologica Scripta* **33**: 187–96.

Erséus, C., & Rota, E. (2003) New findings and an overview of the oligochaetous Clitellata (Annelida) of the North Atlantic deep sea. *Proceedings of the Biological Society of Washington* **116**: 892–900.

Erséus, C., Rota, E., Matamoros, L., & De Wit, P. (2010) Molecular phylogeny of Enchytraeidae (Annelida, Clitellata). *Molecular Phylogenetics and Evolution* **57**: 849–58.

Ewers, R.M., Didham, R.K., Fahrig, L., et al. (2011) A large-scale forest fragmentation experiment: the stability of altered forest ecosystems project. *Philosophical Transactions of the Royal Society B: Biological Sciences* **366**: 3292–302.

Fayle, T.M., Bakker, L., Cheah, C., et al. (2011) A positive relationship between ant biodiversity (Hymenoptera: Formicidae) and rate of scavenger-mediated nutrient redistribution along a disturbance gradient in a SE Asian rain forest. *Myrmecological News* **14**: 5–12.

Fayle, T.M., Turner, E.C., & Snaddon, J.L. (2010) Oil palm expansion into rain forest greatly reduces ant biodiversity in canopy, epiphytes and leaf-litter. *Basic and Applied Ecology* **11**: 337–45.

Fierer, N., Strickland, M.S., Liptzin, D., Bradford, M.A., & Cleveland, C.C. (2009) Global patterns in belowground communities. *Ecology Letters* **12**: 1238–49.

Fisher, B.L. (2010) Antweb. [Online] http://www.antweb.org.

Floren, A., Freking, A., Biehl, M., & Linsenmair, K.E. (2001) Anthropogenic disturbance changes the structure of arboreal tropical ant communities. *Ecography* **24**: 547–54.

Folgarait, P.J. (1998) Ant biodiversity and its relationship to ecosystem functioning: a review. *Biodiversity and Conservation* **7**: 1221–44.

Fonseca de Souza, O.F., & Brown, V.K. (1994) Effects of habitat fragmentation on Amazonian termite communities. *Journal of Tropical Ecology* **10**: 197–206.

Fox, O., Vetter, S., Ekschmitt, K., & Wolters, V. (2006) Soil fauna modifies the recalcitrance-persistence relationship of soil carbon pools. *Soil Biology and Biochemistry* **38**: 1353–63.

Frelich, L.E., Hale, C.M., Scheu, S., et al. (2006) Earthworm invasion into previously earthworm-free temperate and boreal forests. *Biological Invasions* **8**: 1235–45.

Garrido, J.L., Rey, P.J., Cerda, X., & Herrera, C.M. (2002) Geographical variation in diaspore traits of an ant-dispersed plant (*Helleborus foetidus*): are ant community composition and diaspore traits correlated? *Journal of Ecology* **90**: 446–55.

Gaston, K.J. (2000) Global patterns in biodiversity. *Nature* **405**: 220–7.

Gaston, K.J., & Spicer, J.I. (1998) *Biodiversity—an introduction*. Blackwell, Oxford.

Gove, A.D., & Majer, J.D. (2006) Do isolated trees encourage arboreal ant foraging at ground-level? Quantification of ant activity and the influence of season in Veracruz, Mexico. *Agriculture Ecosystems and Environment* **113**: 272–6.

Hale, C.M., Frelich, L.E., Reich, P.B., & Pastor, J. (2005) Effects of European earthworm invasion on soil characteristics in northern hardwood forests of Minnesota, USA. *Ecosystems* **8**: 911–27.

Hartzell, P., Nghiem, J.V., Richio, K.J., & Shain, D.H. (2005) Distribution and phylogeny of glacier ice worms (*Mesenchytraeus solifugus* and *Mesenchytraeus solifugus rainierensis*). *Canadian Journal of Zoology* **83**: 1206–13.

Hastings, A., Byers, J.E., Crooks, J.A., et al. (2007) Ecosystem engineering in space and time. *Ecology Letters* **10**: 153–64.

Hebert, P.D.N., Cywinska, A., Ball, S.L., & DeWaard, J.R. (2003b) Biological identifications through DNA barcodes. *Proceedings of the Royal Society London B* **270**: 313–21.

Hebert, P.D.N., Ratnasingham, S., & DeWaard, J.R. (2003a) Barcoding animal life: cytochrome c oxidase subunit 1 divergences among closely related species. *Proceedings of the Royal Society London B* **270**: S96–S99.

Hedlund, K., Griffiths, B., Christensen, S., et al. (2004) Trophic interactions in changing landscapes: responses of soil food webs. *Basic and Applied Ecology* **5**: 495–503.

Helmus, M.R., Savage, K., Diebel, M.W., Maxted, J.T., & Ives, A.R. (2007) Separating the determinants of phylogenetic community structure. *Ecology Letters* **10**: 917–25.

Hendrix, P.F., Callaham, M.A. Jr., Drake, J.M., et al. (2008) Pandora's box contained bait: the global problem of introduced earthworms. *Annual Review of Ecology, Evolution and Systematics* **39**: 593–613.

Holt, J.A., Coventry, R.J., & Sinclair, D.F. (1980) Some aspects of the biology and pedological significance of mound-building termites in a red and yellow earth landscape near Charters Towers, North Queensland. *Australian Journal of Soil Research* **18**: 97–109.

Hubbell, S.P. (2001) *The Unified Neutral Theory of Biodiversity and Biogeography*. Princeton University Press, Princeton, NJ.

Huhta, V. (1976) Effects of clear-cutting on numbers, biomass and community respiration of soil invertebrates. *Annales Zoologici Fennici* **13**: 63–80.

Huston, M.A. (1994) *Biological Diversity*. Cambridge University Press, Cambridge.

Inward, D.J.G., Beccaloni, G., & Eggleton, P. (2007) Death of an order: a comprehensive molecular phylogenetic study confirms that termites are eusocial cockroaches. *Biology Letters* **3**: 331–5.

James, S.W. (2004) Planetary processes and their interactions with earthworm distributions and ecology. In: C.A. Edwards (eds.) *Earthworm Ecology*, pp. 53–62. CRC Press, Boca Raton, FL.

James, S.W., & Brown, G.G. (2006) Earthworm ecology and diversity in Brazil. In: F.M.S. Moreira, J.O. Siqueira & L. Brussaard (eds.) *Soil biodiversity in Amazonian and other Brazilian ecosystems*, pp. 56–116. CAB International, Wallingford.

James, S.W., Porco, D., Decaëns, T., Richard, B., & Erséus, C. (2010) DNA barcoding reveals cryptic diversity in *Lumbricus terrestris* L., 1758 (Clitellata): resurrection of *L. herculeus* Savigny, 1826. *PLoS ONE* **5**: e15629.

Jeanne, R.L. (1979) A latitudinal gradient in rates of ant predation. *Ecology* **60**: 1211–24.

Jenny, H. (1941) *Factors of Soil Formation: a System of Quantitative Pedology*, pp. 281. McGraw-Hill, New York.

Johns, A.D. (1992) Species conservation in managed tropical forests. In: T.C. Whitmore & J.A. Sayer (eds.) *Tropical deforestation and species extinction*, pp. 15–50. Chapman & Hall, London.

Jones, D.T., Susilo, F.X., Bignell, D.E., Hardiwinoto, S., Gillison, A.N., & Eggleton, P. (2003) Termite assemblage collapse along a land-use intensification gradient in lowland central Sumatra, Indonesia. *Journal of Applied Ecology* **40**: 380–91.

Jonsson, M., Yeates, G.W., & Wardle, D.A. (2009) Patterns of invertebrate density and taxonomic richness across gradients of area, isolation, and vegetation diversity in a lake-island system. *Ecography* **32**: 963–72.

Judas, M. (1988) The species-area relationship of European Lumbricidae (Annelida, Oligochaeta). *Oecologia* **76**: 579–87.

Kairesalo, P. (1978) Ecology of enchytraeids in meadow forest soil in southern Finland. *Annales Zoologici Fennici* **15**: 210–20.

Kambhampati, S., & Eggleton, P. (2000) Phylogenetics and taxonomy. In: T. Abe, D.E. Bignell, & M. Higashi (eds.) *Termites: evolution, sociality, symbioses, ecology*, pp. 1–23. Kluwer Academic Publishers, Dordrecht.

Kapusta, P., Sobczyk, L., Rozen, A., & Weiner, J. (2003) Species diversity and spatial distribution of enchytraeid communities in forest soils: effects of habitat characteristics and heavy metal contamination. *Applied Soil Ecology* **23**: 187–98.

King, R.A., Tibble, A.L., & Symondson, W.O.C. (2008) Opening a can of worms: unprecedented sympatric cryptic diversity within British lumbricid earthworms. *Molecular Ecology* **17**: 4684–98.

Knop, J. (1926) Bakterien und Bakteroiden bei Oligochäten. *Zeitschrift für Morphologie und Ökologie der Tiere* **6**: 587–624.

Koh, L.P., Dunn, R.R., Sodhi, N.S., Colwell, R.K., Proctor, H.C., & Smith, V.S. (2004) Species coextinctions and the biodiversity crisis. *Science* **305**: 1632–4.

Kozak, K.H., Graham, C.H., & Wiens, J.J. (2008) Integrating GIS-based environmental data into evolutionary biology. *Trends in Ecology & Evolution* **23**: 141–8.

Kusnezov, N. (1957) Numbers of species of ants in faunae of different latitudes. *Evolution* **11**: 298–9.

Kutnik, M., Uva, P., Brinkworth, L., & Bagnères, A.G. (2004) Phylogeography of two European *Reticulitermes* (Isoptera) species: the Iberian refugium. *Molecular Ecology* **13**: 3099–113.

Lavelle, P. (1983) The structure of earthworm communities. In: J.E. Satchell (ed.) *Earthworm ecology: from Darwin to vermiculture*, pp. 449–66. Chapman & Hall, London.

Lavelle, P. (1986) Associations mutualistes avec la microflore du sol et richesse spécifique sous les tropiques: l'hypothèse du premier maillon. *Comptes Rendus de l'Académie des Sciences de Paris. Série III—Sciences de la Vie* **302**: 11–14.

Lavelle, P. (1996) Diversity of soil fauna and ecosystem function. *Biology International* **33**: 3–16.

Lavelle, P., Bignell, D.E., Lepage, M., et al. (1997) Soil function in a changing world: the role of invertebrate ecosystem engineers. *European Journal of Soil Biology* **33**: 159–93.

Lavelle, P., Blanchart, E., Martin, A., et al. (1993) A hierarchical model for decomposition in terrestrial ecosystems: application to soils of the humid tropics. *Biotropica* **25**: 130–50.

Lavelle, P., & Pashanasi, B. (1989) Soil macrofauna and land management in Peruvian Amazonia (Yurimaguas, Peru). *Pedobiologa* **33**: 283–91.

Lavelle, P., & Spain, A. (2001) *Soil Ecology*. Kluwer Academic Publishers, Dordrecht.

Lavergne, S., Mouquet, N., Thuiller, W., & Ronce, O. (2010) Biodiversity and climate change: integrating evolutionary and ecological responses of species and communities. *Annual Review of Ecology, Evolution and Systematics* **41**: 321–50.

Lee, K.E. (1959) The earthworm fauna of New Zealand. *CSIRO Bulletin* **130**: 1–9.

Lee, K.E., & Wood, J.T. (1971) *Termites and Soils*. Academic Press, London.

Leniaud, L., Dedeine, F., Pichon, A., Dupont, S., & Bagneres, A.G. (2009) Geographical distribution, genetic diversity and social organization of a new European termite, *Reticulitermes urbis* (Isoptera: Rhinotermitidae). *Biological Invasions* **12**: 1389–402.

Lenz, M., Amburgey, T.L., Zi-Hong, D., et al. (1991) Interlaboratory studies on termite-wood decay fungi associations: II. Response of termites to *Gloeophyllum trabeum* grown on different species of wood (Isoptera: Mastotermitidae, Termopsidae, Rhinotermitidae, Termitidae). *Sociobiology* **18**: 203–54.

Lessard, J.P., Fordyce, J.A., Gotelli, N.J., & Sanders, N.J. (2009) Invasive ants alter the phylogenetic structure of ant communities. *Ecology* **90**: 2664–9.

Loranger, G., Bandyopadhyaya, I., Razaka, B., & Ponge, J.F. (2001) Does soil acidity explain altitudinal sequences in collembolan communities? *Soil Biology and Biochemistry* **33**: 381–93.

Lund, M., Davidson, S.K., Holmstrup, M., et al. (2009) Diversity and host specificity of the *Verminephrobacter*- earthworm symbiosis. *Environmental Microbiology* **12**: 2142–51.

Lundkvist, H. (1983) Effects of clear-cutting on the enchytraeids in a Scots Pine forest soil in central Sweden. *Journal of Applied Ecology* **20**: 873–85.

Lyford, W.H. (1963) The importance of ants to Brown Podzolic Soil Genesis in New England. *Harvard Forest Paper* **7**: 1–18.

Maraun, M., Schatz, H., & Scheu, S. (2007) Awesome or ordinary? Global diversity patterns of oribatid mites. *Ecography* **30**: 209–16.

Martin, A., & Marinissen, J.C.Y. (1993) Biological and physico-chemical processes in excrements of soil animals. *Geoderma* **56**: 331–47.

McGill, B.J., Enquist, B.J., Weiher, E., & Westoby, M. (2006) Rebuilding community ecology from functional traits. *Trends in Ecology & Evolution* **21**: 178–85.

Mermillod-Blondin, F., & Lemoine, D.G. (2010) Ecosystem engineering by tubificid worms stimulates macrophyte growth in poorly oxygenated wetland sediments. *Functional Ecology* **24**: 444–53.

Moreau, C.S., Bell, C.D., Vila, R., Archibald, S.B., & Pierce, N.E. (2006) Phylogeny of the ants: diversification in the age of angiosperms. *Science* **312**: 101–4.

Mueller, U.G., Gerardo, N.M., Aanen, D.K., Six, D.L., & Schultz, T.R. (2005). The evolution of agriculture in insects. *Annual Review of Ecology, Evolution and Systematics* **36**: 563–95.

Nieminen, J.K. (2009) Are spruce boles hot spots for enchytraeids in clear-cut areas? *Boreal Environment Research* **14**: 382–8.

Nobre, T., Eggleton, P., & Aanen, D.K. (2010) Vertical transmission as the key to the colonization of Madagascar by fungus-growing termites? *Proceedings of the Royal Society of London B* **277**: 359–65.

Nobre, T., Koné, N.A., Konaté, S., Linsenmair, K.E., & Aanen, D.K. (2011) Dating the fungus-growing termites' mutualism shows a mixture between ancient codiversification and recent symbiont dispersal across divergent hosts. *Molecular Ecology* **20**: 2619–27.

Nobre, T., Nunes, L., & Bignell, D.E. (2009) Survey of subterranean termites (Isoptera: Rhinotermitidae) in a managed silvicultural plantation in Portugal, using a line-intersection method (LIS). *Bulletin of Entomological Research* **9**: 11–22.

Novo, M., Almodovar, A., & Diaz-Cosin, D.J. (2009) High genetic divergence of hormogastrid earthworms (Annelida, Oligochaeta) in the central Iberian Peninsula: Evolutionary and demographic implications. *Zoologica Scripta* **38**: 537–52.

Novo, M., Almodovar, A., Fernandez, R., Trigo, D., & Diaz Cosin, D.J. (2010) Cryptic speciation of hormogastrid earthworms revealed by mitochondrial and nuclear data. *Molecular Phylogenetics and Evolution* **56**: 507–12.

Nowak, E. (2001) Enchytraeid communities in successional habitats (from meadow to forest). *Pedobiologia* **45**: 497–508.

Nurminen, M. (1965) Enchytraeid and lumbricid records (Oligochaeta) from Spitsbergen. *Annales Zoologici Fennici* **2**: 1–10.

Nurminen, M. (1967) Faunistic notes on North-European enchytraeids (Oligochaeta). *Annales Zoologici Fennici* **4**: 567–87.

Nurminen, M. (1970) Four new enchytraeids (Oligochaeta) from southern Finland. *Annales Zoologici Fennici* **7**: 378–81.

O'Dowd, D.J., Green, P.T., & Lake, P.S. (2003) Invasional meltdown on an oceanic island. *Ecology Letters* **6**: 812–17.

Omodeo, P. (2000) Evolution and biogeography of megadriles (Annelida, Clitellata). *Italian Journal of Zoology* **67**: 179–201.

Peachey, J.E. (1959) Studies on the Enchytraeidae of Moorland Soils. Ph.D. thesis.

Pearce, M.J. (1997) *Termites: Biology and Pest Management*. CAB International, Wallingford.

Perel-Vsevolodova, T.S. (1987) The Altai-Sayan center of endemism of the family Lumbricidae. In: A.M. Bonvicini Pagliai & P. Omodeo (eds.) *On earthworms*, pp. 345–8. Mucchi Editore, Modena.

Pérez-Losada, M., Eiroa, J., Mato, S., & Dominguez, J. (2005) Phylogenetic species delimitation of the earthworms *Eisenia fetida* (Savigny, 1826) and *Eisenia andrei* Bouché, 1972 (Oligochaeta, Lumbricidae) based on mitochondrial and nuclear DNA sequences. *Pedobiologia* **49**: 317–24.

Petersen, H., & Luxton, M. (1982) A comparative analysis of soil fauna populations and their role in decomposition processes. *Oikos* **39**: 287–388.

Pinel, N., Davidson, S.K., & Stahl, D.A. (2008) *Verminephrobacter eiseniae* gen. nov., sp. nov., a nephridial symbiont of the earthworm *Eisenia foetida* (Savigny). *International Journal of Systematic and Evolutionary Microbiology* **58**: 2147–57.

Piper, S.R., MacLean, S.F., & Christensen, B. (1982) Enchytraeidae (Oligochaeta) from taiga and tundra habitats of northeastern U.S.S.R. *Canadian Journal of Zoology* **60**: 2594–609.

Rantalainen, M.L., Fritze, H., Haimi, J., Pennanen, T., & Setälä, H. (2005) Species richness and food web structure of soil decomposer community as affected by the size of habitat fragment and habitat corridors. *Global Change Biology* **11**: 1614–27.

Rao, M., Terborgh, J., & Nuñez, P. (2002) Increased herbivory in forest isolates: implications for plant community structure and composition. *Conservation Biology* **15**: 624–33.

Richard, B., Decaëns, T., Rougerie, R., James, S.W., Porco, D., & Hebert, P.D.N. (2010) Re-integrating earthworm juveniles into soil biodiversity studies: species identification through DNA barcoding. *Molecular Ecology Resources* **10**: 606–14.

Roithmeier, O., & Pieper, S. (2009) Influence of Enchytraeidae (*Enchytraeus albidus*) and compaction on nutrient mobilization in an urban soil. *Pedobiologia* **53**: 29–40.

Romero-Alcaraz, E., & Ávila, J.M. (2000) Effect of elevation and type of habitat on the abundance and diversity of scarabaeoid dung beetle (Scarabaeoidea) assemblages in a Mediterranean area from southern Iberian peninsula. *Zoological Studies* **39**: 351–9.

Rota, E. (1995) Italian Enchytraeidae (Oligochaeta). I. *Italian Journal of Zoology* **62**: 183–231.

Rota, E., & Erséus, C. (2003) New records of Grania (Clitellata, Enchytraeidae) in the Northeast Atlantic (from Tromsø to the Canary Islands), with descriptions of seven new species. *Sarsia* **88**: 210–43.

Rota, E., Healy, B., & Erséus, C. (1998) Biogeography and taxonomy of terrestrial Enchytraeidae (Oligochaeta) in northern Sweden, with comparative remarks on the genus *Henlea*. *Zoologischer Anzeiger* **237**: 155–69.

Rougerie, R., Decaëns, T., Deharveng, L., *et al.* (2009) DNA barcodes for soil animal taxonomy. *Pesquisa Agropecuaria Brasileira* **44**: 789–802.

Rouland-Lefèvre, C. (2000) Symbiosis with fungi. In: T. Abe, D.E. Bignell, & M. Higashi (eds.) *Termites: evolution, sociality, symbioses, ecology*, pp. 289–306. Kluwer Academic Publishers, Dordrecht.

Rouland-Lefèvre, C., & Bignell, D. (2002) Cultivation of symbiotic fungi by termites of the subfamily Macrotermitinae. In: J. Sekbach (ed.) *Symbiosis: mechanisms and model systems*, pp. 731–56. Kluwer, Dordrecht.

Rousset, V., Plaisance, L., Erséus, C., Siddall, M.E., & Rouse, G.W. (2008) Evolution of habitat preference in Clitellata (Annelida). *Biological Journal of the Linnean Society*, **95**: 447–64.

Sanchez de Leon, Y., Zou, X., Borges, S., & Ruan, H. (2003) Recovery of native earthworms in abandoned tropical pastures. *Conservation Biology* **17**: 1–8.

Sanders, N.J., Gotelli, N.J., Heller, N.E., & Gordon, D.M. (2003) Community disassembly by an invasive species. *Proceeding of the National Academy of Sciences* **100**: 2474–7.

Sanderson, M.G. (1996) Biomass of termites and their emissions of methane and carbon dioxide: a global database. *Global Biogeochemical Cycles* **10**: 543–557.

Scheu, S., Albers, D., Alphei, J., *et al.* (2003) The soil fauna community in pure and mixed stands of beech and spruce of different age: trophic structure and structuring forces. *Oikos* **101**: 225–38.

Schmelz, R.M. (2003) Taxonomy of *Fridericia* (Oligochaeta, Enchytraeidae). Revision of species with morphological and biochemical methods. *Abhandlungen Naturwissschaftlicher Verein Hamburg (NF)* **38**: 1–415.

Schmelz, R.M., & Collado, R. (2008) A type-based redescription of *Pachydrilus georgianus* Michaelsen, 1888, the type species of *Marionina* Michaelsen, 1890, with comments on *Christensenidrilus* Dózsa-Farkas & Convey, 1998 (Enchytraeidae, "Oligochaeta," Annelida). *Abhandlungen Naturwissschaftlicher Verein Hamburg (NF)* **44**: 7–22.

Schmelz R.M. and Collado R. (2010) A guide to European terrestrial and freshwater species of Enchytraeidade (Oligochaeta). *Soil Organisms* **82**: 1–176.

Schmitz, O.J. (2010) Spatial dynamics and ecosystem functioning. *PLoS Biology* **8**: e1000378.

Schramm, A., Davidson, S.K., Dodsworth, J.A., Drake, H.L., Stahl, D.A., & Dubilier, N. (2003) *Acidovorax*-like bacterial symbionts in the nephridia of earthworms. *Environmental Microbiology* **5**: 804–9.

Schultz, T.R., & Brady, S.G. (2008) Major evolutionary transitions in ant agriculture. *Proceeding of the National Academy of Sciences of the United States of America* **105**: 5435–40.

Siddall, M.E., Apakupakul, K., Burreson, E.M., Coates, K.A., Erséus, C., Källersjö, M., *et al.* (2001) Validating Livanow: molecular data agree that leeches, branchiobdellidans and *Acanthobdella peledina* are a monophyletic group of oligochaetes. *Molecular Phylogenetics and Evolution* **21**: 346–51.

Siira-Pietikäinen, A., & Haimi, J. (2009) Changes in soil fauna 10 years after forest harvestings: comparison between clear felling and green-tree retention methods. *Forest Ecology and Management* **258**: 332–8.

Siira-Pietikäinen, A., Pietikäinen, J., Fritze, H., & Haimi, J. (2001) Short-term responses of soil decomposer communities to forest management: clear felling versus alternative forest harvesting methods. *Canadian Journal of Forest Research* **31**: 88–9.

Sousa, J.P., Bolger, T., da Gama, M.M., *et al.* (2006) Changes in Collembola richness and diversity along a gradient of land-use intensity: a pan European study. *Pedobiologia*, **50**: 147–56.

Springett, J.A. (1967) An Ecological Study of Moorland Enchytraeidae. Ph.D. Thesis.

Springett, J.A. (1970) The distribution and life histories of some moorland Enchytraeidae (Oligochata). *Journal of Animal Ecology* **39**: 725–37.

Standen, V. (1984) Production and diversity of enchytraeids, earthworms and plants in fertilized hay meadow plots. *Journal of Applied Ecology* **21**: 293–312.

Standen, V. (1988) Oligochaetes in fire climax grassland and coniferous plantations in Papua New Guinea. *Journal of Tropical Ecology* **4**: 38–48.

Standen, V., & Latter, P.M. (1977) Distribution of a population of *Cognettia sphagnetorum* (Enchytraeidae) in relation to microhabitats in a blanket bog. *Journal of Applied Ecology* **46**: 216–29.

Stanton, N.L., & Tepedino, V.J. (1977) Island habitats in soil communities. *Ecological Bulletins* **25**: 511–14.

Stockdill, S.M.J. (1982) Effects of introduced earthworms on the productivity of New Zealand pastures. *Pedobiologia* **24**: 29–35.

Suarez, A.V., Bolger, D.T., & Case, T.J. (1998) Effects of fragmentation and invasion on native ant communities in coastal southern California. *Ecology* **79**: 2041–56.

Sugimoto, A., Bignell, D.E., & MacDonald, J.A. (2000) Global impact of termites on the carbon cycle and atmospheric trace gases. In: T. Abe, D.E. Bignell, & M. Higashi (eds.) *Termites: evolution, society, symbioses, ecology*, pp. 409–35. Kluwer Academic, Dordrecht.

Swift, M.J., Heal, O.W., & Anderson, J.M. (1979) *Decomposition in Terrestrial Ecosystems*. Blackwell, Oxford.

Tagger, S., Périssol, C., Criquet, S., *et al.* (2008) Characterization of an amphimull under Mediterranean evergreen oak forest (*Quercus ilex*): micromorphological and biodynamic descriptions. *Canadian Journal of Forest Research* **38**: 268–77.

Thorne, B.L., Grimaldi, D.A., & Krishna, K. (2000) Early fossil history of the termites. In T. Abe, D.E. Bignell, and M. Higashi (eds.) *Termites: evolution, sociality, symbioses, ecology*, pp. 77–94. Kluwer Academic Publishers, Dordrecht.

Tian, G., Brussaard, L., Kang, B.T., & Swift, M.J. (1997) Soil fauna-mediated decomposition of plant residues under constrained environmental and resource quality conditions. In: G. Cadisch & K.E. Giller (eds.) *Driven by nature: plant litter quality and decomposition*, pp. 125–34. CAB International, Wallingford.

Topoliantz, S., Ponge, J.F., & Viaux, P. (2000) Earthworm and enchytraeid activity under different arable farming systems, as exemplified by biogenic structures. *Plant and Soil* **225**: 39–51.

Tyler, A.N., Carter, S., Davidson, D.A., Long, D.J., & Tipping, R. (2001) The extent and significance of bioturbation on ^{137}Cs distributions in upland soils. *Catena* **43**: 81–99.

Uhía, E., & Briones, M.J.I. (2002) Population dynamics and vertical distribution of enchytraeids and tardigrades in response to deforestation. *Acta Ecologica* **23**: 349–59.

Ulrich, W., & Fiera, C. (2009) Environmental correlates of species richness of European springtails (Hexapoda: Collembola). *Acta Oecologica* **35**: 45–52.

Van Rhee, J.A. (1969) Development of earthworm populations in polder soils. *Pedobiologia* **9**: 128–32.

Van Vliet, P.C.J., Radcliffe, D.E., Hendrix, P.F., & Coleman, D.C. (1998) Hydraulic conductivity and pore size distribution in small microcosms with and without enchytraeids (Oligochaeta). *Applied Soil Ecology* **9**: 277–82.

Van Vliet, P.C.J., West, L.T., Coleman, D.C., & Hendrix, P.F. (1995) The impact of Enchytraeidae (Oligochaeta) on the pore structure of small microcosms. *Acta Zoologica Fennica* **196**: 97–100.

Vasconcelos, H.L., Vilhena, J.M.S., Magnusson, W.E., & Albernaz, A.L.K.M. (2006) Long-term effects of forest fragmentation on Amazonian ant communities. *Journal of Biogeography* **33**: 1348–56.

Vieau, F. (1993) Le termite de saintonge *Reticulitermes santonensis* Feytaud: termite urbain. *Bulletin de la Société Zoologique Française* **118**: 125–33.

Wall, D.H., Bradford, M.A., St. John, M.G., *et al.* (2008) Global decomposition experiment shows soil animal impacts on decomposition are climate-dependent. *Global Change Biology* **14**: 2661–77.

Waller, D.A., Jones, C.G., & LaFage, J.P. (1990) Measuring wood preference in termites. *Entomologia Experimentalis et Applicata* **56**: 117–23.

Wardle, D.A. (2002) *Communities and Ecosystems: Linking the Aboveground and Belowground Components*, pp. 400. Princeton University Press, Princeton, NJ.

Wardle, D.A., Yeates, G.W., Williamson, W., & Bonner, K.I. (2003) The response of a three trophic level soil food web to the identity and diversity of plant species and functional groups. *Oikos* **102**: 45–56.

Webb, C.O., Ackerly, D.D., McPeek, M.A., & Donoghue, M.J. (2002) Phylogenies and community ecology. *Annual Review of Ecology and Systematics* **33**: 475–505.

Webb, C.T., Hoeting, J.A., Ames, G.M., Pyne, M.I., & Poff, N.L. (2010) A structured and dynamic framework to advance traits-based theory and prediction in ecology. *Ecology Letters* **13**: 267–83.

Whitford, W.G. (2000) Keystone arthropods as webmasters in desert ecosystems. In: D.C. Coleman and P.F. Hendrix (eds.) *Invertebrates as webmasters in ecosystems*, pp. 25–41. CABI Publishing, Wallingford.

Whittaker, J.B. (1974) Interactions between fauna and microflora at tundra sites. In: A.J. Holding, O.W. Heal, S.F.J. MacLean & P.W. Flanagan (eds.) *Soil organisms and decomposition in tundra* pp. 183–96. Tundra Biome Steering Committee, Stockholm.

Williams, R.M. (1966) East African termites of genus *Cubitermes* (Isoptera—Termitidae). *Transactions of the Entomological Society of London* **118**: 73–118.

Wilson, E.O. (1974) *The Insect Societies*, pp. x + 548. Belknap Press of Harvard University Press, Cambridge, MA.

Wolters, V. (2000) Invertebrate control of soil organic matter stability. *Biology and Fertility of Soils* **31**: 1–19.

Wood, T.G. (1988) Termites and the soil environment. *Biology and Fertility of Soils* **6**: 228–36.

Wood, T.G., & Sands, W.A. (1978) The role of termites in ecosystems. In: M.V. Brian (ed.) *Production ecology of ants and termites*, pp. 245–92. Cambridge University Press, Cambridge.

Synthesis

Donald R. Strong and Valerie Behan-Pelletier

Soil is the luxuriant tapestry that ensures life on Earth. These six chapters on community structure and biotic assemblages interweave spatial scales, phylogeny, biogeography, and diversity from microbes to among the largest ecosystem engineers, with a central theme of linking biodiversity to ecosystem functioning and ecosystem services. We are confronted with the possibility that research in soil biodiversity can make a difference to human society. We are challenged to rethink the paradigms that underpin much research in soil ecology.

In Chapter 3.1 Justin Bastow describes work on succession in detrital food webs that is involved with diversity and resource processing. His point of departure is to elaborate upon solutions to the "enigma of soil animal diversity," which is seen as an enigma because of the apparent lack of diversity in resources for soil animals. While species of plants are known for their profusion of phytochemical differences above ground, at first blush, detritus is boringly homogenous dead plant material with the occasional live root thrown in; this would make for a fauna with "seemingly unspecialized diets." However, the lack of diversity in resources is only apparent; fixed carbon and nutrients in the soil are actually a complex intertwining of live and dead plant material with abundant dimensions of resource partitioning. Furthermore, soil animals add yet more complexity in the sequential chain of transformations that they perform upon the decomposition of detritus. Bastow's experimental work has made use of the ingenious insight of processing chain decomposition, which made its debut in stream ecosystems; a processing chain of different species of detrital consumers each of which changes the character of the decomposing plant material for the next consumer. The sequential consumption in these chains augments biodiversity with positive, facilitative interactions; they are much richer phenomena than carbon down the Second Law drain. However, there is yet more heterogeneity in soil biotas. The microbial biota increases in diversity as wood is successively decomposed by a series of organisms in the soil. While it has long been known that fungi increase in biomass and diversity while bacteria decrease during decomposition, it has more recently been established that the nematode, earthworm, and arthropod detritivore community structure also changes during this process, and that these changes are followed by a succession of different predator species. Echoing a theme in all contributions to this section, Bastow emphasizes that abiotic conditions insert substantial contingency, and therefore spatial and temporal heterogeneity, into succession of organisms participating in soil decomposition. This is well known from the abiotically influenced contingency of wood, animal dung, and carrion decomposition. The emphasis that Bastow places upon the soil components of carrion decomposition will be a valuable contribution to the knowledge base of soil ecologists.

In chapter 3.2 Matty Berg explores patterns of biodiversity at spatial scales from rhizosphere, to plot, to landscape and to the macroscale in vertical and horizontal separation. Physicochemical factors, to current climate, to plant growth and other environmental drivers contribute differentially at different spatial scales. Spatial heterogeneity in turn is interwoven with genetic traits, such as body size, and dispersal abilities of organisms, contributing to patterns in spatial distribution of soil organisms and their "bewildering diversity." Berg illustrates this spatial dimension with organisms of different body size, from bacteria and nematodes to carabid beetles and earthworms, distributed at different spatial scales. These spatial patterns of soil organisms can be strongly influenced by human management of soil. They also lead to patchiness in soil functions with impacts on ecosystem services. The author advocates a hierarchical approach to study spatial patterns, and leads us down Alice's rabbit-hole from the small-scale of the plot where "horizontal heterogeneity is the rule" much defined by spatial distribution of soil properties, to the fine scale where vertical heterogeneity is at least equally important. He emphasizes that spatial scales are nested and thus the remarkable diversity of soil organisms is in large measure due to spatial heterogeneity and the "nested set of ecological worlds in soil."

Chapter 3.3 by Pedro Antunes, Philipp Franken, Dietmar Schwarz, Matthias Rillig, Marco Cosme, Martha Scott, and Miranda Hart carries on the theme of this section by an exploration of the links between soil biodiversity and human nutrition. This contribution begins with a very useful overview of the daunting statistics on widespread nutrient deficiencies of arid lands and alkaline soils, upon which a substantial proportion of humanity practices subsistence agriculture. Since food was industrialized, in the 19th century, the solution to nutrient deficiencies of soils has been profuse fertilization. An early strategy was to apply manure from the oat-fed horses that towed the carts and barges of coal, which powered the industrial revolution. Then came application of Haber Bosch nitrogen, cheaper yet by an order of magnitude. And then again came heavy fertilization (and the pesticides, hydraulic technology, and high yielding varieties) of the Green Revolution. Additional strategies used by wealthier people are diet diversification, post-harvest fortification of foods, and genetic manipulation to produce crops with boosted levels of human nutrients (biofortification). The incisive point made by the authors is that all of these are "beyond the reach for most of the world's population." They propose an alternative approach through soil ecology and evolution that does not require the high levels of initial investment of the standard industrial food production system. Their proposition is that interactions between plants and endophytic as well as epiphytic microbes, of both roots and shoots, could be fostered that would increase the nutrient value of crops as food. Crops on nutritionally stressed soils could benefit in particular from such an approach. Ironically, fertilization with inorganic fertilizers interferes with a major group of these microbes, cyanobacteria, and despite promise, their use in agriculture has decreased to be "almost non-existent" today. The authors give an annotated catalogue of other kinds of plant microbe interactions that, with the right science and application, are candidates for boosting the quality of food grown on nutrient poor soils: cyanobacteria, rhizobia, mycorrhizal fungi, and non-mycorrhizal fungal endophytes. Their main point is that so little science is being directed at such a potentially promising set of soil interactions with plants that we have no idea if any of these would provide practical benefit. The potential direct benefits for plant vigor and growth of having the right microbes in the right plant associations include increased nutrient uptake, beneficial changes in plant soil chemical relationships, nutrient solubilization and transport through soil, and iron chelation. The indirect effects include improved secondary chemistry that would increase the nutrient quality of plants as human foods. This contribution is distinctly objective in clearly elaborating upon what could be negative effects of new microbe-crop associations: uptake of radioactive isotopes and metals as well as increase in secondary plant chemicals that are toxic to humans. This contribution is an imaginative, novel contribution from soil ecology of potentially great human benefit.

In Chapter 3.4, Gregor Schuurman focuses in on fungus growing termites, the Macrotermitidae, in the arid tropics of Asia and Africa. These animals and their mutualistic, white rot fungi are particularly diverse and have large ecosystem effects through plant litter decomposition in dry areas. One notable aspect of this relationship is its independence of climate, which allows these organisms to achieve the status of the predominant litter decomposers and to attain a collective biomass and rate of plant consumption rivaling the mammalian megafauna. Their massive biomass and high harvesting rates yield major effects upon decomposition, soil carbon, and nutrient recycling rates, soil carbon pools, and nutrient and energy flows. As observed in Chapter 3.6, tropical termites are major ecological engineers of the soil, creating large biogenic structures above- and belowground with their sheets, nests, mounds, burrows, and tunnels. These structures create landscape and resource heterogeneity from tiny to large scales.

In Chapter 3.5, Mark Bradford and Noah Fierer seek to bridge the scientific divide in soil ecosystem science and biogeochemistry between "black box... empirical models" that are based upon physics and chemistry alone and "mechanistic" models built upon biological relationships. Empirical models tend to take the simplest approach; they are regressions of rates of soil processes that are independent of spatial environmental variability. Mechanistic models are based upon microbiological physiology that describes, for example, soil respiration as the product of respiratory rates of enzymes crucially dependent upon variation in soil temperature and moisture. Perhaps implicit, but certainly part and parcel of the empirical models is the assumption that tiny organisms have such great dispersal power as to be functionally ubiquitous. After arriving virtually everywhere, universal-immigrant microbes are seen to have high growth rates and rapid evolutionary adaptation. This would render soil microbial communities to be functionally homogeneous in space and time. It is not surprising therefore that empirical models scale up to regional and global predictions without the messiness that burdens the mechanistic models. While the authors' own research falls on the mechanistic end of the spectrum, they have the objective attitude that biological accuracy has a cost. While mechanistic understanding provides detailed information about what is actually happening in soil among real organisms, it generates large-scale and long-term predictions that are more uncertain than the just-physics-and-chemistry empirical models. The authors' ambition is to lessen the differences between empirical and mechanistic models and they argue for a "fresh look at the paradigms of functional redundancy, similarity and equivalence."

In Chapter 3.6, Lijbert Brussaard Duur K. Aanen, Maria J.I. Briones, Thibaud Decaëns, Gerlinde B. De Deyn, Tom M. Fayle, Samuel W. James, and Tânia Nobre examine the ecological engineering of soil invertebrate communities from a biogeographical and phylogenetic perspective. They begin locally and scale up to a global view. This chapter is one of two in this section that deals with soil ecosystem engineers particularly termites, ants, earthworms, and enchytraeids, and the mutualisms of these with bacteria and fungi. Soil animals join microbes in decomposition and nutrient cycling, but soil animals perform these functions in varied ways that have distinctly different net effects. Their legacies are in mounds, burrows, soil aggregates manifested at coarse scales. In a stunning figure, the authors summarize how soil animals, differing in phylogenetic history differ in geography, climate, latitude, altitude, and the ecosystem consequences: *inter alia*, enchytraeids work Mor and Moder at high latitudes, earthworms and ants enter the picture, especially to work the Mull of grasslands at mid latitudes, and the whole menagerie are at work in moist, low latitude forests and savannas. This figure also anticipates the ecosystem results of human driven global change: invasive species, dewatering, warming, fragmentation, deforestation, overgrazing, and knock-on effects such as desertification, as it will play out through soil animals. The authors demonstrate the contributions of both phylogeny and biogeography to any functional trait-based approach to ecosystem functioning.

Together these chapters transcend the "black box" view of soil inhabited by an unknown and undefined set of functional groups. On spatial scales from the smallest unit of ecological soil, the minuscule aggregate of mineral, fixed carbon, other nutrients, and associated microbes to biomes, on timescales from recent effects of management to phylogenetic history, authors show how soil diversity from microbes to ecosystem engineers together provide the living basis for functioning of ecosystems. These are the players and the processes in soil ensuring ecosystem services.

SECTION 4

Global Changes

SECTION EDITORS: **Richard D. Bardgett and T. Hefin Jones**

Introduction

Richard D. Bardgett and T. Hefin Jones

Human activities have had a substantial and often irreversible impact on Earth's ecosystems, to the extent that humans have either transformed, or influenced in some way, much of the world's land surface. The most noticeable impact of humans is the transformation of land for agriculture and forestry, which has been occurring for centuries and is probably the main force behind biodiversity loss worldwide (Millennium Ecosystem Assessment 2005). However, the Earth's terrestrial ecosystems and their biodiversity have been, and continue to be, affected by several other global change phenomena. These include climate change, atmospheric nitrogen deposition, the invasion of alien species into new territories, and urbanization, which is occurring at an ever-increasing rate worldwide. All of these global change phenomena, and their interactions, can have strong direct and indirect impacts on soils, their biodiversity, and their capacity to deliver ecosystem services. Indeed, the extent to which humans have impacted on soils is highlighted by the fact that degraded soils, which are no longer capable of delivering ecosystem services, now cover some 15–17% of the Earth's land surface (ISRIC 1990).

This section includes a series of chapters that explore the impact of different global change phenomena, namely climate change, nitrogen enrichment, urbanization, and agriculture, on soil and its capacity to deliver ecosystem services. The list of global change factors covered is not exhaustive, in that we do not cover topics such as invasive species and their impacts on soils, which can be profound (Wardle et al. 2011), or effects on soils of arable farming, deforestation or persistent organic pollutants. Also, we consider global change factors singularly, rather than simultaneously, although as we highlight in our synthesis, that our ability to predict future responses of soils to global change will increasingly require understanding of the simultaneous effects of multiple global change drivers. However, our intention is not to provide an exhaustive catalogue of how soils and ecosystem services are affected by global change, but rather to illustrate, using selected examples, the capacity for human activities to impact on soils and ecosystem services, and where appropriate, the potential to manage soils to mitigate these impacts.

To this end, the first chapter (4.1), by Nick Ostle and Susan Ward, considers how climate change can impact on soils and ecosystem services. Specifically, the chapter focuses on the role of soils in the carbon cycle and climate mitigation, and how this is potentially affected by climate change, both directly, for instance through direct impacts of changes in temperature and water availability on soil processes, and indirectly via changes in plant productivity and community structure which cascade belowground. Ostle and Ward also discuss the potential for predicting the future responses of soils to climate change using models, and identify gaps in our knowledge. In the next chapter (4.2), Manning illustrates how terrestrial ecosystems and their soils are also strongly affected by nitrogen enrichment, which is of high relevance because anthropogenic activities have substantially increased global rates of nitrogen fixation and deposition. In this chapter Manning considers how nitrogen enrichment affects soils, their biota, and vegetation, and the consequences of this for ecosystem services.

Next, in Chapter 4.3, Mitchell Pavao-Zuckerman covers the topic of urbanization, soils, and ecosystem services. As Pavao-Zuckerman highlights, we are living in an increasingly urbanized world, with more than half of the world's population living in cities. This presents many challenges for soils, and in this chapter Pavao-Zuckerman highlights the impacts on, constraints for, and potential to develop ecosystem services in cities through management of urban soils. Finally, in Chapter 4.4, Phil Murray, Felicity Crotty, and Nick van Eekeren consider the topic of land use for agriculture and its impacts on soils and ecosystem services. As mentioned earlier, the conversion of land for agriculture has had substantial, and often strongly negative and irreversible, impacts on soils and the ecosystem services that they deliver. Moreover, there are currently considerable challenges and tensions surrounding the need to manage soils in order to feed a burgeoning world population, whilst sustaining their long-term fertility and ability to deliver other ecosystem services, such as carbon storage and the storage of clean water. In this chapter, Murray et al. consider some of these issues in the context of grasslands, which cover around a quarter of the Earth's land surface, and the trade-offs that exist with their management for food production and the delivery of other ecosystem services through their soils.

As highlighted at the start of this introduction, human activities have, and are increasingly having, substantial impacts on soils with far reaching consequences for the services that they can deliver. We hope that these chapters provide an insight into some of the ways that soils and their services are affected by human-induced global change, and how changes in the way that we manage soils might help mitigate human impacts in the future.

References

ISRIC (1990) Global Assessment of Human-induced Soil Degradation. http://www.isric.org/projects/global-assessment-human-induced-soil-degradation-glasod

Millennium Ecosystem Assessment (2005) *Ecosystems and human well-being: current state and trends*. Island Press, Washington DC.

Wardle, D.A., Bardgett, R.D., Callaway, R.M. and Van der Putten, W.H. (2011) Terrestrial ecosystem responses to species gains and losses. *Science* **332**, 1273–7.

CHAPTER 4.1

Climate Change and Soil Biotic Carbon Cycling

Nicholas J. Ostle and Susan E. Ward

4.1.1 Introduction

Climate change is now considered as a major threat to global ecosystems, with a predicted minimum increase in world temperatures of 0.2°C every decade over the next 50 years (IPCC 2007) as fossil fuel derived CO_2 emissions rise at an average rate of 2.5% a year (Friedlingstein *et al.* 2010). Changes in air and soil temperature, the magnitude, frequency, and distribution of precipitation, growing season length, drought, flood, and elevated carbon dioxide (CO_2) concentrations have all been predicted as climate change proceeds. These changes will converge with pressures on land use as demands for fuel, water, and food increase with human population growth (Ingram *et al.* 2008). Together, these powerful drivers of change will have consequences for soil biota and dependent ecosystem functions, including plant productivity, decomposition, and the regulation of the three primary greenhouse gases (GHG): CO_2, methane (CH_4), and nitrous oxide (N_2O) (Fig. 4.1.1). Ultimately, the terrestrial biospheric feedback of these natural biogenic greenhouse gases will be largely dependent on plant–soil interactions that control organic matter sequestration and decomposition processes.

The Earth's soils are the product of multiscale and dynamic phenomena including climate, underlying geology, local biodiversity, biotic activity, time, and, in many instances, the actions of human populations (i.e. for agriculture, industry, urbanization and natural resource management, harvest and extraction). There is a well-established relationship between the health of the soil and the sustainability of human civilizations that rely on it as a crucial resource for the provision of food, fuel, and fibre, and as a regulator of ecosystem nutrient and carbon dynamics (e.g. Montgomery 2007; Lal 2004, 2009). The structure and activity of soil decomposer communities, including microbes and mesofauna, are critical in the breakdown of complex organic matter

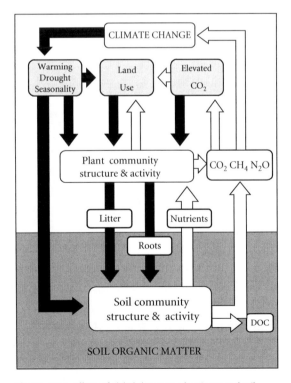

Figure 4.1.1 Effects of global change on plant inputs and soil processes with feedbacks to ecosystem greenhouse gas emissions.

Soil Ecology and Ecosystem Services. First Edition. Edited by Diana H. Wall *et al.*
© 2012 Oxford University Press. Published 2012 by Oxford University Press.

inputs (plant and animal derived), accumulation of soil carbon stocks, and in the regulation of ecosystem GHG emissions. There is, however, growing evidence of powerful feedbacks between climate change and soil processes that underpin ecosystem services of carbon sequestration, nutrient cycling, primary productivity, and water quality regulation (Bardgett et al. 2008; Quinton et al. 2010). These feedbacks include the acceleration of soil carbon cycling and emissions of ecosystem green house gases CO_2, CH_4, and N_2O, which is of particular concern in ecosystems that contain large stores of organic matter belowground in the soil.

The purpose of this chapter is to examine the reported effects of climate change on processes crucial to the plant–soil–atmosphere carbon cycle and the production and consumption of soil greenhouse gases. We examine the evidence for *direct* (i.e. temperature, water availability) and *indirect* (i.e. plant changes, elevated CO_2) effects of climate change on plant–soil interactions and soil biotic carbon (C) cycling. We have focused on the role of the soil biotic community as a whole with some specific reference to soil microbes (bacteria, fungi, and archaea) and mesofauna (oligochaetes, micro-arthropods, nematodes). We also discuss the potential for predicting future soil biological responses to climate change using mathematical models and identify some gaps in knowledge that need to be considered to better understand the role and resilience of soil biotic C cycling in the provision of life-supporting ecosystem services in the future.

4.1.2 Climate change and plant–soil interactions

Plant–soil interactions are at the core of global ecological and biogeochemical cycles, with climate change having considerable potential to influence their role in terrestrial carbon cycling. Direct effects of climate change on soil C processes include atmospheric warming and alterations in precipitation patterns that regulate moisture levels belowground. The indirect effects of climate change include shifts in soil C cycling as a result of changes in growing season duration, plant species occurrence, community composition, plant distributions, and productivity (Heimann & Reichstein 2008). The combination of instantaneous direct and, relatively longer-term, indirect climate change effects on key ecosystem C processes (i.e. photosynthesis, productivity, and soil decomposition) will be strongly influenced by interactions between plant-soil

Figure 4.1.2 Schematic of dynamic climate change effects on soil biota and feedbacks to plant community composition their inputs and ecosystem green house gas emissions.

biology. Specifically, climate change effects will determine the size, composition, and activity of soil biological communities and their feedbacks to plant community composition and productivity (Figure 4.1.2). The continuous feedback between climate and soil processes is mediated by vegetation responses, and is a critical step in the adaptation of ecosystems to climate change. Providing an improved understanding of the impacts of climate change on soil biodiversity and carbon biogeochemistry poses a significant challenge in the development of protection, mitigation, and adaptation strategies for the future (Trumbore & Czimczik 2008; Ostle *et al.* 2009a).

Natural and man-made variations between soils at local, regional, and global scales can be substantial making it difficult to generalize as to their possible reactivity to climate change. For example, it is unlikely that a South American Rendzina soil will react in the same way as a boreal peat, or that soil communities currently constrained by drought will respond in the same way as those constrained by water-logged conditions. But, whatever the environmental constraints, it is clear that climate acts both directly and indirectly on soil components of terrestrial ecosystems to influence subsequent rates of biologically mediated soil organic matter processing, nutrient flows, carbon cycling, and green house gas feedbacks to the atmosphere (van der Putten *et al.* 2004). In the following section, the nature of direct and indirect climate change effects on soil biological processes such as respiration and organic matter mineralization/decomposition that govern ecosystem carbon cycling are examined.

4.1.3 Direct effects

The direct effects of warming and changes in soil water conditions on soil biota and their role in decomposition can be instantaneous, but their consequences for soil C cycling are likely to be cumulative and long term.

4.1.3.1 Temperature

Temperature exerts a powerful control over the biological properties and functions of all ecosystems, and soil is no exception to this. Consequently, warming is a strong driver of soil biotic abundance and its role in soil organic matter decomposition. Currently, most scientific evidence shows that global warming will result in accelerated soil respiration to release CO_2, and increased exports of dissolved organic carbon (Craine *et al.* 2010; Briones *et al.* 2010). Soil respiration rates are thought to be generally more sensitive to temperature than primary production (Fisher *et al.* 2010), so that as climate warming progresses it is predicted that the net transfer of carbon from soil to atmosphere will increase, thereby creating a positive feedback to climate change (Schimel *et al.* 1994). While it is accepted that temperature is an important determinant of instantaneous rates of soil organic matter decomposition (Dorrepaal *et al.* 2009; Carrera *et al.* 2009), the nature of the longer-term relationship between temperature and heterotrophic respiration, and its potential to feedback to climate change, are far from clear (e.g. Trumbore 2006). A major cause of this uncertainty is the "difficult-to-study" complexity and diversity of soil biotic communities and soil organic matter quality.

There is growing evidence that the temperature dependence of decomposition is influenced by the presence and abundance of different soil trophic groups, and by the quantity and chemical quality of organic matter substrates (Davidson & Janssens 2006). For example, studies have shown that the presence of oligochaete enchytraeid worms within organic soils significantly increased soil respiration rates with warming (Cole *et al.* 2002), with consequences for accelerated nutrient mineralization, and that this interaction also led to greater releases of "old" soil carbon as CO_2 and dissolved organic carbon (DOC) (Briones *et al.* 2004). Results from controlled laboratory experiments suggest that the temperature sensitivity of decomposition increases as the quality of consumed organic carbon declines (Fierer *et al.* 2005), which is consistent with "kinetic theory" that predicts greater temperature sensitivity of recalcitrant, difficult to decompose carbon pools (e.g. Knorr *et al.* 2005). However, other analyses of laboratory experiments have found that this relationship is not necessarily universal (Fang *et al.* 2005). There is also potential for interactions between soil biota, carbon substrates, and the physical and chemical protection of organic matter which can also influence warming effects (Kardol et al. 2010).

This picture is further complicated by uncertainty about how reactive different microbial groups and mesofauna species are to temperature change. Data from field experiments are fairly consistent. For example, in temperate grasslands the impacts of warming on soil animal populations can be significant (Briones et al. 2009) with predictable feedbacks to carbon cycling (Wall et al. 2008). In peatland soils climate-induced changes in enchytraeid populations have been shown to accelerate decomposition rates (Briones et al. 2007). Observations from a long-term warming experiment in the McMurdo Antarctic dry valleys showed a significant sustained reduction in soil nematode populations over 8 years (Simmons et al. 2009). However, there is also conflicting evidence as to whether increases in carbon mineralization commonly observed in warming experiments in the field (Lou et al. 2001; Melillo et al. 2002) will be sustained as substrate availability reduces and decomposer communities adapt (Kirschbaum 2004). For example, in a long-term warming of a subarctic heath Rinnan et al. (2011) found that decreases in soil bacterial community growth had no effects on its temperature adaptation after 7 and 17 years due to decreased availability of labile substrate. Other work (Heinemeyer et al. 2006), suggests that there is also potential for interactions between warming (surface) and mycorrhizal fungi respiration. The possibility for thermal adaptation of soil microbes to enable them to respire more and more complex soil compounds is also possible but contentious (Bradford et al. 2008; Hartley et al. 2008). However, there is a lack of knowledge at the decomposer functional level that contributes to unreliable model predictions of soil carbon cycling and feedbacks to climate change (Kirschbaum 2006). Resolving the link between climate change, soil biodiversity, carbon substrates (quality and availability), and soil GHG emissions represents a major research challenge for the future.

4.1.3.2 Water

The water status of soils is predicted to be one of the most powerful determinants of how climate change will influence land surface to atmosphere feedbacks (e.g. Seneviratne et al. 2006). It is likely that climate change related increases in the frequency of extreme weather events, such as drought and freeze-thawing, will have even greater effects on soil organisms and their activities than overall changes in temperature and precipitation at the ecosystem scale. It is well established that both drought and freeze-thaw cycles can have substantial direct effects on microbe and fauna community composition and physiology (e.g. Schimel & Mikan 2005; Briones et al. 2010) and can also contribute to increased vulnerability to soil erosion (Quinton et al. 2010), with consequences for ecosystem-level carbon cycling and nutrient flows (Schimel et al. 2007). Increased frequency and intensity of drought in already dry ecosystems, for example, may result in moisture-limiting conditions for faunal and microbial activity, creating a negative feedback to decomposition and soil carbon loss as respiration (Taylor & Wolters 2005). This view is supported by studies in forest ecosystems which report significant falls in litter isoenzyme diversity and soil bacterial and fungal biomass during dry periods (Nardo et al. 2004; Krivtsov et al. 2006), and by a manipulation experiment in dry Californian annual grassland, where the addition of water increased soil hydrolytic enzyme activity (Henry et al. 2005). However, in contrast, increased drought and drying in wetlands and peatlands can favor microbial activity, by lowering the water table and introducing oxygen into the previously anaerobic peat. This has been shown to influence the abundance and distribution of mesofauna in peatlands as animals are forced to move away from the peat surface as it dries (Briones et al. 1998). Severe drying can also increase the activity of phenol oxidases (Freeman et al. 2004a; Zibilske & Bradford 2007), which play a pivotal role in the breakdown of complex organic matter and the cycling of phenolic compounds that interfere with microbially derived extracellular enzymes (Albers et al. 2004). The "enzymic latch mechanism" (Freeman et al. 2001), proposes that changes in the activity of extracellular phenol oxidases can directly affect the retention of carbon in soil *via* the breakdown of otherwise highly recalcitrant organic matter compounds and by releasing extracellular hydrolase enzymes from phenolic inhibition. Because peatlands and wetlands are one of the largest stocks of terrestrial carbon (Ward et al. 2007), such enhanced breakdown of recalcitrant organic matter under drying could have major implications for the global carbon cycle (Freeman et al. 2004b).

While the increase in O_2 availability that accompanies drought promotes decomposition in moist or waterlogged soils, the same cannot be said for methanogenesis. In peatlands, water table depth is generally a strong predictor of methane emissions (Roulet & Moore 1995), and while this is generally assumed to be due to toxic effects of O_2, there is also evidence that methanogens are sensitive to desiccation (Fetzer et al.1993). Also, toxic effects on methanogens of the oxidized products of denitrification have been noted (Kluber & Conrad 1998), while net methane emissions are also suppressed under drought conditions by the action of methanotrophic bacteria (Freeman et al. 2002). Drying also exerts a strong regulatory effect on soil nitrous oxide (N_2O) fluxes with emissions depending on the severity of drought conditions i.e. modest summer drought scenarios may have little effect on soil N_2O emissions, whereas more extreme drought can increase N_2O emissions substantially (Dowrick et al. 1999).

Climate change-related reductions in snow cover will also impact soil microbial communities and decomposition processes. It has been estimated that 25% of Earth's permafrost could thaw by 2100 due to climate warming, releasing considerable amounts of otherwise protected organic matter for microbial decomposition (Anisimov et al. 1999), thus creating a positive feedback to climate change. Permafrost soils store an estimated $1{,}672 \times 10^{15}$ g of C (Tarnocai et al. 2009) that has accumulated and stabilized under prevalent freezing temperatures. However, as predicted climate change progresses permafrost will thaw and this C will be vulnerable to greater biological activity and release as CO_2 and CH_4 (Schuur et al. 2008; Schuur et al. 2009). There is now strong evidence that permafrost thawing will also make previously inert (old) soil C stocks more available to decomposition processes (Nowinski et al. 2010) further amplifying GHG feedbacks to the atmosphere. Resultant accelerated decomposition could also increase nitrogen availability to microbes and plants, exacerbating the direct effects of warming on plant–soil interactions (Mack et al. 2004; Nowinski et al. 2008). In time, the mixing of minerals into former permafrost soils could act to stabilize organic C (Striegl et al. 2005; Kawahigashi et al. 2006), but the capacity for physical protection of soil organic matter in permafrost regions is poorly understood. Despite these important recent discoveries, surprisingly little is known about permafrost biogeochemistry and how plant-soil interactions that determine GHG emissions will evolve with warming and changes in hydrology. Key questions remain about the extent to which permafrost C is stabilized by freezing, the effects of hydrological changes, and associated shifts in macronutrient limitation on the soil microbial processes that ultimately govern Arctic CO_2 and CH_4 fluxes. Also, because snow is an important insulator of soil biological processes, reductions in snow cover in alpine and arctic regions could increase soil freezing, with consequences for root mortality, nutrient cycling, and microbial processes of decomposition (Groffman et al. 2001). Strong microbial responses to freeze-thaw have been detected in several studies, resulting in increased microbial activity and greenhouse gas emissions (Christensen & Tiedje 1990; Sharma et al. 2006), altered microbial substrate use (Schimel & Mikan 2005), and the expression of denitrifying genes leading to greater N_2O release. However, recent studies in subalpine forest in Colorado indicate that reduced snow cover can suppress rates of soil respiration due to a unique and highly temperature sensitive soil microbial community that occurs beneath snow (Monson et al. 2006). Such responses could have substantial consequences for winter soil microbial activity, carbon storage, and CO_2 efflux in alpine and arctic regions.

4.1.4 Indirect effects

Climate change can also have marked indirect effects on soil biotic communities and their activity through its influence on plant growth and vegetation composition, with consequent impacts on soil conditions and resources due to differences in the quality and quantity of plant inputs to the soil. These plant-mediated effects of climate change on soil organisms and their function operate via a variety of mechanisms that produce changes in plant productivity and community composition to influence the quantity and quality of organic resource inputs.

4.1.4.1 Plant community productivity

The first mechanism concerns the indirect effects of rising atmospheric CO_2 concentrations on the soil decomposer community as a result of increased plant photosynthesis and productivity. Experimental evidence suggests that the effects of higher CO_2 concentrations on photosynthesis and plant productivity vary significantly between plant species (Matamala et al. 2003; Ainsworth & Long 2005; Long et al. 2006), but studies have also shown that elevated CO_2 augments the transfer of photosynthate carbon to mycorrhizal fungi (Treseder 2004; Johnson et al. 2005; Högberg & Read 2006) and heterotrophic microbes (Zak et al. 1993). It is well established that elevated CO_2 increases plant photosynthesis, especially under nutrient-rich conditions (Long et al. 2004), and this in turn increases the flux of carbon to roots, their symbionts, and soil fauna and bacteria via root exudation of easily degradable sugars, organic acids, and amino acids in grasslands (Ostle et al. 2003, 2007), forests (Drake et al. 2011) and peatlands (Fenner et al. 2007). The consequences of increased carbon flux from roots to soil communities will vary substantially with plant identity, soil food web interactions, soil fertility, and a range of other ecosystem properties (Wardle 2002), but there is strong evidence that fresh cascades of rhizodeposition will accelerate both soil C and nitrogen (N) cycling by the decomposer community (e.g. Phillips et al. 2011). This can result in increases in soil carbon loss by respiration and in drainage waters as dissolved organic carbon due to stimulation of microbial abundance and activity (Freeman et al. 2004b; Heath et al. 2005), and enhanced mineralization of recent and old soil organic carbon, a phenomenon known as "priming" (Fontaine & Barot 2005; Kuzyakov 2006). There is also potential for the stimulation of microbial biomass and immobilization of soil N, thereby limiting N availability to plants, creating a negative feedback that constrains future increases in plant growth and carbon transfer to soil in forests (Norby et al. 2010; Korner et al. 2005) and grasslands (Newton et al. 2010). Elevated CO_2 experiments in grasslands have also shown associated increases in symbiotic clover N fixation (Zanetti et al. 1996). There is also some evidence that elevated CO_2 influences changes in root exudation to promote methanogenesis and N_2O emissions and hence, carbon loss from soil as methane (Baggs et al. 2003). However, the mechanisms involved are poorly understood (Ström et al. 2005).

4.1.4.2 Plant community composition

The second mechanism concerns indirect effects of climate change on soil communities via shifts in the functional composition and diversity of vegetation over longer timescales. Plant species can be broadly classified into a number of functional types on the basis of ecological, physiological, and biogeochemical characteristics or traits (e.g. photosynthetic activity, litter quality, root turnover) that have different regulatory influences on soil biotic processes (DeDeyn et al. 2008). Evidence suggests that elevated CO_2 can alter plant community structure (Owensby et al. 1999), but there is far stronger evidence that warming and changing precipitation regimes influence the distribution of plant species and functional groups at both local and global scales (Prentice et al. 1992; Woodward et al. 2004). For example, recent changes in precipitation patterns have markedly affected vegetation composition in tropical rainforest (Engelbrecht et al. 2007) and African savanna (Sankaran et al. 2005), and warming is leading to rapid replacement of Canadian tundra by boreal forest (Danby & Hik 2007) and pan-arctic shrub encroachment in arctic tundra (Epstein et al. 2004). Such changes in vegetation composition can strongly regulate carbon exchange by affecting uptake of CO_2 by photosynthesis and by modifying the soil physical environment, for example, by changes in root architecture and rooting depth (Jackson et al. 1996). Meta analysis of the International Tundra Experiment (ITEX) warming experiment in the Tundra showed a rapid response of plants to two growing seasons of warming, with increased growth of shrubs and graminoids at the expense of mosses and lichens, resulting in reduced species diversity and evenness (Walker et al. 2006). There is also evidence that changes in Alaskan Arctic vegetation could significantly alter surface albedo properties to amplify feedbacks of CO_2 (Chapin et al. 2005). But, in contrast, a high arctic unproductive Canadian evergreen shrub heath showed no strong effects on vascular plant cover or species diversity, even after 15 years of warming, suggesting that unproductive sites would only be affected by a substantial climate change (Hudson & Henry 2010).

A key mechanism by which climate-driven shifts in vegetation composition influences soil decomposers, and hence carbon-cycle feedback, is through changes in the quality and quantity of organic matter entering the soil as plant inputs. Leaf litter and rhizodeposition quality is known to differ consistently across plant functional groups (Dorrepaal et al. 2005; Rasmussen et al. 2008) and correlates strongly with rates of decomposition and hence heterotrophic respiration. Slow-growing plants, such as evergreen shrubs, produce poor quality litter which is low in nutrients and rich in recalcitrant compounds, such as lignin and phenolic acids, and hence decompose slowly due to retardation of microbial activity. In contrast, fast-growing plants, such as graminoids and N-fixers, produce relatively high quality litter that is more readily decomposed through microbial activity (Wardle 2002). Therefore, climate-driven increases in the dominance of evergreen shrubs with recalcitrant litter, as is occurring in the arctic, might constitute a negative feedback on carbon exchange and global warming due to reduced soil heterotrophic respiration (Cornelissen et al. 2007). In grassland ecosystems, increased dominance of legumes over grasses, evidenced in elevated carbon dioxide experiments (Ross et al. 2004; Hanley et al. 2004), could induce a positive feedback on microbial activity and carbon mineralization due to enhanced soil nutrient availability and the decomposition of nutrient rich litter. In temperate grasslands, small increases in legume cover have been shown to alter soil physicochemical conditions and biotic processes resulting in up to 10% increases in C sequestration in 3 years (DeDeyn et al. 2011). As noted above, ecosystem-level shifts in substrate quality could have implications for the temperature sensitivity of decomposition and hence complicate further our ability to predict the magnitude of carbon-cycle feedbacks (Fierer et al. 2005).

Plant functional groups also differ markedly in their mycorrhizal status (Read et al. 2004) and their mechanisms of nutrient uptake, including acquisition of different chemical forms of inorganic and organic nitrogen (McKane et al. 2002; Weigelt et al. 2005; Harrison et al. 2007). Therefore, climate-driven changes in vegetation composition will alter nutrient competition between plant species, and between plants and soil microbes, with potential consequences for ecosystem carbon cycling. Our understanding of the importance for carbon exchange of such feedbacks between climate change, vegetation, and soil microbial functioning represents an important and ongoing research challenge.

4.1.4.3 Seasonality and phenology

As well as changes in plant species distributions and community composition, changes in atmospheric temperatures and precipitation patterns can also result in significant alterations to plant phenological processes (Nord & Lynch 2009). Changes in the timing of germination, spring-time emergence, flowering, seed dispersal, and senescence have the potential to feedback to soil biotic processes responsible for carbon cycling as the quantity and timing of organic matter inputs to soil are altered. Using satellite observations, Delbart et al. (2008) identified a significant lengthening of the growing season which revealed an 8-day advance in leaf appearance date recorded between 1982 and 1991 in northern latitudes and across the Eurasian Boreal zone. Similarly, the average first flowering date of 385 British plant species has advanced by 4.5 days in the 1990s compared with the previous four decades with flowering most especially sensitive to the temperature in the previous month (Fitter & Fitter 2002). In another study, the effects of warmer autumn months in northern temperate ecosystems can be seen on both saprobic and mycorrhizal soil fungi. Gange et al. (2007) analyzed 52,000 individual fungal fruiting records, collected from nearly 1,400 localities, and found that first fruiting dates were earlier by 8.6 days per decade and that overall fruiting periods have increased from 33.2 days in the 1950s to 74.8 days in the 2000s. More recently, similar patterns in earlier and longer fungal fruiting activity have been observed in Scandinavia (Kauserud et al. 2010). Changes in the timing of phenological phenomena that influence primary productivity and the quantity and quality of plant derived inputs to the soil have the potential to significantly influence soil biotic diversity and function as climate change progresses.

4.1.4.4 Multifactor effects

Most experimental studies on the effects of climate change on soil have examined single factors, such as elevated atmospheric CO_2, warming, or drought. However, there is considerable potential for interactions between these factors and others (e.g. land use-management, nutrient availability) to have additive or antagonistic effects on soil biota and their function (Shaw *et al.* 2002; Mikkelsen *et al.* 2008). In Mediterranean shrub ecosystems the effects of warming and drought on soil enzyme activities have been shown to be due to the direct effects on soil temperature and soil water content, and not to changes in litter or soil organic matter quantity and quality (Sardans *et al.* 2008). In peatland, the combined and positive effect of increased temperature and elevated CO_2 on decomposition of peat was found to be greater than when these factors operated alone (Fenner *et al.* 2007), creating an even stronger positive feedback on carbon loss from soil as DOC and microbial respiration. On the other hand, Wan *et al.* (2007) found that in old-field grasslands there were no interactive effects of warming, moisture conditions and elevated CO_2 on soil microbial respiration rates. In another grassland ecosystem a factorial elevated CO_2 × N addition experiment showed feedbacks between N limitation and plant productivity responses to CO_2 with implications for belowground biotic processes (Reich *et al.* 2006). Added to this inconsistency is the knowledge that other organisms and trophic groups that influence soil microbes directly (e.g. microbial-feeding fauna and ecosystem engineers) or indirectly through altering vegetation diversity and productivity (e.g. herbivores, plant pathogens, and parasites) will also respond differently to multiple climate change factors (Wardle 2002; Bardgett 2005).

4.1.5 Making predictions

Changes in temperature and moisture resulting from climate change will clearly have strong impacts on both the ecology and biogeochemistry of soils across the globe. Predicting the magnitude of these impacts on ecosystem and global C cycling and potential feedbacks to atmospheric GHG concentrations remains a priority. The inherent complexity of soil food webs and trophic interactions, however, poses significant obstacles to our ability to predict effects of multiple climate change drivers on soil biotic communities and soil C dynamics. A number of mathematical plant–soil carbon models have been developed over the past 50 years to predict land use and global change effects on ecosystem nutrient dynamics and GHG emissions. Currently, the most widely applied soil C models have multiple soil organic matter pools to simulate C dynamics (e.g. DNDC Li *et al.* 1994; CENTURY Del Grosso *et al.* 2006; ROTH-C Coleman & Jenkinson 1996; SUNDIAL Bradbury *et al.* 1993). They generally include the minimum of soil biogeochemical processes needed to estimate soil C and N turnover and have been applied to arable croplands to grasslands, forests, and peatlands (McGill 1996; Smith *et al.* 1998; Smith 2001; Smith 2002; Peltoniemi *et al.* 2007). Most soil models have adopted the representation of soil C in multiple pools of organic matter that decompose according to first-order rate kinetics (Paustian 1994), but others have attempted to include more detailed understanding of soil food webs (Hunt *et al.* 1991; de Ruiter & van Faassen 1994; de Ruiter *et al.* 1993, 1995).

The further development of mathematical plant–soil interaction models now offers soil scientists an opportunity to develop and test hypotheses regarding climate change effects on biological processes that govern both terrestrial carbon cycling and ecosystem GHG emissions (Ostle *et al.* 2009b). A number of modeling approaches span a broad range of dynamic and spatial dimensions from global circulation models (GCMs) that operate at the 2° global grid cell scale to soil carbon process models that can be parameterized at the plot, core or microsite scale (Figure 4.1.3). Linking between these are dynamic global vegetation models (DGVMs) of varying degrees of complexity that represent the state-of-the-art for studying the impacts of change on plant–soil interactions and their feedbacks to the climate system, e.g. CLM-CN, IGSM, LPJ, BIOME-BGC, CENTURY, DNDC, HYBRID, SDGVM, TRIFFID, ORCHIDEE (Schimel *et al.* 1996; Friend *et al.*

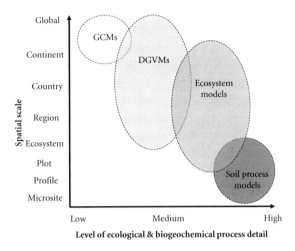

Figure 4.1.3 Process and prediction resolution of soil models from point to global grid scale.

1997; Woodward et al. 1998; Cox 2001; Sitch et al. 2003; Krinner et al. 2005; Thornton & Rosenbloom 2005; Miehle et al. 2006; Thornton et al. 2007; Xu & Prentice 2008; Sokolov et al. 2008; Thornton et al. 2009). These DGVMs can be operated whilst coupled to global atmospheric circulation models (GCMs) to enable explicit simulation of the feedbacks between the biosphere and atmosphere (Friedlingstein et al. 2010). As well as simulating the instantaneous biogeochemical processes of photosynthesis and plant respiration, DGVMs can also simulate longer-term impacts of climate change on vegetation cover and soil carbon storage (Cox et al. 2000; Levy et al. 2004; Sato et al. 2007; Sitch et al. 2003; Woodward & Lomas 2004). Most DGVMs typically utilize the concept of "plant functional types" (PFTs; numbering between 3 and 20) to classify global vegetation diversity that also determine soil carbon dynamics and its reactivity to climate. Each PFT represents a broad class of vegetation type such as deciduous forest or grassland and is parameterized for a core set of physiological processes. All of these models draw on established ecological understanding and field measurements to develop, parameterize, and test their predictive certainty (see Sitch et al. 2008 for a detailed breakdown of DGVM model compositions). All current DGVMs use process based descriptions of soil C dynamics some with a single C soil pool, others with multiple C pools similar to those found in soil models (e.g. Parton et al. 1988; Jones et al. 2005). None represent soil biotic communities explicitly.

4.1.6 Conclusions

The evidence is clear and growing that soil biology plays a crucial role in terrestrial biosphere carbon cycling and that soil biotic communities are sensitive to climate change. *Ipso facto* the ecological response of this active and vital component of terrestrial ecosystems will be a key determinant of terrestrial feedbacks to climate change in the future. The effects of climate change on the interactions between plant community function and the diversity and activity of soil communities will, therefore continue to have a powerful effect on ecosystem carbon cycling at a global scale. Increasingly powerful and complex mathematical modeling approaches are simulating and predicting future climate change with greater certainty. At the same time, it is apparent that the direct and indirect effects of climate change on plant–soil interactions need to be considered carefully in further experimental research. Together the knowledge of fundamental soil biotic processes and their response to climate variations need to be explicitly addressed within large-scale models to reduce uncertainty in biosphere carbon-climate predictions.

The severity of effects of human influenced global change (e.g. land use, climate change, nitrogen deposition, invasive species, ozone, pollution) on the survival of species, habitats, and the function of natural and managed ecosystems, is raising concern among scientists, politicians, and environmental groups alike. It is evident that rapid and chronic long-term shifts in environmental conditions, due to human activities, are occurring simultaneously across a range of spatial and temporal scales, in some instances threatening the existence of valued and finite natural capital (Costanza *et al.* 1997; Millennium Ecosystem Assessment 2005; Janssens & Vicca 2010).

Climate has been, is, and will continue to be a primary regulator of soil biotic communities and their role in global carbon cycling. A key challenge in the coming years will be to use scientific evidence for decision making to protect, conserve and manage the biosphere and its resources, including the soil, to sustain the delivery of life supporting functions and ecosystem services that underpin food security, fuel provision, and water quality. The challenge remains to gather evidence from global gradients, long-term and multifactorial experiments to underpin decision making on human actions to mitigate and adapt to climate change. Crucial to this will be our understanding of how to sustain the natural capital held within soil biodiversity and to predict its influence on the Earth's carbon cycle and ecosystem services that we rely on.

References

Ainsworth, E.A., & Long, S.P. (2005) What have we learned from 15 years of free-air CO_2 enrichment (FACE)? A meta-analytic review of the responses of photosynthesis, canopy. *New Phytologist* **165**: 351–71.

Albers, D., Migge, S., Schaefer. M., & Scheu, S. (2004) Decomposition of beech leaves and spruce needles in pure and mixed stands of beech and spruce. *Soil Biology and Biochemistry* **36**: 155–64.

Anisimov, O.A., Nelson, F.E., & Pavlov, A.V. (1999) Predictive scenarios of permafrost development under conditions of global climate change in the XXI century. *Earth Cryology* **3**: 15–25.

Baggs, E.M., Richter, M.,Cadisch, G., *et al.* (2003) Denitrification in grass swards is increased under elevated atmospheric CO_2. *Soil Biology and Biochemistry* **35**: 729–132.

Bardgett, R.D. (2005) *The Biology of Soil. A Community and Ecosystem Approach.* Oxford University Press, Oxford.

Bardgett, R.D., Freeman, C., & Ostle, N.J. (2008) Microbial contributions to climate changethrough carbon cycle feedbacks. *ISME Journal* **2**: 805–14.

Bradbury, N.J., Whitmore, A.P., Hart, P.B.S., & Jenkinson, D.S. (1993) Modelling the fate of nitrogen in crop and soil in the years following application of ^{15}N-labelled fertilizer to winter wheat. *Journal of Agricultural Science* **121**: 363–79.

Bradford, M.A., Davies, C.A, Frey, S.D., *et al.* (2008) Thermal adaptation of soil microbial respiration to elevated temperature. *Ecology Letters* **11**: 1316–27.

Briones, M.J.I., Garnett, M.H., & Ineson, P. (2010) Soil biology and warming play a key role in the release of "old C" from organic soils. *Soil Biology and Biochemistry* **42**: 960–7.

Briones, M.J.I., Ineson, P., & Poskitt, J. (1998) Climate change and Cognettiasphagnetorum: effects on carbon dynamics in organic soils. *Functional Ecology* **12**: 528–35.

Briones, M.J.I., Ostle, N.J., & Garnett, M.H. (2007) Invertebrates increase the sensitivity of non-labile soil carbon to climate change. *Soil Biology and Biochemistry* **39**: 816–18.

Briones, M.J.I., Ostle, N.J., McNamara, N.R., *et al.* (2009) Functional shifts of grassland soil communities in response to soil warming. *Soil Biology and Biochemistry* **41**: 315–22.

Briones, M.J.I., Poskitt, J., & Ostle, N. (2004) Influence of warming and enchytraeid activities on soil CO_2 and CH_4 fluxes. *Soil Biology and Biochemistry* **36**:1851–9.

Carrera, N., Barreal, M.E., Gallego, P.P., *et al.* (2009) Soil invertebrates control peatland C fluxes in response to warming. *Functional Ecology* **23**: 637–48.

Chapin III, F.S., Sturm, M., Serreze, M.C., *et al.* (2005) Role of land-surface changes in Arctic summer warming. *Science* **310**: 657–60.

Christensen, S., & Tiedje, J.M. (1990) Brief and vigorous N_2O production by soil at spring thaw. *European Journal of Soil Science* **41**: 1–4.

Cole, L., Bardgett, R.D., Ineson, P., & Adamson, J.K. (2002) Relationships between enchytraeid worms (oligochaeta), climate change, and the release of dissolved organic carbon from blanket peat in northern England. *Soil Biology & Biochemistry* **34**: 599–607.

Coleman, K., & Jenkinson, D.S. (1996) ROTHC 26.3- A Model for the turnover of carbon in soil. In: D.S. Powlson, P. Smith, & J.U. Smith. (eds.), *Evaluation of Soil Organic Matter Models Using Existing, Long-Term Datasets*, pp. 237–46. Springer-Verlag, Heidelberg.

Cornelissen, J.H.C., van Bodegom, P.M., Aerts, R., *et al.* (2007) Global negative vegetation feedback to climate warming responses of leaf litter decomposition rates in cold biomes. *Ecology Letters* **10**: 619–27.

Costanza, R., d'Arge, R., deGroot, R., et al. (1997) The value of the world's ecosystem services and natural capital. *Nature* **387**: 253–60.

Cox, P.M. (2001) *Description of the "TRIFFID" dynamic global vegetation model*. Hadley Centre Technical Note 24, Met Office, UK.

Cox, P.M., Betts, R.A., Jones, C.D., Spall, S.A., & Totterdell, I.J. (2000) Acceleration of global warming due to carbon-cycle feedbacks in a coupled climate model. *Nature* **408**: 184–7.

Craine, J.M, Fierer, N., & McLauchlan, K.K. (2010) Widespread coupling between the rate and temperature sensitivity of organic matter decay. *Nature Geoscience* **3**: 854–7.

Danby, R.K., & Hik, D.S. (2007) Variability, contingency and rapid change in recent subarctic alpine tree line dynamics. *Journal of Ecology* **95**: 352–63.

Davidson, E.A., & Janssens, I.A. (2006) Temperature sensitivity of soil carbon decomposition and feedbacks to climate change. *Nature* **440**: 165–73.

De Deyn, G.B., Cornelissen, J.H.C., & Bardgett, R.D. (2008) Plant functional traits and soil carbon sequestration in contrasting biomes. *Ecology Letters* **11**:516–31.

De Deyn, G.B., Shiel, R.S., Ostle, N.J., et al. (2011) Additional carbon sequestration benefits of grassland diversity restoration. *Journal of Applied Ecology* **48**: 600–8.

de Ruiter, P.C., Neutel, A.M., & Moore, J.C. (1995) Energetics and stability in belowground food webs. In: G.A. Polis & K.O. Winemiller (eds.) *Food Webs, Integration of Patterns and Dynamics*, pp. 201–10. Chapman and Hall, New York.

de Ruiter, P.C., & Van Faassen, H.G. (1994) A comparison between an organic matter dynamics model and a food web model simulating nitrogen mineralization in agro-ecosystems. *European Journal of Agronomy* **3**: 347–54.

de Ruiter, P.C., Van Veen, J.A., Moore, J.C., Brussaard, L., Hunt, H.W. (1993) Calculation of nitrogen mineralization in soil food webs. *Plant and Soil* **157**: 263–73.

Del Grosso, S.J., Parton, W.J., Mosier, A.R., Walsh, M.K., Ojima, D.S., & Thornton, P.E, (2006) DAYCENT national-scale simulations of nitrous oxide emissions from cropped soils in the United States. *Journal of Environmental Quality* **35**: 1451–60.

Delbart, N., Picard, G., Le Toans, T., et al. (2008) Spring phenology in boreal Eurasia over a nearly century time scale. *Global Change Biology* **14**: 603–14.

Dorrepaal, E., Cornelissen, J.H.C., Aerts, R., Wallen, B., & van Logtestijn, R.S.P. (2005) Are growth forms consistent predictors of leaf litter quality and decomposability across peatlands along a latitudinal gradient? *Journal of Ecology* **93**: 817–28.

Dorrepaal, E., Toet, S., van Logtestijn, R.S.P., et al. (2009) Carbon respiration from subsurface peat accelerated by climate warming in the subarctic. *Nature* **460**: 616–19.

Dowrick, D.J., Hughes, S., Freeman, C., Lock, M.A., Reynolds, B.R., & Hudson, J.A. (1999) Nitrous oxide emissions from a gully mire in mid-Wales UK, under simulated summer drought. *Biogeochemistry* **44**: 151–62.

Drake, J.E., Gallet-Budynek, A., & Hofmockel, K.S. (2011) Increases in the flux of carbon belowground stimulate nitrogen uptake and sustain the long-term enhancement of forest productivity under elevated CO_2. *Ecology Letters* **14**: 349–57.

Engelbrecht, B.M., Comita, L.S., Condit, R., Kursar, T.A., Tyree, M.T., Turner, B.L., Hubbell, S.P. (2007) Drought sensitivity shapes species distribution patterns in tropical forests. *Nature* **447**: 80–2.

Epstein, H.E., Beringer, J., Gould, W.A., et al. (2004) The nature of spatial transitions in the Arctic. *Journal of Biogeography* **31**: 1917–33.

Fang, C.M., Smith, P., Moncrieff, J.B., & Smith, J.U. (2005) Similar response of labile and resistant soil organic matter pools to changes in temperature. *Nature* **433**: 881.

Fenner, N., Freeman, C., Lock, M.A., Harmens, H., Reynolds, B., & Sparks, T. (2007) Interactions between elevated CO_2 and warming could amplify DOC exports from peatland catchments. *Environmental Science & Technology* **41**: 3146–52.

Fetzer, S., Bak, F., & Conrad, R. (1993) Sensitivity of methanogenic bacteria from paddy soil to oxygen and desiccation. *FEMS Microbiology Ecology* **12**: 107–15.

Fierer, N., Craine, J.M., McLauchlan, K., & Schimel, J.P. (2005) Litter quality and the temperature sensitivity of decomposition. *Ecology* **86**: 320–6.

Fisher, R., McDowell, N., Purves, D., et al. (2010) Assessing uncertainties in a second-generation dynamic vegetation model caused by ecological scale limitations. *New Phytologist* **187**: 666–81.

Fitter, A.H., & Fitter, R.S.R. (2002) Rapid changes in flowering time in British plants. *Science* **296**: 1689–91.

Fontaine, S., & Barot, S. (2005) Size and functional diversity of microbe populations control plant persistence and long-term soil carbon accumulation. *Ecology Letters* **8**: 1075–87.

Freeman, C., Fenner, N., Ostle, N.J., et al. (2004b) Export of dissolved organic carbon from peatlands under elevated carbon dioxide levels. *Nature* **430**: 195–8.

Freeman, C., Nevison, G.B., Kang, H., Hughes, S., Reynolds, B., & Hudson, J.A. (2002) Contrasted effects of simulated drought on the production and oxidation of methane in a mid-Wales wetland. *Soil Biology and Biochemistry* **34**: 61–7.

Freeman, C., Ostle, N., & Kang, H. (2001) An enzymic "latch" on a global carbon store—a shortage of oxygen locks up carbon in peatlands by restraining a single enzyme. *Nature* **409**: 149.

Freeman, C., Ostle, N.J., Fenner, N., & Kang, H. (2004a) A regulatory role for phenol oxidase during decomposition in peatlands. *Soil Biology and Biochemistry* **36**: 1663–7.

Friedlingstein, P., Houghton, R.A., Marland, G., et al. (2010) Update on CO_2 emissions. *Nature Geosciences* **3**: 811–12.

Friend, A.D., Stevens, A.K., Knox, R.G., & Cannell, M.G.R. (1997) A process-based, terrestrial biosphere model of ecosystem dynamics (Hybrid v3.0). *Ecological Modelling* **77**: 233–55.

Gange, A.C., Gange, E.G., Sparks, T.H., et al. (2007) Rapid and recent changes in fungal fruiting patterns. *Science* **316**: 71.

Groffman, P.M., Driscoll, C.T., Fahey, T.J., Hardy, J.P., Fitzhugh, R.D., & Tierney, G.L. (2001) Colder soils in a warmer world: a snow manipulation study in northern hardwood forest. *Biogeochemistry* **56**: 135–50.

Hanley, M.E., Trofmov, S., & Taylor, G. (2004) Species-level effects more important than functional group-level responses to elevated CO_2: evidence from simulated turves. *Functional Ecology* **18**: 304–13.

Harrison, K.A., Bol, R., & Bardgett, R.D. (2007) Preferences for uptake of different nitrogen forms by co-existing plant species and soil microbes in temperate grasslands. *Ecology* **88**: 989–99.

Hartley, I.P., Hopkins, D.W., Garnett, M.H., et al. (2008) Soil microbial respiration in arctic soil does not acclimate to temperature. *Ecology Letters* **12**: 1092–100.

Heath, J., Ayres, E., Possell, M., et al. (2005) Rising atmospheric CO_2 reduces soil carbon sequestration. *Science* **309**: 1711–13.

Heimann, M., & Reichstein, M. (2008) Terrestrial ecosystem carbon dynamics and climate feedbacks. *Nature* **451**: 289–92.

Heinemeyer, A., Ineson, P., Ostle, N., et al. (2006) Respiration of the external mycelium in the arbuscularmycorrhizal symbiosis shows strong dependence on recent photosynthates and acclimation to temperature. *New Phytologist* **171**: 159–70.

Henry, H., Juarez, J.D., Field, C.B., & Vitousek, P.M. (2005) Interactive effects of elevated CO_2, N deposition and climate change on extracellular enzyme activity and soil density fractionation in a Californian annual grassland. *Global Change Biology* **11**: 1808–15.

Högberg, P., & Read, D.J. (2006) Towards a more plant physiological perspective on soil ecology. *Trends in Ecology & Evolution* **21**: 548–54.

Hudson, J.M.G., & Henry, G.H.R. (2010) High arctic plant community resists 15 years of experimental warming. *Journal of Ecology* **98**: 1035–41.

Hunt, H.W., Trlica, M.J., Redente, E.F., et al. (1991) Simulation model for the effects of climate change on temperate grassland ecosystems. *Ecological Modelling* **53**: 205–46.

Ingram, J.S.I., Gregory, P.J., & Izac, A.M. (2008) The role of agronomic research in climate change and food security policy. *Agriculture Ecosystems & Environment* **126**: 4–12.

IPCC (2007) *Climate Change: Fourth Assessment Report of the Intergovernmental Panel on Climate Change*. Cambridge University Press, Cambridge.

Jackson, R.B., Canadell, J., Ehleringer, J.R., Mooney, H.A., Sala, O.E., & Schulze, E.D. (1996) A global analysis of root distributions for terrestrial biomes. *Oecologia* **108**: 389–411.

Janssens, I.A., & Vicca, S. (2010) Biogeochemistry: Soil carbon breakdown. *Nature Geoscience* **3**: 823–4.

Johnson, D., Kresk, M., Stott, A.W., et al. (2005) Soil invertebrates disrupt carbon flow through fungal networks. *Science* **309**: 1047.

Jones, C., McConnell, C., Coleman, K., et al. (2005) Global climate change and soil carbon stocks; predictions from two contrasting models for the turnover of organic carbon in soil. *Global Change Biology* **11**: 154–66.

Kardol, P., Cregger, M.A., Campany, C.E., et al. (2010) Soil ecosystem functioning under climate change: plant species and community effects. *Ecology* **91**: 767–81.

Kauserud, H., Heegaard, E., Semenov, M.A., et al. (2010) Climate change and spring-fruiting fungi. *Proceedings of the Royal Society B Biological Sciences* **277**: 1169–77.

Kawahigashi, M., Kaiser, K., Rodionov, A. & Guggenberger, G. (2006) Sorption of dissolved organic matter by mineral soils of the Siberian forest tundra. *Global Change Biology* **12**, 1868–77.

Kirschbaum, M.U.F. (2004) Soil respiration under prolonged soil warming: are rate reductions caused by acclimation or substrate loss? *Global Change Biology* **10**: 1870–7.

Kirschbaum, M.U.F. (2006) The temperature dependence of organic-matter decomposition still a topic of debate. *Soil Biology and Biochemistry* **38**: 2510–18.

Kluber, H.D., & Conrad, R. (1998) Effects of nitrate, nitrite, NO and N_2O on methanogenesis and other redox processes in anoxic rice field soil. *FEMS Microbiology Ecology* **25**: 301–18.

Knorr, W., Prentice, I.C., House, J.I., & Holland, E.A. (2005) Long-term sensitivity of soil carbon turnover to warming. *Nature* **433**: 298–301.

Korner, C., Asshoff, R., Bignucolo, O., et al. (2005) Carbon flux and growth in mature deciduous forest trees exposed to elevated CO_2. *Science* **309**: 1360–2.

Krinner, G., Viovy, N., de Noblet-Ducoudre, N., et al. (2005) A dynamic global vegetation model for studies of the coupled atmosphere-biosphere system. *Global Biogeochemical Cycles* **19**: GB1015.

Krivtsov, V., Bezginova, T., Salmond, R., *et al.* (2006) Ecological interactions between fungi, other biota and forest litter composition in a unique Scottish woodland. *Forestry* **79**: 201–16.

Kuzyakov, Y. (2006) Sources of CO_2 efflux from soil and review of partitioning methods. *Soil Biology and Biochemistry* **38**: 425–48.

Lal, R. (2004) Soil carbon sequestration to mitigate climate change. *Geoderma* **123**: 1–22.

Lal, R. (2009) Soil degradation as a reason for inadequate human nutrition. *Food Security* **1**: 45–57.

Levy, P.E., Cannell, M.G.R., & Friend, A.D. (2004) Modelling the impact of future changes in climate, CO_2 concentration and land use on natural ecosystems and the terrestrial carbon sink. *Global Environmental Change-Human and Policy Dimensions* **14**: 21–30.

Li, C., Frolking, S., & Harriss, R. (1994) Modelling carbon biogeochemistry in agricultural soils. *Global Biogeochemical Cycles* **8**: 237–54.

Long, S.P., Ainsworth, E.A., Leakey, A.D.B., *et al.* (2006) Food for thought: Lower-than-expected crop yield stimulation with rising CO_2 concentrations. *Science* **312**: 1918–21.

Long, S.P., Ainsworth, E.A., Rogers, A., *et al.* (2004) Rising atmospheric carbon dioxide: Plants face the future. *Annual Review of Plant Biology* **55**: 591–628.

Lou, Y., Wan, S., & Hui, D. (2001) Acclimatization of soil respiration to warming in tall grass prairie. *Nature* **413**: 622–5.

Mack, M. C., Schuur, E. A. G., Bret-Harte, M. S., Shaver, G. R. & Chapin, F. S. (2004) Ecosystem carbon storage in arctic tundra reduced by long-term nutrient fertilization. *Nature* **431**, 440–3.

Matamala, R., Gonzalez-Meler, M.A., Jastrow, J.D., *et al.* (2003) Impacts of fine root turnover on forest NPP and soil C sequestration potential. *Science* **302**: 1385–7.

McGill, W.B. (1996) Review and classification of ten soil organic matter (SOM) models. In:D.S. Powlson, P. Smith, & J.U. Smith (eds) *Evaluation of Soil Organic Matter Models Using Existing, Long-Term Datasets NATO ASI Series I, Vol.38*, pp. 111–33. Springer-Verlag, Heidelberg.

McKane, R.B., Johnson, L.C., Shaver, G.R., Nadelhoffer, K.J., Rastetter, E.B., Fry, B., *et al.* (2002) Resource-based niches provide a basis for plant species diversity and dominance in arctic tundra. *Nature* **415**: 68–71.

Melillo, J.M., Steudler, P.A., Aber, J.D., *et al.* (2002) Soil warming and carbon-cycle feedbacks to the climate system. *Science* **298**: 2173–5.

Miehle, P., Livesley, S.J., Li, C., Feikema, P.M., Adams, M.A., & Arndt, S.K. (2006) Quantifying uncertainty from large-scale model predictions of forest carbon dynamics. *Global Change Biology* **12**: 1421–34.

Mikkelsen, C., Beier, S., Jonasson, M., *et al.* (2008) Experimental design of multifactor climate change experiments with elevated CO_2, warming and drought: the CLIMAITE project. *Functional Ecology* **22**: 185–95.

Millennium Ecosystem Assessment (2005) *Ecosystems and Human Well-being: Synthesis*. Island Press, Washington, DC.

Monson, R.K., Lipson, D.L., Burns, S.P., *et al.* (2006) Winter soil respiration controlled by climate and microbial community composition. *Nature* **439**: 711–14.

Montgomery, D.R. (2007) Soil erosion and agricultural sustainability. *Proceedings of the National Academy of Sciences of the United States of America* **104**: 13268–72.

Nardo, C.D., Cinquegrana, A., Papa, S., Fuggi, A., & Fioretto, A. (2004) Laccase and peroxidase isoenzymes during leaf litter decomposition of Quercus ilex in a Mediterranean ecosystem. *Soil Biology and Biochemistry* **36**: 1539–44.

Newton, P.C.D., Lieffering, M., Bowatte, W.M.S.D., *et al.* (2010) The rate of progression and stability of progressive nitrogen limitation at elevated atmospheric CO_2 in a grazed grassland over 11 years of free air CO_2 enrichment. *Plant and Soil* **336**: 433–41.

Norby, R.J., Warren, J.M., Iversen, C.M., *et al.* (2010) CO_2 enhancement of forest productivity constrained by limited nitrogen availability. *Proceedings of the National Academy of Sciences of the United States of America* **107**: 19368–73.

Nord, E.A.,& Lynch, J.P. (2009) Plant phenology: a critical controller of soil resource acquisition. *Journal of Experimental Botany* **60**: 1927–37.

Nowinski, N. S., Taneva, L., Trumbore, S. E. & Welker, J. M. (2010) Decomposition of old organic matter as a result of deeper active layers in a snow depth manipulation experiment. *Oecologia* **163**: 785–92.

Nowinski, N. S., Trumbore, S. E., Schuur, E. A. G.,Mack, M. C.& Shaver, G. R. (2008) Nutrient addition prompts rapid destabilization of organic matter in an arctic tundra ecosystem. *Ecosystems* **11**: 16–25.

Ostle, N., Briones, M.J.I., Ineson, P., Cole, L., Staddon, P., & Sleep, D. (2007) Isotopic detection of recent photosynthate carbon flow into grassland rhizosphere fauna. *Soil Biology and Biochemistry* **39**: 768–77.

Ostle, N., Whiteley, A.S., Bailey, M.J., Sleep, D., Ineson, P., & Manefield, M. (2003) Active microbial RNA turnover in a grassland soil estimated using a $^{13}CO_2$ spike. *Soil Biology and Biochemistry* **35**: 877–85.

Ostle, N.J., Levy, P.E., Evans, C.D., & Smith, P. (2009a) UK land use and soil carbon sequestration. *Land Use Policy* **26**: 274–83.

Ostle, N.J., Smith, P., Fisher, R., *et al.* (2009b) Integrating plant-soil interactions into global carbon cycle models. *Journal of Ecology* **97**: 851–63.

Owensby, C.E., Ham, J.M., Knapp, A.K., et al. (1999) Biomass production and species composition change in a tallgrass prairie ecosystem after long-term exposure to elevated atmospheric CO_2. *Global Change Biology* **5**: 497–506.

Parton, W.J., Stewart, J.W.B., & Cole, C.V. (1988) Dynamics of C, N, P and S in grassland soils—A model. *Biogeochemistry* **5**: 109–31.

Paustian, K. (1994) Modelling soil biology and biochemical processes for sustainable agricultural research. In: C.E. Pankhurst, B.M. Doube, V.V.S.R. Gupta, & P.R. Grace (eds.) *Soil Biota. Management in Sustainable Farming Systems*, pp. 182–93. CSIRO Information Services, Melbourne.

Peltoniemi, M., Thürig, E., Ogle, S., et al. (2007) Models in country scale carbon accounting of forest soils. *Silva Fennica* **41**: 575–602.

Phillips, R.P., Finzi, A.C., & Bernhardt, E.S. (2011) Enhanced root exudation induces microbial feedbacks to N cycling in a pine forest under long-term CO_2 fumigation. *Ecology Letters* **14**: 187–94.

Prentice, I.C., Cramer, W., Harisson, S.P., Leemans, R., Monserud, R.A., Solomon, A.M. (1992) A global biome model based on plant physiology and dominance, soil properties and climate. *Journal of Biogeography* **19**: 117–34.

Quinton, J.N., Govers, G., Van Oost, K., et al. (2010) The impact of agricultural soil erosion on biogeochemical cycling. *Nature Geosciences* **3**: 311–14.

Rasmussen, C., Southard, R.J., & Horwath, W.R. (2008) Litter type and soil minerals control temperate forest soil carbon response to climate change. *Global Change Biology* **14**: 2064–80.

Read, D.J., Leake, J.R., & Perez-Moreno, J. (2004) Mycorrhizal fungi as drivers of ecosystem processes in heathland and boreal forest biomes. *Canadian Journal of Botany* **82**: 1243–63.

Reich, P.B., Hobbie, S.E., Lee, T., et al. (2006) Nitrogen limitation constrains sustainability of ecosystem response to CO_2. *Nature* **440**: 922–5.

Rinnan, R., Michelsen, A., & Baath, E. (2011) Long-term warming of a sub-arctic heath decreases soil bacterial community growth but has no effects on its temperature adaptation. *Applied Soil Ecology* **47**: 217–20.

Ross, D.J., Newton, P.C.D., & Tate, K.R. (2004) Elevated CO_2 effects on herbage and soil carbon and nitrogen pools and mineralization in a species–rich, grazed pasture on a seasonally dry sand. *Plant & Soil* **260**: 183–96.

Roulet, N.T., & Moore, T.R. (1995) The effect of forestry drainage practices on the emissions of methane from northern peatlands. *Canadian Journal of Forest Research* **25**: 491–9.

Sankaran, M., Hanan, N.P., Scholes, R.J., et al. (2005) Determinants of woody cover in African savannas. *Nature* **438**: 846–9.

Sardans, J., Penuelas, J., & Estiarte, M. (2008) Changes in soil enzymes related to C and N cycle and in soil C and N content under prolonged warming and drought in a Mediterranean shrubland. *Applied Soil Ecology* **39**: 223–35.

Sato, H., Itoh, A., & Kohyama, T. (2007) SEIB–DGVM: A new Dynamic Global Vegetation Model using a spatially explicit individual-based approach. *Ecological Modelling* **200**: 279–307.

Schimel, D.S., Braswell, B.H., Holland, E.A., et al. (1994) Climatic, edaphic and biotic controls over storage and turnover of carbon in soils. *Global Biogeochemical Cycles* **8**: 279–93.

Schimel, D.S., Braswell, B.H., McKeown, R., Ojima, D.S., Parton, W.J., & Pulliam, W. (1996) Climate and nitrogen controls on the geography and timescales of terrestrial biogeochemical cycling. *Global Biogeochemical Cycles* **10**: 677–92.

Schimel, J.P., Balser, T.C., & Wallenstein, M. (2007) Microbial stress-response physiology and its implications for ecosystem function. *Ecology* **88**: 1386–94.

Schimel, J.P., & Mikan, C. (2005) Changing microbial substrate use in Arctic tundra soils through a freeze-thaw cycle. *Soil Biology and Biochemistry* **37**: 1411–18.

Schuur, E. A. G. et al. (2008) Vulnerability of permafrost carbon to climate change: implications for the global carbon cycle. *Bioscience* **58**: 701–14.

Schuur, E. A. G. et al. (2009) The effect of permafrost thaw on old carbon release and net carbon exchange from tundra. *Nature* **459**: 556–9.

Seneviratne, S.I., Luethi, D., Litschi, M., & Schar, C. (2006) Land-atmosphere coupling and climate change in Europe. *Nature* **443**: 205–9.

Sharma, S., Szele, Z., Schilling, R., Munch, J.C., & Schloter, M. (2006) Influence of freeze-thaw on the structure and function of microbial communities in soil. *Applied and Environmental Microbiology* **72**: 48–54.

Shaw, M.R., Zavaleta, E.S., Chiariello, N.R., Cleland, E.E., Mooney, H.A., & Field, C.B. (2002) Grassland responses to global environmental changes suppressed by elevated co_2. *Science* **298**: 1987–90.

Simmons, B.L., Wall, D.H., Adams, B.J., et al. (2009) Long-term experimental warming reduces soil nematode populations in the McMurdo Dry Valleys, Antarctica. *Soil Biology and Biochemistry* **41**: 2052–60.

Sitch, S., Huntingford, C., Gedney, N., et al. (2008) Evaluation of the terrestrial carbon cycle, future plant geography and climate-carbon feedbacks using 5 Dynamic Global Vegetation Models (DGVMs). *Global Change Biology* **14**: 2015–39.

Sitch, S. Smith, B., Prentice, I.C., et al. (2003) Evaluation of ecosystem dynamics, plant geography and terrestrial carbon cycling in the LPJ dynamic global vegetation model. *Global Change Biology* **9**: 161–85.

Smith, P. (2001) Soil organic matter modelling. In: R. Lal (ed.) *Encyclopedia of Soil Science*. Marcel Dekker Inc., New York.

Smith, P. (2002) Soil and the environment: Role of soil in models for climate change. In: D. Hillel, C. Rosenzweig, D. Powlson, K. Scow, M. Singer, & D. Sparks (eds.) *Encyclopedia of Soils in the Environment*. Academic Press, London.

Smith, P., Andrén, O., Brussaard, L., Dangerfield, M., Ekschmitt, K., Lavelle, P., & Tate, K. (1998) Soil biota and global change at the ecosystem level: describing soil biota in mathematical models. *Global Change Biology* **4**: 773–84.

Sokolov, A.P., Kicklighter, D.W., Melillo, J.M., Felzer, B.S., Schlosser, C.A., & Cronin, T.W. (2008) Consequences of considering carbon–nitrogen interactions on the feedbacks between climate and the terrestrial carbon cycle. *Journal of Climate* **21**: 3776–96.

Striegl, R. G., Aiken, G. R., Dornblaser, M. M., Raymond, P. A. & Wickland, K. P. (2005) A decrease in discharge-normalized DOC export by the Yukon River during summer through autumn. *Geophysical Research Letters* **32**, L21413, doi:10.1029/2005GL024413.

Ström, L., Mastepanov, M., & Christensen, T.R. (2005) Species-specific effects of vascular plants on carbon turnover and methane emissions from wetlands. *Biogeochemistry* **75**:65–82.

Tarnocai, C. et al. (2009) Soil organic carbon pools in the northern circumpolar permafrost region. *Global Biogeochemical Cycles* **23**, GB2023, doi:10.1029/2008GB003327.

Taylor, A.R., & Wolters, V. (2005) Responses of oribatid mite communities to summer drought: The influence of litter type and quality. *Soil Biology and Biochemistry* **37**: 2117–30.

Thornton, P.E., Doney, S.C., Lindsay, K., et al. (2009) Carbon-nitrogen interactions regulate climate-carbon cycle feedbacks: results from an atmosphere-ocean general circulation model. *Biogeosciences Discussions* **6**: 3303–54.

Thornton, P.E., Lamarque, J.F., Rosenbloom, N.A., & Mahowald, N.M. (2007) Influence of carbon-nitrogen cycle coupling on land model response to CO_2 fertilization and climate variability. *Global Biogeochemical Cycles* **21**: 1–15.

Thornton, P.E., & Rosenbloom, N.A. (2005) Ecosystem model spin-up: Estimating steady state conditions in a coupled terrestrial carbon and nitrogen cycle model. *Ecological Modelling* **189**: 25–48.

Treseder, K.K. (2004) A meta-analysis of mycorrhizal responses to nitrogen, phosphorus, and atmospheric CO_2 in field studies. *New Phytologist* **164**: 347–55.

Trumbore, S. (2006) Carbon respired by terrestrial ecosystems—recent progress and challenges. *Global Change Biology* **12**: 141–53.

Trumbore, S.E., & Czimczik, C.I. (2008) Geology—An uncertain future for soil carbon. *Science* **321**: 1455–6.

van der Putten, W.H., de Ruiter, P.C., Bezemer, T.M., et al. (2004) Trophic interactions in a changing world. *Basic and Applied Ecology* **5**: 487–94.

Walker, M.D., Wahren, C.H., Hollister, R.D., et al. (2006) Plant community responses to experimental warming across the tundra biome. *Proceedings of the National Academy of Sciences of the United States of America* **103**: 1342–6.

Wall, D.H., Bradford, M.A., St John, M.G., et al. (2008) Global decomposition experiment shows soil animal impacts on decomposition are climate-dependent. *Global Change Biology* **14**: 2661–77.

Wan, S., Norby, R.J., Ledford, J., et al. (2007) Responses of soil respiration to elevated CO_2, air warming, and changing soil water availability in a model old-field grassland. *Global Change Biology* **13**: 2411–24.

Ward, S.E., Bardgett, R.D., McNamara, N.P., Adamson, J.K., & Ostle, N.J. (2007) Long-term consequences of grazing and burning on northern peatland carbon dynamics. *Ecosystems* **10**:1069–83.

Wardle, D.A. (2002) *Communities and Ecosystems: Linking the aboveground and belowground components*. Princeton University Press, Princeton, NJ.

Weigelt, A., Bol, R., & Bardgett, R.D. (2005) Preferential uptake of soil N forms by grassland plant species. *Oecologia* **142**: 627–35.

Woodward, F.I., & Lomas, M.R. (2004) Vegetation-dynamics—simulating responses to climate change. *Biological Reviews* **79**: 643–70.

Woodward, F.I., Lomas, M.R., & Betts, R. (1998) Vegetation-climate feedbacks in a greenhouse world. *Philosophical Transactions of the Royal Society B* **353**: 29–39.

Woodward, F.I., Lomas, M.R., & Kelly, C.K. (2004) Global climate and the distribution of plant biomes. *Proceedings of the Royal Society B Biological Sciences* **359**: 1465–76.

Xu R., & Prentice, I.C. (2008) Terrestrial nitrogen cycle simulation with a dynamic global vegetation model. *Global Change Biology* **14**: 1745–164.

Zak, D.R., Pregitzer, K.S., Curtis, P.S., Teeri, J.A., Fogel, R., & Randlett, D.L. (1993) Elevated atmospheric CO_2 and feedback between carbon and nitrogen cycles. *Plant and Soil* **151**: 105–17.

Zanetti, S., Hartwig, U.A., Luscher, A., et al. (1996) Stimulation of symbiotic N_2 fixation in Trifoliumrepens L. under elevated atmospheric pCO_2 in a grassland ecosystem. *Plant Physiology* **112**: 575–83.

Zibilske, L.M., & Bradford, J.M. (2007) Oxygen effects on carbon, polyphenols, and nitrogen mineralization potential in soil. *Soil Science Society of America Journal* **71**: 133–9.

CHAPTER 4.2

The Impact of Nitrogen Enrichment on Ecosystems and Their Services

Peter Manning

4.2.1 Nitrogen—the Earth's most limiting resource?

Despite its global ubiquity as an unreactive gas, nitrogen (N), which makes up ~78% of the atmosphere, is in short supply in most terrestrial ecosystems (Galloway et al. 2008). Only reactive N (Nr) is usable by plants and its natural pathway of entry, N fixation by free living and symbiotic bacteria is constricted by a range of biotic and abiotic limitations. As a result, soil N availability falls short of plant demand, and primary productivity is limited by N availability in almost every biome (Le Bauer & Treseder 2008). Even tropical systems, which are traditionally thought of as phosphorus (P) limited, have been shown to be N limited. This may be strongest in the post-disturbance phase (Le Bauer & Treseder 2008) that will become more prevalent as human disturbance advances into the natural ecosystems of the tropics.

The constraint of N availability on biomass production has had major implications for agricultural practices and the evolution of life. Throughout the history of human agriculture, N fertilization has played a major role in determining food availability, with a wide range of practices emerging as a means of alleviating N limitation and ultimately enhancing food supply (Smil 1991). Evolution has also developed a myriad of solutions for coping with low and variable supplies of this essential nutrient. These solutions include conservative nutrient use strategies and associations with mycorrhizal fungi in plants (Aerts & Chapin 2000) as well as a wide diversity of nutrient foraging mechanisms in soil microbes, including the reorganization of soil structure, which facilitates resource capture and the "mining" of recalcitrant organic matter for N (Craine et al. 2007).

4.2.1.1 Nitrogen enrichment of terrestrial ecosystems—causes and trends

The long history of nutrient limitation in terrestrial ecosystems has seen significant disruption in recent times. It is estimated that inputs of Nr into terrestrial ecosystems have more than doubled since pre-industrial times (Vitousek et al. 1997) because of human activity. Foremost amongst these is the development of the Haber–Bosch process at the end of the 19th century. A whole new pathway of reactive N input was opened up through industrial means of fixing unreactive N to form ammonia; this synthetic process has boosted agricultural productivity immensely and played a major role in the human population explosion. It is estimated that at least 2 billion people would not be alive today without the use of chemical nitrogen fertilizers (Smil 1991). Between 1860 and 1995, human Nr creation rose from 15 to 156 Tg N; by 2005 it had risen again to 187 Tg (Galloway et al. 2008). The application of this N has been notoriously inefficient with only ~50% of fertilizer N being converted into harvested products, and the remainder ending up in either air, water, soils, or vegetation (Galloway et al. 2008).

This direct and intentional fertilization has been accompanied by a rise in a more diffuse form of enrichment, the dry and wet deposition of N.

Soil Ecology and Ecosystem Services. First Edition. Edited by Diana H. Wall *et al.*
© 2012 Oxford University Press. Published 2012 by Oxford University Press.

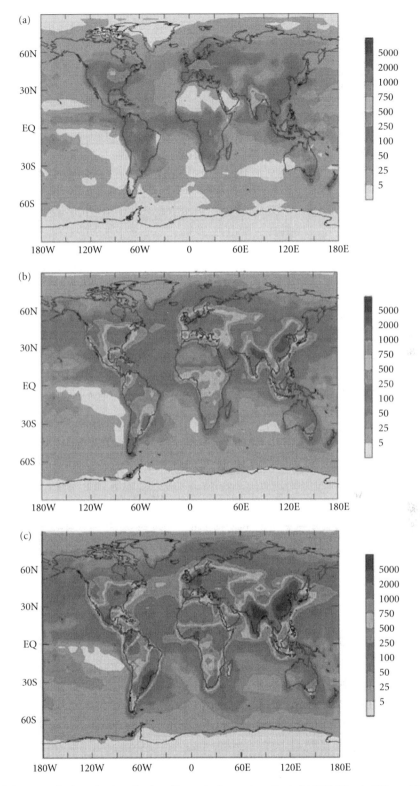

Figure 4.2.1 Global patterns of estimated and predicted total inorganic nitrogen deposition in (a) 1860, (b) early 1990s, and (c) 2050. Values are in mg N m^{-2} year^{-1}. (From Galloway *et al.* 2004, with permission from Springer Science + Business Media B.V.)

Nitrogen deposition originates from a variety of sources, including fossil fuel and biomass burning and volatilization from livestock production and fertilizers (Vitousek et al. 1997). It comes in two main forms: reduced (NH_y) (e.g. NH_4) and oxidized (NO_x), (e.g. NO and NO_2). The total of NH_y and NO_x deposits rose from 34 Tg N year^{-1} in 1860 to 100 Tg in 1995 (Galloway et al. 2004). However, this deposition is spatially uneven, with some areas receiving almost no deposition and others subject to rates in excess of 30 kg N ha^{-1}. Most ammonia is deposited within 5km of the source, but often deposition occurs considerable distances from the source, with upland (high precipitation) areas downwind of urban, industrial and agricultural regions often receiving the highest deposits.

The third major source of increased N inputs is through the planting of leguminous crops, a process termed cultivation induced biological nitrogen fixation (C-BNF). Legumes rarely form monocultures in the wild, but in modern agriculture they often do, and given the large land area occupied by these fields, C-BNF is estimated to have been 40 Tg N in 2005 (Galloway et al. 2008). Organic N inputs also come from manure application, and as many livestock are fed on crops and on fertilized pastureland, some of this will originate from industrially fixed sources.

This dramatic fertilization of both human dominated and natural ecosystems shows little sign of halting. While fertilizer use and fossil fuel burning may be plateauing in the developed world, the emerging world is anticipated to dramatically increase both in the coming century (Galloway et al. 2004; Fig. 4.2.1). Further, rapidly rising meat and dairy production and consumption is expected (Bouwman et al. 2005). Some of the highest emissions of reactive N, and subsequent deposition are reported from animal husbandry. Even if this somewhat profligate consumption of Nr is controlled, the demands of a rising population, and the reduction in land availability indicates that producing sufficient food is unlikely without the widespread use of N fertilizers. Predictions for 2050 are that N deposition will rise to 200 Tg year^{-1} (Galloway et al. 2004).

Because N limits productivity in so many ecosystems, evidence is showing that any increase in its supply can have massive impact on ecosystem composition and functioning (Knorr et al. 2005; Le Bauer and Treseder 2008; Bobbink et al. 2010; Lee et al. 2010). From an ecosystem services perspective, N enrichment has a range of impacts, and the perception of these will differ depending on the land use in question. In agricultural systems for example, the primary service of food production will be increased, possibly causing detriment to other ecosystem services, such as soil carbon (C) storage. This local gain may also come at a cost to the primary services of neighboring ecosystems. Leached fertilizer and N deposition may hamper clean water provision in freshwaters and drift and deposition from Nr sources may cause declines in plant diversity in semi-natural and natural ecosystems that are often used as conservation and amenity lands.

To understand the impacts of N enrichment on ecosystem services such as food and clean water provision, and the maintenance of aesthetic and cultural value it is vital to understand the mechanisms through which N alters the physical, chemical, and biological processes that underlie them (Millennium Ecosystem Assessment 2005). Given that ecosystem responses to N are also highly variable, it is also important to identify the ways in which N impacts are modified by ecosystem properties such as soil characteristics and species composition. This chapter takes an ecosystem approach to the impacts of N enrichment on soils and their ecosystem services, with an emphasis on the somewhat neglected role of compositional change in driving long-term ecosystem responses. First, the immediate effects of N enrichment on soil chemistry and biology will be discussed followed by a description of how species differences in response to N alter the community structure of plants and soil organisms and how these N-induced changes modify ecosystem responses to N. Finally, I relate these changes to ecosystem properties to the ecosystem services they underpin.

4.2.2 Direct impacts of nitrogen enrichment on soil chemistry and plant and microbial metabolism

The range of oxidized, reduced, and organic nitrogen inputs into ecosystems is diverse, with each form differing, sometimes significantly, in its effect on soil properties. Upon entering the system, N may follow a number of pathways. The relative importance of these pathways depends on the form of deposited N and on a number of ecosystem properties including soil chemistry, precipitation rates, and the abundance of various biota. The processes and changes are complex and are reviewed in more detail elsewhere (e.g. Vitousek *et al.* 1997; Bobbink *et al.* 1998; Booth *et al.* 2005; Stevens *et al.* 2011 and the forthcoming European Nitrogen Assessment), but the major routes are outlined here.

Oxidized N forms tend to be the most labile and mobile as nitrate (NO_3^-) converts to nitrite (NO_2^-) and then gaseous nitrogen by denitrification; nitrates also leach readily from the soil, with potential consequences for the eutrophication of freshwaters (Vitousek *et al.* 1997). This movement of nitrate can also bring cations along with it, potentially depleting calcium (Ca) and magnesium (Mg) from the soil and causing the leaching of aluminum (Al) (Vitousek *et al.* 1997). Reduced nitrogen inputs to soils often cause acidification as both biological uptake of ammonium and nitrification produce hydrogen ions. Deposition of nitric acid (HNO_3) can also cause acidification. Acidification can also lead to the loss of trace nutrients and the release of Al, which may come to contaminate drinking water supplies. It is assumed that organic N inputs have fewer immediate effects on the soil as they are generally less mobile than NH_y and NO_x, and are initially only available to specialized soil decomposers. Unless soil mineralization rates are very high, organic N inputs will have more of a gradual effect on soils and ecosystem properties; and yet high levels of organic N inputs still have the potential to cause eutrophication of terrestrial ecosystems. Pastures receiving high manure inputs for example, share many common properties with grasslands subject to inorganic N addition, including raised primary productivity, high leaching rates and reduced plant diversity.

Immobilization (uptake) by microbes is a common fate of added N, and in N limited systems, it will be the fate of most N inputs. In such cases, the turnover rate of the microbial biomass will greatly determine the degree of ecosystem response; where slow, the N may remain sequestered, and other changes may not be manifested. Plant uptake is another major pathway for N inputs, where once acquired, it is either utilized to produce more biomass, to make defensive compounds or stored in what is known as "luxury uptake," thus lowering tissue C:N ratio (Aerts & Chapin 2000).

Nitrogen saturation occurs where N input of any form exceeds plant and microbial uptake and sorption sites are saturated. The capacity to predict saturation is limited, but it is likely at sites that have a history of high N inputs, e.g. via fertilization. Once a system is saturated, N outputs from the system, often in the form of nitrate leaching, are likely to be very high, and thus increasing major risks for lake, stream, and groundwater pollution.

4.2.3 Effects of nitrogen enrichment on plants and the soil biota

It is clear that the enrichment of any N limited ecosystem will favor species typically found in more naturally fertile conditions, but this belies a great underlying complexity that makes the prediction of community response to N enrichment inexact (Wardle 2002; Pennings *et al.* 2005; Tylianakis *et al.* 2008). This section discusses the effects of N enrichment on the biota of ecosystems and the ecosystem functions they control. I focus on N limited, as opposed to saturated ecosystems, which will see relatively little change in their community composition under further N enrichment.

4.2.3.1 Plants

A large body of literature describes the often-pronounced changes that occur in plant communities as a response to N enrichment. Perhaps the most widely reported consequences of N enrichment are declines in plant species richness and evenness (Suding *et al.* 2005; Bobbink *et al.* 2010). There are exceptions to this trend (e.g. in the case of the beta diversity of nutrient poor sites) (Chalcraft *et al.* 2008); however, the vast majority of experimental studies report declines (Bobbink *et al.* 2010) with rare species and legumes being particularly likely to undergo local extinction (Suding *et al.* 2005).

Despite the general consistency in overall plant community response there is currently much debate

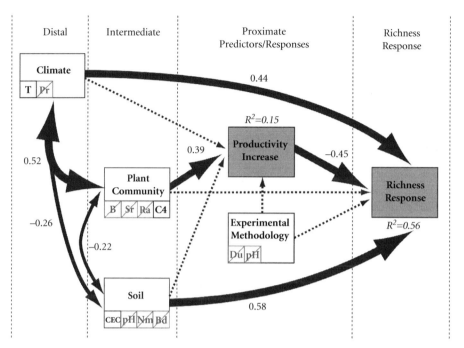

Figure 4.2.2 Structural equation model of plant community responses to N enrichment. Numbers correspond to partial regression coefficients for unidirectional arrows and correlations for bidirectional arrows. Arrow sizes approximate the strength of relationship (non-significant pathways are dashed). Predictors are crossed out if not included in the final model. Also shown is the amount of variance explained (R^2) in the productivity increase, and the richness response. Distal predictors included regional climate properties (T, temperature; Pr, precipitation). Intermediate predictors included plant community properties (B, standing biomass; Sr, species richness; Ra, proportion of species that were rare; FG, relative abundance of different functional groups), and soil properties (CEC, soil cation exchange capacity; pH; Nm, monthly net N mineralization averaged over the growing season; Bd, bulk density). All intermediate properties are based on average conditions in control plots. Proximate predictors include the relative increase in standing aboveground biomass following fertilization and specific experimental conditions (Du, duration of study; Ap, application rate of fertilizer). (Redrawn from Clark et al. 2007., with permission from John Wiley & Sons.)

over the mechanisms responsible, with proposed causes including ammonium toxicity, acidification, light exclusion, increased susceptibility to secondary stress factors, and changes to plant–soil feedback (Bobbink et al. 1998; Suding et al. 2008; Bobbink et al. 2010). It is likely that each of these mechanisms operates under certain conditions, but that its relative importance depends upon the dominant form of N deposition and environmental context (Pennings et al. 2005; Stevens et al. 2011).

Where NH_y is the dominant form of N deposition and nitrification rates are low, NH_4 may accumulate to toxic levels (Stevens et al. 2011). While species found in such soils may be naturally tolerant of NH_4, they are likely to differ in their tolerance within the community, with the least tolerant being lost first. This may mean that the identity of the species lost may differ depending on community context. In acid grasslands and heathlands, Kelijn et al. (2008) found that typically rare plant species were absent in sites with high NH_4 concentrations.

In weakly acid soils (pH ~5) with low buffering capacity and NH_y inputs, nitrification and plant uptake of NH_4 will acidify the soil, making conditions unsuitable for some species (Bobbink et al. 1998; Stevens et al. 2011). The multisite path analysis of Clark et al. (2007; Fig. 4.2.2) supports this view; they found that cation exchange capacity, which can be viewed as an indicator for potential acidification, showed a strong relationship with species declines in response to N.

In soils where nitrification rates and buffering capacity are high, NH_4 will not accumulate, acidification will be weak, and the cause of community change is more likely to be related to fertilization. Nitrogen fertilization stimulates the growth of almost any plant species when N is limiting. At the community level, this stimulation is observed as an increase in shoot and root biomass with a relative decrease in belowground allocation (Xia & Wan 2008; Lee et al. 2010), which masks a great variation in plant species responses. The overall gain in aboveground productivity shows a strong relationship with species declines (Clark et al. 2007) as canopy-forming species monopolize light whereby shorter and less shade-tolerant species decline. The effects of light exclusion appear to be particularly strong for understory species and seedlings (Hautier et al. 2009), with those of smaller seeded species being less able to tolerate N induced canopy closure (Manning et al. 2009). Such losses may be highest where only a few species already present in the community with high growth rates are able to dominate under the more fertile conditions, and other species sharing similar traits are unable to disperse into the community.

An increase in N also has the potential to increase the N content of plant material (Xia & Wan 2008). This change in litter quality may be the first step in generating a positive feedback on N supply that further favors nitrophilous species (Fig. 4.2.3). Litters with lower C:N ratio tend to decompose faster and accelerate N mineralization rates (Aerts & Chapin 2000; Manning et al. 2008), thus further increasing N supply to plants and microbes. As increases in the N supply rate often favor plant species with inherently high quality litters and fast turnover of tissues (Chapin 1980), N supply rates are increased further. One of the best known examples of such a change is in the Dutch heaths, where very high levels of N deposition have displaced dwarf-shrubs, while favoring nitrophilous grasses (Aerts & Berendse 1988). The dwarf-shrubs possess slow growth rates and a relative preference for organic N sources, while having high tissue concentrations of lignin and plant secondary compounds that make their plant litter slow to decompose, thus resulting in the slow nutrient mineralization rates at which they compete well. In contrast, the grass species are more rapidly growing, have a relative preference for inorganic N uptake and produce more decomposable plant materials that result in faster rates of soil nutrient turnover (Aerts & Chapin 2000). While described here as a specific example, this dichotomy represents a widespread plant strategy trade-off (Hobbie 1992), an important exception being the N fixing legumes, which typically perform well at low nutrient supply, but have a high tissue N content (Craine et al. 2002). Further, similar patterns of plant composition mediated changes in function in response to N enrichment have been observed in a range of ecosystems including grasslands (Wedin & Tilman 1996). These can be viewed as two positive feedbacks, and hence alternate ecosystem states that can be disrupted by a change in nutrient supply or loss (Fig. 4.2.3). Initially, these feedbacks may stabilize the system rendering it less susceptible to change, but if pushed beyond a threshold level of N supply, the system may be pushed into the alternate state (Suding et al. 2008). Legume rich systems may form an exception to this trend for positive feedbacks in compositional responses to N enrichment. The idea is unexplored, but it is possible that the decline in N fixation counteracts N enrichment, dampening the shift towards a saturated and N leaky system.

Switches in dominance between functionally distinct species are likely to have a large impact on services underpinned by ecosystem functions such as nutrient cycling and C storage. However, changes to ecosystem service supply may even be observed in cases of rare and subdominant species loss. It is not unusual for these species to have a high aesthetic appeal, and so these local extinctions represent an erosion of ecosystem services relating to recreational and spiritual value. Extinctions may also affect ecosystem stability in a way that is not immediately apparent in short-term experiments. Nitrogen-induced species losses were seen to reduce ecosystem stability in a long-term grassland experiment where the productivity of species poor plots was less drought resistant (Tilman & Downing 1994).

4.2.3.2 Saprotrophic fungi and bacteria

Another commonly observed response to N enrichment is a shift from fungal to bacterial dominance in the soil decomposer community (e.g. Frey et al. 2004). This can also be accompanied by an increase

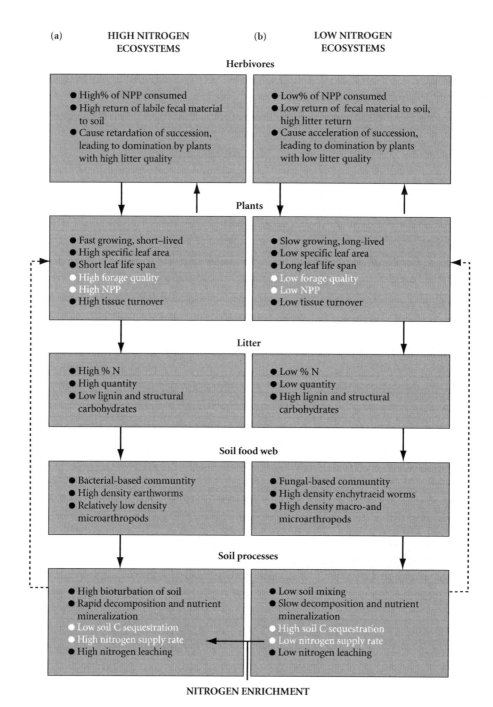

Figure 4.2.3 Simplified conceptual model showing high nitrogen and low nitrogen ecosystem states. Plant and soil communities can form self-reinforcing ecosystem states in which low or high rates of nutrient cycling are maintained. There is evidence that nitrogen enrichment may transform low nitrogen ecosystems into those with more rapid rates of C and N cycling, with changes to several ecosystem services, which are shown in white type, occurring as a result. (Figure adapted from Wardle *et al.* 2004, reprinted with permission from AAAS.)

in bacterial feeding nematodes and their predators (Wardle 2002). These bacterial food webs that come to dominate in fertile conditions are characterized by higher rates of nutrient loss, faster nutrient cycling, and lower levels of soil organic matter than the fungal dominated food webs they replace (van der Heijden *et al.* 2008; Fig. 4.2.3). The shift from fungi to bacterial dominance in response to N shares a number of parallels with the changes in plant communities that were described previously, and it will often occur alongside it (Wardle 2002; Fig. 4.2.3). In both cases, N enrichment shifts the community towards a functional group that utilizes inorganic N forms and is more readily consumed by grazing animals and decomposers, thus accelerating the turnover of biomass and nutrients.

This compositional shift, along with an alteration of enzyme production, explains (and is also explained by) the often-strong alteration of organic N and C turnover that N enrichment causes. Fog (1988) described the tendency for high quality materials to decompose faster following N deposition and for recalcitrant materials to decompose slowly in terms of the microbial community and its activities. Under N limited conditions microbes produce enzymes (e.g. phenol oxidase) that aid in the "mining" of recalcitrant soil C for N (Frey *et al.* 2004; Knorr *et al.* 2005; Craine *et al.* 2007), while under higher N availability cellulosic decomposers and the production of enzymes (e.g. cellulase) promote the degradation of rapid turnover materials (Manning *et al.* 2008). This difference between limited and higher available N is linked to the often observed change in fungal to bacterial ratios, as relatively speaking, fungi are superior at the former activity while bacteria perform better at utilizing inorganic and simple organic materials.

The pattern for faster turnover of rapidly decomposing materials and slower decomposition of more recalcitrant materials has proven remarkably consistent across the world's ecosystems, and has now been observed in tropical forest soils (Cusak *et al.* 2010; Fig. 4.2.4a) and in temperate zone forests and grasslands that constitute the bulk of N enrichment studies. These patterns need to be incorporated with knowledge of compositional shifts in plant communities when considering the effects of N enrichment on soil C sequestration. General patterns have not yet been identified, but it is likely that in non-woody vegetation there will be a shift towards species with more rapidly decomposing litters that will result in either a net neutral effect or an overall reduction in C stocks (Manning *et al.* 2008), while in forests, the compositional change will be much slower. Here, a shift from fine root to stem production will cause a direct stimulation of woody litter that is likely to increase inputs of recalcitrant materials and enhance C sequestration (Magnani *et al.* 2007). Where compositional change occurs, it is likely that biomass (and C stock) increases will be seen aboveground while reduced root allocation and higher turnover of soil organic matter will reduce belowground C stocks.

4.2.3.3 Mycorrhiza

The relationship between plants and mycorrhizal fungi hangs in a delicate balance between mutual benefit and parasitism. Under conditions of high N and P availability, plants have less to gain from allocating resources belowground and to a mycorrhizal partner, and root colonization is often reduced in both arbuscular (AMF) and ectomycorrhizal fungi (Johnson *et al.* 2003; Treseder 2004). Where the association remains it may become less mutualistic, with more parasitic fungi increasing in abundance (Johnson 1993).

These changes may play a part in plant community responses to N enrichment, but can also affect soil properties and the ecosystem services related to them. AMF exude glycoproteins such as glomalin that directly contribute to the stable soil C fraction and enhance soil aggregation. Declining AMF abundance following N enrichment, which averages 15% in field studies (Treseder 2004; Fig. 4.2.4b) but may be stronger in P rich soils (Johnson *et al.* 2003), can therefore result in an ecosystem that has weaker soil aggregation and lower C and N sequestration (Wilson *et al.* 2009). While there is relatively little direct evidence for this relationship at present, and Wilson *et al.* (2009) found that N fertilization *increased* soil AMF hyphae, Manning *et al.* (2006) found that N

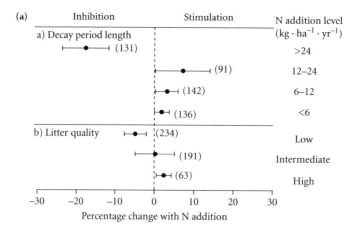
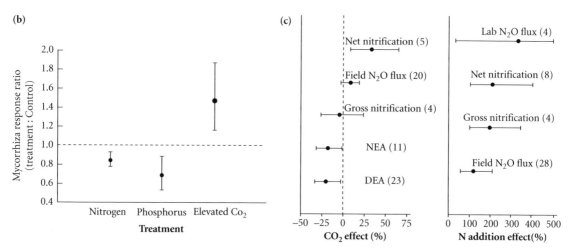

Figure 4.2.4 Meta-analyses of N addition impacts upon ecosystem properties. a) Response of litter mass remaining to N additions when the data were grouped by i) the length of the litter decay period (in months), and ii) initial litter quality (low, intermediate, high). (Redrawn from Knorr et al. 2005, with permission from the Ecological Society of America.) b) The effects of CO_2, phosphorus, and nitrogen addition on mycorrhizal abundance A response ratio >1 indicates an increase in abundance relative to the control, and <1 indicates a decrease. (Redrawn from Treseder 2004, with permission from John Wiley & Sons.) c) Shows the effects of CO_2 and nitrogen enrichment on nitrification and denitrification measures. (Redrawn from Barnard et al. 2005, with permission from the American Geophysical Union.) On a), b), and c) bars show 95% confidence interval of the effect size, in a) and c) based on the number of experiments indicated in parentheses.

enrichment reduced AMF root colonization directly but also triggered compositional change towards less mycorrhizal species. These declines in AMF abundance closely corresponded with a reduction in mineral associated C. Legumes are also a strongly mycorrhizal functional group that commonly declines in response to N enrichment (Suding et al. 2005), and so both direct (reduced allocation) and composition mediated changes to function that operate via mycorrhiza may occur frequently in nature.

4.2.3.4 Multitrophic impacts above- and belowground

By influencing the physiology and community structure of plants and microbes N enrichment has bottom-up effects on both aboveground and belowground food webs. Aboveground there are often changes to biomass availability and the quality of this material in terms of both nutritional content and secondary defense compounds (Chen &

Ruberson 2010). While the latter can lead to negative effects, the general trend is towards increased aboveground herbivory in N enriched systems (Tylianakis *et al.* 2008). The increased quality and quantity of plant materials may also explain the increased severity of pathogen infection that tends to occur in response to N enrichment, and the transfer of food quantity and quality effects from lower trophic levels may explain the general increase in parasitoid effects on hosts under N fertilization. Changes in plant community composition will also undoubtedly influence invertebrate community structure, e.g. via the extinction of larval food plant species. At higher trophic levels natural enemies might also be affected by changes in vegetation structure but neither of these areas has been investigated in any great detail (Chen & Ruberson 2010).

Belowground there is also evidence that the stimulation of plant litter inputs by N can increase fauna abundances at higher trophic levels of the food web (e.g. Collembola via bottom up control; Manning *et al.* 2006). However, observed effects in these poorly understood food webs are very variable (Tylianakis *et al.* 2008). Bottom-up effects will only be seen where N has increased the quantity and quality of plant and microbial food resources, and the strength of N effect on any particular functional feeding group may be modified by its position in the food web, with top-down and bottom-up regulated organisms showing divergent responses (Wardle 2002).

4.2.4 Net effects on ecosystem services

How does the suite of changes described in the previous section ultimately affect the net levels of higher-level ecosystem functions and services? This section outlines a number of functions and services such as total C sequestration, food and forage production, clean water supply, human health, and greenhouse gas emissions and the effect that N enrichment has upon them.

4.2.4.1 Greenhouse gases and carbon sequestration

The overall effect of N enrichment on the C balance of ecosystems, and therefore its capacity to modify climate change, is difficult to quantify. Nitrogen boosts plant growth, but the quality, quantity, and subsequent decomposition of litter inputs can also be strongly affected. What is the net effect of these changes at the ecosystem level and how does it vary between ecosystems? It is hypothesized that overall C storage will either decline or remain largely unchanged in ecosystems limited by other factors (e.g. deserts), and in those with a low standing biomass and fast turnover of N (e.g. in agricultural grasslands). However, it may be increased in systems with a high standing biomass and slow turnover of plant biomass that produces recalcitrant litter materials (e.g. forests). In the former case, biomass increases and the faster turnover and decomposition of this biomass will effectively cancel out. In the latter, a shift to aboveground allocation of lignified materials (e.g. from fine roots to tree trunks), and a slower decomposition of these materials, will lead to a net gain in C sequestration. This allocation shift may also occur in non-forest systems, but the differences in the decomposability between aboveground and belowground parts are less pronounced here, and so it may have less impact on C dynamics. Largely neutral effects of N addition on C balance have been seen in temperate steppe environments (Niu *et al.* 2010), while evidence for the latter is provided by the multisite comparison of Magnani *et al.* (2007), who found that the overall C balance of European forests had shifted towards significant C sequestration in areas of high N deposition. However, these results have proven rather controversial, with others proposing that important covariates (e.g. site fertility) were omitted from their analysis, and that the real overall C sequestration response is much smaller (see communications arising from Magnani *et al.* 2007).

When considering the effects of N on global climate change, the role of other greenhouse gases should not be forgotten. Nitrous oxide is also a powerful greenhouse gas and a significant contributor to global climate change; the IPCC (2007) estimate that the atmospheric concentration of nitrous oxides has risen 44 ppb since 1750, causing a 0.15 W m^{-2} increase in radiative forcing. Further N enrichment is likely to increase emissions of this gas further, as experimental field studies show a consistent rise in nitrification and N$_2$O emission in response to N addition (Barnard *et al.* 2005; Fig. 4.2.4c).

4.2.4.2 Food, biomass, and forage production, clean water supply, and human health

The immediate benefits seen in food, biomass and forage production following N addition are the main reason that N inputs have increased so much in recent years. Nitrogen enrichment unquestionably increases aboveground productivity (Le Bauer & Treseder 2008; Xia and Wan 2008; Lee *et al.* 2010) and gives great benefits to agricultural yields. While sustenance and economic development of world's growing population is probably impossible without further application and release of Nr, this application must be managed carefully if other ecosystem services are not to be severely damaged. It is well known that much N is applied inefficiently, and that the same benefits could be realized with much less environmental impact if N applications were applied more efficiently (e.g. through the use of slow-release fertilizers, precision agriculture and the planting of more nutrient efficient crop genotypes) (Good *et al.* 2004).

If not well managed, N addition to N saturated sites and soils with a low retention capacity results in substantial N losses to freshwater and groundwater, which can cause eutrophication and substantial damage to freshwater ecosystems and their services (Vitousek *et al.* 1997). Recent evidence suggests that the extent of this leaching could be modified by plant community composition. Dijkstra *et al.* (2007) found that N leaching losses were lower in species rich grasslands, presumably as a result of more complete resource use, while Phoenix *et al.* (2008) found that N leaching in limestone grassland mesocosms was lowest in species-rich and perennial grass dominated communities and high in those dominated by forbs and sedges. Such findings may help to inform management, e.g. the planting of species rich mixtures when creating buffer vegetation strips between agricultural fields and freshwaters, a common practice in agri-environment schemes.

The effects of Nr increases on human health are also still poorly understood, but there is a large and growing literature to support the hypotheses that tropospheric ozone production from the production of nitrous oxides can exacerbate pulmonary disease. In addition, high nitrate in drinking water can influence cancer risk and cause methemoglobinemia. However, by providing improved nutrition for billions Nr has no doubt prevented illness via improved immune responses (Galloway *et al.* 2008).

4.2.4.3 Biodiversity conservation and nitrogen management

Nitrogen enrichment generally causes biodiversity declines in both terrestrial and aquatic habitats and so, ecosystem services provided by biodiversity conservation, such as genetic resources and recreation and aesthetic benefits, are adversely affected by N enrichment (Millennium Ecosystem Assessment 2005). This puts their management with respect to N at odds with food and biomass production, and possibly C sequestration.

This trade-off further stresses the importance of improving the management of N inputs so that the benefits of N enrichment seen in biomass and food production are retained, while the negative impacts on biodiversity and water quality are minimized. The management of N sources as well as sites should be carefully considered to formulate optimal management of Nr. Large benefits can be realized through the improved management of animal husbandry, increased efficiency of fertilizer application and better sewage treatment. Where N enrichment has already occurred or is inevitable, there are a number of management options available that may help shift the ecosystem back towards a more slowly cycling and fungal dominated state. These include traditional measures such as the removal of mown biomass but also the addition of organic C, which increases the C:N ratio of the soil environment and so can induce microbial communities to immobilize N. Several studies have reported successful reduction of soil N availability and a reduction in nitrophillous plant species as a result of C addition in the form of sucrose, sawdust, starch, and cellulose (e.g. Eschen *et al.* 2007).

New opportunities may also arise through the manipulation of plant species composition to modify and limit N impacts so that the delivery of multiple ecosystem services is maximized. For instance, it may be possible to hamper the transition towards leaky fast turnover systems by removing or preventing the invasion of species that accelerate the transition (e.g. the nitrophilous grasses of heaths). Similarly, the encouragement of species that produce recalcitrant litter materials may be an option if higher C sequestration under high N deposition is desired.

4.2.5 Conclusion and future directions

The nature of the most pressing issues in N enrichment research is changing. The major mechanisms through which N enrichment alters ecosystems have probably now been identified in temperate zone ecosystems. Major challenges remain in understanding how N enrichment impacts higher level ecosystem services. Soil scientists tend to study a subset of direct effects in controlled studies that may give very different results to whole ecosystem studies (e.g. see Barnard et al. 2005). This understanding must now be linked, and the relative importance of each mechanism must be estimated in a range of ecosystems. The fact that direct effects have been the focus of soil science's interest in N enrichment is perhaps unsurprising; it is easier (and cheaper) to do short, small scale lab incubation experiments than to set up large, long-term field experiments. But a more holistic approach is necessary if we are to understand the impacts of N enrichment at the whole ecosystem scale, at which ecosystem services are utilized. Researchers must adopt a holistic but mechanistic approach to N impacts (e.g. by considering changes in composition and in the quality of litter inputs) (Manning et al. 2006, 2008) when investigating the effects of N on C turnover as well as its immediate effects on the soil. Full budgeting of ecosystem N and C, and estimates of N turnover in various pools are also required; the fate of ~65% of Nr is largely unknown (Galloway et al. 2008). General rules regarding the modification of N response by ecosystem properties have been identified by the use of multisite studies and meta analysis (e.g. Suding et al. 2005; Barnard et al. 2005; Knorr et al. 2005; Magnani et al. 2007; Lee et al. 2010), thus highlighting the future potential of this approach. Understanding the fate of Nr and the relative importance of the mechanisms through which it alters ecosystems that would enable better and new management strategies to be devised (e.g. by altering the properties that control ecosystem response).

Greater study of N impacts in the emerging world is also needed. Population and economic growth in areas such as China, India, and Brazil means huge increases in N inputs are almost certain in the next few decades (Galloway et al. 2004), yet studies of N impacts in these regions are sorely lacking. Many biodiversity hotspots coincide with regions where N inputs are expected to increase (Phoenix et al. 2006) making research of the ecological impacts of N enrichment a priority in these areas. The fundamental trade-off in ecosystem service supply, that between food production and a range of other ecosystem services, also needs to be investigated more thoroughly. At what level of N application is overall benefit to humanity maximized, and how does the spatial configuration of N application and our relative valuation of each service affect this estimate? Such questions are unlikely to yield their answers easily, and highly integrated interdisciplinary research seems the only possible solution.

Finally, it is important that researchers studying N enrichment ensure that its impacts are not ignored, relative to more prominent environmental issues such as climate change. There is considerable evidence that the effects of N enrichment are as strong and in some cases stronger, than those of CO_2 enrichment and global climate change (although the number of direct comparisons is limited, Treseder 2004; Barnard et al. 2005; Lee et al. 2010; Fig. 4.2.4b,c). There is little awareness of this within the scientific community, let alone the public. These changes will not occur in isolation, however, and the impacts of N enrichment should not be viewed singly. A growing number of studies report very strong interactions between N enrichment and other environmental change factors, including CO_2 enrichment and fire frequency (e.g. Dijkstra et al. 2007). Identifying general patterns in these interactions will greatly aid in our prediction and management of future ecosystem services.

References

Aerts, R., & Berendse, F. (1988) The effect of increased nutrient availability on vegetation dynamics on wet heathland. *Vegetatio* **76**: 63–9.

Aerts, R., & Chapin III, F.S. (2000) The mineral nutrition of wild plants revisited: a re-evaluation of processes and patterns. *Advances in Ecological Research* **30**: 1–67.

Barnard, R., Leadley, P.W., & Hungate, B.A. (2005) Global change, nitrification, and denitrification: A review. *Global Biogeochemical Cycles* **19**: GB1007.

Bobbink, R., Hicks, K., Galloway, J., et al. (2010) Global assessment of nitrogen deposition effects on terrestrial plant diversity: a synthesis. *Ecological Applications* **20**: 30–59.

Bobbink, R., Hornung, M., & Roelofs, J.G.M. (1998) The effects of air-borne nitrogen pollutants on species diversity in natural and semi-natural European vegetation. *Journal of Ecology* **86**: 717–38.

Booth, M.S., Stark, J.M., & Rastetter, E. (2005) Controls on nitrogen cycling in terrestrial ecosystems: A synthetic analysis of literature data. *Ecological Monographs* **75**: 139–57.

Bouwman, A.F., Ven der Hoek, K.W., Eikhout, B., & Soenarie, I. (2005) Exploring changes in world ruminant production systems. *Agricultural Systems* **84**: 121–53.

Chalcraft, D.R., Cox, S.B., Clark, C., et al. (2008) Scale-dependent responses of plant biodiversity to nitrogen enrichment. *Ecology* **89**: 2165–71.

Chapin, F.S. (1980) The mineral nutrition of wild plants. *Annual Review of Ecology and Systematics* **11**: 233–60.

Chen, Y., & Ruberson, J.R. (2010) Effects of nitrogen fertilisation on tritrophic interactions. *Arthropod-Plant Interactions* **4**: 81–94.

Clark, C.M., Cleland, E.E., Collins, S.L., et al. (2007) Environmental and plant community determinants of species loss following nitrogen enrichment. *Ecology Letters* **10**: 596–607.

Craine, J.M., Morrow, C., & Fierer, N. (2007) Microbial nitrogen limitation increases decomposition. *Ecology* **88**: 2105–13.

Craine, J.M., Tilman, D., Wedin, D., Reich, P., Tjoelker, M., Knops, J. (2002) Functional traits, productivity and effects on nitrogen cycling of 33 grassland species. *Functional Ecology* **16**: 563–74.

Cusak, D.F., Torn, M.S., McDowell, W.H., & Silver, W.L. (2010) The response of heterotrophic activity and carbon cycling to nitrogen additions and warming in two tropical soils. *Global Change Biology* **16**: 2555–72.

Dijkstra, F.A., West, J.B., Hobbie, S.E., Reich, P.B., & Trost, J. (2007) Plant diversity, CO2, and N influence inorganic and organic N leaching in grasslands. *Ecology* **88**: 490–500.

Eschen, R., Mortimer, S.R., Lawson, C.S., et al. (2007) Carbon addition alters vegetation composition on ex-arable fields. *Journal of Applied Ecology* **44**: 95–104.

Fog, K. (1988) The effect of added nitrogen on the rate of decomposition of organic-matter. *Biological Reviews* **63**: 433–62.

Frey, S.D., Knorr, M., Parrent, J.L., & Simpson, R.T. (2004) Chronic nitrogen enrichment affects the structure and function of the soil microbial community in temperate hardwood and pine forests. *Forest Ecology and Management* **196**: 159–71.

Galloway, J.N., Dentener, F.J., Capone, D.G., et al. (2004) Nitrogen cycles: past, present, and future. *Biogeochemistry* **70**: 153–226.

Galloway, J.N., Townsend, A.R., Erisman, J.W., et al. (2008) Transformation of the nitrogen cycle: Recent trends, questions and potential solutions. *Science* **320**: 889–92.

Good, A.G., Shrawat, A.K., & Muench, D.G. (2004) Can less yield more? Is reducing nutrient input into the environment compatible with maintaining crop production? *Trends in Plant Science* **9**: 1360–85.

Hautier, Y., Niklaus, P.A., & Hector, A. (2009) Competition for light causes plant biodiversity loss after eutrophication. *Science* **324**: 636–8.

Hobbie, S.E. (1992) Effects of plant species on nutrient cycling. *Trends in Ecology and Evolution* **7**: 336–9

IPCC (2007) *Contribution of Working Group I to the Fourth Assessment Report of the Intergovernmental Panel on Climate Change*. Solomon, S., Qin, D., Manning, M., et al (eds.). Cambridge University Press, Cambridge.

Johnson, N.C. (1993) Can fertilisation of soil select less mutualistic mycorhizae? *Ecological Applications* **3**: 749–57.

Johnson, N.C., Rowland, D.L., Cordiki, L., Egerton-Warburton, L.M., & Allen, E.B. (2003) Nitrogen enrichment alters mycorrhizal allocation at five mesic to semiarid grasslands. *Ecology* **84**: 1895–908.

Kleijn, D., Bekker, R.M., Bobbink, R., De Graaf, M.C.C., Roelofs, J.G.M. (2008) In search for key biogeochemical factors affecting plant species persistence in heathland and acidic grasslands: a comparison of common and rare species. *Journal of Applied Ecology* **45**: 680–7.

Knorr, M., Frey, S., & Curtis, P. (2005) Nitrogen additions and litter decomposition: a meta-analysis. *Ecology* **86**: 3252–7.

LeBauer, D.S., & Treseder, K.K. (2008) Nitrogen limitation of net primary productivity in terrestrial ecosystems is globally distributed. *Ecology* **89**: 371–9.

Lee, M., Manning, P., Rist, J., Power, S.A., & Marsh, C. (2010) A global comparison of grassland biomass responses to CO_2 and nitrogen enrichment. *Philosophical Transactions of the Royal Society Series B* **365**: 2047–56.

Magnani, F., Mencuccini, M., Borghetti, M., et al. (2007) The human footprint in the carbon cycle of temperate and boreal forests. *Nature* **447**: 848–450.

Manning, P., Houston, K., & Evans, T. (2009) Shifts in seed size across experimental nitrogen enrichment and plant density gradients. *Basic and Applied Ecology* **10**: 300–8.

Manning, P., Newington, J.E., Robson, H.R., et al. (2006) Decoupling the direct and indirect effects of nitrogen deposition on ecosystem function. *Ecology Letters* **9**: 1015–24.

Manning, P., Saunders, M., Bardgett, R.D., et al. (2008) Direct and indirect effects of nitrogen deposition on litter decomposition. *Soil Biology and Biochemistry* **40**: 688–98.

Millennium Ecosystem Assessment (2005) *Ecosystems and Human Well-being: Biodiversity Synthesis*. World Resources Institute, Washington, DC.

Niu, S., Wu, M., Han, Y., et al. (2010) Nitrogen effects on net ecosystem carbon exchange in a temperate steppe. *Global Change Biology* **16**: 144–55.

Pennings, S.C., Clark, C.M., Cleland, E.E., et al. (2005) Do individual plant species show predictable responses to nitrogen addition across multiple experiments? *Oikos* **110**: 547–55.

Phoenix, G.K., Hicks, K., Cinderby, S., et al. (2006) Atmospheric nitrogen deposition in world biodiversity hotspots: the need for a greater global perspective in assessing N deposition impacts. *Global Change Biology* **12**: 470–6.

Phoenix, G.K., Johnson, D., Grime, J.P.G., & Booth, R.E. (2008) Sustaining ecosystem services in ancient limestone grassland: importance of major component plants and community composition. *Journal of Ecology* **96**: 894–902.

Smil, V. (1991) Population growth and nitrogen: an exploration of a critical existential link. *Population and Development Review* **17**: 569–601.

Stevens, C.J., Manning, P., van den Berg, L.J.L., et al. (2011) Ecosystem responses to reduced and oxidised nitrogen inputs in European terrestrial habitats. *Environmental Pollution* **159**(3): 665–76.

Suding, K.N., Ashton, I.W., Bechtold, H., Bowman, W.D., Mobley, M.L., & Winkleman, R. (2008) Plant and microbe contribution to community resilience in a directionally changing environment. *Ecological Monographs* **78**: 313–29.

Suding, K.N., Collins, S.L., Gough, L., et al. (2005) Functional- and abundance-based mechanisms explain diversity loss due to N fertilization *Proceedings of the National Academy of Sciences of the United States of America* **102**: 4387–92.

Tilman, D., & Downing, J.A. (1994) Biodiversity and stability in grasslands. *Nature* **367**: 363–5.

Treseder, K.K. (2004) A meta-analysis of mycorrhizal responses to nitrogen, phosphorus, and atmospheric CO_2 in field studies. *New Phytologist* **164**: 347–55.

Tylianakis, J.M., Didham, R.K., Bascompte, J., & Wardle, D.A. (2008) Global change and species interactions in terrestrial ecosystems. *Ecology Letters* **11**: 1351–63.

van der Heijden, M.G.A., Bardgett, R.D., & van Straalen, N.M. (2008) The unseen majority: soil microbes as drivers of plant diversity and productivity in terrestrial ecosystems. *Ecology Letters* **11**: 296–310.

Vitousek, P.M., Aber, J.D., Howarth, R.W., et al. (1997) Human alteration of the global nitrogen cycle: sources and consequences. *Ecological Applications* **7**: 737–50.

Wardle, D.A. (2002) *Communities and Ecosystems: Linking the Abovegound and Belowground Components*, pp. 277–81. Monographs in Population Biology 34, Princeton University Press, Princeton, NJ.

Wardle, D.A., Bardgett, R.D., Klironomos, J.N., Setälä, H., van der Putten, W.H., & Wall, D.H. (2004) Ecological linkages between aboveground and belowground biota. *Science* **304**: 1634–7

Wedin, D.A., & Tilman, D. (1996) Influence of nitrogen loading and species composition on the carbon balance of grasslands. *Science* **274**: 1720–3.

Wilson, G.W., Rice, C.W., Rillig, M.C., Springer, A., & Hartnett, D.C. (2009) Soil aggregation and carbon sequestration are tightly correlated with the abundance of arbuscular mycorrhizal fungi: results from long-term field experiments. *Ecology Letters* **12**: 452–61.

Xia, J.Y., & Wan, S.Q. (2008) Global response patterns of terrestrial plant species to nitrogen addition. *New Phytologist* **179**: 428–39.

CHAPTER 4.3

Urbanization, Soils, and Ecosystem Services

Mitchell A. Pavao-Zuckerman

4.3.1 Introduction to urbanization and soils in cities

We are living in an ever increasingly urbanized world, where greater than half of the world's population currently lives in an urban landscape and population growth models predict this trend continuing throughout this century (Millennium Ecosystem Assessment 2005). While this marks a dramatic transformation for the human population with respect to economics, politics, and social interactions, there are also highly significant environmental changes that accompany urbanization that impact human well-being through direct and indirect influences on the ecosystem services that cities provide. Despite the seeming disconnection that cities have from nature, there are very real linkages between urban populations and ecological patterns and processes that become very apparent through the lens of ecosystem services.

The notion that cities provide ecosystem services can at first thought seem contradictory. Cities for many years have been considered extractive systems reliant upon the landscapes around them (Cronon 1991), with ecological footprints that far exceed non-urban ecosystems (Rees 1997). However, even within the densest urban centers, there remains the presence of remnant ecosystem patches as well as living systems that were designed to function within the city. Urban theoretical perspectives in the last decade have developed around the consideration of cities as urban ecosystems (Pickett et al. 2001). Thus, cities can be viewed as ecosystems that do provide ecosystem services (Bolund & Hunhammar 1999). This has the potential to complicate our view of an urban landscape, because at the same time that they potentially generate ecosystem services at the local level, cities have significant negative impacts on ecosystem services from a regional perspective. In this chapter, I will highlight the impacts on, constraints for, and potential to develop ecosystem services in cities using soil-based services as examples. First, I review the impacts cities have on soils, distinguishing between direct and indirect effects that are important for soil ecosystem services. Then I discuss specific urban soil-based ecosystems services within the framework of the Millennium Ecosystem Assessment. I conclude with a discussion of management implications that are important for both the study and promotion of urban soil ecosystem services.

4.3.2 Urbanization effects on soils

Cities have many environmental impacts on their immediate environments that can alter ecosystem services. One of the universal influences urbanization has on soils is that soil formation factors within cities are often greatly diverted from more "natural" trajectories. Moreover, the influence of management, land-use, and development on the nature of urban soils is often dominated by the incorporation of anthropogenic material and fill in the "parent material" of urban soils, something that contributes to the high degree of spatial heterogeneity found in urban soils and urban ecosystems (Effland & Pouyat 1997; Pickett & Cadenasso 2009). Within the context of ecosystem services, it is useful to describe

Soil Ecology and Ecosystem Services. First Edition. Edited by Diana H. Wall *et al.*
© 2012 Oxford University Press. Published 2012 by Oxford University Press.

urban effects on soils as either 1) direct (resulting from the physical process of urbanization and land development), or 2) indirect (the cascading effects that are the result of the presence and functioning of a city), because this dichotomy provides different ways to manage cities and the urban environment to affect the provision of ecosystem services.

4.3.2.1 Direct effects

Direct impacts on soils are the result of changes to physical properties of soils, a shift in inputs to soils, or the inclusion of imported materials in soil profiles. Soil compaction is a universal trait of urban soils. Usually the result of the physical clearing and grading of land by heavy machinery, soils can also become compacted in cities due to the proximity to built surfaces, which constrain the volume in which soils can exist. Compaction has many indirect influences on soil properties, from water holding capacity to soil biota. Plant communities in cities have been shown to be more invasible in cities due to changes in soil structure, such as compaction and hydrologic drought in riparian areas (Sung *et al.* 2011). These belowground–aboveground linkages can expand the reach of direct soil effects into the ecosystem at-large.

Anthropogenic influences due to management and artifacts of development often change the character of urban soils relative to non-urban soils in the same region. Soil formation can be conceptualized as an ecological process that results from the interactions of "formation factors" which include: climate, organisms, parent material, topography, and time that weather parent material into soil mineral particles and add organic matter to soils (Fig. 4.3.1; Jenny 1941). While the "organism" factor was initially considered to be insufficient to include the actions of humans, recent theoretical conceptualizations of soil formation have sought a more explicit role for human action (i.e. Amundson & Jenny 1991) and include humans as a subset of organisms in a state factor model. Other theoretical advances that have altered the state factor model include the actions of humans in the formation of soils and soil structure as key components of urban ecosystems (Fig. 4.3.1) (Effland & Pouyat 1997; Pickett & Cadenasso 2009). Effland and Pouyat (1997) achieve this by adding an explicit anthropic factor, with people mediating the other state factors in the formation of urban soils, as well as contributing novel anthropogenic parent materials to soils. Pickett and Cadenasso (2009) take an approach similar to Jenny's (1941), where the actions of people in cities serve to modify the more 'natural' soil formation trajectories. The main point with either conceptual approach is to highlight that the end result of urbanization produces regionally novel soils, the implications being that urban soils are unique in structure and function relative to non-urban soils, and this difference is the result of human management and unintended consequences of development.

Soil sealing is a direct impact that can disconnect an urban soil from its surrounding ecosystem. Soil sealing can be defined as either the physical or chemical change to soil such that water infiltration is impeded (Scalenghe & Marsan 2009) or the actual physical covering of soil and functionally locking it away from the environment (a broader disconnection from the ecosystem than the former definition) (Wessolek 2008). The net result of soil sealing is to limit, cease, or modify the interaction between a soil and its environment and forming factors. Sealed soils have manipulated properties that impact their ability to provide services that are thermodynamic, chemical, and hydrologic in nature. At its most extreme, soil sealing effectively removes a soil aerially from providing ecosystem services within an urban environment. Much literature on soil sealing treats the process as follows: finding correlates between indices of sealing (such as percent impervious cover) and altered ecosystem services (such as hydrologic flow rates) and assuming that once sealed, a soil will have its ability to provide services degraded. It is also often assumed that increasing the pervious cover in a city (and thus reducing the seal soil cover) will lead to an increase in green space, open space, plant cover, etc. However, not all reductions in pervious cover add patches that will increase the provisioning of ecosystem services in cities.

4.3.2.2 Indirect effects

While direct urban impacts will shift soil structure in ways that result in a novel ecosystem, other

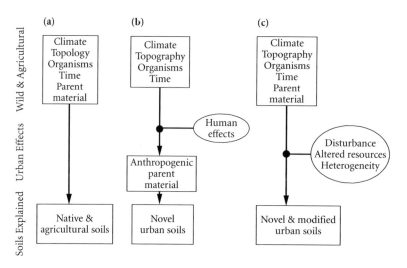

Figure 4.3.1 Three frameworks for the development of urban soils given the relationship with state factors of soil formation: a) Jenny (1940); b) Effland and Pouyat (1997); and c) Pickett and Cadenasso (2009). The latter emphasizes that each of the five state factors may be affected by human actions and structures in urban ecosystems, which can be thought of as "filters" of human actions (either physical disturbances, altered resource availability, and/or spatial heterogeneity Pavao-Zuckerman and Byrne (2009)). State factors interact with these anthropogenic filters to either yield new urban soils or modify existing soils in urban ecosystems. (From Pickett & Cadenasso 2009, with permission from Springer Science + Business Media B.V.)

impacts on ecosystem services result from indirect influences of cities on ecosystem structure and function. Indirect effects are the result of cascades of influence from a direct impact though physical changes in the urban environment. Indirect effects such as urban heat islands, alterations in hydrology, chemical inputs, food web shifts, and biogeochemical cycling impacts tend to be universal in cities but vary by degree between different cities and bioregions. The urban heat island effect is a robust landscape pattern found in most urban regions. Urban temperatures are generally elevated relative to the surrounding landscape, sometimes as much as 7–12°C (Oke 1982; Spronken-Smith & Oke 1998). Differences in the storage of heat in construction materials along urban-rural gradients generate this pattern, where structures absorb heat during the day and radiate them back at night, increasing urban temperatures. Urban heat islands can result in reduced soil moisture contents and shifts in the rates of biogeochemical cycling (McDonnell et al. 1997). Pouyat et al. (1997) for example found that urban heat islands may contribute to enhanced rates of leaf decomposition in urban forests. Furthermore, urban heat islands alter biogeochemical cycling by affecting above–belowground linkages due to shifts in litter inputs to soils. Warmer urban temperatures can result in a change in plant phenological properties, with leaf out occurring earlier and leaf fall occurring later in the season (Breuste et al. 1998), resulting in a potential for asynchrony between plant C inputs to soils and soil detritivore activity. Plant physiological responses to the heat island can result in elevated rates of plant productivity (Ziska et al. 2004), which may lead to enhanced litter inputs from the plant canopy.

The built environment can also dramatically alter hydrologic regimes in cities through an increase in the spatial cover of impervious surfaces, which tends to decrease infiltration capabilities within cities and result in an increase in surface runoff (Paul & Meyer 2001). Moreover, the increase in surface runoff impacts the physical structure of urban streams, causing banks to become deeply incised with a deeper depth to the water table (Pickett et al. 2001). Soils may also become hydrophobic (White & McDonnell 1988) in cities, further reducing the infiltration of water in soils. The net result for soils is that urban ecohydrology is often disconnected from precipitation inputs, with cascading effects to soil biology and soil functioning.

Soil biology demonstrates complex responses to specific environmental conditions that are a function of city size, age, density, history, and the ecological template that urbanization occurs within. Urbanization indirectly impacts food webs through shifts in resource availability, modification of species composition, and alteration of feeding behaviors and species interactions (Faeth *et al.* 2005). Management can indirectly influence soil food webs and their function with bacterial or fungal pathways responding to plant species and functional type (Vauramo & Setälä 2010). As with aboveground urban communities, the urban soil community reflects complex combinations of refuge patches of native and endemic taxa and introduced and exotic species (Vilisics & Hornung 2009). However, the contribution of exotic species to local soil fauna communities varies. While all isopods reported in forests in Baltimore are introduced exotics (Hornung & Szlavecz, 2003) and there are no naive earthworm species found in the New York City region (Steinberg *et al.* 1997), in Baltimore, native species comprise nearly half the earthworms sampled (Szlavecz *et al.* 2006). While these trends are partly the result of interactions of history, biogeography, climate, and mechanisms of dispersal, current management decisions can have indirect impacts on the distribution of soil communities. For example, earthworms respond to environmental conditions brought about by differing mulching techniques (Byrne *et al.* 2008). Alterations in microbial (Wang *et al.* 2011) and faunal (Pavao-Zuckerman & Coleman 2007; Szlavecz *et al.* 2006) communities have important consequences for decomposition and biogeochemical cycles in cities.

Another important control point in urban biogeochemical cycles is an increase in nutrient deposition and flux of pollutants to urban soils (Kaye *et al.* 2006). Soil chemistry responds fairly predictably to measures of urbanization such as density and traffic volume, with increased soil concentrations of heavy metals and salts in the urban core (Pickett *et al.* 2001). Nitrogen deposition to soils also occurs fairly predictably along urban gradients, with some modification given local urban morphology (Pouyat *et al.* 2008). Again, the response of soil organisms is important, as soil community composition can shift and biomasses are reduced in response to the presence of heavy metals in soils (Pouyat *et al.* 1994) and nitrogen inputs (Treseder *et al.* 2004). Nutrient cycling rates will respond differently depending on the size of the city or the biomes that these cities occur in, which is likely a function of the nature of inputs to soils or how severe physical modifications (heat islands, compaction) turn out to be. Pouyat *et al.* (2010) report that for three cities that differ over two orders of magnitude in size (New York City, Baltimore, and Asheville) rates of litter decomposition were either elevated or suppressed in urban forests, while urban impacts on the rate of N-mineralization didn't vary with city size. Residential and municipal scale decision-making about water management and plant cover in arid cities can lead to a convergence of soil carbon accumulation rates in cities that exist in very different climatic regimes (Pouyat *et al.* 2009), and enhancement of microbial function and enzyme activity (Green & Oleksyszyn 2002).

4.3.3 Examples of ecosystem services in cities

Despite these direct and indirect urban effects, soils do provide ecosystem services in urban landscapes. To illustrate examples of links between soils and potential urban ecosystem services, I will use the MEA framework of supporting, provisioning, regulating, and cultural services (see Fig. 4.3.2), where:

- Supporting ecosystem services can be thought of as the patterns and processes that underlie the other, more directly economically relevant, ecosystem services.
- Provisioning ecosystem services are the goods and products obtained from ecosystems, often conceived as either some plant product with economic value (e.g. food, fiber) or potable water.
- Regulating ecosystem services are benefits that come from ecosystem processes that regulate, mediate, and control the boundaries

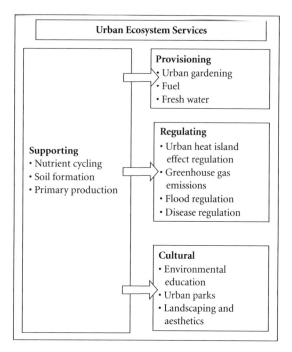

Figure 4.3.2 Ecosystem services described for urban ecosystems. Supporting services are the foundation for provisioning, regulating, and cultural ecosystem services in cities. (Modified from the Millennium Ecosystem Assessment.)

of environmental regimes (such as those related to climate, disease, flooding, etc.).
- Cultural ecosystem services are non-material benefits from ecosystems that include aesthetics, education, and spiritual values.

4.3.3.1 Supporting ecosystem services

Urbanization affects the physical, chemical, and biological properties of soils in a way that fundamentally alters the nature of the belowground component of urban ecosystems, which ultimately shift ecosystem functions and processes related to biogeochemical cycling (Carreiro et al. 1999; Zhu & Carreiro 1999; Kaye et al. 2005; Pavao-Zuckerman & Coleman 2005). Kaye et al. (2006) suggest that cities have a fundamentally different biogeochemistry than 'natural' systems because human actions alter control points, inputs, and outputs in urban ecosystems. Nutrient cycling in urban soils is often greatly mediated by the interaction of direct and indirect impacts on urban soils that link physical, chemical, and biological structures to the emerging function (McDonnell et al. 1997).

Cities have been shown to have a negative impact on local and regional estimations of biodiversity (Savard et al. 2000; McKinney 2002). Soil biodiversity has not been studied to a very great extent in an urban context, where generally a much greater emphasis is placed on the diversity of birds and plants, and to a lesser extent insects and smaller vertebrates. Strong and robust patterns in urban biodiversity have been found among avifauna and plants that relate to urban geography, demography, and socioeconomics. A "luxury effect" has been described where increased plant diversity is strongly and positively correlated with metrics of economic status (Hope et al. 2003). The urban gradient approach has been successful for describing the influence of urbanization on soil organisms because the selection of field sites in these studies holds a natural "template" constant such that differences in taxa, abundance, and diversity can be attributed to direct and indirect urban environmental factors. Gradient studies have demonstrated lower fungal abundances along urban–rural gradients, but the response of microfaunal abundances varies depending on the city and setting (Ohtonen et al. 1992; Pouyat et al. 1994). Importantly, for linking soil biodiversity to other ecosystem services, urbanization may not influence diversity indices per se, but actually impacts biodiversity via a functional shift in soil community structure, which is in turn linked to altered biogeochemical processes in urban soils (Pavao-Zuckerman & Coleman 2005; Pavao-Zuckerman & Coleman 2007; Vauramo & Setälä 2010). Strong linkages have been demonstrated between management practices and the resulting soil biodiversity, which are important for the local-scale management of ecosystem services (Cheng et al. 2008).

4.3.3.2 Provisioning ecosystem services

Urban gardening (through allotments, community gardens, etc.) is increasing in prevalence as people try to locally produce some food products and reduce externalities of industrial agriculture. However, the physical and chemical changes that accompany urbanization often result in impaired conditions for plant growth. Heavy metal contamination in urban soils can be a concern, especially as studies have reported a transference of metals from soils to roots, stems, and leaves (although fruiting bodies may be less of a concern (Finster *et al.* 2004)). It is important to note that while risk may be minimal for exposure to heavy metals from consumption of products from urban gardens, subpopulations (e.g. children) may be at greater risk for exposure (Hough *et al.* 2004). Soils are an integral component of urban gardening, yet serve as a potential point for human health risks from a potentially beneficial ecosystem service (Wakefield *et al.* 2007).

Urban soils also play an important role in storm water management for the purposes of water quality improvement. Low-impact development and green infrastructure are two environmental design elements that have important soil components in their design and function. For example, rain gardens (or bioretention cells—depressions that are graded and planted to collect and retain storm water) are considered an urban best-management practice for mitigating the flux of pollutants from urban landscapes into ground water and surface bodies of water (Carpenter & Hallam 1999). The design of rain gardens with respect to physiochemical soil properties such as hydraulic conductivity, bulk, density, particle size distribution, etc. has a significant impact on their ability to retain nitrogen, phosphorus, and sediments in storm water, ultimately linking to the ability to provide clean water as an ecosystem service (Carpenter & Hallam 1999). On-site waste treatment systems (i.e. septic systems) also rely upon specific soil physical parameters for effective drainage and waste treatment (Anderson & Otis 2000). Poor siting practices including imprecise soil surveys (especially a problem in urban areas) and compaction and damage to system elements can cause irreparable damage that impairs functioning, creating a potential for pollutants and pathogens to leach off site and impair waters (Anderson & Otis 2000).

4.3.3.3 Regulating ecosystem services

Soils play an important role in regulating the hydrologic regime in cities, where even a very general property, such as percent pervious surface, can be linked to flood intensities (Ogden *et al.* 2000). Ogden *et al.* (2000) link other specific urban soil properties, such as soil saturated hydraulic conductivity to the incidence of flash floods. Cities also have indirect impacts on local hydrologic cycles, where impervious surfaces and a lack of soil moisture can create feedbacks in urban energy cycles that have the end result of increased precipitation intensity in cities (Ashley *et al.* 2005).

Linkages between cities and global climate change come about from different management strategies that either sequester C or increase the flux of greenhouse gasses from soils in cities. Soil carbon stocks may increase in certain urban land uses through time, particularly those that result in higher levels of primary productivity, such as conversion of land into golf courses and other recreational sites (Golubiewski 2006). Urban landscapes may contribute enough to greenhouse gas fluxes that regional estimates of emissions may be underestimates (Townsend-Small *et al.* 2011). The net balance of greenhouse gas emissions needs to consider management approaches that have emissions related to fertilizer and fossil fuel use, or irrigation practices which may alter soil conditions to promote efflux of CO_2 or N_2O (Kaye *et al.* 2004, 2005; Pataki *et al.* 2006).

4.3.3.4 Cultural ecosystem services

Urban soils are an integral component of an active learning process in urban environmental education (Johnson & Catley 2009). While cultural ecosystem services are often difficult to place a value upon, their role in bridging gaps between people and nature cannot be underemphasized. Participation in urban restoration projects has been suggested as an important way to develop relationships with one's local ecology and progress towards "ecological citizenship" (Light 2006). Incorporation of urban soils into education and outreach programs and linking urban soils to participatory urban restoration and gardening experiences are a key way to ground urban residents in their local ecology (Thomashow 2001) and help to overcome the extinction of experience that is often assumed to accompany urbanization of a population.

4.3.4 Management for urban ecosystem services

While our understanding of ecosystem services in urban areas itself is in relative infancy, we do know a good deal about the management of soil for the provision of ecosystem services, even if that specific nomenclature is not utilized. Agroecology and ecological forestry have been important to the development of soil ecology as a discipline, with management decisions and best management practices (BMPs) implemented as experimental treatments from a research perspective. The provision of ecosystem services in this context (crop or timber production, reduction in erosion, retention of nutrients, economic and cultural livelihoods, etc.) is frequently linked to the ecology of soils, diversity of soil communities, and general soil quality. This knowledge base extends beyond systems focused on production to heavily disturbed systems, where the success or failure of ecological restoration can be linked to the degree to which soil ecological knowledge is applied in design (Heneghan et al. 2008). One caveat with working in urban ecosystems is that notions of reference, baseline, and functional goals often have to be modified for the particular context and case in question (Pavao-Zuckerman 2008). City planners and managers are often aware of urban environmental challenges at both local and regional scales, and the framework of ecosystem services can be used as an integrating concept to guide planning, management, or research in urban ecosystems. More efforts need to be made to conduct research on urban ecosystem services that integrate the ecological sciences with the practice of planning, design, management, and restoration. Opportunities for management intervention for urban soils to improve ecosystem services can be considered either 1) at localized scale, where interventions improve soil quality for ecosystem services, or 2) at the scale of the urban landscape, where interventions increase the surface area of soils for ecosystem services.

4.3.4.1 Local scale improvements

Local scale interventions to promote ecosystem services often incorporate a mechanism to improve soil organic matter contents, as this is a key variable affecting soil biology and biogeochemical function, while often organic matter contents are depressed in cities. The use of organic soil amendments is a fairly successful strategy for ameliorating the impact of soil compaction on plant roots (Kozlowski 1999). Byrne et al. (2008) found that the use of organic mulches (rather than rock) promoted the biomass of soil faunal groups, with likely feedbacks into nutrient cycling rates. Zhu et al. (2004) suggest that management techniques to increase soil carbon pools, such as leaving grass clippings on site after mowing, can facilitate the use of soils in retention basins to remove nitrogen from surface waters through stimulation of denitrification. Organic wastes (e.g. food scraps, yard waste, sewage) are a readily available resource in cities that could be accumulated at the household or municipal scale for urban soil improvement, and innovative projects such as Growing Power, Inc. are using such aggregations of compost to promote urban agriculture despite locally poor soil quality (Sievert 2006).

Management of soil organisms (either targeting exotic species or promoting endemics) may also help restore soil ecosystem services in cities. Heneghan et al. (2006) point to an interaction between exotic earthworms and an invasive plant (European buckthorn), where efforts that solely target the removal of the plant to meet goals of native plant restoration in Chicago ignore feedbacks between the earthworms, plant litter, and soil nitrogen. It is the removal of the invasive earthworm, or ameliorating their soil impacts, that facilitates the removal of the invasive plant, and thus the restoration of ecosystem services provided by native plant species. Fungal populations are often reduced in cities (Pouyat et al. 1994) suggesting that the beneficial effects of mycorrhizal fungi for plants may be missing. Handel et al. (1997) report that depressed mycorrhizal inocula contributes to the failed establishment and root growth of plants used in landfill restoration. And while inoculation with mycorrhizal fungi can help promote the establishment of plants on degraded soils, it can also have additive impacts on ecosystem services by promoting microbial enzyme activity and aggregate stability in restored soils (Alguacil et al. 2005).

Most efforts to improve soils for ecosystem services are done at a very local scale, and models often assume a linear increase in the benefit that these elements will provide. However, these returns are likely to be non-linear on the ground as the percent urban cover that they impact increases. Gober *et al.* (2010) found a non-linear increase and ultimate saturation of the benefit of using greenspace to mitigate the urban heat island effect. Non-linearities can be driven by the heterogeneous and unequal (and often inequitable) distribution of capital and plant cover and diversity in the urban landscape (Gober *et al.* 2010). Further research to extrapolate and link local scale improvements of soil quality to the improved provision of ecosystem services at the city or urban landscape scale is needed.

4.3.4.2 Urban scale intervention

In addition to localized improvements of soils quality, environmental policy and regulations can promote healthier soils and a reduction in the proportion of sealed soils in cities. For example, new legislation in England targets driveways in an effort to unseal soils and promote ecosystem services such as drainage, flood reduction, and pollution abatement by requiring the use of permeable paving when gardens are paved over (Davies *et al.* 2011a). Tree planting programs can mitigate the effects of soil-sealing and other urban soil impacts, with several large-scale efforts are underway, such as the Million Trees Projects in New York City (http://www.milliontreesnyc.org/) and Los Angeles (http://www.milliontreesla.org/), which seek to use urban reforestation to increase ecosystem services in cities. One caveat for such projects and extrapolation to soil improvements is that the allometric models that have been used to estimate the growth of trees in cities and the ecosystem services they provide were developed for non-urban trees and may misestimate tree growth on the ground (McHale *et al.* 2009). Further research should redefine these models for urban environments and also make direct measures of the soil feedbacks (i.e. carbon sequestration) related to ecosystem services.

In addition to unsealing soils and reforestation efforts, novel ecosystems are being created or restored in cities to increase the percent cover of ecologically functional space in the urban landscape. For example, wetlands are often created for the purposes of nutrient retention (Hogan & Walbridge 2007), and incorporation of soil knowledge into these designs will help reach performance goals as they relate to ecosystem services. Green roofs provide another opportunity to bring new ecosystems into cities, and promote soil ecosystem services of nutrient retention, storm flow reduction, and carbon sequestration (Oberndorfer *et al.* 2007). Felson and Pickett (2005) promote designed experiments, where designers and ecologists can integrate ecological research and landscape design elements together to better test the efficacy of BMPs and green infrastructure in cities. Ecologists and designers could collaborate in a replicated study/installation where design elements make use of novel structured soils to help integrate plants into storm water management (Bartens *et al.* 2009) in comparison to "controls" that use *in situ* urban soils, providing both the green infrastructure that might be desired in a city and also the ability to robustly study these installations from an ecological perspective.

4.3.4.3 Urban assessment for ecosystem services

Cities are the least well characterized ecosystem type in the Millennium Ecosystem Assessment (2005) which is due in part to a relative lack of study of urban ecosystems at a broad and integrated scale. As efforts are made to expand our understanding of the ecology of urban areas, it is important to recognize the need to improve the spatial characterization of urban soils for ecosystem services as part of this expansion of data sets. Fordyce *et al.* (2005), for example, illustrate the importance of national scale mapping of urban geochemistry for broader purposes of restoration, human health risk assessment related to toxins and pathogens, and soil links to water quality. That said, even when large-scale inventories are made, many regional estimates of ecosystem services fail to consider the possibility that cities contribute positively to ecosystem services. Davies *et al.* (2011b) illustrate the importance of scale in inventories by mapping an ecosystem service (carbon storage) at the scale of a city, and show that national estimates underestimate the urban contribution to carbon storage by an order of magnitude. So, while mapping and inventories may help to better estimate ecosystem services, current efforts to provide integrated urban data sets (including the US National Science Foundation Urban Long-Term Research Area sites and National Ecological Observatory Network) should consider the appropriate scales for measurement and assessment.

Complicating the discussion of urban soil ecosystem services is that the demand for urban ecosystem services has been discussed and studied to a much lesser degree than environmental impacts of urbanization (McDonald 2009). Accounting for demand will help with environmental decision making on the ground, as the spatial arrangement of demand, land use, population density, etc. have important implications for how ecosystem services might be managed in cities and urban regions. In cities we may see tradeoffs between approaches that look to either 1) conservation planning, purchasing open space for conservation and parks or 2) designing novel patches. Building from the interdisciplinary foundation of the ecosystem service concept will be an important step in moving forward with monitoring studies and application of the idea from a planning and management perspective. For example, Pataki et al. (2011) suggest that addressing potential tradeoffs in ecosystem services in cities will require both integrated studies of multiple biogeochemical cycles and the incorporation of assumptions via cost–benefit analyses within a decision support framework.

Finally, when considering urban ecosystem services, it is important to recognize the distinction between urban and more natural ecosystems. Cities reflect the integration of physical, biological, economic, and cultural processes on the landscape. Research and management of urban ecosystems should therefore proceed from an interdisciplinary social-ecological system framework (Alberti et al. 2003; Colding 2007). From this perspective, an overlooked "service" that soils provide is support for the physical structures in the built environment that give an urban ecosystem part of its structure. Anderson and Otis (2000) describe consideration of soil texture, chemical composition, and hydrologic properties that relate to the construction and longevity of buildings and roads and the stability of soils that support such structures. The character of urban soils has been linked to building stability and can result in negative impacts on building structures through time (essentially an ecosystem disservice) (Gould et al. 2002). Consideration of soils for the purposes of construction shifts focus from the more commonly discussed ecosystem services we are concerned with preserving (biodiversity, nutrient cycling, etc.); however, when dealing with urban systems it is an appropriate element to be discussed when considering a full accounting of the costs and benefits related to ecosystem services in the context of urbanization, urban density, and tradeoffs between "sealed" soils vs. open spaces that can provide services.

4.3.5 Summary

Despite being systems that are characterized by densely built environments and that draw heavily on surrounding ecosystems for goods and services, cities are places that can provide ecosystem services. There exist potential trade-offs between the use of urban space for what we usually consider the services of cities (trade, housing, entertainment) and the ecosystem services we might desire a city to have to promote sustainability and resilience. Moreover, cities often negatively impact the ability of an ecosystem to provide services. Soils are the foundation of ecosystems, and can serve that role in cities, acting as a foundation for reconciling the trade-offs between environmental impacts and the desired ecological components we seek from sustainable cities. However, much place-specific studies are required to advance the concept of urban ecosystem services, and to determine the limits and opportunities of what we might expect from urban soils. Working from an interdisciplinary perspective and in collaboration with planners, designers, and city managers will bring soils to the forefront of ecological consideration and simultaneously provide new opportunities for study and for generating urban landscapes that yield what ecosystem services they can.

References

Alberti, M., Marzluff, J.M., Shulenberger, E., Bradley, G., Ryan, C., & Zumbrunnen, C. (2003) Integrating humans into ecology: Opportunities and challenges for studying urban ecosystems. *Bioscience* 53: 1169–79.

Alguacil, M.M., Caravaca, E., & Roldan, A. (2005) Changes in rhizosphere microbial activity mediated by native or allochthonous AM fungi in the reafforestation of a Mediterranean degraded environment. *Biology and Fertility of Soils* **41**: 59–68.

Amundson, R., & Jenny, H. (1991) The place of humans in the state factor theory of ecosystems and their soils. *Soil Science* **151**: 99–109.

Anderson, D.L., & Otis, R.J. (2000) Integrated wastewater management in growing urban environments. In: Brown, R.B. (ed.) *Managing soils in an urban environment*, pp. 199–250. American Society of Agronomy, Crop Science Society of America, Soil Science Society of America, Madison, WI.

Ashley, R.M., Balmforth, D.J., Saul, A.J., & Blanksby, J.D. (2005) Flooding in the future—predicting climate change, risks and responses in urban areas. *Water Science and Technology* **52**: 265–73.

Bartens, J., Day, S.D., Harris, J.R., Wynn, T.M., & Dove, J.E. (2009) Transpiration and root development of urban trees in structural soil stormwater reservoirs. *Environmental Management* **44**: 646–57.

Bolund, P., & Hunhammar, S. (1999) Ecosystem services in urban areas. *Ecological Economics* **29**: 293–301.

Breuste J., Feldmann H. & Uhlmann O. (eds.) (1998) *Urban ecology*. Springer Verlag, Berlin.

Byrne, L.B., Bruns, M.A., & Kim, K.C. (2008) Ecosystem properties of urban land covers at the aboveground-belowground interface. *Ecosystems* **11**: 1065–77.

Carpenter, D.D., & Hallam, L. (1999) Influence of planting soil mix characteristics on bioretention cell design and performance. *Journal of Hydrologic Engineering* **15**: 404–16.

Carreiro, M.M., Howe, K., Parkhurst, D.F., & Pouyat, R.V. (1999) Variation in quality and decomposability of red oak leaf litter along an urban-rural gradient. *Biology and Fertility of Soils* **30**: 258–68.

Cheng, Z., Grewal, P.S., Stinner, B.R., Hurto, K.A., & Hamza, H.B. (2008) Effects of long-term turfgrass management practices on soil nematode community and nutrient pools. *Applied Soil Ecology* **38**: 174–84.

Colding, J. (2007) "Ecological land-use complementation" for building resilience in urban ecosystems. *Landscape and Urban Planning* **81**: 46–55.

Cronon, W. (1991) *Nature's metropolis: Chicago and the great west*. W.W. Norton & Co., Inc. New York.

Davies, L., Kwiatkowski, L., Gaston, K.J., et al. (2011a) Urban. The UK National Ecosystem Assessment Technical Report. UK National Ecosystem Assessment, UNEP-WCMC, Cambridge.

Davies, Z.G., Edmondson, J.L., Heinemeyer, A., Leake, J.R., & Gaston, K.J. (2011b) Mapping an urban ecosystem service: quantifying above-ground carbon storage at a city-wide scale. *Journal of Applied Ecology* **48**(5): 1125–34.

Effland, W.R., & Pouyat, R.V. (1997) The genesis, classification, and mapping of soils in urban areas. *Urban Ecosystems* **1**: 217–28.

Faeth, S.H., Warren, P.S., Shochat, E., & Marussich, W.A. (2005) Trophic dynamics in urban communities. *BioScience* **55**: 399–407.

Felson, A.J., & Pickett, S.T.A. (2005) Designed experiments: new approaches to studying urban ecosystems. *Frontiers in Ecology and the Environment* **3**: 549–56.

Finster, M.E., Gray, K.A., & Binns, H.J. (2004) Lead levels of edibles grown in contaminated residential soils: a field survey. *Science of the Total Environment* **320**: 245–57.

Fordyce, F.M., Brown, S.E., Ander, E.L., et al. (2005) GSUE: urban geochemical mapping in great Britain. *Geochemistry-Exploration Environment Analysis* **5**: 325–36.

Gober, P., Brazel, A., Quay, R., et al. (2010) Using watered landscapes to manipulate urban heat island effects: How much water will it take to cool Phoenix? *Journal of the American Planning Association* **76**: 109–21.

Golubiewski, N.E. (2006) Urbanization increases grassland carbon pools: Effects of landscaping in Colorado's front range. *Ecological Applications* **16**: 555–71.

Gould, R., Bedell, P.R., & Muckle, J.G. (2002) Construction over organic soils in an urban environment: four case histories. *Canadian Geotechnical Journal* **39**: 345–56.

Green, D.M., & Oleksyszyn, M. (2002) Enzyme activities and carbon dioxide flux in a Sonoran desert urban ecosystem. *Soil Science Society of America Journal* **66**: 2002–8.

Handel, S.N., Robinson, G.R., Parsons, W.F.J., & Mattei, J.H. (1997) Restoration of woody plants to capped landfills: Root dynamics in an engineered soil. *Restoration Ecology* **5**: 178–86.

Heneghan, L., Miller, S.P., Baer, S., et al. (2008) Integrating soil ecological knowledge into restoration management. *Restoration Ecology* **16**: 608–17.

Heneghan, L., Steffen, J., & Fagen, K. (2006) Interactions of an introduced shrub and introduced earthworms in an Illinois urban woodland: Impact on leaf litter decomposition. *Pedobiologia* **50**: 543–51.

Hogan, D.M., & Walbridge, M.R. (2007) Best management practices for nutrient and sediment retention in urban stormwater runoff. *Journal of Environmental Quality* **36**: 386–95.

Hope, D., Gries, C., Zhu, W.X., et al. (2003) Socioeconomics drive urban plant diversity. *Proceedings of the National Academy of Sciences of the United States of America* **100**: 8788–92.

Hornung, E., & Szlavecz, K. (2003) Establishment of a Mediterranean isopod (*Chaetophiloscia sicula*, Verhoeff, 1908) in a North American temperate forest. *Crustaceana Monographs* **2**: 181–9.

Hough, R.L., Breward, N., Young, S.D., et al. (2004) Assessing potential risk of heavy metal exposure from consumption of home-produced vegetables by urban populations. *Environmental Health Perspectives* **112**: 215–21.

Jenny, H. (1941) *Factors of soil formation: a system of quantitative pedology*. McGraw-Hill, New York.

Johnson, E.A., & Catley, K.M. (2009) Urban soil ecology as a focal point for environmental education. *Urban Ecosystems* **12**: 79–93.

Kaye, J.P., Burke, I.C., Mosier, A.R., & Guerschman, J.P. (2004) Methane and nitrous oxide fluxes from urban soils to the atmosphere. *Ecological Applications* **14**: 975–81.

Kaye, J.P., Groffman, P.M., Grimm, N.B., Baker, L.A., & Pouyat, R.V. (2006) A distinct urban biogeochemistry? *Trends in Ecology & Evolution* **21**: 192–9.

Kaye, J.P., McCulley, R.L., & Burke, I.C. (2005) Carbon fluxes, nitrogen cycling, and soil microbial communities in adjacent urban, native and agricultural ecosystems. *Global Change Biology* **11**: 575–87.

Kozlowski, T.T. (1999) Soil compaction and growth of woody plants. *Scandinavian Journal of Forest Research* **14**: 596–619.

Light, A. (2006) Ecological citizenship: the democratic promise of restoration. In: Platt, R.H. (ed.) *The humane metropolis: people and nature in the 21st-century city*, pp. 169–82. University of Massachusetts Press, Amherst & Boston, MA.

McDonald, R. (2009) Ecosystem service demand and supply along the urban-to-rural gradient. *Journal of Conservation Planning* **5**: 1–14.

McDonnell, M.J., Pickett, S.T.A., Groffman, P.M., *et al.* (1997) Ecosystem processes along an urban-to-rural gradient. *Urban Ecosystems* **1**: 21–36.

McHale, M.R., Burke, I.C., Lefsky, M.A., Peper, P.J., & McPherson, E.G. (2009) Urban forest biomass estimates: is it important to use allometric relationships developed specifically for urban trees? *Urban Ecosystems* **12**: 95–113.

McKinney, M.L. (2002) Urbanization, biodiversity, and conservation. *BioScience* **52**: 883–90.

Millennium Ecosystem Assessment (2005) *Ecosystems and Human Well-being: Synthesis*. Island Press, Washington, DC.

Oberndorfer, E., Lundholm, J., Bass, B., *et al.* (2007) Green roofs as urban ecosystems: Ecological structures, functions, and services. *Bioscience* **57**: 823–33.

Ogden, F.L., Sharif, H.O., Senarath, S.U.S., Smith, J.A., Baeck, M.L., & Richardson, J.R. (2000) Hydrologic analysis of the Fort Collins, Colorado, flash flood of 1997. *Journal of Hydrology* **228**: 82–100.

Ohtonen, R., Ohtonen, A., Luotonen, H., & Markkola, A.M. (1992) Enchytraeid and nematode numbers in urban, polluted Scots pine (*Pinus sylvestris*) stands in relation to other soil biological parameters. *Biology and Fertility of Soils* **13**: 50–4.

Oke, T.R. (1982) The energetic basis of the urban heat-island. *Quarterly Journal of the Royal Meteorological Society* **108**: 1–24.

Pataki, D.E., Alig, R.J., Fung, A.S., *et al.* (2006) Urban ecosystems and the North American carbon cycle. *Global Change Biology* **12**: 2092–102.

Pataki, D.E., Carreiro, M.M., Cherrier, J., *et al.* (2011) Coupling biogeochemical cycles in urban environments: ecosystem services, green solutions, and misconceptions. *Frontiers in Ecology and the Environment* **9**: 27–36.

Paul, M.J., & Meyer, J.L. (2001) Streams in the urban landscape. *Annual Review of Ecology and Systematics* **32**: 333–65.

Pavao-Zuckerman, M.A. (2008) The nature of urban soils and their role in ecological restoration in cities. *Restoration Ecology* **16**: 642–9.

Pavao-Zuckerman, M.A., & Byrne, L.B. (2009) Scratching the surface and digging deeper: exploring ecological theories in urban soils. *Urban Ecosystems* **12**: 9–20.

Pavao-Zuckerman, M.A., & Coleman, D.C. (2005) Decomposition of chestnut oak (Quercus prinus) leaves and nitrogen mineralization in an urban environment. *Biology and Fertility of Soils* **41**: 343–9.

Pavao-Zuckerman, M.A., & Coleman, D.C. (2007) Urbanization alters the functional composition, but not taxonomic diversity, of the soil nematode community. *Applied Soil Ecology* **35**: 329–39.

Pickett, S.T.A., & Cadenasso, M.L. (2009) Altered resources, disturbance, and heterogeneity: a framework for comparing urban and non-urban soils. *Urban Ecosystems* **12**: 23–44.

Pickett, S.T.A., Cadenasso, M.L., Grove, J.M., *et al.* (2001) Urban ecological systems: linking terrestrial ecological, physical, and socioeconomic components of metropolitan areas. *Annual Review of Ecology and Systematics* **32**: 127–57.

Pouyat, R.V., Carreiro, M.M., Groffman, P.M., & Pavao-Zuckerman, M.A. (2010) Investigative approaches of urban biogeochemical cycles: New York and Baltimore Cities as case studies. In: M.J. McDonnell, A. Hahs, & J. Breuste (eds.) *Comparative ecology of cities and towns*, pp. 329–51. Cambridge University Press, New York.

Pouyat, R.V., McDonnell, M.J., & Pickett, S.T.A. (1997) Litter decomposition and nitrogen mineralization in oak stands along an urban-rural land use gradient. *Urban Ecosystems* **1**: 117–31.

Pouyat, R.V., Parmelee, R.W., & Carreiro, M.M. (1994) Environmental effects of forest soil-invertebrate and fungal densities in oak stands along and urban-rural land use gradient. *Pedobiologia* **38**: 385–99.

Pouyat, R.V., Yesilonis, I., & Golubiewski, N.E. (2009) A comparison of soil organic carbon stocks between residential turf grass and native soil. *Urban Ecosystems* **12**: 45–62.

Pouyat, R.V., Yesilonis, I.D., Szlavecz, K., *et al.* (2008) Response of forest soil properties to urbanization gradients in three metropolitan areas. *Landscape Ecology* **23**: 1187–203

Rees, W.E. (1997) Urban ecosystems: the human dimension. *Urban Ecosystems* **1**: 63–75.

Savard, J.L., Clergeau, P., & Mennechez, G. (2000) Biodiversity concepts and urban ecosystems. *Landscape and Urban Planning* **48**: 131–42.

Scalenghe, R., & Marsan, F.A. (2009) The anthropogenic sealing of soils in urban areas. *Landscape and Urban Planning* **90**: 1–10.

Sievert, L.N. (2006) Urban watershed management: the Milwaukee River experience. In: Platt, R.H. (ed.) *The humane metropolis: people and nature in the 21st-century city*, pp. 141–53. University of Massachusetts Press, Amherst & Boston, MA.

Spronken-Smith, R.A., & Oke, T.R. (1998) The thermal regime of urban parks in two cities with different summer climates. *International Journal of Remote Sensing* **19**: 2085–104.

Steinberg, D.A., Pouyat, R.V., Parmelee, R.W., & Groffman, P.M. (1997) Earthworm abundance and nitrogen mineralization rates along an urban-rural land use gradient. *Soil Biology and Biochemistry* **29**: 427–30.

Sung, C.Y., Li, M.H., Rogers, G.O., Volder, A., & Wang, Z.F. (2011) Investigating alien plant invasion in urban riparian forests in a hot and semi-arid region. *Landscape and Urban Planning* **100**: 278–86.

Szlavecz, K., Placella, S.A., Pouyat, R.V., Groffman, P.M., Csuzdi, C., & Yesilonis, I. (2006) Invasive earthworm species and nitrogen cycling in remnant forest patches. *Applied Soil Ecology* **32**: 54–62.

Thomashow, M. (2001) *Bringing the biosphere home: learning to perceive global environmental change.* MIT Press, Boston, MA.

Townsend-Small, A., Pataki, D.E., Czimczik, C.I., & Tyler, S.C. (2011) Nitrous oxide emissions and isotopic composition in urban and agricultural systems in southern California. *Journal of Geophysical Research-Biogeosciences* **116**: G01013.

Treseder, K.K., Masiello, C.A., Lansing, J.L., & Allen, M.F. (2004) Species-specific measurements of ectomycorrhizal turnover under N-fertilization: combining isotopic and genetic approaches. *Oecologia* **138**: 419–25.

Vauramo, S., & Setälä, H. (2010) Urban belowground food-web responses to plant community manipulation—Impacts on nutrient dynamics. *Landscape and Urban Planning* **97**: 1–10.

Vilisics, F., & Hornung, E. (2009) Urban areas as hot-spots for introduced and shelters for native isopod species. *Urban Ecosystems* **12**: 333–45.

Wakefield, S., Yeudall, F., Taron, C., Reynolds, J., & Skinner, A. (2007) Growing urban health: Community gardening in South-East Toronto. *Health Promotion International* **22**: 92–101.

Wang, M.E., Markert, B., Shen, W.M., Chen, W.P., Peng, C., & Ouyang, Z.Y. (2011) Microbial biomass carbon and enzyme activities of urban soils in Beijing. *Environmental Science and Pollution Research* **18**: 958–67.

Wessolek, G. (2008) Sealing of soils. In: Marzluff, J.M., Shulenberger, E., Endlicher, W., Alberti, M., Bradley, G., Ryan, C., Simon, U., & Zumbrunnen, C. (eds.) *Urban ecology: and international perspective on the interaction between humans and nature*, pp. 161–79. Springer, NY.

White, C.S., & McDonnell, M.J. (1988) Nitrogen cycling processes and soil characteristics in an urban versus rural forest. *Biogeochemistry* **5**: 243–62.

Zhu, W.X., & Carreiro, M.M. (1999) Chemoautotrophic nitrification in acidic forest soils along an urban-to-rural transect. *Soil Biology and Biochemistry* **31**: 1091–100.

Zhu, W.X., Dillard, N.D., & Grimm, N.B. (2004) Urban nitrogen biogeochemistry: status and processes in green retention basins. *Biogeochemistry* **71**: 177–96.

Ziska, L.H., Bunce, J.A., & Goins, E.W. (2004) Characterization of an urban-rural CO_2/temperature gradient and associated changes in initial plant productivity during secondary succession. *Oecologia* **139**: 454–8.

CHAPTER 4.4

Management of Grassland Systems, Soil, and Ecosystem Services

Phil Murray, Felicity Crotty, and Nick van Eekeren

4.4.1 Introduction

Grasslands occupy almost a quarter of the land surface of the Earth (Harvey 2001) including pasture, prairie, rangelands, savannah, and steppe. The types of grasslands are varied and cover a wide range of management regimes, from intensively managed, highly productive to totally unmanaged. Globally and regionally, grasslands provide a number of key ecosystem services including support (e.g. water and nutrient cycling), provisioning (e.g. food production), regulating (e.g. climate regulation), cultural (e.g. recreational), and biocontrol (e.g. source of predatory organisms) services. Many of these services are mediated or provided by the soil biota. Grassland ecosystems tend to have a relatively stable and permanent plant cover, which in turn, provides a habitat for a large and diverse invertebrate fauna and microbes.

The perennial nature of grasslands means that plant–soil interactions are extremely important in regulating soil processes and the ecosystem services. The vegetation cover contributes an abundant nutrient supply and also buffers the soil environment from temperature extremes. A key feature of grasslands is the high turnover of shoot and root biomass, that consequently results in the retention of a large pool of labile organic matter at the soil surface. This allows grasslands to support a relatively stable and numerous soil biota that contribute substantially to effective soil functioning, including the maintenance of fertility. In addition, unlike in cropping systems, there is often a regular substantial input of carbon and nutrients in the form of animal returns as dung and urine.

The dynamics of the carbon (C) and nitrogen (N) cycle within grasslands directly affect the quality of the plants and the soil. A large proportion of the processes in soils are mediated by the soil biota. These include the comminution and incorporation of litter into soil, the building and maintenance of structural porosity, the aggregation in soils through burrowing activities, the control of microbial communities and activities, and the improvement of plant production (Lavelle et al. 2006). The interactions between groups of organisms and physical and chemical processes shape the soil as a habitat and influence the nature of the soil food web with consequences for the vegetation it supports.

Plant net primary production provides food for the soil biota, whether it is through the living plant itself, the herbivory channel, or through dead plant matter (the detrital channel). In turn, the soil biota improves soil structure, water regulation and nutrient cycling, and increases root production, and water and nutrient uptake, which results in greater plant production. In this way, the quantity and quality of litter and root exudates are increased, and the cycle continues. The challenge for sustainable grassland management is to allow this cycle to function with an optimum use of inputs (nutrients and water) (Fig. 4.4.1).

The influence of management on the grass sward cannot be decoupled from its effects on soil biota. Agricultural grasslands are generally managed to maximize production of products such as milk, meat, leather, and other goods. The ability of different grassland systems to maximize production is determined by the carrying capacity of the grass-

Soil Ecology and Ecosystem Services. First Edition. Edited by Diana H. Wall et al.
© 2012 Oxford University Press. Published 2012 by Oxford University Press.

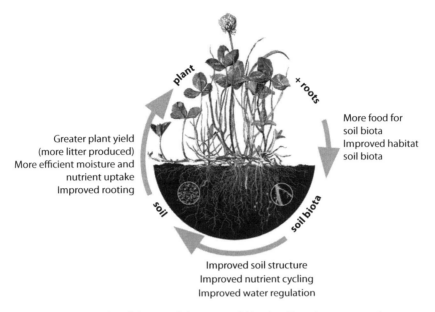

Figure 4.4.1 Cyclic interactions between above/belowground plant, roots, soil biota (root biota, decomposers, and ecosystem engineers), and soil properties (chemical and physical). (From van Eekeren et al. 2007.)

land (i.e. the number of animals per unit area). This, in turn, is determined by the productivity of the plant species present and the soil physical properties. How these plants are managed, and which species are present, can have a major influence on the soil biota and the ecosystem services they provide. From this we can make better use of the ecosystem services associated with soil biota and can help to limit inadvertent negative effects of management measures on soil biota and ecosystem functioning. This is described within a conceptual framework of cyclical interactions proposed by van Eekeren et al. (2007) (Fig. 4.4.1).

4.4.2 Plant–soil interactions

The quantity and quality of litter and root exudates (including secondary metabolites) of various herb and grass species have been shown to be major determinants of soil food web structure and soil biodiversity (Bardgett 2005). Grazing can affect soil biota through defoliation, dung and urine return, and the physical presence of animals (causing treading and excreta return in patches). These three mechanisms can again have an indirect effect on soil biota through their effect on botanical composition. The plant diversity of grasslands affects biological quality through various mechanisms: 1) the quantity and quality of resources allocated to the soil, 2) the extent to which different plant species deplete nutrients and water from soils, and 3) the modification and formation of habitats for soil biota (Wardle 2002). Wardle and Nicholson (1996) found that in grassland soils most plant species had a stimulatory effect on the microbial biomass. They showed that the biomass of saprophytic microorganisms is predominantly set by plant primary production, although it is also influenced by the number of plant species present. In a study by Proulx et al. (2010), plant diversity increased stability across trophic levels, at the community or ecosystem level. Scherber et al. (2010), however, showed that although plant diversity does impact the soil biota and their functions, the effect weakens with increasing trophic position. It is thought that the combination of diversity and rapid C flux makes the grassland soil ecosystem highly resilient. Although the soil communities appear to be relatively resilient to plant biodiversity loss, they are highly susceptible to any perturbation that affects the soil structure itself (e.g. cultivation).

According to Whipps (1990), 35–80% of the net fixed C in perennial grasses is transferred belowground. Grassland plants are able to sequester large amounts of C in the soil with an overall average sequestration rate in European grasslands of ~60 g m^{-2} year^{-1} (Janssens *et al.* 2005). The vegetation type and root architecture also contribute to soil C maintenance (De Deyn *et al.* 2008; Fissore *et al.* 2008). The variable ability of some grasses to deposit C at depth, whether by root turnover or exudation, may enable a greater degree of C storage due to the differences in the processes that govern C dynamics between topsoils and subsoils (Salomé *et al.*, 2009). New developments in grass breeding for deeper roots may allow managers to take advantage of this phenomenon and raise the possibility of managing grasslands for increased subsoil sequestration (Abberton *et al.* 2008).

There is increasing evidence that there is a need for a diversity of plant functional groups that can influence soil processes e.g. legumes have been identified as key species in facilitating a number of ecosystem processes. The inclusion of legumes in grassland systems can promote C and N storage (De Deyn *et al.* 2009), and where red clover (*Trifolium pretense*) was promoted, there were associated improvements in soil structure, respiration, and increased soil organic matter content (De Deyn *et al.* 2010). Thus, it appears that the presence of key species and the properties they impart are more important than plant diversity *per se*.

Legumes can have important effects on the soil biota. For example, Elgersma and Hassink (1997) reported a greater active soil microbial biomass and measured a higher N-mineralization in grass-clover than in grass-only swards. Ryan *et al.* (2000) reported that white clover roots had a higher infection with mycorrhizal fungi than ryegrass roots, whereas in a field experiment, De Vries *et al.* (2006) measured a higher fungal and bacterial biomass in grass-only swards than in grass-clover. Mytton *et al.* (1993) found higher drainage rates in white clover than in perennial ryegrass. Moreover, in field experiments a higher biomass of earthworms was found in clover-only than in grass-only swards (van Eekeren *et al.* 2009a). Results suggest that mixed grass-clover swards combines the positive effects of clover-only on the ecosystem service of nutrient supply with the positive effects of grass-only on soil structure maintenance.

4.4.3 Ecosystem services provided by the soil biota

Grasslands provide a number of key ecosystem services, many of which are regulated by the soil biota. These include plant production, nutrient and C cycling, maintenance of soil structure, and water regulation (Brussaard *et al.* 1997). The interactions between groups of organisms and the physical and chemical processes shape the soil as a habitat. This feeds back to affect the diversity of soil biota themselves as well as impacting ecosystem processes.

4.4.3.1 Soil structure maintenance

In intensively managed grassland systems, livestock directly affect the structure of the soil in two ways, either through compaction or poaching, which occurs when soil and vegetation on poorly drained or waterlogged sites are damaged by livestock. Grazing pressure and traffic load are the main causes of soil compaction and can affect herbage composition and plant cover. This can have a deleterious impact on the soil biota; for example, Bouwman and Arts (2000) in a grassland found a shift in trophic groups of nematodes when the soil was compacted. This shift was mainly caused by an increase in herbivorous nematodes following an increase of root biomass in the topsoil.

There are a number of ways the soil biota can ameliorate compaction. Earthworms are commonly referred to as ecosystem engineers (*sensu* Jones *et al.* 1994) due to their wide ranging impacts on the soil ecosystem. In a study of upland grassland in the UK, Cole *et al.* (2006) concluded that macrofauna, particularly earthworms, have a more profound effect on soil structure than the microflora. Earthworms, in particular, affect soil structure through producing fecal casts, promoting humification, and creating pores. In a study examining a 20-year absence of earthworms and other soil invertebrates (due to pesticide treatment in perennial ryegrass pasture in the UK) (Clements *et al.* 1991) soil bulk density had increased, while there was a decrease in penetrability, soil organic matter content, initial infiltration rate, and soil moisture content. Earthworm activity in soil has also been related to improved nutrient cycling and enhanced plant productivity (Bhadauria & Saxena 2010).

4.4.3.2 Water regulation

Water regulation is closely related to soil structure maintenance. The importance of soil biota for water regulation was shown by the increased waterlogging of Scottish grassland soils where flatworm predation had significantly reduced earthworm populations (Haria *et al.* 1998). Water infiltration through macropores and stable crumb formation are two key soil processes strongly affected by earthworms. The presence of earthworms can reduce surface runoff due to the enhancement of soil porosity caused by burrowing, but there is then the possibility that this may lead to increased bypass flow and nutrient leaching.

Grass rooting depth is important for the drought resistance of grasslands. Earthworm burrows provide pathways for root penetration throughout the soil profile. In particular, the common deep-burrowing, surface-feeding earthworm species *Lumbricus terrestris* and *Aporrectodea longa* generally make vertical burrows and are able to penetrate hard pans and aid root growth.

4.4.3.3 Nutrient cycling

All groups of soil biota are involved in nutrient cycling. Bacteria and fungi contribute to this supporting service through nutrient mineralization and immobilization. In many systems plant litter is deposited on the surface of the soil where it is subsequently mineralized and incorporated. In cropped grassland systems (such as for hay or silage), the amount of litter deposited is limited because of the way the grass is cut and removed. In grazed pastures, a greater part of the herbage consumed by livestock is directly deposited on the ground through excretion of both dung and urine. This results in a more heterogeneous distribution of C and N within the pasture. In more intensively managed systems there is a greater use of liquid slurries; however, these animal returns may lead to increased N losses through leaching and volatilization. Soil organic matter content and soil biological activity can be enhanced through inorganic as well as organic fertilizers, especially when initial N levels are low. Inorganic fertilizers (containing only mineral N) feed the plant and soil micro-organisms directly and the entire soil biota indirectly by increased root biomass and exudates, and plant litter. Inorganic fertilizers, however, involve high fossil fuel energy consumption and are easily lost from the soil by nitrate leaching and denitrification. In grassland soils, organic fertilization compared to inorganic fertilization, has been shown to increase the organic C, the total N, the activity of decomposers, and the supply of nutrients via the soil food web.

Studies on the effect of specific quality aspects of organic fertilizers (higher C:N and lower mineral N:total N) on soil biota are rare. Griffiths *et al.* (1998) observed that the number of protozoa responded more quickly to the application of pig slurry than cattle slurry, and explained this by the greater proportion of readily-available C in pig slurry compared to that of cattle slurry. Van Eekeren (2009b) measured different organic fertilizers and found a higher bacterial activity and the highest amount of mineralizable N with normal manure slurry compared to inorganic fertilizers. This suggests a positive effect on the supply of nutrients and water regulation.

Nitrate leaching and denitrification are the reasons why N is predominantly lost from grassland pastures. Denitrification is a microbially mediated process involving the reduction of NO_3^- to N_2, under anaerobic conditions, the intermediaries being NO_2^-, NO, and N_2O. For denitrification to take place there needs to be a readily available supply of substrates and low oxygen concentrations. Efforts to increase denitrification through to N_2 may prove useful in reducing some of the negative impacts of excess N in soils. There are also natural nitrification inhibitors present in some grass plants; for example, the tropical pasture grass *Brachiaria* spp. inhibits *Nitrosomonas* function and therefore suppresses N_2O emissions.

4.4.3.3.1 Microorganisms
The soil microbial biomass is mostly composed of bacteria and fungi. A number of factors, such as soil

moisture, pH, and substrate availability, affect the relative abundance of bacteria and fungi. Bacteria and fungi represent a large pool of N and C within the soil and increases in soil fertility can induce shifts from fungal- to bacterial-based energy channels in soil food webs (Bardgett & McAlister 1999). Intensively managed systems tend to have decreased microbial biomass and favor the bacterial-based energy channel. This has important consequences for the soil food web. Grasslands with higher N input have lower fungal to bacterial ratios, which affect the level of N leaching (De Vries et al. 2006).

The soil microbial population represents a large store of phosphorus (P) in organic forms that is a potential source of inorganic P for crops—either directly or by replenishing the inorganic pools. There is a general understanding that some organic P will be mineralized as cells die; the rate, precise mechanisms and controlling factors have been poorly investigated. For N mineralization the role of faunal grazing as an accelerating mechanism for N release has been demonstrated (e.g. Bonkowski 2004) but this has not been demonstrated for P, although it seems logical that it should occur.

4.4.3.3.2 Symbiosis

In grassland systems symbiosis between plants and soil biota are extremely important for plant productivity, especially where essential nutrients are limited. Phosphorus is typically very immobile in soil, and the supply to roots or microbes is extremely slow. Mycorrhizal fungi form mutualistic associations with plant roots, where the fungus derives C from the host plant and forms extensive mycelial networks through the soil, absorbs P and transfers this directly to the plant. Since such hyphae are typically of the order of a few microns in diameter, they are able to explore and exploit considerably larger volumes of soil per unit C than plant roots. For example, Van der Heijden et al. (1998) found positive mycorrhizal effects on shoot P concentrations and shoot biomass of grasses such as *Bromus* spp. and *Festuca* spp. The filamentous structure of the mycelium provides an extensive surface area, where 1 cm^3 of soil can contain 1 km of fungal hyphae with a surface area of more than 300 cm^2. Some groups of soil biota (e.g. Collembola) potentially disrupt mycorrhizal linkages with host plants and consequently may have important impacts on herbage production (Jonas et al. 2007).

Legumes have long been important in grassland agriculture due to their high feeding value and their ability to form symbiotic relationships with N-fixing bacteria. In agricultural grasslands, this symbiotic association between legumes and some bacteria (including members of the orders *Rhizobiales* and *Burkholderiales*) are significant and provide around 100 kg ha^{-1} N equivalent, increasing the yields of herbage obtained without additional mineral N fertilization of the crop. Considerable effort is being focused on exploiting the potential of white clover in animal production systems which can meet the financial and environmental requirements likely to prevail. In the UK by far the most important forage legume is white clover. It is included in 75% of grassland seed mixtures. Although the presence of white clover in swards is desirable, both for its N-fixing capability and its enhanced feeding value for livestock, white clover also makes a significant contribution to the botanical composition in a maximum of 20% of UK swards. The failure of clover to thrive in pasture systems is, in part, due to the impact of pest and diseases on the seedling crop grassland production.

4.4.3.3.3 Microfauna

Nematodes and protozoa affect nutrient cycling processes indirectly through grazing on soil microbial biomass and nutrient excretion. Griffiths (1989) observed that the N content of ryegrass increased by 14% when nematodes or protozoa were added to microcosms with bacteria. Where grassland is characterized by N limitation (e.g. tallgrass prairie), herbivorous nematode densities respond positively to increased root inputs that occur following fertilization (Todd et al. 1999). Although plants may directly affect soil pH, their influence on soil water content may be of greater importance. Plants may change soil water status through increased evapotranspirational losses due to increased plant growth when subjected to fertilizer addition, or through climate change effects.

Not only microbivorous nematodes are involved in nutrient cycling, but also plant parasitic nematodes and herbivorous nematodes. In experiments with clover, low levels of root infestation by clover cyst nematodes (*Heterodera trifolii*) positively influenced the rhizosphere microbial community in the soil (Yeates *et al*. 1998b). Transfer of N from legumes to companion grasses is generally caused by the death and decomposition of the root tissue which is relatively high in N. Root herbivory by nematodes, however, has been found to increase the root growth of white clover and perennial ryegrass; such herbivory of white clover roots may enhance the flux of clover N to the soil, which is subsequently recycled and taken up by the neighboring ryegrass plants (Bardgett *et al*. 1999). Similar results have also been demonstrated for larger invertebrates. For example, the larvae of the clover weevil (*Sitona* spp.) have been shown to facilitate the transfer of N from clover to companion ryegrass (Murray & Hatch 1994).

4.4.3.3.4 Macrofauna
Macrofauna affect nutrient cycling processes directly, through fragmentation and the transport of organic and mineral particles, and indirectly through regulating microbial communities and stimulating microbial activity. The role of the soil fauna in decomposition has received considerable attention over the last three decades, but is still not adequately defined. The animals in the soil may exert a major influence on the decomposition through interactions with the microflora. The available data on faunal contribution to N mineralization indicate a larger role than their metabolic rate would suggest. This may be due to their low production efficiency, and low N to C requirement. This means that excess N is returned to the soil in excreta in forms that are readily available to the microbial biomass. Subsequently, this biomass provides substrate for the microbivorous fauna, particularly protozoans and nematodes, resulting in the rapid turnover of a small but important pool.

In grassland, earthworms contribute significantly to organic matter fragmentation through breaking down the sod. Hoogerkamp *et al*. (1983) observed that the introduction of earthworms to grasslands in reclaimed polders resulted in the development of dark-colored top soil within 3 years after introduction. In an experiment by Clements *et al*. (1991), the most apparent effect of a pesticide treatment that excluded earthworms was the accumulation of litter. Next to fragmentation, the transport and mixing of organic and mineral particles is an important function of earthworms. In a glasshouse study with perennial ryegrass and rock phosphate, the presence of earthworms resulted in a higher yield, not only through the better mixing of rock phosphate by earthworm activity but also due to increased availability of P in worm casts (Mackay *et al*. 1982).

4.4.3.4 Herbage production

For farmers, grass production (quantity and quality) is the ultimate ecosystem service, in which the maintenance of soil structure, water regulation, and particularly nutrient supply play prominent roles. The section on nutrient supply provides several examples of positive effects of soil biota on root growth, shoot N and P content, and yield. It appears that the positive effects of soil biota on grass production under field circumstances are most obvious for earthworms, particularly when earthworms are introduced to soils where they were previously absent (Hoogerkamp *et al*. 1983).

Soil biota can, however, adversely impact grass production. Many of the invertebrates found in the soil are phytophagous and there are only limited data on the effects they may have on the growth and development of pasture plants. Such plants can cope well with aboveground herbivory, through traits such as placement of meristems near the ground and/or the ability to hold nutrient reserves in organs such as stolons and roots, that enable rapid regrowth following foliar herbivory. They may, however, be less well adapted to belowground pressures; and plant responses to belowground herbivory can be variable and will depend on the ability of the plant to compensate for the functions lost.

Root herbivores are known to have many effects on their host plants, for example, they damage and consume plant roots, sever roots segments, and increase the turnover rate of root tissue. A general loss of root material may be a particular handicap to plant performance in terms of nutrient and water absorbing capacity, but damage to more specialized structures, such as the root nodules of legumes represents both a direct loss of plant material and an impairment of the N-fixing capability. Belowground herbivory may also affect the plant composition through the selective suppression of dominant plant species (De Deyn et al. 2003).

4.4.4 Impact of management intensity of grassland systems

Many managed grasslands are established with a limited number of species. For example, a mixture predominantly of *Lolium perenne* and *Trifolium repens* is used in the UK. As pastures age, they tend to become more botanically diverse and less productive. If there is a desire to restore productivity, there are a number of options including fertilization or lime ($CaCO_3$) addition.

In intensively managed pastures, such as those commonly found in dairy systems, the carrying capacity is high. These systems are comprised of fast growing, highly nutritious forage species with a high N fertilizer input providing a large amount of animal feed. This results in a significant return of nutrients to the soil in feces and urine. These systems are managed so that plant diversity is low and that the litter that is returned to the soil has a high N and low C content, and that there is a high rate of root turnover. These traits, in turn, lead to the development of bacterial based food webs with a fast turnover of organic matter and nutrient mineralization, and with a low C sequestration. In intensive grassland systems, the importance of soil organisms has often been ignored, as physical manipulation of the soil and nutrient supply are increasingly provided by human inputs rather than by natural processes (Brussaard et al. 1997). At the other end of the scale, in semi-natural systems, plant species diversity tends to be high, with slower growing and less productive species and with poorer feeding value. In these systems the use of N fertilizers is low and there is a smaller carrying capacity. There is also a smaller amount of natural return of animal waste. The plants in these grasslands tend to have a low N and high C content, and their litter returns reflect this. The soils in these systems tend to have soil food webs dominated by fungi and micro-arthropods such as mites and Collembola (Wardle et al. 2004). With densities in lowland grassland reaching over 200,000 m^2, they have a large impact on decomposition processes. Despite this, decomposition tends to be slower and leads to high C sequestration.

Most grasslands are somewhere in between these two extremes. This is illustrated by a study by Rutgers et al. (2008) in the Netherlands (Table 4.4.1): agricultural land use represented the most fertile soil conditions while heathland and mixed forest represented the most infertile conditions. Bacterial biomass, and the abundance of nematodes and earthworms (bold-faced numbers in Table 4.4.1), are higher in agricultural and semi-natural grassland, whereas fungi and especially micro-arthropods are more abundant in heathland and mixed forests.

Studies have reported changes in the composition of soil biota on sites with an arable cropping history followed by the establishment of perennial pasture or abandonment of cultivation. Yeates et al. (1998a) found that earthworm populations increased when a perennial pasture was established on sites formerly under arable cropping. In a long-term study comparing a ley-arable crop rotation with permanent grassland, most soil biota were found to recover quickly from declines during the arable phase of the crop rotation, reaching overall abundance and activity levels in the ley phase comparable to those observed in permanent grassland. The physiological diversity of the bacteria community and the anecic earthworm abundance were, however, significantly reduced in the ley-arable crop rotation compared to permanent grassland. This may impair the ecosystem services of the supply of nutrients and water regulation. The combination of no disturbance and a stable supply of resource in permanent grassland probably keep these ecosystem services at a high level (van Eekeren et al. 2010).

4.4.5 Trade-offs between ecosystem services

In general the effect of management intensity on soil biota and ecosystem services are not independent of one other. For agriculture, the problem is typically posed as trade-off between provisioning services—for example, production of agricultural goods such as food, fiber, or bioenergy—and regulating services such as water purification, soil conservation, or C sequestration. The potential for a

Table 4.4.1 Soil biota in sandy soils with different land uses in the Netherlands (Rutgers et al. 2008). Bold numbers indicate higher values for biotic groups in certain ecosystems

Land use	Unit (no. of locations)	Arable farms (34)	Dairy farms (87)	Semi-natural grassland (10)	Heathland (10)	Mixed forest (20)
Soil chemical parameters						
SOM[b]	g kg dry soil^{-1}	75	64	93	73	57
C:N[b]	N	20	14	18	34	25
pH-KCl[b]		5.2	5.2	4.5	3.2	3.2
P-Al[b]	mg P$_2$O$_5$ 100g soil^{-1}	56	54	27	2	3
P-total[b]	mg P$_2$O$_5$ 100g soil^{-1}	105	149	144	41	19
Soil biological parameters						
Earthworm number[a]	n m^{-2}	38	**187**	133	0	9
Earthworm taxa[a]	N	2.0	4.6	6.8	0	0.7
Enchytraeid number[a]	n 10^3 m^{-2}	22	24	13	14	21
Enchytraeid taxa[a]	N	8.1	8.2	14	6.2	4.7
Micro-atropod number[a]	n 10^3 m^{-2}	23	46	101	**157**	**150**
Micro-atropod taxa[a]	n m^{-2}	22	26	24	22	59
Nematode number[b]	n 100g soil^{-1}	3717	**4926**	**5054**	2053	730
Nematode taxa[b]	N	27	32	36	22	25
Bacterial biomass[b]	µg C g dry soil^{-1}	88	146	204	75	47
Bacterial activity[b]	Thy incorp.	67	66	17	4	2
Bacterial diversity	N DNA bands	68	51	—	—	25
Fungal biomass[b]	µg C g dry soil^{-1}	—	22	24	**53**	—
CLPP, ES50[b][c]	µg dry soil	1415	637	324	9293	39712
CLPP slope[b][d]		0.55	0.57	0.35	0.41	0.60

[a] Data for dairy farm land use were collected in grassland only.
[b] Data for dairy farm land use were collected in grassland as well as the arable land of the same dairy farms.
[c] CLPP ES50: the amount of soil extract needed to convert 50% of all substrates in ECO plates. This is a measure for the physiological activity of the bacterial community. A low ES50 indicates a high activity.
[d] CLPP slope: a measure of the physiological diversity of the bacterial community. A low slope indicates a high diversity.

"win–win" management increases with an increased awareness of ecosystem services and effective methods for evaluating these services. Even for the supporting ecosystem services discussed in this chapter we see trade-offs between other ecosystem services. Improving water infiltration can require more earthworms, which are stimulated by a higher quantity and quality of resource (more production) and for water retention you need organic matter. For a better uptake of water and nutrients by grasses you need a higher density and deeper roots, but with lower fertilization there are more roots. Fine grass roots with their rhizosphere have an important positive effect on soil structure and the latter will also improve water infiltration.

The impact on soil structure is different, in that the intensively managed pastures are often regularly ploughed and reseeded. This will help maintain structure. As the pastures age there is likely to be increased treading and machinery movements that will increase compaction. As the degree of management reduces further, and the stocking densities reduce further, there is less compaction pressure. The biota which are the ecosystem engineers (e.g. earthworms, termites, etc.) can often overcome the management pressures depending on the climate and soils.

In production-orientated grassland systems the goal is to produce as much product as possible, whether this is herbage, milk, or meat, and the management practices employed by the farmer are to achieve this. High productivity is, however, almost always at the expense of other ecosystem services. If high levels of inorganic fertilizer are used, then there is the potential for increased losses in diffuse ground pollution or gaseous emissions. The stocking rate is usually higher on the more productive swards due to the increased availability of forage. This may have a significant influence on microbial communities in grassland soils because of the returns from grazing animals. If animals are housed for longer periods, then there is the problem of how to best utilize the manure and slurries that may cause potential pollution and also impacts the soil biota.

In more extensive systems, the effects are reduced, but at the expense of production. In an experiment where there was a cessation of fertilizer input, De Deyn et al. (2010) found that plant species diversity increased as did the rate of soil C and N accumulation. There was, however, a concomitant reduction in plant biomass both above- and belowground. The increasing use of N impacts the relative abundance of fungi and bacteria in the soil, with implications for nutrient cycling and C storage.

4.4.6 Conclusions

The wide range of management practices in grasslands impacts ecosystem services at different temporal and spatial scales. In many temperate grasslands the level of management intensity is governed by two factors, carrying capacity and fertilizer usage. The two are almost always inextricably linked.

The functional relationships between the intensity of grassland management and ecosystem services are complex and, by definition, interactive and additive. Production per unit area declines as the intensity of management decreases, from highly fertilized and stocked to unfertilized and loosely grazed and unfertilized. The effect of the management intensity is different and often conflicting for the other ecosystem services discussed in this chapter. To improve water infiltration you need more earthworms, which are stimulated by a higher quantity and quality of resource (more production). For water retention you need organic matter. For improved uptake of water and nutrients by grass you need a higher density of roots at a greater depth. With lower fertilization, however, there are more roots. Fine grass roots and consequently the rhizosphere around them, have an important positive effect on soil structure. The latter will also improve water infiltration. The impact on soil structure is different, in that the intensively managed pastures are often regularly ploughed and reseeded which will help maintain structure. As the pastures age, there is likely to be increased treading and machinery movements which will increase compaction. As the degree of management reduces further and the stocking densities reduce further there is less compaction pressure and the biota, the ecosystem engineers (e.g. earthworms, termites, etc.), can often overcome the management pressures, although there are exceptions.

There are few studies where N has been added to bare soil and it is therefore difficult to differentiate the direct effects of fertilizer addition from those indirect ones manifested through the plants, for example, plant derived nutrients and litter inputs. There has been relatively little consideration given to how plant-mediated changes in soil abiotic conditions influence soil communities. While it is important to determine the direct effects of nutrient addition and plants on soil communities, it is perhaps more important to take a more holistic approach. We need to consider how increased plant growth may influence the soil physical environment and how such environmental changes may in turn affect the soil biota. For example, Murray *et al.* (2006) describe how fertilizer addition to upland grassland resulted in increased plant growth, with concomitant increase in evapotranspiration, which in turn created a drier soil environment that impacted the soil communities.

The introduction of a legume appears to play a more important role in how soil biota function, rather than any change in the composition of the C that is derived from that plant. The introduction of such species may potentially reduce the overall productivity of the sward, but does promote other ecosystem services including soil structure, water retention, biodiversity, and C and N storage. Therefore we can ask whether we should sacrifice a degree of production to enhance delivery of the ecosystem services (which can be more difficult to price). The trade-offs required are dependent on the goals of the land managers, and it may be that the same goals do not apply for each individual field, farm, or region. In all these scenarios there are optima which may be achieved, usually at an intermediate level. It is therefore our contention that a moderate management intensity will deliver the best use of the supporting ecosystem services discussed, while at the same time having the least compromise on actual grassland production.

References

Abberton, M.T., Marshall, A.H., Humphreys, M.W., Macduff, J.D., Collins, R.P., Marley, C.L. (2008) Genetic improvement of forage species to reduce the environmental impact of temperate livestock grazing systems. *Advances in Agronomy* **98**: 311–55.

Bardgett, R.D. (2005) *The Biology of Soil: a community and ecosystem approach*. Oxford University Press, New York.

Bardgett, R.D., Denton, C.S., & Cook, R. (1999) Belowground herbivory promotes soil nutrient transfer and root growth in grassland. *Ecology Letters* **2**: 357–60.

Bardgett, R.D., & McAlister, E. (1999) The measurement of soil fungal: bacterial biomass ratios as an indicator of ecosystem self-regulation in temperate meadow grasslands. *Biology and Fertility of Soils* **29**: 282–90.

Bonkowski, M. (2004) Protozoa and plant growth: the microbial loop in soil revisited. *New Phytologist* **162**: 617–31.

Bouwman, L.A., & Arts, W.B.M. (2000) Effects of soil compaction on the relationships between nematodes, grass production and soil physical properties. *Applied Soil Ecology* **14**: 213–22.

Brussaard, L., Behan-Pelletier, V.M., Bignell, D.E., *et al.* (1997) Biodiversity and ecosystem functioning in soil. *Ambio* **26**: 563–70.

Clements, R.O., Murray, P.J., & Sturdy, R.G. (1991) The impact of 20 years absence of earthworms and 3 levels of N-fertilizer on a grassland soil environment. *Agriculture Ecosystems & Environment* **36**: 75–85.

Cole, L., Bradford, M.A., Shaw, P.J.A., & Bardgett, R.D. (2006) The abundance, richness and functional role of soil meso- and macrofauna in temperate grassland—A case study. *Applied Soil Ecology* **33**: 186–98.

De Deyn, G.B., Cornelissen, J.H.C., & Bardgett, R.D. (2008) Plant functional traits and soil carbon sequestration in contrasting biomes. *Ecology Letters* **11**: 516–31.

De Deyn, G.B., Quirk, H., Yi, Z., Oakley, S., Ostle, N.J., & Bardgett, R.D. (2009) Vegetation composition promotes carbon and nitrogen storage in model grassland communities of contrasting soil fertility. *Journal of Ecology* **97**: 864–75.

De Deyn, G.B., Raaijmakers, C.E., Zoomer, H.R., *et al.* (2003) Soil invertebrate fauna enhances grassland succession and diversity. *Nature* **422**: 711–13.

De Deyn, G.B., Shiel, R.S., Ostle, N.J., *et al.* (2010) Additional carbon sequestration benefits of grassland diversity restoration. *Journal of Applied Ecology* **48**: 600–8.

De Vries, F.T., Hoffland, E., van Eekeren, N., Brussaard, L., & Bloem, J. (2006) Fungal/bacterial ratios in grasslands with contrasting nitrogen management. *Soil Biology & Biochemistry* **38**: 2092–103.

Elgersma, A., & Hassink, J. (1997) Effects of white clover (*Trifolium repens* L.) on plant and soil nitrogen and soil organic matter in mixtures with perennial ryegrass (*Lolium perenne* L.). *Plant and Soil* **197**: 177–86.

Fissore, C., Giardina, C.P., Kolka, R.K., *et al.* (2008) Temperature and vegetation effects on soil organic carbon quality along a forested mean annual temperature gradient in North America. *Global Change Biology* **14**: 193–205.

Griffiths, B.S. (1989) Enhanced nitrification in the presence of bacteriophagous protozoa. *Soil Biology & Biochemistry* **21**: 1045–51.

Griffiths, B.S., Wheatley, R.E., Olesen, T., Henriksen, K., Ekelund, F., & Ronn, R. (1998) Dynamics of nematodes and protozoa following the experimental addition of cattle or pig slurry to soil. *Soil Biology & Biochemistry* **30**: 1379–87.

Haria, A.H., McGrath, S.P., Moore, J.P., Bell, J.P., & Blackshaw, R.P. (1998) Impact of the New Zealand flatworm (*Artioposthia triangulata*) on soil structure and hydrology in the UK. *Science of the Total Environment* **215**: 259–65.

Harvey, G. (2001) *The forgiveness of nature: the story of grass*. Johnathan Cape Ltd, London.

Hoogerkamp, M., Rogaar, H., & Eysakkers, H.J.P. (1983) Effects of earthworms on grassland on recently reclaimed polder soils in the Netherlands. In: J.E. Satchell (ed.), *Earthworm Ecology: from Darwin to Vermiculture*, pp. 85–105. Chapman and Hall, London.

Janssens, I.A., Freibauer, A., Schlamadinger, B., et al. (2005) The carbon budget of terrestrial ecosystems at country-scale—a European case study. *Biogeosciences* **2**: 15–26.

Jonas, J.L., Wilson, G.W.T., White, P.M., & Joern, A. (2007) Consumption of mycorrhizal and saprophytic fungi by Collembola in grassland soils. *Soil Biology & Biochemistry* **39**: 2594–602.

Jones, C.G., Lawton, J.H., & Shachak, M. (1994) Organisms as ecosystem engineers. *Oikos* **69**: 373–86.

Lavelle, P., Decaens, T., Aubert, M., et al. (2006) Soil invertebrates and ecosystem services. *European Journal of Soil Biology* **42**: S3–S15.

Mackay, A.D., Syers, J.K., Springett, J.A., Gregg, P.E.H. (1982) Plant availability of phosphorus in super-phosphate and a phosphate rock as influenced by earthworms. *Soil Biology & Biochemistry* **14**: 281–7.

Murray, P.J., Cook, R., Currie, A.F., et al. (2006) Interactions between fertilizer addition, plants and the soil environment: Implications for soil faunal structure and diversity. *Applied Soil Ecology* **33**: 199–207.

Murray, P.J., & Hatch, D.J. (1994) Sitona weevils (Coleoptera, Curculionidae) as agents for rapid transfer of nitrogen from white clover (*Trifolium-Repens* L) to perennial ryegrass (*Lolium-Perenne* L). *Annals of Applied Biology* **125**: 29–33.

Mytton, L.R., Cresswell, A., & Colbourn, P. (1993) Improvement in soil structure associated with white clover. *Grass and Forage Science* **48**: 84–90.

Proulx, R., Wirth, C., Voigt, W., et al. (2010) Diversity promotes temporal stability across levels of ecosystem organization in experimental grasslands. *PLoS One* **5**: e13382. doi: 10.1371/journal.pone.0013382.

Rutgers, M., Mulder, C., Schouten, A.J., et al. (2008) *Soil Ecosystem profiling in the Netherlands with ten references for biological soil quality. Report 607604009*. RIVM, Bilthoven.

Ryan, M.H., Small, D.R., & Ash, J.E. (2000) Phosphorus controls the level of colonisation by arbuscular mycorrhizal fungi in conventional and biodynamic irrigated dairy pastures. *Australian Journal of Experimental Agriculture* **40**: 663–70.

Salomé, C., Nunan, N., Pouteau, V., Lerch, T.Z., & Chenu, C. (2009) Carbon dynamics in topsoil and in subsoil may be controlled by different regulatory mechanisms. *Global Change Biology* **16**: 416–26

Scherber, C., Eisenhauer, N., Weisser, W.W., et al. (2010) Bottom-up effects of plant diversity on multitrophic interactions in a biodiversity experiment. *Nature* **468**: 553–6.

Todd, T.C., Blair, J.M., & Milliken, G.A. (1999) Effects of altered soil-water availability on a tallgrass prairie nematode community. *Applied Soil Ecology* **13**: 45–55.

van der Heijden, M.G.A., Klironomos, J.N., Ursic, M., et al. (1998) Mycorrhizal fungal diversity determines plant biodiversity, ecosystem variability and productivity. *Nature* **396**: 69–72.

van Eekeren, N., Bos, M., de Wit, J., Keidel, H., & Bloem, J. (2010) Effect of individual grass species and grass species mixtures on soil quality as related to root biomass and grass yield. *Applied Soil Ecology* **45**: 275–83.

van Eekeren, N., van Liere, D., De Vries, F., Rutgers, M., de Goede, R., & Brussaard, L. (2009a) A mixture of grass and clover combines the positive effects of both plant species on selected soil biota. *Applied Soil Ecology* **42**: 254–63.

van Eekeren, N., de Boer, H., Bloem, J., et al. (2009b) Soil biological quality of grassland fertilized with adjusted cattle manure slurries in comparison with organic and inorganic fertilizers. *Biology and Fertility of Soils* **45**: 595–608.

van Eekeren, N., Murray, P.J., & Smeding, F.W. (2007) Soil biota in grassland, its ecosystem services and the impact of management. In: *Permanent and Temporary Grassland: plant, environment and economy. Proceedings of the 14th Symposium of the European Grassland Federation*, pp. 247–58. Ghent, Belgium, 3–5 September 2007. Belgian Society for Grassland and Forage Crops, Merlbeke Belgium,

Wardle, D.A. (2002) *Communities and Ecosystems: Linking the Aboveground and Belowground Components. Monographs in Population Biology 34*. Princeton University Press, Princeton, NJ.

Wardle, D.A., Bardgett, R.D., Klironomos, J.N., Setala, H., van der Putten, W.H., & Wall, D.H. (2004) Ecological linkages between aboveground and belowground biota. *Science* **304**: 1629–33.

Wardle, D.A., & Nicholson, K.S. (1996) Synergistic effects of grassland plant species on soil microbial biomass and activity: Implications for ecosystem-level effects of enriched plant diversity. *Functional Ecology* **10**: 410–16.

Whipps, J.M. (1990) Carbon economy. In: Lynch, J.M. (ed.) *The Rhizosphere*, pp. 59–98. Wiley Interscience/John Wiley and Sons, Chichester,

Yeates, G.W., Saggar, S., Denton, C.S., Mercer, C.F., 1998b. Impact of clover cyst nematode (*Heterodera trifolii*) infection on soil microbial activity in the rhizosphere of white clover (*Trifolium repens*)—A pulse-labelling experiment. *Nematologica* **44**: 81–90.

Yeates, G.W., Shepherd, T.G., & Francis, G.S. (1998a) Contrasting response to cropping of populations of earthworms and predacious nematodes in four soils. *Soil & Tillage Research* **48**: 255–64.Soil Ecology and Ecosystem Services. First Edition. Edited by Diana H. Wall *et al.*

Synthesis

Richard D. Bardgett and T. Hefin Jones

The chapters in this section clearly demonstrate the potential for human activities to have profound impacts on soils, their biodiversity, and their capacity to deliver ecosystem services. Moreover, they amply illustrate how decisions about the future management of soils for ecosystem services will need to take into account how they are being, and will be, affected by multiple global change phenomena in the future, and how soil management might actually be changed in order to mitigate global change, for example, through carbon storage in soil. In other words, future soil management strategies need to be developed in the context of global change. Across the chapters of this section, certain themes emerge. First, it is evident that feedbacks between plant and soil biological communities are central to understanding how ecosystem processes and the services that they underpin are affected by global change. As has been highlighted elsewhere, aboveground and belowground components of ecosystems have traditionally been considered in isolation from one another. However, as illustrated in these chapters, it is now widely appreciated that biotic interactions between aboveground and belowground communities play a fundamental role in regulating the response of terrestrial ecosystems to human-induced global change. This is amply illustrated in Chapters 4.1 and 4.2, where the effects of climate change and nitrogen deposition on ecosystem services is shown to often be indirect, being mediated by feedbacks between plant and soil biological communities. And, as shown in Chapter 4.3 and 4.4, the delivery of ecosystem services in urban environments and managed grasslands are strongly dependent on interactions between plants and soil biological communities.

Second, it is becoming increasingly recognized that global change phenomena have both direct and indirect effects on soils and their biodiversity, and that such effects create feedbacks that influence ecosystem functioning. This understanding is probably most developed in relation to climate change. For example, as detailed in Chapter 4.1, there are now many studies that show that climate change can impact directly on soil organisms and ecosystem carbon cycling through changes in temperature, precipitation, and extreme climatic events, and indirectly via climate-driven changes in plant productivity and community composition. This in turn alters soil physicochemical conditions, the supply of carbon to soil, and the structure and activity of soil communities involved in decomposition processes and carbon loss from soil. Likewise, as shown in Chapter 4.2, several studies demonstrate that nitrogen enrichment, which has increased substantially as a result of anthropogenic activity, can significantly affect soils and the ecosystem services that they deliver through strong direct effects on soil biota and chemistry, and through indirect effects on vegetation change.

Third, global change effects on soils need to be fully integrated into future land management strategies. This is essential for two reasons: first, in order to minimize future negative impacts of global change on soils and their capacity to deliver a range of ecosystem services, and, second, to potentially mitigate global change factors themselves, for instance through soil carbon storage which can contribute to climate change mitigation. These general issues are touched on in all chapters, but as discussed in Chapter 4.4, central to this is the need to understand the likely trade-offs among ecosystem services that will results from different land management strategies in the face of global change.

Finally, as highlighted in our introduction, the majority of studies to date that have explored effects of global change on soils, their biodiversity, and functioning have considered single factors, as also done in this series of chapters. However, there is much potential for interactions between global change factors to have additive or antagonistic effects on soils and the ecosystem services that they deliver. Remarkably little is known about the combined effects of global changes on soils, but they will likely have the potential to amplify, suppress, or perhaps even neutralize global change driven effects on soils. Such unanticipated effects of multiple drivers acting simultaneously create major challenges in predicting future responses to global environmental change, and future experimental studies that simultaneously vary two or more global change drivers are hence required to improve our understanding of how soils and their services will respond to global change.

SECTION 5

Sustainable Soils

SECTION EDITORS: **Johan Six and Jeffrey E. Herrick**

Introduction

Johan Six and Jeffrey E. Herrick

The contribution of soil biota to sustainability: it depends. Or does it?

The six chapters in the final section of the book address the role of soil biota in a wider context of maintaining soil resources for *sustaining* the Earth's capacity to provide critical ecosystem services. This is quite possibly the most important, and frequently overlooked, function that soil biota provide. Without soil biota, the soil resource as we know it would not exist, and soil recovery in most ecosystems would proceed at orders of magnitude less than a snail's pace. Net primary productivity would be dramatically reduced by much slower rates of nutrient cycling, and therefore lower plant nutrient availability and nutrient effects would be exacerbated by water limitations as the lack of a stable soil structure would maintain relatively low plant available water holding capacities and even lower infiltration rates.

Soil organisms are responsible for ensuring that the nutrients necessary to produce the ear of corn that we will consume next year will be made available from last year's roots and residues. They are also responsible for sustaining the soil's capacity to provide us with food in the decades and centuries to come. While fertilizer additions can temporarily maintain production by substituting for biologically-mediated nutrient cycling, only soil biota can create and maintain stable soil structure. Soil structural stability increases resistance to soil erosion. Higher infiltration rates associated with better soil structure effectively decrease the erosivity of storms, further reducing soil loss and degradation. The rate and extent of recovery following degradation (resilience; sensu Holling & Meffe 1996; Seybold *et al.* 1999) are similarly supported by soil biotic activity.

A world without soil biota is not necessarily inconsistent with local increases in provisioning services (Millennium Ecosystem Assessment 2005), as illustrated by the success of hydroponics and biologically simplified fertigation systems on sandy soils; but, these systems, at least in their current form, are unsustainable in the absence of significant external inputs. At the same time, the sheer complexity and diversity of soil biota, and their resistance and resilience in response to disturbance, has, ironically enough, made it virtually impossible to identify a universally consistent role of specific groups of soil organisms for sustaining ecosystem services.

In the following chapters, however, the authors only implicitly acknowledge that if the big elephants (all soil biota) were to leave all of the rooms (ecosystems) in the global house, the house would collapse. Instead, they explicitly focus on the precise nature of the elephants and the characteristics (other than their size and ubiquitousness) that define their contributions to sustaining soil resources and related ecosystem services under particular sets of conditions. For decades, wildlife biologists were puzzled by the reported, but unconfirmed, existence of a "dwarf" elephant, hidden deep in the forests of Central Africa. The discovery that the "dwarfs" were simply juvenile forest-savanna elephant hybrids was disappointing to some, but trivial in comparison to the virtually simultaneous recognition that forest and savanna elephants play a functionally similar role in maintaining herbaceous "gaps" in their respective habitats (Western 2002). The study of soil biology may be at a similar stage: while we dither about the existence and importance of individual species and functional groups, we frequently fail to simply recognize what life would be like in the absence of them all.

The primary objective of this section of the book is not to recognize the importance of soil biota for sustaining soil resources and related ecosystem services; rather it is to define the frontiers of our understanding of the complexity of the diverse roles that soil biota play in sustainable ecosystems, and to identify critical knowledge gaps. Nearly all of the chapters in this section effectively describe the limitations to our understanding of the contributions of specific groups of soil biota. They identify the contradictions found in the results of reductionist and more holistic approaches alike, and attribute the inconsistencies to site- and disturbance-specific interactions with the soil biotic community. Each chapter also identifies what we *do* know, what we *might* know, and how this new knowledge could help increase our ability to support the development of sustainable management systems that take full advantage of the contributions of the soil biotic community to a sustainable soil resource and, in general, sustainable ecosystem functioning.

Chapter 5.3 by Karlen begins with a quote attributed to Plato where he describes degraded soils as "the skeleton of a sick man." Karlen continues this chapter with a description of the re-emergence of the concept of soil health and associated emphasis on soil biota during the past several decades, and then continues with a brief description of two approaches to assessment and monitoring. In Chapter 5.2, Barrios, Sileshi, Shepherd, and Sinclair deepen the discussion of soil health monitoring systems with an emphasis on the value of local knowledge, and the potential for systematic, coordinated strategies. This is also the only chapter in the book addressing the unique role of trees in fostering the provision of soil-based ecosystem services. This is an important and often underappreciated issue, particularly in light of the authors' assertion that "nearly half of all agricultural land has more than 10% tree cover."

Soil erosion is one of the primary and most persistent results and causes of soil degradation. In Chapter 5.1, Van Oost and Bakker consider the extent of soil loss and the challenges of estimating the effects of these losses on crop production. They include a key figure illustrating the generalized relationship between incremental soil loss and relative crop yield. Interpreted in the context of other chapters in this section, this figure suggests that studies of soil biota–management–crop productivity relationships should focus on nutrient cycling on relatively undegraded soils, while at later stages of degradation they should focus on soil water infiltration and holding capacity. The figure also supports the importance of careful soil profile characterization for these studies, such as the long-term farming system comparison used to parameterize the erosion model simulation for three farming systems described by Barrios *et al.* in Chapter 5.2. Chapter 5.4 by Cavigelli, Maul, and Szlavecz clearly describes the complex relationships between soil biota and ecosystem services, and describes a recently published set of six steps necessary to "determine if organic management supports biodiversity that is functionally significant with respect to pest control." A similar set of steps could be usefully applied to many questions about the importance of soil biota for other issues. The last two chapters address resilience and recovery processes. In Chapter 5.5, Grandy, Fraterrigo, and Billings apply ecological theory to understanding long-term soil dynamics. This chapter, together with Baer, Heneghan and Eviner's (Chapter 5.6) review of the contribution of soil ecology to restoration, emphasizes the importance of recognizing hysteresis, whereby degradation and restoration processes can be quite different, and occur over very different time periods.

Together, these chapters provide a summary of current knowledge and remaining challenges to identify, measure, and generalize the importance of soil biota for sustaining soil resources and their related ecosystem services in agroecosystems.

References

Holling, C.S., & Meffe, G.K. (1996) Command and control and the pathology of natural resource management. *Conservation Biology* **10**: 328–37.

Millennium Ecosystem Assessment (2005) *Ecosystems and human well-being: current state and trends*. Island Press, Washington DC.

Seybold, C.A., Herrick, J.E., & Brejda, J.J. (1999) Soil resilience: A fundamental component of soil quality. *Soil Science* **164**: 224–34.

Western, D. (2002) *In the Dust of Kilimanjaro*. Island Press, Covelo, CA.

CHAPTER 5.1

Soil Productivity and Erosion

Kristof Van Oost and Martha M. Bakker

5.1.1 Introduction

Ever since the wide-scale cultivation of soil by humans, accelerated erosion has been an environmental concern. Human-induced or accelerated erosion deteriorates soils and negatively impacts their productivity. The impact is difficult to measure, as erosion occurs gradually over long time spans along with other productivity-determining processes. In addition, spatial variability of soil erosion is high, which complicates global quantitative estimates. Lastly, the relative importance of erosion-induced productivity losses for agricultural land use is largely unknown, as many other factors determine primary productivity (yield), and even more determine economic productivity (revenues).

John Boardman stated in his reflections on the limitations of current approaches in soil erosion science that some of the most important questions—such as "how serious is soil erosion," "what are the costs of soil erosion" and "why and where is it happening" still haven't been answered (Boardman 2006). These questions are difficult to answer because they require interdisciplinary approaches and funding. In this chapter we discuss some of these problems by presenting an integrative compilation of existing knowledge. We review literature recently published on these topics, to obtain an overview of the current state of the art. We try to identify problems in assessment techniques and gaps in current knowledge. We start by making an inventory of accelerated erosion rates, and we place these within the context of both soil formation rates and erosion rates under natural conditions. We discuss how soil erosion and formation rates are closely related and how we use soil formation rates as a benchmark with which accelerated erosion rates are compared. Soil functions can generally be judged not to deteriorate as long as soil erosion does not exceed natural rates (Verheijen *et al.* 2009). Next, we try to translate accelerated soil losses into yield reductions based on published literature, and identify the relative importance of these productivity losses for agriculture. Does erosion affect farmers' decision-making? If so, is the problem perhaps self-regulatory or is the physical impact too small to give rise to a feed-back to land use? We compare past agricultural conditions to present situations and also sketch an outlook for future developments. We conclude that, with the ongoing developments in agriculture, the erosion-induced onsite productivity losses as well as the downstream effects of erosion may become an important environmental threat.

5.1.2 Soil gain versus soil loss, and accelerated versus natural erosion

5.1.2.1 Soil gain

Soils are formed from the mechanical and (bio-)chemical weathering of parent material *in situ* or from unconsolidated deposits that have been transported by erosional processes. Mechanical weathering involves the physical disintegration of bedrock into smaller mineral particles and occurs as a result of, among other factors, temperature fluctuations, hydrological processes, and the action of animals and plants. The chemical transformation of parent material as a result of dissolution, oxidation, or hydrolysis processes may also contribute to the weathering of bedrock. The composition, age, and surface area of minerals, and the hydrological context (solute temperature, flux, and composition) are the main determinants of potential chemical weathering rates (Anderson et al. 2007). As soils are exposed to weathering, the more easily decomposable particles weather away first which leads to a gradual enrichment in more resistant minerals. The release of inorganic nutrients by these weathering processes has direct consequences for soil fertility and plant production; soils on old substrates are typically more depleted in bedrock-derived nutrients and are less fertile than soils formed on younger landforms (Vitousek et al. 1999).

A soil system is in equilibrium in terms of soil stock when soil erosion (i.e. the removal of soil at the surface) is balanced by the formation of new soil. Soil formation and soil erosion in natural systems are roughly at a steady state over longer periods of time for mature residual soils formed from uniform bedrock parent material (Phillips 2010). This steady state is the result of the feedback between soil thickness and weathering rate at the base of the regolith (Heimsath et al. 1999). High rates of weathering occur in shallow soils, and rates decline exponentially when soils thicken. Although this notion of steady-state soil thickness is only valid under the specific conditions mentioned previously, typical rates of soil formation provide a valuable reference against which accelerated soil erosion may be evaluated.

Soil formation rates can be obtained from a range of techniques such as geochemical mass-balances or in situ produced cosmogenic nucleides such as ^{10}Be or ^{26}Al (e.g. Heimsath et al. 1997). At present, relatively few observations are available, and estimates vary over four orders of magnitude (Table 5.1.1). Although the sample size is limited, the soil formation rates can be considered representative as they are in the same order of magnitude as independently derived global and continental estimates (Montgomery 2007a). The main physical and chemical factors controlling soil formation are temperature and moisture, biological activity, relief, geological substrate, and time (Jenny 1941). The observations that are available suggest that soil formation rates in (semi-) arid conditions are substantially lower than those in humid areas. Soil formation rates also decrease with depth as exposure to weathering agents decreases. For that reason current soil formation rates are slower on stable cratons, i.e. on landforms which have remained undisturbed for long periods of time and that contain old and deeply weathered soils, than in areas where relatively fresh, unweathered bedrock is available. Although recent progress has been made in elucidating the main small-scale controls on weathering, a comprehensive, process-based model of weathering that can predict large-scale patterns and rates of soil formation is currently lacking.

While weathering transforms parent material into soil, the physical process of erosion may mobilize and transport previously weathered minerals by a range of agents (e.g. runoff, ice, wind, raindrop, gravity, etc.), and upon deposition, create landforms from which soil formation begins. For example, alluvial soils are formed in lowland areas where the capacity of rivers to carry sediments gradually drops which results in the deposition of the previously eroded soil particles. Under the absence of a dense vegetation cover and dry conditions, wind can also erode and transport loose sediments that, once deposited, contribute to the local input of minerals and nutrients.

Soils developing on depositional landforms have the ability to remain young when they periodically receive fresh nutrients by ongoing deposition. For example, the soils in the Nile river valley are very productive, even after thousands of years of intensive agriculture (Montgomery 2007b). Other depositional landforms were created by large erosional disturbances in the geologic past. Loess soils are derived from the accumulation of wind-blow silt and have long been regarded as among the most fertile in the world (e.g. Catt 2001). At present, a substantial area of productive cropland is situated on soils that were formed by depositional processes. Due to their low topographic potential, depositional soils are also less vulnerable to physical soil degradation processes than their upland counterparts.

Natural deposition is still occurring today. It is estimated that, at present, 1.0–2.1 Gt year^{-1} of sediment is emitted into the global atmosphere by wind erosion, and the Sahara is by far the largest source contributing 40–60% of global emissions (Maher et al. 2010). A large part of this flux is deposited into the North Atlantic but substantial deposition of

Table 5.1.1 Reported rates of soil formation, dust deposition, natural, and accelerated erosion.

Method	Soil gain/loss (mm year^{-1})	Comment	Reference
Soil gain/weathering			
Mass-balance/cosmogenic nucleides	0.036	Compilation of data from 188 studies	Montgomery 2007a
Cosmogenic ^{10}Be	0.02–0.1	Middle European rivers, assuming steady-state	Schaller et al. 2001
Mass balance	0.03–0.1		Wakatsuki & Rasyidin 1992
Soil gain/dust deposition			
	0–0.03	Averages for N and S Europe	Verheijen et al. 2009
	0.02–0.2	California & great plains	Verheijen et al. 2009
Soil loss/natural erosion			
Sedimentary volumes	0.03	Average for Phanerozoic	Wilkinson & McElroy 2007
Sedimentary volumes	0.1	Pliocene Epoch	Wilkinson & McElroy 2007
Model-based river export	0.1	Pre-Anthropocene	Syvitski et al. 2005
Soil loss/accelerated erosion			
Plot extrapolation	1.2	Developed ag	Pimentel et al. 1995
Plot extrapolation	2-3	Undeveloped ag	Pimentel et al. 1995
Plot extrapolation	3.9	Global cropland	Montgomery 2007a
Model-based	1.0	Global cropland	Van Oost et al. 2007

mineral particles on the continent is observed with typical rates of 0.0008 and 0.015 mm year^{-1} for North and South Europe, respectively (Verheijen et al. 2009). Locally, higher rates of dust deposition can be observed for soils in the vicinity of semi-arid areas, such as the Great Plains in the US where average rates of $c.$0.2 mm yr^{-1} are reported (Verheijen et al. 2009). The data presented here nevertheless show that, although the local variation in dust deposition can be high, the current relative contribution of dust deposition in soil formation at the global scale is probably small when compared to weathering and erosion processes (Table 5.1.1).

5.1.2.2 Natural soil erosion

Soil erosion is a natural process that occurs over geologic timescales and, as we described earlier, also contributes to soil formation elsewhere. It occurs due to the transport by water, wind, and ice, or by gravitational forces (e.g. creep, landslides). Similar to soil formation rates, rates of erosion under natural conditions offer a valuable benchmark with which accelerated erosion rates can be compared. At long timescales, natural erosion rates are approximately balanced by rates of accumulation, so that long-term natural erosion rates can be inferred from preserved volumes of sedimentary rocks (Wilkinson 2005; Willenbring & von Blanckenburg 2010). Geologic fluxes have ranged over an order of magnitude and equal $c.$0.1 mm year^{-1} for the most recent Epoch (Wilkinson & McElroy 2007). Natural erosion rates can also be derived from estimates of the global flux of sediment to the oceans under pre-human conditions. It is estimated that about 21 Pg of sediments are annually exported to the ocean by global rivers under natural conditions and this corresponds to a mean erosion rate of $c.$0.1 mm yr^{-1} (Syvitski et al. 2005). Note that this river-flux derived estimate is very similar to the rate derived from sedimentary rock volumes. In contrast to sedimentary rock volumes, river flux data can be analyzed at a much higher spatial resolution, and many investigations have related climate, slope, relief, vegetation, and substrate characteristics to erosion. Although there is great variability in the factors controlling river sediment fluxes, these studies suggest that the dominant controls on natural erosion are climate, topography, and geology (e.g. Ludwig et al. 1996). As a result, natural erosion occurs mainly in orogenic belts and is therefore highly localized (Wilkinson & McElroy 2007).

5.1.2.3 Accelerated erosion

Accelerated erosion happens when the protective cover is removed by humans for the purpose of food, feed, or fiber production. Protective cover primarily concerns vegetation, but can also include a naturally formed protective soil crust that forms in the absence of abundant natural vegetation. The removal of this cover results in an increased exposure of the soil surface to the energetic input of climate, and depending on its cohesive properties, soil becomes detached and redistributed. Removal of vegetation mostly happens by harvesting crops, but also by grazing and slash-and-burn practices; protective soil crusts are typically removed during plowing practices.

Accelerated erosion primarily occurs on annually ploughed and harvested land, and to a lesser extent on land planted with permanent crops and grasslands (Cerdan et al. 2010). Much of the eroded sediment is retained on land (Van Oost et al. 2007); therefore, accelerated erosion also affects the rate of soil accumulation. Depending on the properties of the eroded soil and the pace of deposition, sediment deposition can either improve or degrade soils upon which it is deposited. The rates of accelerated erosion reported in the literature vary widely because land cover is only one of the determinants of soil erosion, along with topography, rainfall erosivity, soil cohesive properties, and management techniques such as contour plowing. Due to the very high spatial and temporal variability of controlling factors, quantifying rates of accelerated erosion over larger areas is not straightforward. Data-based assessments rely on the extrapolation of results derived from plot-scale observations, which often result in overestimations. For example, Pimentel et al. (1995) estimated the average rate of accelerated erosion at 17 Mg ha^{-1} year^{-1} for Europe and North America, and at 30–40 Mg ha^{-1} year^{-1} for Africa, Latin-America and Asia. These rates correspond to an average surface lowering on cropland of 1.2 mm year^{-1} and 2.2–3 mm year^{-1} respectively. In another study, based on 448 observations, a typical global cropland erosion rate was estimated at c.4 mm year^{-1} (Montgomery 2007a). These rates appear to be quite drastic when one considers that the majority of cropland is located in relatively flat, slowly eroding areas. More recently, accelerated erosion has been parameterized over continental and global scales in conceptually simplified ways that captures the essence of empirical observations. Using these models in combination with a range of geographically explicit parameters at high spatial resolutions, Cerdan et al. (2010) obtained an average cropland erosion rate of 0.36 mm year^{-1} for Western Europe, which is a factor of ten lower than the estimates by Pimentel and Montgomery. Van Oost et al. (2007) used a spatially distributed soil erosion model which they calibrated so that modeled erosion rates for the USA and Europe were similar to those obtained through independent estimates. Their estimate of accelerated water erosion was 22 Pg year^{-1} on croplands and 11 Pg year^{-1} on pasture and rangelands (or 1.0 and 0.3 mm year^{-1}, respectively).

While it is certain that large uncertainties are associated with both methods, it is clear that spatial explicit estimations result in substantially lower estimates than those derived from data-based approaches. A possible explanation for this is that the erosion studies, on which data-based approaches rely, are usually set up in areas where erosion is known to be a problem, and this may result in a biased estimate (Cerdan et al. 2010). The simulated global distribution of accelerated erosion on crop, pasture- and rangelands obtained by Van Oost et al. (2007), is shown in Fig. 5.1.1. At a global scale, the map indicates a very high spatial variability with Africa and Asia representing c.65% of the total erosion. In contrast, Europe and North America contribute little and represent <15% of the total accelerated erosion. However, erosion is a spatially varying process across a range of scales. While the regional average value may be low, severe erosion and soil degradation may still occur in specific locations. For example, Cerdan et al. (2010) estimate that c.70% of the erosion occurs over <15% of the total area of Europe.

5.1.2.4 Synthesis

The global estimates presented here confirm the contention that for a range of climate and geologic forcings, soil formation and erosion approximately balance each other under natural conditions. However, there is an order of magnitude discrepancy between present rates of accelerated erosion on

Figure 5.1.1 Spatially distributed estimates of accelerated soil erosion on crop-, pasture-, and rangeland. (Based on Van Oost et al. 2007.)

agricultural land (0.5–1.5 mm year^{-1}) and rates of soil formation and erosion under natural conditions (0.01–0.1 mm year^{-1}). Despite the possibly exaggerated view generated by the plot estimates, accelerated erosion rates are still a magnitude higher than natural erosion and formation rates. In contrast, erosion rates under conservation practices (e.g. terracing, reduced tillage) are substantially lower than those obtained under conventional agriculture and are in the same order of magnitude as natural erosion and soil formation.

Based on these numbers, the question of whether accelerated erosion under conventional agriculture impairs the soil's ability to provide ecosystem goods and services is fully justified. Faced with a global population predicted to reach 9 billion by 2050, and a global grain demand expected to double in the same timeframe (Tilman et al. 2002; Trewavas 2002), the issue of accelerated erosion and long-term sustainability of conventional agriculture should therefore be evaluated carefully.

5.1.3 Erosion's effect on agricultural productivity

5.1.3.1 How erosion affects agricultural productivity

Erosion affects a crop's phenological development via a multitude of mechanisms. It removes the fertile topsoil, leads to a residual enrichment of subsurface layers with clay or stone, and can decrease the depth of rooting. These three mechanisms can lead to a reduced uptake of water and nutrients. Next, we discuss these mechanisms in more detail.

1. Soil erosion removes the topsoil of a profile. With perturbation of the soil (by tillage or by biological activity) the subsoil mixes with the remaining topsoil, but as subsoil material generally contains less organic matter and key nutrients such as N and P, and is typically enriched in clays, the resulting soil is poorer and more clayey than the original soil. Ongoing erosion causes the subsoil to become increasingly incorporated in the top layer which, depending on the subsoil properties, will gradually reduce the fertility of the soil. Although an increased application of fertilizers and continued input of photosynthetically derived plant material may to a large extent compensate for these losses, this comes with an additional economical cost and quite often also with undesired off-site eutrophication of water bodies.

2. The selectivity of erosion processes also has to be considered: soil detachment and transport depends on the density and weight of the soil particles and the transport capacity of the wind or water. The fine sand and silt

fraction of the soil and particulate organic matter are generally more easily eroded and transported by erosion and this may lead to a residual enrichment of clay and stones (Lal 2001). When fine particles and organic matter are removed, the water and nutrients available to plants also decrease: the coarse fraction has no ability to absorb water and nutrients, while clay may absorb nutrients and water too strongly so that they are no longer available to the plant.

3. The nutrient and water resource for a plant is also determined by the total amount of soil available for the rooting system. When soil losses lead to the rooting zone approaching an impenetrable layer, the available soil is reduced. Particularly the water availability is affected by this mechanism, as water uptake generally happens in the lower regions of the rooting zone. Roots may extend laterally, but this leads to stronger competition between plants for water.

Most of the time, these mechanisms act simultaneously, but generally either nutrient deficit or water deficit is limiting the crop's growth. This limiting factor is often referred to as the *regressor* (Christensen & McElyea 1988; Bakker *et al.* 2004).

5.1.3.2 Estimating erosional impacts on agricultural productivity

5.1.3.2.1 Complicating factors
Erosional effects on productivity are difficult to assess for different reasons. The obvious complication by the variability in soil types, erosion rates, and crop types relates to 1) the agricultural management that partly masks the effects, 2) the long time spans over which erosion has an impact, and 3) the differences in temporal development of responses. We discuss these three complicating factors next, and try to conclude with some numbers to give an indication of the impact of erosion on agricultural productivity.

The first complicating factor is that erosion-induced income losses result from two types of costs: the costs of extra inputs and labor, and the income loss due to lower yields. In an ideal experimental set-up, one of these two is measured in response to soil erosion, while the other is kept constant. Such experiments which are discussed in a moment unfortunately have disadvantages that often make it preferable to look at real-world impacts of erosion. In reality, however, most farmers will distribute the costs over the two types: they apply more inputs and labor, and at the same time harvest fewer crops. Therefore, uncontrolled empirical observations of, for instance, yield responses to soil loss reflect only part of the story; the other part, the increased application of labor and inputs, is not observed.

Another complication is the relatively long time span over which erosional effects on productivity become clear. Erosion may have an effect over a time span of 30 years, a period in which many other productivity-affecting processes, such as the improvement of cultivars, strongly confounds the impact of erosion. Therefore, the effect of erosion on productivity is measured indirectly, using different techniques that all have their drawbacks. Artificially desurfacing a plot and comparing the yields on that plot to yields from a plot left intact, i.e. the experimental setup, is a popular technique (e.g. Dormaar *et al.* 1986; Larney *et al.* 2000); however, the observed effects are probably much stronger than when erosion occurs gradually over a longer time span. When erosion occurs gradually, its impact on the nutrient balance in the soil is much smaller, for reasons explained in Section 5.1.3.1, as is the subsequent effect on crop productivity. Other indirect measurements are complicated by the fact that erosion coincides with other productivity-determining factors. For example, areas prone to erosion have generally less favorable hydrological circumstances than areas where deposition occurs. The method of evaluating yields along a catena, with erosion occurring upslope at the convex part and deposition occurring downhill in the concave part, probably also leads to overestimation of erosional impacts. After all, the higher yields in the downhill, concave parts may partly be attributed to the better available soil moisture compared to the upslope, convex parts. Deposition of soil material at the bottom of catenas also influences the agricultural productivity at these locations and this should be accounted for.

In fine-textured soils, deposition typically results in the development of deep Ap horizons which are rich in nutrients and can support higher yields than non-eroding soils (Quine & Zhang 2002; Heckrath et al. 2005). Deposition can also have a detrimental effect when, due to selective transport of the fines, depositional soils become enriched in silt and sand and retain less water and nutrients than their eroding counter-parts. An assessment of erosion-induced changes in agricultural production should therefore integrate the effects on both erosion and depositional sites along a catena (Pennock 1997).

Probably the least biased method of evaluating erosional effects on productivity is the comparison of plots that are similar in all ways except for the time span over which they have been subject to erosion (i.e. the period over which they were cultivated). This application of the ergodicity principle leads to estimates that are 2.5 times less damaging than studying yields along a catena, and six times less damaging than the desurfacing experiments (Bakker et al. 2004). Most likely, these are the most realistic estimates.

A third complication is raised by the multitude of mechanisms through which erosion affects a crop's phenological development. When Bakker et al. (2004) combined the results of many studies they found no systematic difference in magnitude of response between the different mechanisms, but they did find that the development of water deficit-induced yield reductions over time was quite different from that of the nutrient deficit-induced yield reductions. These different temporal developments can be explained as follows: consider a newly cultivated plot that becomes subject to erosion from the moment cultivation starts (Fig. 5.1.2). We may assume that at the onset of cultivation the soil is relatively deep and has a certain nutrient stock in the topsoil. The first few centimeters of erosion will not have much affect on the water availability, as the soil depth is hardly reduced. The nutrient stock, however, is seriously reduced, which will affect crop growth. As nutrients decrease in depth, each additional centimeter of soil loss will have a smaller impact. The resulting response curve will be concave: with each incremental soil loss, the response will be smaller. When erosion continues, at some point the soil depth becomes significantly reduced, and water deficit starts to become the regressor of crop growth. As with decreasing depth, more water-uptaking roots (i.e. the roots that penetrate deepest into the soil) become affected, and each additional centimeter of soil loss will have a stronger impact. The resulting response curve in this case will be convex: with each incremental soil loss, the response will be larger.

Hence, the phase of cultivation, together with the (resulting) soil characteristics, determines whether the response curve of crop yield to soil erosion is convex or concave (Bakker et al. 2004). In areas where soils are shallow, for example in Mediterranean regions, the response curve is generally convex: each incremental soil loss will show a stronger impact than the previous soil loss. In areas where soils are generally deep but nutrient-poor, for example in tropical regions, the response curve is generally concave: most damage is done during the removal of the first centimeters of topsoil, after which the impact levels off. It is important to know the temporal context to evaluate an observed impact of erosion on the crop yield. A relatively small impact can still be dangerous when water deficit is the regressor, because the effect may worsen, while

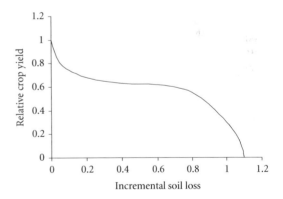

Figure 5.1.2 Erosion-crop productivity relations where a soil with an initial depth of 1.2 m becomes completely eroded over time. The initial soil depth reductions affect the nutrient availability, but as nutrients are concentrated towards the surface, this impact becomes negligible after the first 10–15 cm of soil loss. This is the concave part of the curve. For a while the crop yield may stay at this reduced level, until a significant portion of water-uptaking roots become hindered by the impermeable layer (i.e. when approximately 60 cm of soil depth is left). The shallower the soil becomes, the more roots are hindered in their water uptake, until at some point (e.g. when only 10 cm is left) the crop can no longer grow.

a severe impact is not expected to extend much further into the future when nutrient deficit is the regressor (Bakker et al. 2004). Moreover, in theory, nutrient deficits may be easier to overcome than water deficits because synthetic nutrients are logistically easier to distribute and apply than water. In practice, however, many parts in Africa are still suffering from nutrient deficits because of poor infrastructure and a limited knowledge on the specific nutrient deficits of African soils (Voortman et al. 2003).

5.1.3.2.2. Estimates of the impact
Taking into account the previous discussions, meaningful estimates of the effects of soil erosion on crop productivity are rare and certainly not universal. Bakker et al. (2004) estimated the effect of every 10 cm of soil loss to be around 4.3%, based on an analysis of experiments using the method of plot comparisons. A similar number (3–4.3%) was found for loess soils of southeast Poland by Rejman and Iglik (2010). In a later study, Bakker et al. (2007) found that in Germany, France, and Greece each mm reduction in soil water holding capacity (the prevalent regressor in these countries), resulted in a wheat yield reduction of 21 kg ha^{-1} year^{-1}. When an average water holding capacity of 150 mm m^{-1} is assumed, 10 cm of soil loss results in 0.31 ton ha^{-1} year^{-1}, which is about 5% of the average yield of these countries (i.e. 6.5 ton ha^{-1} year^{-1}). Den Biggelaar et al. (2004) also inventoried many comparing plot exercises and found higher reductions of around 7% for Europe and 17% for North America. It must be mentioned that the reference yields used in this study (relative to which these percentages are computed) seem rather low: 6.2 ton ha^{-1} year^{-1} for maize in North America and 3.5 ton ha^{-1} year^{-1} for wheat in Europe. Should these yield reductions be expressed as fractions of the average yield of the continents, which are roughly twice as high, the reported reduction rates would be around 3.5% for Europe and 8.5% for Northern America. These estimates cited above were based on modernized agriculture, where even on controlled plots, technological substitutes (e.g. synthetic fertilizers, irrigation techniques) conceal part of the effects of erosion. Stocking (2003), on the other hand, conducted experiments in Africa among other places, and found that for many tropical soils, yields dropped to zero after a few centimeters of soil loss. It is not clear whether these experiments concern the topsoil desurfacing experiments, as this may partly explain the strong effect. Also Mbagwu et al. (1984) and Lal (1987) found severe yield reductions in Nigeria (reductions >95%), but again these were obtained by the desurfacing experiments. Lal, however, also performed comparing-plot experiments in Nigeria, which resulted in reductions of 10–20% per 10 cm of soil loss. Den Biggelaar's inventory of comparing-plot experiments conducted in Africa reported reductions of around 30% per 10 cm of soil loss. Although significantly lower than the desurfacing experiment results, these reductions are still much stronger than for the intensive European and North-American agriculture. This is partly due to the fact that the erosion effect is less compensated by other inputs, but also because these numbers are inflated by the low reference yields. Lal also noted that the numbers reported concerned the first centimeters of topsoil reduction, and that reductions would probably decrease with incremental soil loss, as is indeed the case when nutrient deficit is the regressor.

At the time of the studies, as reported by Den Biggelaar (i.e. 1960–1990), Asia and Latin America were still at a much lower developmental stage than they are now, which resulted in a high impact of erosion on crop yields (similar to reductions reported in African sites). Now that these continents have modernized their agriculture to a great extent, impacts of erosion have probably become less significant. Recent erosion-induced yield reductions reported by Wang et al. (2009) demonstrated a loss of 16%, but again, as these were obtained from artificial desurfacing experiments they are likely to be overestimations. Should we "rescale" these to comparing plot estimates, a division by six (see Section 5.1.3.2.1) results in a reduction of around 3%.

These estimates should of course be interpreted in relation to the erosion rates, in order to provide an overall estimate of erosion's impact on productivity. As erosion rates and yield reductions are not independent (i.e. both are dependent on the crop grown), simply substituting the 10 cm of soil loss for actual soil depth reductions based on Table 5.1.1 will not give a reliable estimate. As Den Biggelaar already made an elaborate estimate of accumulated yield reductions over time for several important crops, we limit ourselves to the following brief conclusions: for Europe and North America the erosion-induced reductions seem to be quite small, and long periods of quite severe erosion must pass before yields are significantly affected. In Africa, the relative impact of erosion is likely to be much more severe. Not because erosion rates are higher there, but because nutrient losses are to a lesser extent replaced by fertilizers, and because yields are already low and reductions are felt much harder. Accurate and up-to-date numbers are lacking for Asia, but with the ongoing modernization of agriculture in this region the impact of erosion per cm of soil loss is likely to be similar to that of Europe and North America. The soil erosion rates, however, are generally higher (see also Fig. 5.1.1), which may result in a more severe overall impact of erosion on agricultural productivity.

5.1.4 The importance of erosion-induced productivity losses for agriculture

The fact that accelerated erosion occurs implies that erosion-prone land is indeed cultivated despite the risk of deteriorating the soil's productivity. Naturally, farmers' preference will go to level, less erosive areas, where soils are workable, fertile, well-drained, and have high infiltration rates and high water storage capacity. The decision to move to suboptimal areas is driven by the relative scarcity of suitable land. When farmers move to suboptimal areas, different constraints are weighed against each other, whereby erosion risk often weighs less than other constraints. The nature of the land scarcity (real scarcity or a poor competitive disposition of arable land relative to other land uses) and the trade-offs made between constraints may differ among regions and also over time. Next we briefly discuss the development of (relative) land scarcity and elaborate on the importance of erosion risk relative to other constraints for arable cultivation. We conclude with an outlook for the future, in which we emphasize the importance of governance constructs to discourage the cultivation of erosion-prone land.

At the onset of sedentary agriculture, people settled in locations where crop cultivation was expected to be successful. Mostly, settlements were started in flat and fertile alluvial plains, where clearance of the natural vegetation did not result in erosion of the soil (Van Andel & Runnels 1995). But with the success of sedentary agriculture, societies grew and the area used for the production of food, feed, and fiber had to expand. Until halfway through the previous century, cropland generally grew in proportion to the world population, which was roughly exponential. Although technological progress occurred gradually over centuries, in the course of the 20th century yields boosted as a result of technological developments were often referred to as the Green Revolution. For decades the resulting yield increases allowed the growing demand for agricultural products to be met without significant area expansion. More recently, this compensation of demand increase by yield increase is becoming largely cancelled by the increased consumption of meat, requiring much larger areas to produce the same amount of calories and proteins (Tilman et al. 2002; Keyzer et al. 2005).

With the expansion of the world population, suitable land became a scarce commodity in many regions. Economic theory says that in such a case, land will be used for that purpose and bring the greatest benefits to the owner, which means that urban land use, gardening, and intensive dairying will normally outcompete field crops. So when urban land use, gardening, and dairying grew in response to economic growth, arable farming was pushed towards more marginal areas, in accordance with Ricardo's and Von Thünen's 19th-century land-use theories. Arable farming in turn outcompeted extensive grazing and natural forests as crops yielded higher benefits per hectare. Furthermore, arable cultivation is relatively intense compared to extensive forms of grazing, forestry and permanent-crop cultivation, as it requires annual tillage, irrigation, fertilizer and biocide application, etc. As a result it is generally more cost-effective to place cropland in level areas, and (extensive) cattle breeding, orchards, and forests on steeper slopes.

This may suggest that arable farming only moves to erosion-prone areas when all better suited areas are occupied by land uses with a higher purchasing power. When we look at Fig. 5.1.3, we see that this is not the case: arable fields are located in locations

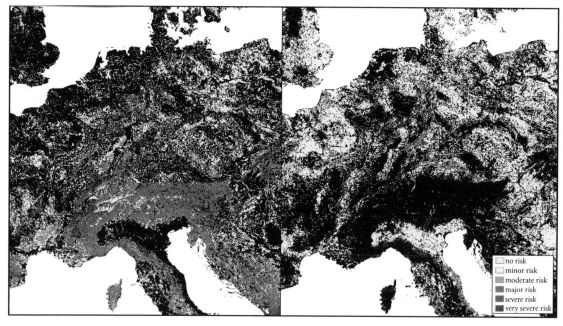

Figure 5.1.3 Erosion and land use. The figure shows two overlays of land use and potential erosion risk in areas with arable cultivation and permanent crops is shown in different shades of gray (all other land cover types are black), and it can be seen that, although the majority is located on areas with a small erosion risk (light gray), quite some fields are located in locations with high erosion risk (dark gray). On the left, the erosion risk in areas covered by grassland and natural vegetation cover is shown (other land use types are black). This figure suggests that the arable land that is currently located at high erosion potential areas could also have been located in less erosion-prone areas (the light gray areas).

with high erosion risk (see for instance southern Germany) despite the "availability" of some non-erosion-prone land (i.e. not or extensively used; see for instance northern Germany). This suggests that income losses related to erosion are relatively small (or perceived as such) compared to other constraint-related income losses. We may assume that the non-erosion-prone areas supposedly available for arable cultivation exhibit other constraints (e.g. remoteness, drought, water-logging, low temperatures, and low solar radiation) that weigh more heavily on the farmer's income balance than those costs related to erosion. The estimates in the previous section indeed suggest that costs of erosion-induced productivity losses in Europe are small.

5.1.4.1 Past impacts of erosion on land use

In the past (roughly before the industrial and the green revolution) things may have been different in two respects. One, is that cultivation of erosion-prone areas happened on a much wider scale than today; second, is that the technological means to compensate for erosion-induced fertility losses were largely absent. The result was that soil erosion increased, causing the food supply to fall short after a while, which, under specific conditions, may have contributed to the collapse of a civilization (Montgomery 2007b). Easter Island is a famous, though not uncontested example (Diamond 2004), but also the Roman Empire is known for having had problems with maintaining soil productivity (Van Andel *et al.* 1990). Some scientists also relate the collapse of the Maya civilization to soil erosion (Jacob & Hallmark 1996).

Beginning with the first aspect: wide-scale cultivation of erosion-prone areas happened in ancient civilizations because the degrees of freedom to look for more suitable land were severely limited compared to modern times. In terms of constraints: the constraint of overland transport costs dominated land-use decision making. When a village or city was located in a small valley, it was often more cost-effective to expand uphill than going to the next valley despite the erosion risk. Particularly in Mediterranean Europe, where large deltas were not abundantly available, occasional severe soil erosion occurred in the hilly cropland surrounding the main cities and towns (Van Andel et al. 1990).

As for the second aspect, whenever erosion-prone land was cultivated, erosion-induced productivity losses were hard to compensate, as artificial fertilizers were largely non-existent. According to Montgomery, food production in the Roman Empire at some stage suffered so much from soil erosion that armed forces weakened and increasingly lost battles with invading armies. Even when erosion does not drastically affect food production, it can have important impacts on the land use system. In cases where the overall pressure of land is not very high, but suitable soils are nevertheless scarce, erosion may lead to shifting cultivation patterns. Shifting cultivation is currently still practiced in tropical regions, but only a few studies identified erosion as the driving factor (Gafur et al. 2000). A relatively recent European example of land-use change because of soil erosion was found in Lesbos, a Greek island in the Aegean Sea, in the middle of the last century (Bakker et al. 2005). Reconstruction suggested that soil erosion of cereal fields reduced the average soil depth from 98 cm to 72 cm, after which the cereal fields moved to a new location with deeper soils and less steep slopes. In modern, high-input agricultural systems, the impact of erosion on crop productivity is probably too small to affect the land-use decision making, for reasons explained earlier. Indeed, an analysis of Greece, France, and Germany in the period 1970–2000 showed that even though yields were related to soil erosion (Bakker et al. 2007), no relationship observed between the erosion-induced yield losses and a decline in the agricultural area (Bakker et al. 2011).

5.1.4.2 Future impacts of erosion on land use

The knowledge that the impact of erosion was different in the past justifies an outlook to possible changes in the future. One may expect that the ever-decreasing share of transport costs on farmers' income balances may stimulate a further migration to less erosion-prone areas. This would be a favorable development from the point of view of soil conservation. But there are other reasons to be less optimistic. One is that cropland is increasingly marginalized compared to other land uses; second is the increased use of fertilizers and irrigation systems which only masks the loss of soil and which may give rise to a much more serious problem than on-site productivity losses: the off-site effects.

Regarding the marginalization of cropland, the demand for products such as fruits, vegetables, meat, and dairy products is increasing rapidly due to global economic growth. At the same time agricultural land is consumed by urban sprawl, particularly in countries where zoning policies are absent or failing. Because urbanization happens mostly at the fringes of existing cities, which are—for historical reasons—located on fertile lands, the soils that are being built up are generally highly productive agricultural land. Apart from the area that is built up, urbanization also has an impact on the valuation of other land cover types in the vicinity. Nature and traditional (often low-yielding) agriculture are valued highly by urban citizens for recreation purposes. As a consequence, such areas are often protected and cannot be used for commercial production. The result of the expansion of land uses with a higher purchasing-power than cropland (i.e. urban areas, intensive dairying, horticulture, and recreation areas) is that the latter is being pushed towards more marginal areas. Currently in Western Europe the average slope of cropland is increasing because of this mechanism, despite an overall decline of the cropland area (Bakker & Veldkamp 2011).

Secondly, crop farmers also increasingly choose to expand in areas with high potential productivity, which are often semi-arid areas. They tolerate the constraints inherent to these areas such as drought and erratic (i.e. erosive) rainfall, because irrigation and fertilization are relatively cheap. Even sloping land is no longer considered much of a limitation, witnessing the decreasing role of slope in land-use decision-making (Bakker & Veldkamp 2011). This is again a matter of a relative shift in the importance of constraints. Technological developments the disadvantages of drought and erosion risk have apparently started to weigh less than the advantages of higher potential productivity in warmer and sunnier areas.

The combination of cropland moving to marginal places and the increasing ease by which onsite productivity losses are compensated gives rise to serious concerns, especially when the area for food, feed, and fiber production expands as is projected by the most recent Food and Agriculture Organization (FAO) prognoses. In intensive, high-input agriculture, the onsite productivity losses are too small to discourage cultivation of erosion-prone land. In low-input systems in the past, a natural feedback mechanism existed that led erosion-prone land to be taken out of cultivation after a certain time. Today, cultivation continues despite the erosion it causes. The offsite effects brought additional concern as in disastrous events, such as floods and mudflows, but also damage to fragile ecosystems in down-slope streams and lakes because of the inflow of fertilizers and biocides.

Without a social or governance structure that discourages the cultivation of erosion-prone land or stimulates erosion-mitigation measures, offsite damage is likely to increase in the future. When costs of offsite damage are recouped from farmers, the price of cultivating erosion-prone areas may start to approach other constraint-induced costs, and land-use or management changes may be the result. Putting such a policy into practice is cumbersome, as costs are generally difficult to estimate. While the costs for cleaning the mud from a down-slope road may still be fairly easy to assess, damage to fragile ecosystems in down-slope streams and lakes is much harder to estimate, let alone trace back to the exact field the pollutants originate from. In the case of more dramatic events, whereby excessive erosion results in the flooding of a downstream area, one can question the meaningfulness of holding a group of farmers liable for the possible victims.

A perhaps more feasible and certainly more farmer-friendly policy would be to encourage the cultivation of non-erosion-prone land rather than to penalize farmers for cultivating erosion-prone land. Governments can try to reduce costs of other constraints by, for instance, draining a water-logged area, or opening up a potentially suitable area by constructing roads. In addition, governments can stimulate the application of measures to mitigate soil erosion, such as conservation tillage, contour plowing, planting cover crops during the winter, etc. In addition, governments should try to stimulate production increases on current, non-erosion-prone cropland, so that further expansion of cropland can be avoided as much as possible. Yield gaps are still large in many parts of the world, which, to a large extent, can be reduced by improving education, infrastructure, and investment climates.

5.1.5 Summary

In this chapter we compiled existing knowledge on 1) the extent of accelerated soil erosion, 2) the effects of accelerated soil erosion on soil productivity, and 3) the relative importance of erosion-induced productivity losses for agriculture. In general, soil erosion leads to a reduced water- and nutrient-holding capacity, and productivity losses are likely to occur on most soils if they erode for several centuries or decades at present rates. In contrast to accelerated erosion, soil formation is a very slow process, with typical rates that are one order of magnitude lower. While it is clear that soils cannot renew the eroded surface under conventional agriculture, we showed that assessing the direct effect of erosion on crop productivity is difficult for a multitude of reasons. Soil erosion-productivity relationships can at best be derived indirectly, but controlling for other confounding effects remains difficult. As a result, meaningful estimates of the effects of soil erosion on crop productivity are rare and certainly not universal. However, when taking into account the complicating factors described above, the evidence available in the literature strongly suggests that for Europe and North America, the erosion-induced productivity losses are relatively small, on the order of 4% per 10 cm of soil loss. Although spatial variation in accelerated erosion and initial soil depth is likely to be very high, a global average accelerated erosion rate of 0.5–2 mm yr^{-1} and soil depth of 130 cm suggests that the effect of erosion is relatively small, and by far cancelled out by the increases in crop productivity during the last decades in the developed world (Wilkinson & McElroy 2007). In low-input agricultural systems, however, where yields are lower and nutrient losses are less replaced by fertilizers, the relative impact of erosion is without any doubt much more severe, up to 30% per 10 cm of soil loss. In such systems, erosion should be considered a serious threat to overall soil productivity.

Finally, we discussed the implications of erosion-induced productivity losses for agriculture. Current land-use patterns suggest that high land pressure in combination with small productivity impacts make erosion a secondary consideration in land-use decision-making. In the past and in low-input systems, the technological means to compensate for erosion-induced productivity losses are largely absent which results in a much stronger impact of soil erosion on land use. The trade offs of increased use of fertilizers and soil conservation techniques cause erosion-induced yield losses to weigh less in the decision-making, which may lead to offsite effects of erosion becoming a much bigger environmental threat than any onsite productivity losses.

References

Anderson, S.P., von Blanckenburg, F., & White, A.F. (2007) Physical and chemical controls on the Critical Zone. *Elements* **3**: 315–19.

Bakker, M.M., Govers, G., Kosmas, C., Vanacker, V., Van Oost, K., & Rounsevell, M.D.A. (2005) Soil erosion as a driver of land-use change. *Agriculture, Ecosystems & Environment* **105**: 467–81.

Bakker, M.M., Govers, G., & Rounsevell, M.D.A. (2004) The crop productivity-erosion relationship: an analysis based on experimental work. *Catena* **57**: 55–76.

Bakker, M.M., Hatna, E., Mucher, S., & Kuhlman, T. (2011) Changing environmental characteristics of European cropland. *Agricultural Systems* **104**: 522–32.

Bakker, M.M., Rounsevell, M.D.A., Govers, G., Jones, R.A., & Rounsevell, M.D.A. (2007) The effect of soil erosion on crop yields in Europe. *Ecosystems* **10**: 1209–19.

Bakker, M.M., & Veldkamp, A. (2011) Changing relationships between land use and environmental characteristics and their consequences for spatially explicit land-use change prediction. *Journal of Land Use Science*, in press.

Boardman, J. (2006) Soil erosion science: Reflections on the limitations of current approaches. *Catena* **68**(2–3): 73–86.

Catt, J.A. (2001) The agricultural importance of loess. *Earth-Science Reviews* **54**(1–3): 213–29.

Cerdan, O., Govers, G., Le Bissonnais, Y., et al. (2010) Rates and spatial variations of soil erosion in Europe: A study based on erosion plot data. *Geomorphology* **122**: 167–77.

Christensen, L.A., & McElyea, D.E. (1988) Toward a general method of estimating productivity-soil depth response relationships. *Journal of Soil and Water Conservation* **43**: 199–202.

Den Biggelaar, C., Lal, R., Wiebe, K., & Breneman, V. (2004) The global impact of soil erosion on productivity. *Advances in Agronomy* **81**: 1–48.

Diamond, J. (2004) Twilight at Easter. *The New York review of books* **51**(5). [online] http://www.nybooks.com/articles/archives/2004/mar/25/twilight-at-easter/

Dormaar, J.F., Lindwall, C.W., & Kozub, G.C. (1986) Restoring productivity to an artificially eroded dark brown chernozemic soil under dryland conditions. *Canadian Journal of Soil Sciences* **66**: 273–85.

Gafur, A., Borggaard, O.K., Jensen, J.R., & Petersen, L. (2000) Changes in soil nutrient content under shifting cultivation in the Chittagong Hill Tracts of Bangladesh. *Geografisk Tidsskrift* **100**: 27–36.

Heckrath, G., Djurhuus, J., Quine, T.A., Van Oost, K., Govers, G., & Zhang, Y. (2005) Tillage erosion and its effect on soil properties and crop yield in Denmark. *Journal of Environmental Quality* **34**: 312–24.

Heimsath, A.M., Dietrich, W.E., Nishiizumi, K., & Finkel, R.C. (1997) The soil production function and landscape equilibrium. *Nature* **388**: 358–61.

Heimsath, A.M., Dietrich, W.E., Nishiizumi, K., & Finkel, R.C. (1999) Cosmogenic nuclides, topography, and the spatial variation of soil depth. *Geomorphology* **27**: 151–72.

Jacob, J.S., & Hallmark, C.T. (1996) Holocene stratigraphy of Cobweb Swamp, a Maya wetland in northern Belize. *Geological Society of America Bulletin* **108**: 883–91.

Jenny, H. (1941) *Factors of soil formation: a system of quantitative pedology*. McGraw-Hill Book Company, Inc., New York.

Keyzer, M.A., Merbis, M.D., Pavel, I.F.P.W., & van Wesenbeeck, C.F.A. (2005) Diet shifts towards meat and the effects on cereal use: can we feed the animals in 2030? *Ecological Economics* **55**: 187–202.

Lal, R. (1987) Erosion-crop productivity relationships for Africa. *Soil Science Society of America Journal* **59**: 661–7.

Lal, R. (2001) Soil degradation by erosion. *Land Degradation & Development* **12**: 519–39.

Larney, F.J., Olson, B.M., Janzen, H.H., & Lindewall, C.W. (2000) Early impacts of topsoil removal and soil amendments on crop productivity. *Agronomy Journal* **92**: 948–56.

Ludwig, W., Probst, J.L., & Kempe, S. (1996) Predicting the oceanic input of organic carbon by continental erosion. *Global Biogeochemical Cycles* **10**: 23–41.

Maher, B.A., Prospero, J.M., Mackie, D., Gaiero, D., Hesse, P.P., & Balkanski, Y. (2010) Global connections between aeolian dust, climate and ocean biogeochemistry at the present day and at the last glacial maximum. *Earth-Science Reviews* **99**: 61–97.

Mbagwu, J.S.C., Lal, R., & Scott, T.W. (1984) Effects of desurfacing of Alfisols and Ultisols in Southern Nigeria: I. Crop performance. *Soil Science Society of America Journal* **48**: 828–33.

Montgomery, D.R. (2007a) Soil erosion and agricultural sustainability. *Proceedings of the National Academy of Sciences of the United States of America* **104**: 13268–72.

Montgomery, D.R. (2007b) *Dirt, The erosion of Civilizations.* University of California Press, Berkeley and Los Angeles, CA.

Pennock, D. (1997) Effects of soil redistribution on soil quality: Pedon, landscape, and regional scales. *Developments in Soil Science Volume* **25**: 167–85.

Phillips, J.D. (2010) The convenient fiction of steady-state soil thickness. *Geoderma* **156**: 389–98.

Pimentel, D., Harvey, C., Resosudarmo, P., et al. (1995) Environmental and economic costs of soil erosion and conservation benefits. *Science* **267**: 1117–23.

Quine, T.A., & Zhang, Y. (2002) An investigation of spatial variation in soil erosion, soil properties, and crop production within an agricultural field in Devon, United Kingdom. *Journal of Soil and Water Conservation* **57**: 55–65.

Rejman, J., & Iglik, I. (2010) Topsoil reduction and cereal yields on loess soils of southeast Poland. *Land Degradation & Development* **21**: 401–5.

Schaller, M., von Blanckenburg, F., Hovius, N., & Kubik, P.W. (2001) Large-scale erosion rates from in situ-produced cosmogenic nuclides in European river sediments. *Earth and Planetary Science Letters* **188**: 441–58.

Stocking, M. (2003) Tropical soils and food security: the next 50 years. *Science* **302**: 1356–9.

Syvitski, J.P.M., Vorosmarty, C.J., Kettner, A.J., & Green, P. (2005) Impact of humans on the flux of terrestrial sediment to the global coastal ocean. *Science* **308**: 376–80.

Tilman, D., Cassman, K.G., Matson, P.A., Rosamond, N., & Polasky, S. (2002) Agricultural sustainability and intensive production practices. *Nature* **418**: 671–7.

Trewavas, A. (2002) Malthus foiled again and again. *Nature* **418**: 668–70.

Van Andel, T.H., & Runnels, C.N. (1995) The earliest farmers in Europe. *Antiquity* **69**: 481–500.

Van Andel, T.H., Zangger, E., & Demitrack, A. (1990) Land-use and soil-erosion in prehistoric and historical Greece. *Journal of Field Archaeology* **17**: 379–96.

Van Oost, K., Quine, T.A., Govers, G., et al. (2007) The impact of agricultural soil erosion on the global carbon cycle. *Science* **318**: 626–9.

Verheijen, F.G.A., Jones, R.J.A., Rickson, R.J., & Smith, C.J. (2009) Tolerable versus actual soil erosion rates in Europe. *Earth-Science Reviews* **94**: 23–38.

Vitousek, P.M., Chadwick, O.A., Crews, T.E., Fownes, J.H., Hendricks, D.M., & Herbert, D. (1999) Soil and ecosystem development across the Hawaiian Islands. *GSA Today* **7**: 1–8.

Voortman, R.L., Sonneveld, B.G.J.S., & Keyzer, M.A. (2003) African land ecology: Opportunities and constraints for agricultural development. *Ambio* **32**: 367–73.

Wakatsuki, T., & Rasyidin, A. (1992) Rates of weathering and soil formation. *Geoderma* **52**: 251–63.

Wang, Z.Q., Liu, B.Y., Wang, X.Y., Gao, X.F., & Liu, G. (2009) Erosion effect on the productivity of black soil in Northeast China. *Science in China, Series D: Earth Sciences* **52**: 1005–21.

Wilkinson, B.H. (2005) Humans as geologic agents: A deep-time perspective. *Geology* **33**: 161–4.

Wilkinson, B.H., & McElroy, B.J. (2007) The impact of humans on continental erosion and sedimentation. *Geological Society of America Bulletin* **119**: 140–56.

Willenbring, J.K., & von Blanckenburg, F. (2010) Long-term stability of global erosion rates and weathering during late-Cenozoic cooling. *Nature* **465**: 211–14.

CHAPTER 5.2

Agroforestry and Soil Health: Linking Trees, Soil Biota, and Ecosystem Services

Edmundo Barrios, Gudeta W. Sileshi, Keith Shepherd, and Fergus Sinclair

5.2.1 Introduction

A significant and increasing proportion of the Earth's land area is covered by crop and range lands. Agricultural landscapes hold a large proportion of the world's biodiversity but the relative contribution of each land management type to conservation of biodiversity and the maintenance of ecosystem service delivery is poorly understood (Jackson *et al.* 2005). Ecosystem services can be classified into those associated with the provision of goods (e.g. food, fibers, and fresh water), those that support and regulate ecosystem function (e.g. climate regulation, disease control, soil formation, and nutrient cycling), and those cultural services that are not associated with material benefits (e.g. recreation, spiritual, and aesthetic value) (MEA 2005). Agricultural ecosystems both require and generate ecosystem services and may enhance or degrade natural capital through time depending on how they are managed. Soil health is a key indicator of the state of natural capital, and is considered here as an integrative property that reflects the capacity of soil to respond to agricultural management by maintaining both the agricultural production and the provision of other ecosystem services (Kibblewhite *et al.* 2008).

Soil organisms contribute to a wide range of ecosystem services that are essential to the functioning of natural and managed ecosystems (Wall 2004). Evidence has shown that there is a strong link between organisms above- and belowground (Wardle *et al.* 2004), highlighting the impact that land use and management can have on the provision of soil-based ecosystem services. Little research has been conducted on the role of soil biota in high input agriculture because natural processes regulating soil structure, nutrient supply, and pests and diseases have been largely replaced by soil tillage, artificial fertilizers, and biocides (Barrios 2007). Recent concern about sustaining soil function in intensive agriculture has created a new demand for agricultural practices that are less dependent on external inputs, tighten nutrient cycles, and are productive while enhancing rather than degrading natural capital (Swift *et al.* 2004).

Agroforestry is now broadly defined, in scale-neutral terms, as the interaction of agriculture and trees (Sinclair 2004). The field and landscape scales that are a focus here, involve land use practices that combine trees with crops and/or animals in some form of spatial arrangement or temporal sequence that results in significant ecological and economic interactions among trees and agricultural components (Sinclair 1999; Fig. 5.2.1). Recent global estimates indicate that nearly half of all agricultural land has >10% tree cover, an area of about 1 billion ha that is home to more than 500 million people (Zomer *et al.* 2009). The perennial nature of most trees has a profound impact on soil properties, and hence on the abundance, diversity, and function of the soil biota, underpinning soil health.

Soil Ecology and Ecosystem Services. First Edition. Edited by Diana H. Wall *et al.*
© 2012 Oxford University Press. Published 2012 by Oxford University Press.

Figure 5.2.1 Trees in agricultural fields: a) bean crop growing under diverse naturally regenerating trees (pruned and free-growing) in the Quesungual slash-and-mulch agroforestry system (Honduras); b) Shaded coffee system including *Erythrina poepiggiana* (pruned) that is often combined with naturally generated *Cordia alliadora* grown for timber (Costa Rica); c) cassava intercropped with beans growing in combination with pruned *Gliricida sepium* trees (foreground) and pigeonpea growing under *Schizolobium amazonicum* timber trees (background) (Brazil); d) maize crop under *Faidherbia albida* that is known for shedding their leaves in the wet season when the crop is growing, reducing competition while contributing to nutrient cycling (Tanzania). (Photo credits: a), c) Edmundo Barrios; b) Philippe Vaast; d) World Agroforestry Centre image database.)

This chapter first discusses the potential of trees to modify the soil and its impact on soil biota. The exploration of the linkages between the biological activity of soil organisms in agroforestry systems and their impact on soil-based ecosystem services and soil health follows next. Then recent advances in soil health monitoring systems and approaches to harnessing the complementary nature of local and scientific knowledge are discussed. We conclude by highlighting the role of agroforestry practices in adaptive and multifunctional land management with a view to enhancing soil health and agricultural sustainability, as well as recommendations for future research.

5.2.2 How trees influence soil properties and biota

The integration of trees into agricultural landscapes has the potential to generate a number of improvements in the soil as a habitat for soil organisms and also for crop growth. Trees modify the soil environment in many ways: leaves intercept rainfall, transpire water taken up by roots from the soil, and provide shade to the understory and soil, and dead or pruned leaves and branches provide soil cover and nutrient inputs to soils. These processes affect the temperature, moisture, erosion, and nutrient content of the soil as well as influencing soil biota. Martius *et al.* (2004) showed that soil macrofauna biomass in Amazonian forests strongly correlated with canopy closure, consistent with the tree canopy protecting the soil macrofauna from high temperature variation and drought stress. Similarly, research in coffee agroforestry systems of Southern Mexico demonstrated the impact of shading on lowering soil temperature, resulting in reduced water losses through evapotranspiration and maintenance of suitable soil moisture for crop growth (Lin 2010). Soil cover by tree litter and pruning biomass in Quesungual slash and mulch agroforestry practices in sub-humid Western Honduras has also been related to increased duration of soil moisture availability during critical periods resulting in sustained crop yield increases (Castro *et al.* 2009). Furthermore, studies by Pauli *et al.* (2010) which determined the spatial relationships among tree distribution, mulch cover, and earthworm casts in the same agroforestry context showed that production and distribution of earthworm casts (an indicator of biological activity) was closely related to the spatial arrangement of trees as shown (Fig. 5.2.2). These results emphasize the role of trees in fostering conditions for increased biological activity.

Soil improvements by trees can also occur by increased supply and availability of nutrients for crops and soil biota (Buresh & Tian 1998). The increased supply of nutrients through a "deep capture" of subsoil nutrients by tree roots returns these nutrients to the surface soil as litter (Rowe *et al.* 1999). This mechanism can also recycle fertilizer applied by farmers, thus improving nutrient use efficiency and the returns to fertilizer application. Published values for leguminous trees in different agroforestry systems show average annual additions of dry matter biomass of up to 20 t ha^{-1} year^{-1} (Young 1997). The size of annual biomass additions is largely influenced by climate, soil fertility, tree species and tree management regime.

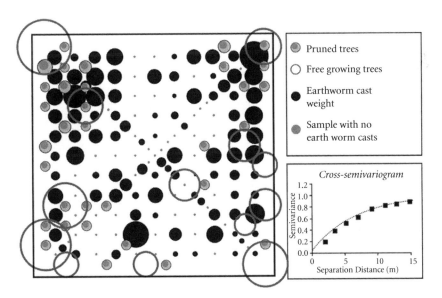

Figure 5.2.2 Comparison of earthworm casts and tree distribution in the Quesungual Slash-and-Mulch Agroforestry System. The size of the open circles represents the size of the tree canopy. The size of the light circles indicates the number of pruned trees found within each sampling cell (range of values: 1–3 pruned trees). The size of the dark circles represents the weight of earthworms casts (range of values: 0.6–10.6 g). The smaller graphics at the bottom right show the cross-semivariogram for the spatial relationship between tree distribution and earthworm cast distribution. An exponential model variogram provided the best fit to data (nugget = 0.03; sill = 1.22; range = 10.36 m). (Adapted from Pauli *et al.* 2010.)

Alley cropping in the Nigerian savanna-forest transition has reported about 5 and 7 t ha^{-1} year^{-1} of pruning biomass from *Gliricidia sepium* and *Leucaena leucocephala* respectively (Kang et al. 1999). In the Colombian Andes, pruning biomass contributions from *Indigofera constricta* and *Calliandra calothyrsus* planted fallows added about 9 t ha^{-1} year^{-1} to soil as mulch following pruning while *Tithonia diversifolia* contributed close to 15 t ha^{-1} year^{-1} (Barrios & Cobo 2004). In eastern Zambia, a drier environment, Sileshi and Mafongoya (2006a,b, 2007) recorded wide variations in pruning biomass contributions within the *Leucaena* genus (e.g. *L. palida, L. esculenta, L. collinsi* and *L. diversifolia* contributing 4.4, 3.4, 2.9, and 2.2 t ha^{-1} year^{-1} respectively during intercropping with maize), while *Acacia angustissima, G. sepium, Senna siamea* and *C. calothyrsus* contributed 3.3, 2.9, 2.2, and 1.4 t ha^{-1} year^{-1} respectively.

The contribution of agroforestry trees to soil nutrients through biomass additions and their utilization by intercropped plants has been reviewed by Palm (1995). One important highlight from that review is that while the nutrient concentration of pruning additions of some agroforestry trees is sufficient for most nutrients to meet crop demands, there is a general exception for phosphorus. Published values indicate that leguminous trees in alley cropping systems can contribute as much as 358 kg nitrogen (N) ha^{-1}, 28 kg phosphorus (P) ha^{-1}, 232 kg potassium (K) ha^{-1}, 144 kg calcium (Ca) ha^{-1}, and 60 kg magnesium (Mg) ha^{-1} (Palm 1995). Considerable interest in planted fallows using *T. diversifolia* has been generated because of its particular ability to accumulate nutrients, including P, in its biomass (Jama et al. 2000). Slash and mulch management of *T. diversifolia* in the Colombian Andes accumulated up to 417 kg N ha^{-1}, 85 kg P ha^{-1}, 928 kg K ha^{-1}, 299 kg Ca ha^{-1}, and 127.6 kg Mg ha^{-1} after 27 months (Barrios & Cobo 2004).

5.2.3 Agroforestry systems increase abundance of soil biota

Agroforestry trees have the potential to promote positive changes in the abundance, diversity, and function of soil organisms through their impact on soil as habitat for soil biota. There are few studies of tree—soil biota interactions in agroforestry systems, and most agroforestry studies reported in the literature focus on changes in the abundance of soil macrofauna with limited consideration of changes in diversity and function. For instance, studies in slash and mulch agroforestry practices in Honduras showed that total soil macrofauna densities were 52% (dry season) and 80% (wet season) higher than in the natural forest (Pauli et al. 2011). These figures are about five times greater than those found in the highlands of Central Honduras (271 individuals m^{-2}) (Ericksen & McSweeney 1999), close to twice the density of *Theobroma grandiflorum, Bactris gasipaes* (peach palm), and *Bertholetia excelsa* (Brazil nut) agroforestry (1059 individuals m^{-2}), and comparable to density values reported for coffee, *Schizolobium amazonicum* agroforestry (2054 individuals m^{-2}) and coffee, *Hevea brasiliensis* (rubber) agroforestry (2122 individuals m^{-2}) for the western Brazilian Amazon (Barros et al. 2002). Differences in abundance of soil organisms can be even greater when contrasting the impact of agroforestry systems to that of continuous cropping without trees (Table 5.2.1).

Agroforestry systems consistently generated substantial increases in the mean abundance of all soil organisms studied compared to the continuous cropping control (Table 5.2.1). The response ratio (RR), the ratio of the mean value of the agroforestry practice to that of the control (continuous cropping) (Hedges et al. 1999), was used to synthesize and compare different soil biota in soils under agroforestry and continuous cultivation without trees. While agroforestry systems consistently generated substantial increases in the mean abundance of soil organisms studied, some groups of organisms showed greater response than others. For example, millipedes and centipedes with RR near six appeared to benefit most from trees, followed by earthworms, ants, and mites with RR near three, springtails and beetles with RR near two. Termites and parasitic nematodes with RR near one appeared to be largely unaffected. While these results highlight a general pattern of trees promoting an increase in beneficial soil organisms, the limited number of studies and soil organisms suggests caution with generalizations regarding other soil organisms. Further, the paucity of studies which relate increases in

Table 5.2.1 Comparison of mean densities (individuals per m^2) of different soil biota in soils under agroforestry and continuous cultivation without trees, with the calculated response ratios (RR)[a]

	Agroforestry	Monocrop	RR	References
Soil macrofauna				
Earthworms	54.4	17.6	3.1	Dangerfield 1993; Tian et al. 1997, 2000; Sileshi & Mafongoya 2006a, b; Fonte et al. 2010
Beetles	20.9	9.6	2.2	Dangerfield 1993; Sileshi & Mafongoya 2006a,b
Centipedes	2.7	0.5	5.6	Dangerfield 1993; Sileshi & Mafongoya 2006a,b
Millipedes	8.1	1.3	6.1	Dangerfield 1993; Sileshi & Mafongoya 2006a,b
Termites	90.7	81.0	1.1	Dangerfield 1993; Sileshi & Mafongoya 2006a,b
Ants	23.2	8.6	2.7	Dangerfield 1993; Sileshi & Mafongoya 2006a,b
Soil mesofauna				
Collembola	3890.1	2000.7	1.9	Adejuyigbe et al. 1999
Mites	5100.7	1860.1	2.7	Adejuyigbe et al. 1999
Soil microfauna				
Non-parasitic nematodes	2922	1288	2.3	Kang et al. 1999
Parasitic nematodes	203.7	211.5	1	Kang et al. 1999

[a] The response ratio (RR), is the ratio of the mean value of the agroforestry practice to that of the control (continuous cropping). In this table RR quantifies both the direction and magnitude of changes in soil biota abundance: if trees do not have any effect on abundance RR = 1. If agroforestry trees favour soil biota the value of RR will be larger than unity, and vice versa if trees do not favour soil biota.

abundance with diversity and functional attributes limits inferences about possible functional benefits that the trees may promote.

Studies by Sileshi and Mafongoya (2007) found that soil biota responded differently to the application of organic resources of different quality. While earthworms and beetles were more abundant under legumes producing fast decomposing "high-quality" biomass, millipedes predominated under legumes producing slow decomposing "low-quality" biomass, and spiders and centipedes were not influenced by biomass quality. Studies by Barrios et al. (2005) compared coppiced planted fallows which showed that earthworm abundance beneath *I. constricta* was five times that of the values beneath *T. diversifolia*. Although both species had very similar plant tissue qualities the latter generated the greatest biomass and received the greatest nutrient inputs. These results suggest that while plant tissue quality measures provide a good prediction of nutrient release patterns, there could be additional factors influencing changes in the abundance of soil biota. Further, the limited number of studies considering soil biota/plant tissue quality interactions, particularly in the tropics, has limited the development of a predictive understanding. Nevertheless, the notion that the functional characteristics of dominant plants rather than diversity, may be a key driver of soil biodiversity and function (Hooper et al. 2005), suggests considerable opportunities to optimize tree/soil biota interactions in agroforestry systems.

5.2.4 Soil biological processes and soil-based ecosystem services

The relationships between the soil biological community, the biological processes they generate, and the provision of ecosystem goods and services in

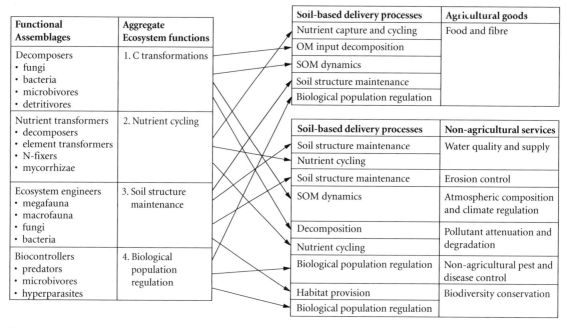

Figure 5.2.3 Conceptual framework of linkages between soil biota, biologically-mediated processes and the provision of soil-based ecosystem goods and services. (Adapted from Kibblewhite et al. 2008.)

agricultural soils have been recently synthesized (Fig. 5.2.3).

Soil organisms can be grouped into four functional assemblages (Kibblewhite *et al.* 2008): 1) decomposers, 2) nutrient transformers, 3) ecosystem engineers, and 4) biocontrollers, each composed of several functional groups. Functional attributes of these assemblages can be similarly grouped into four aggregated ecosystem functions that include carbon (C) transformations, nutrient cycling, soil structure maintenance, and population regulation. The decomposition of organic matter, where organic C in litter and other organic inputs are transformed through the consecutive fragmentation and enzymatic activity of a diverse suit of decomposer organisms, results in the release of CO_2 and the synthesis of soil organic matter (SOM) (Barrios 2007). While strongly linked to decomposition, the cycling of nutrients is largely mediated by soil microorganisms whose activity levels are regulated by food web interactions within the soil community (Susilo *et al.* 2004). The maintenance of soil structure is fostered by the combined action of plant roots and soil organisms known as "soil ecosystem engineers" that continuously modify the soil by forming "biological" aggregates, pores, and channels, thus altering soil physical properties and creating microhabitats for other soil organisms (Six *et al.* 2002). The biological control of pest and diseases takes place through the action of a wide range of soil organisms that regulate the populations of soil-borne diseases and pests largely through competition, predation, and parasitism (Susilo *et al.* 2004). These aggregated ecosystem functions participate in more than one soil-based delivery process. One or several soil-based delivery processes in turn are needed for the provision of ecosystem goods and services in agricultural landscapes. This framework is used to examine soil-based ecosystem services in agroforestry systems.

5.2.5 Tree–soil biota interactions foster the provision of soil-based ecosystem services

Trees and soil biota interact in a number of positive ways through facilitation and synergies. Facilitation is simply understood as diverse benefits provided by one species to other species (e.g. what trees provide to soil organisms and crop plants). Synergies occur when interacting species perform better together than individually (e.g. symbiosis between nitrogen fixing bacteria and leguminous trees). The enhancement of agricultural production has been the focus of attention for many decades; however, agricultural sustainability concerns have increasingly shifted attention to ecosystem services responsible for life support (i.e. C transformations and nutrient cycling) and regulation of ecosystem processes (i.e. soil structure maintenance and biological population regulation) (Swift et al. 2004; Barrios 2007). This section highlights tree–soil biota interactions in agroforestry systems that contribute to the provision of soil-based ecosystem services of life support and regulation.

5.2.5.1 Carbon transformations and nutrient cycling

A major contribution of agroforestry trees to soil-based ecosystem services occurs as a result of aboveground and belowground organic inputs that provide C-substrates and nutrients needed for the soil organisms involved in C transformations and nutrient cycling. C transformations occur during the decomposition of organic inputs as a result of the collective action of decomposer organisms that fragment organic inputs (e.g. earthworms, millipedes, termites and mites). This transformation in turn facilitates the enzymatic action by fungi and bacteria that results in the release of nutrients to the soil matrix, loss of C to the atmosphere, largely as CO_2, and the synthesis of SOM (Barrios 2007). C transformations and nutrient cycling take place through coordinated interaction of decomposers and nutrient transformers (Kibblewhite et al. 2008) and are treated here as a functional continuum.

Increased nutrient availability in agroforestry systems is often associated with higher levels of SOM under trees than away from trees (Buresh & Tian 1998). Nevertheless, while increases in total SOM are closely related with increases in soil water availability, this is not the case for soil nutrient availability because nutrient release is dependent on the biologically active portion of SOM (i.e. microbial biomass, light fraction SOM). The addition of biomass to soil from tree legumes such as *G. sepium* biomass through prunings (Barrios et al. 1996a,b), and also through litter and root turnover in *L. leucocephala* alley cropping (Vanlauwe et al. 1996) and planted tree fallows (Barrios et al. 1997), significantly contribute to increased light fraction SOM. Nevertheless, the relative contribution to light fraction SOM varies significantly amongst tree species. For example, the contribution of *Sesbania sesban* to the light fraction SOM was five times greater than species such as *C. calothrysus, Flemingia macrophylla, G. sepium, L. leucocephala,* and *S. siamea* (Barrios et al. 1997). The amount of N in the light fraction SOM was significantly correlated with N mineralization in the whole soil and with the yield of maize grown after the fallow phase (Barrios et al. 1998). Similarly, P in the light fraction SOM has been correlated with the amount of readily available P in the soil (Phiri et al. 2001).

The relative contribution of organic inputs to nutrient and CO_2 release through mineralization processes and SOM synthesis is strongly regulated by plant-soil biota interactions. The quality of organic inputs influences the decomposer biota composition and thus regulates the magnitude and rate of nutrient release (Wardle et al. 2004). Organic resource quality is an indicator of chemical composition and has been operationally defined by the concentrations of total N, lignin, and soluble polyphenols (Palm et al. 2001). Organic inputs of "high quality" (e.g. low lignin + polyphenol/N ratio) will decompose faster than those of "low quality" (e.g. high lignin + polyphenol/N ratio) and thus contribute relatively more to soil nutrient availability than to SOM formation and effects on microclimate. Studies comparing the effects of different agroforestry trees as planted fallows have concluded that high litter quality, and the ability of symbiotic microorganisms to fix N_2, characterize trees with the highest potential for increasing soil N availability (Barrios et al. 1997). However, if increased nutrient availability following organic inputs is not synchronized with crop demand, nutrient use efficiency can be low and lead to higher nutrient losses to the environment. Early increases in soil N availability after the addition of high quality residues of *I. constrictu* and *T. diversifolia* resulted in 20% and 17% N recovery by crop plants, respectively, whereas slower N release by the low quality residues of *C. calothyrsus* considerably increased plant N recovery to over 47% (Cobo *et al.* 2002).

Tree biomass also serves as a substrate for the synthesis of SOM. Regular organic inputs through leaf litter, tree prunings, and root turnover will have long term impacts on soil carbon and nutrient stocks and thus agroecosystem sustainability. While low quality organic resources are often associated with larger relative contributions to SOM (Palm *et al.* 2001), repeated applications of pruning of high quality biomass, such as those in the *Gliricidia*-maize intercropping systems in Malawi and Zambia also build SOM (Beedy *et al.* 2010). Root turnover is likely to be an important source of organic matter that significantly contributes to SOM synthesis and soil carbon storage in these contexts (Makumba *et al.* 2007). The physical protection of SOM during soil aggregation ensures sustained increases in SOM because soil aggregates generated by ecosystem engineers prevent rapid loss of SOM potentially reducing greenhouse gas emissions (Six *et al.* 2002). According to Castro *et al.* (2009) the Quesungual slash and mulch agroforestry practice was a net CH_4 sink (-102 mg CH_4 m^{-2} year^{-1}) compared to slash and burn agriculture that was a net source of 150 mg CH_4 m^{-2} year^{-1}. While both land uses were net sources of N_2O and CO_2, the overall global warming potential for slash and burn was nearly four times higher than that of slash and mulch. These results are consistent with other studies at the same location showing increased earthworm activity and reduced C loss with mulching rather than burning (Fonte *et al.* 2010; Pauli *et al.* 2010). Davidson *et al.* (2008) in the Brazilian Amazon also showed that burning generated five times higher CO_2-equivalent emissions than mulching, which contributes to global warming mitigation efforts. These results highlight important opportunities to design and manage agroforestry practices to include mixtures of trees that generate residues of different qualities, promoting SOM synthesis, nutrient release, and reduction of greenhouse gas emissions in ways that optimize the ecosystem services from organic matter decomposition, including nutrient supply, water quality and supply, as well as climate regulation.

5.2.5.2 Symbiotic interactions and nutrient cycling

Biological nitrogen fixation (BNF) constitutes a key nutrient input to agroecosystems (Giller 2001). The contribution of leguminous trees to building up N in degraded soils through BNF is well recognized as an important component of the ecosystem service of nutrient cycling (Barrios 2007). There are significant differences in estimates of BNF in trees, ranging from high rates up to 472 kg N_2 ha^{-1} year^{-1} in *L. leucocephala*, *G. sepium*, *C. calothyrsus* to low rates <50 kg N_2 ha^{-1} year^{-1} in *Acacia melanoxylon* and *A. holoserica* (Giller 2001). However, actual BNF rates under field conditions are often lower than the potential maximum as they are considerably affected by soil and climatic conditions. The high variability in percentage of total plant N derived from the atmosphere among tree provenances of *L. leucocephala* (37–74%) and *Faidherbia albida* (6–37%) found by Sanginga *et al.* (1990), and *G. sepium* (Sanginga *et al.* 1994), suggest opportunities for optimizing this symbiotic tree–soil biota interaction. Nevertheless, the precise quantification of the amount of N_2 fixed in trees continues to be limited by methodological constraints (Giller 2001).

Arbuscular mycorrhizal fungi (AMF) associated with trees can complement the nutrient capture function of deep roots by increasing the recovery of nutrients from the subsoil when allowing exploration of a larger soil volume. They may also reduce nutrient loss through leaching and associated pollution, while also increasing uptake of less mobile nutrients like phosphorus. For example, the remarkable ability of *T. diversifolia* to accumulate large quantities of P in its biomass, as well as all other nutrients, seems to be related to the unusual specificity of its mycorrhizal associations (Sharrock *et al.* 2004). The lack of adequate P nutrition is usually a key limiting factor to BNF and therefore the combined action of both symbionts should be encouraged. However, legumes known to be able to fix N_2 under low soil P availability should also be targeted for agroforestry (Sprent 1999). Improved nutrition of trees hosting AMF and nitrogen fixing bacteria would encourage greater nutrient input and greater quantities of nutrients being recycled. The optimization of these interactions is of particular significance for resource poor farmers with limited access to fertilizers and in most tropical soils where N and/or P are limited.

5.2.5.3 Soil structure maintenance

Soil aggregates resulting from the arrangement of soil primary particles and SOM bound by organic and inorganic agents constitute the structural units within the soil. Soil structure is thus a dynamic property reflecting the balance between aggregate forming factors and those that disrupt them (Six et al. 1998). The formation of "biological aggregates" and their stabilization is the result of the activity of fungi, bacteria, plant roots, and macrofauna (Six et al. 2002). Studies by Kang et al. (1994) reported that surface casting by *Hyperiodrilus africanus* was higher under trees including *Dactilenia (Acioa) barteri*, *Alchornea cordifolia*, *G. sepium*, and *L. leucocephala* than in a control plot without trees. In that study, while casting activities under *D. barteri* and *G. sepium* were of similar magnitude (26.4 and 24.4 Mg ha^{-1} year^{-1} respectively) the content of water-stable aggregates in worm casts varied with tree species, being highest under *D. barteri* and lowest under *G. sepium*. Recent studies in the Quesungual slash and mulch agroforestry practice (Fig. 5.2.2) show that the spatial distribution of casts was closely related to the spatial arrangement of trees and mulch (Pauli et al. 2010). This highlights the role of trees in promoting biological activity that contributes to soil structure maintenance. Another study in the same locality which compared the impact of slash and mulch with slash and burn showed that the mean soil erosion rate after three years was about six times higher with burning than mulching (Castro et al. 2009). Additionally, higher soil mesoporosity (30%) and macroporosity (19%) for mulching versus burning are consistent with increases in biological activity of various soil organisms that generate pores and channels of different sizes and shapes. These porosity differences significantly influence the plant available soil water storage capacity, water infiltration, surface runoff, and soil erosion.

Arbuscular mycorrhizal fungi have received considerable attention because of their contribution to the formation and maintenance of soil structure through hyphal enmeshment of soil aggregates and deposition of glomalin, an AMF-specific glycoprotein strongly linked to water stable aggregation (Rillig 2004). The dynamics of physical protection of SOM in soil aggregates has received considerable attention because of its importance for soil carbon sequestration. When soil aggregates break into smaller pieces upon wetting, erosion rates increase, and SOM is readily exposed to microbial action that results in C loss to the atmosphere (Barrios 2007). Therefore, the overall potential effect of agroforestry on soil erosion control and soil C sequestration are clearly underpinned by tree–soil biota interactions that foster increases in the proportion of soil aggregates that are stable upon wetting during rainfall events, and in the magnitude and diversity of soil porosity that allows a balance between infiltration and soil water storage for plants and soil organisms. These results highlight important opportunities for the design of agroforestry practices that incorporate tree species diversity and mulch management, to promote soil biological diversity and activity that optimize the aggregate dynamics required for soil erosion control, C sequestration, and the supply of good quality water.

5.2.5.4 Control of pests and diseases

The control of soil-borne pest and diseases through biological regulation is an ecosystem service of great economic, human health, and environmental importance because global annual crop losses are near 30% and commonly controlled with application of biocides toxic for humans and the environment (Oerke & Dehne 2004). The relationship between the soil biota, soil fertility, and plant health is strengthened in agroforestry systems as trees improve soil fertility, foster above- and below-ground biodiversity, and support the development of complex food webs that keep pests and diseases under control through a combination of predation, parasitism, and competition. Several soil organisms such as collembolans, ants, beetles, centipedes, spiders, and predatory mites and nematodes act as biocontrol agents (Sileshi *et al.* 2001). Termite damage to maize in eastern Zambia was reported to be consistently lower in maize-tree (*L. leucocephala*, *G. sepium*, and *S. sesban*) associations compared to monoculture maize (Sileshi *et al.* 2005). The increased damage in monoculture maize was attributed to low soil organic matter, low soil fertility and, water stress. Fungus-growing termites preferentially feed on crop residues, mulches, and soil organic matter; however, if these are not available, they will eat live plants. The addition of large quantities of leaf litter or pruning biomass in maize-tree associations could increase labile pools of soil organic matter and water availability, and improve soil fertility, which could increase crop vigour and reduce termite damage (Sileshi *et al.* 2005). According to studies in the same area where the soils are heavily infested with *Striga asiatica* under conventional management, maize infestation was negligible following planted tree fallows of *S. sesban* and *S. siamea* that increase soil N availability (Barrios *et al.* 1998). Nevertheless, it is important to highlight that single species agroforestry systems (e.g. planted fallows) also have a high potential to face similar pest and disease problems found in crop monocultures as shown in some *S. sesban* planted fallows (Sileshi *et al.* 2008). Therefore, agroforestry systems that include different tree species, especially if they represent different plant functional types, are likely to increase the diversity of niches suitable for biological control agents. Current understanding about factors affecting plant-soil biodiversity interactions that could influence the effectiveness of biological control agents is particularly limited for agroforestry systems and suggest opportunities for future agroforestry system design to ensure adequate plant and soil biodiversity levels that would allow the tree–soil biota interactions required for biological control of soil-borne pests and diseases.

5.2.6 Soil health monitoring systems

Evidence on the benefits of trees on soil biota and ecosystem services is fragmented and strongly biased towards small-scale plot experiments. However, new advances in remote sensing, georeferenced field surveys, and proximal soil sensing (e.g. Sanchez *et al.* 2009) are providing new opportunities for vegetation and soil measurement and monitoring at multiple scales referred to here as land health surveillance. Developments in information and communication technology are also providing unprecedented opportunities for engaging local communities in systematic data collection (Ballantyne *et al.* 2010).

5.2.6.1 Land health surveillance

Land health is the capacity of land to sustain delivery of essential ecosystem services. Land health surveillance implies large area monitoring of land health using standardized measurement protocols that permit meta-analysis, and where appropriate, use of statistical approaches for sampling of populations to avoid sampling bias (Shepherd *et al.* 2008). The Africa Soil Information Service (AfSIS) provides an example of the application of these principles and provides opportunity for systematic study of tree–soil interactions from continental to plot (or stand) scales. AfSIS deploys a randomized set of "sentinel" sites, spatially randomized within major Köppen climatic zones in non-desert portions of sub-Saharan Africa (Fig. 5.2.4). A sentinel site is a 10 × 10 km block of land, within which a spatially stratified, randomized ground sampling scheme is implemented. Tree and shrub density are measured and soil samples taken in 100-m^2 sub-plots, which are nested within 1000-m^2 plots, in turn nested within 1-km^2 diameter clusters. The position of the clusters within the 2.5 × 2.5 km tiles is also randomized. Soil samples from each plot are characterized using infrared spectroscopy (Shepherd & Walsh 2007) as a front-line soil screening tool (Fig. 5.2.4). Conventional reference analyses of soil chemical, physical, and biological properties are done on a random subset of samples and related to infrared spectral signatures to infer values for the entire set of samples. The georeferenced soil properties are mapped through hierarchical statistical modeling of the soil data from the combined set of sentinel sites to satellite data (e.g. Landsat, Modis) and other GIS data (e.g. digital elevation models) with continental coverage.

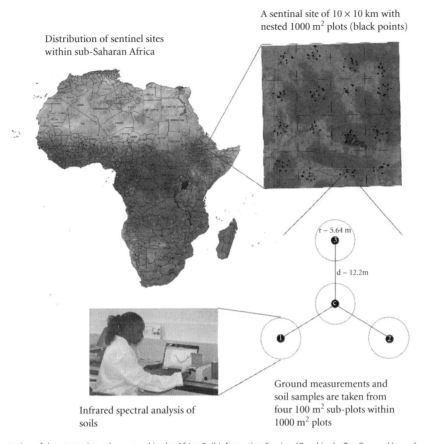

Figure 5.2.4 Illustration of the measuring scheme used in the Africa Soil Information Service. (Graphics by Tor-Gunnar Vagen.)

The richness and consistency of the data sets provide unprecedented opportunities for exploring tree–soil interactions at different scales. For example, collection of data on soil microbial and faunal diversity is being piloted using DNA sequencing (e.g. Fierer & Jackson 2006; Wu et al. 2009). The soil biodiversity data can then be related to other data collected at the different scales, such as soil chemical and physical properties, soil erosion status, vegetation characteristics, woody biomass density, land form, climate, etc.

The population based sampling frame permits statistical distributions of key soil biodiversity indicators to be established, and these can be used to develop norms conditioned on factors such as climate zone, topography, land cover classification, historic land cover, geology, landscape position, and static or slowly changing soil variables. Comparison of indicator values against norms can be used as a statistical basis for developing indicators of degradation. This represents a major advance since it is currently difficult to interpret soil biodiversity data in terms of soil functional capacity. Furthermore, risk factors associated with biodiversity loss could be quantitatively established and verified through monitoring changes in prevalence of degradation over time (i.e. incidence). The surveillance approach has potential to greatly increase the efficiency of research in terms of knowledge gained per unit research investment. For example, the combination of probability sampling, co-located measurements, and use of standardized protocols enables characterization of whole populations and the meta-analysis of results at different scales. This contrasts with

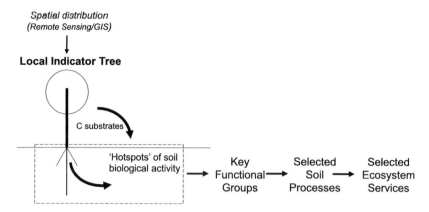

Figure 5.2.5 Integrative approach to identify hotspots of soil biological activity. (Adapted from Barrios 2007.)

existing approaches in which studies typically do not sample a known population of soil spatial units and results can rarely be combined to provide multiscale insights or generalizable conclusions. Surveillance can provide a practical, evidence-based approach for considering soil biodiversity and other land health indicators when planning and evaluating land management interventions.

There is potential for land users to participate in centrally coordinated land health surveillance systems, and in doing so increase the quality of the information they are able to access. Land users could make simple georeferenced observations on land quality using a standardized protocol and submit this data through mobile phone technology to centralized databases. Ways to avoid sampling bias would have to be found, but there may be opportunity for researchers to direct sampling efforts to locally recognized degradation hot spots by using adaptive sampling schemes. Systems whereby communities take soil samples from pre-defined georeferenced locations may also be possible. Further, land users could utilize the same technology to tap into information systems that provide highly location-specific information on land and climatic conditions and access interpreted results from their observations. Local observations could be used to improve recommendations through Bayesian updating (Pearl 1988) of prior information supplied from regional environmental databases, returning the improved estimate to the user.

5.2.6.2 Integrating local knowledge about soil health

The increasing attention paid to local knowledge in recent years is recognition that the knowledge of people who have closely interacted with their environment for a long time can offer many insights about the sustainable management of natural resources (Barrios et al. 2006). Participatory research approaches that encourage the integration of local and scientific knowledge could be useful to reduce the uncertainty of plant–soil biota interaction studies at the landscape scale by adding relevance and legitimacy to the process. Barrios (2007) proposed an approach to integrate local knowledge for the identification of soil biota "hotspots" in the landscape that are presumably responsible for a large proportion of the provision of soil-based ecosystem services. In short, local knowledge about native plants as indicators of soil health is consistently considered a key source of information for land use decision-making across farming communities of Latin America and Africa (Barrios et al. 2006). The presence of native plants indicating healthy soils informs and assists farmers to make decisions during establishment of new agricultural plots. Similarly, Barrios (2007) proposed to use local indicator trees to identify healthy soils where "hotspots" of soil biological activity are likely to be concentrated (Fig. 5.2.5).

These "hot spots" include the rhizosphere, biogenic structures (i.e. soil aggregates), soil C pools (i.e. light fraction SOM), and organic detritus (i.e. litter), where key functional assemblages can be studied to focus on soil biological processes that underpin the provision of soil-based ecosystem services. Given the difficulty of studying soil biota at the landscape scale, greater knowledge about indicator plant–soil biota interactions combined with spatial information obtained from remote sensing about indicator plants, could guide inferences about the role of soil biota and function in the provision of soil-based ecosystem services. The general consensus that soil biological processes are not randomly distributed but largely aggregated near C substrates, and that greater knowledge about tree–soil biota interactions have great potential to improve understanding of the impacts of soil biota at larger scales, are consistent with this approach (Wardle *et al.* 2004).

5.2.7 Conclusions and recommendations

Agroforestry systems have the potential to facilitate the transition to multifunctional agriculture that successfully addresses the challenge of optimizing crop productivity while maintaining the provision of other ecosystem services. In order to realize this potential, however, there is considerable need for greater understanding of how to optimize tree–soil biota interactions that improve agroecosystem function and soil health.

The promotion of agroforestry systems including multiple tree species (e.g. multistrata agroforestry systems) has been highlighted here as a strategy to enhance the sustained provision of soil-based ecosystem services. Combining trees and crops that can coexist while generating sufficient organic inputs of different quality is seen as a way to preserve soil cover and increase the diversity and persistence of active soil biota. There is a need to further study the impact of spatial arrangements and management that minimize competition and favors complementarities and facilitative interactions among trees and associated crops in terms of biomass production, nutrient and water use efficiency, and how these in turn influence the abundance, diversity, and activity of key soil biota. Tree–soil biota interactions both respond and influence ecosystem properties, and so, a greater understanding of the feedbacks involved is necessary to link experimental results at smaller scales with those at large scales. Agroforestry practices embrace manageable levels of complexity that would help address fundamental questions about the role of interacting above- and belowground biodiversity in increasing functional resilience to disturbance or climate change. The use of gradients of physical factors and agricultural intensification as the basis of landscape experimental design would be helpful to gain greater understanding of tree–soil biota interactions under different disturbance regimes and how they influence agroecosystem function and the provision of ecosystem services.

A better understanding of tree–soil biota interactions would provide opportunities to design systems that maximize complementarities, facilitation, and synergies that result in the sustained provision of ecosystem services. Major challenges to the measurement of ecosystem services and the interpretation of data include the particularly limited number of published quantitative field studies, the diversity of applied methods, the difficulty in sampling and identification of some taxa, the spatial biases created by some sampling methods, and the different spatial and temporal scales at which ecosystem services are delivered. The focus proposed by Kibblewhite *et al.* (2008) on four aggregate ecosystem functions and key functional groups or assemblages constitutes a practical approach to address the difficulty of studying all soil biodiversity as part of soil health evaluation. The application of common methodologies for sampling and characterizing soil biota (Moreira *et al.* 2008) may allow greater comparability among studies in agroforestry systems. Furthermore, the strategic use of molecular tools, analysis of stable isotopes, and spectroscopic techniques will increase the ability to identify and characterize "hotspots" of biological activity and facilitate the study of linkages between key soil biota and ecosystem functions at different temporal and spatial scales.

The Land Health Surveillance approach used in the AfSIS project, provides a robust experimental framework to systematically analyze and integrate information at different spatial and temporal scales, and thus provide a comprehensive evaluation of how changes in tree density and diversity influences soil health in agricultural landscapes. Research efforts are also needed for the development of local soil health monitoring systems that inform land users about their land's capacity to provide ecosystem services (Barrios 2007). The empowerment of local communities, and agricultural research and extension institutions, to conduct local monitoring can generate valuable information. Such data, combined with new approaches for the economic valuation of ecosystem services, may be used during negotiations for payments of ecosystem services that reward good management practices and thus become a further incentive mechanism for sustainable land management and development.

Acknowledgements

We are grateful to Richard Coe for statistical advice.

References

Adejuyigbe, C.O., Tian, G., & Adeoye, G.O. (1999) Potential of woody fallows in restoration of soil microarthropods in a degraded tropical soil. *Agroforestry Systems* **47**: 263–72.

AfSIS. African Soil Information System. [Online] http://www.africasoils.net (accessed 14 February 2011).

Ballantyne, P., Maru, A., & Porcari, E.M. (2010) Information and Communication Technologies—Opportunities to Mobilize Agricultural Science for Development. *Crop Science* **50**: S-63–9.

Barrios, E. (2007) Soil biota, ecosystem services and land productivity. *Ecological Economics* **64**: 269–85.

Barrios, E., Buresh, R.J., & Sprent, J.I. (1996a) Organic matter in soil particle size and density fractions from maize and legume cropping systems. *Soil Biology & Biochemistry* **28** (2): 185–93.

Barrios, E., Buresh, R.J., & Sprent, J.I. (1996b) Nitrogen mineralization in density fractions of soil organic matter from maize and legume cropping systems. *Soil Biology & Biochemistry* **28** (10/11): 1459–65.

Barrios, E., Kwesiga, F., Buresh, R.J., & Sprent, J.I. (1997) Light fraction soil organic matter and available nitrogen following trees and maize. *Soil Science Society of America Journal* **61**(3): 826–31.

Barrios, E., Kwesiga, F., Buresh, R.J., Sprent, J.I., & Coe, R. (1998) Relating preseason soil nitrogen to maize yield in tree legume-maize rotations. *Soil Science Society of America Journal* **62**(6): 1604–9.

Barrios, E., & Cobo, J.G. (2004) Plant growth, biomass production and nutrient accumulation by slash/mulch agroforestry systems in tropical hillsides of Colombia. *Agroforestry Systems* **60**: 255–65.

Barrios, E., Cobo, J.G., Rao, I.M., et al. (2005) Fallow management for soil fertility recovery in tropical Andean agroecosystems in Colombia. *Agriculture, Ecosystems and Environment* **110**: 29–42.

Barrios, E., Delve, R.J., Bekunda, M., et al. (2006) Indicators of soil quality: A South-South development of a methodological guide for linking local and technical knowledge. *Geoderma* **135**: 248–59.

Barros, E., Pashanasi, B., Constantino, R., & Lavelle, P. (2002) Effects of land-use system on the soil macrofauna in western Brazilian Amazonia. *Biology and Fertility of Soils* **35**: 338–47.

Beedy, T.L., Snapp, S.S., Akinnifesi, F.K., & Sileshi, G.W. (2010) Impact of *Gliricidia sepium* intercropping on soil organic matter fractions in a maize-based cropping system. *Agriculture, Ecosystems and Environment* **138**: 139–46.

Buresh, R.J., & Tian, G. (1998) Soil improvement by trees in sub-Saharan Africa. *Agroforestry Systems* **38**: 51–76.

Castro, A., Rivera, M., Ferreira, O., et al. (2009) Quesungual slash and mulch agroforestry system improves crop water productivity in hillside agroecosystems of the sub-humid tropics. In: E. Humphreys, & R.S. Bayot (eds.) *Increasing the productivity & sustainability of rainfed cropping systems of poor smallholder farmers*, pp. 89–97. Proceedings of the CGIAR challenge program on water and food international workshop on rainfed cropping systems. CGIAR Challenge Program on Water & Food, Battaramulla, Sri Lanka.

Cobo, J.G., Barrios, E., Kass, D., & Thomas, R.J. (2002) Nitrogen mineralization and crop uptake from surface-applied leaves of green manure species on a tropical volcanic-ash soil. *Biology and Fertility of Soils* **36**(2): 87–92.

Dangerfield, J.M. (1993) Characterization of soil fauna communities. In: M.R. Rao, & R.J. Scholes (eds.) *Report on characterization of an experimental field at KARI farm, Muguga, Kenya*, pp. 51–67. ICRAF, Nairobi.

Davidson, E.A., de Abreu Sá, T.D., Carvalho, C.J.R., et al. (2008) An integrated greenhouse gas assessment of an alternative to slash-and-burn agriculture in eastern Amazonia. *Global Change Biology* **14**: 998–1007.

Ericksen, P., & McSweeney, K. (1999) Fine-scale analysis of soil quality for various land uses and landforms in central Honduras. *American Journal of Alternative Agriculture* **14**: 146–57.

Fierer, N., & Jackson, R.B. (2006) The diversity and biogeography of soil bacterial communities. *Proceedings of the National Academy of Sciences of the United States of America* **103**: 626–31.

Fonte, S.J., Barrios, E., & Six, J. (2010) Earthworms, soil fertility and aggregate-associated soil organic matter dynamics in the Quesungual agroforestry system. *Geoderma* **155**: 320–8.

Giller, K.E. (2001) *Nitrogen fixation in tropical cropping systems* (2nd Edn.) CAB International Publishing, Wallingford.

Hedges, L.V., Gurevitch, J., Curtis, P.S. (1999) The meta-analysis of response ratios in experimental ecology. *Ecology* **80**(4): 1150–6.

Hooper, D.U., Chapin, F.S., Ewel, J.J., *et al.* (2005) Effects of biodiversity on ecosystem functioning: a consensus of current knowledge. *Ecological Monographs* **75**(1): 3–35.

Jackson, L., Bawa, K., Pascual, U., & Perrings, C. (2005) *agroBIODIVERSITY: A new science agenda for biodiversity in support of sustainable agroecosystems.* DIVERSITAS Report No 4, Paris.

Jama, B., Palm, C.A., Buresh, R.J., *et al.* (2000) *Tithonia diversifolia* as a green manure for soil fertility improvement in western Kenya: a review. *Agroforestry Systems* **49**: 201–21.

Kang, B.T., Akinnifesi, F.K., & Pleysier, J.L. (1994) Effect of agroforestry woody species on earthworm activity and physicochemical properties of worm casts. *Biology and Fertility of Soils* **18**(3): 193–9.

Kang, B.T., Caveness, F.E., Tian, G., & Kolawole, G.O. (1999) Long-term alley cropping with four species on an Alfisol in southwest Nigeria—effect on crop performance, soil chemical properties and nematode population. *Nutrient Cycling in Agroecosystem* **54**: 145–55.

Kibblewhite, M.G., Ritz, K., & Swift, M.J. (2008) Soil health in agricultural systems. *Philosophical Transactions of the Royal Society B Biological Sciences* **363**: 685–701.

Landsat. [Online] http://landsat.gsfc.nasa.gov/ (accessed 14 February 2011).

Lin, B.B. (2010) The role of agroforestry in reducing water loss through soil evaporation and crop transpiration in coffee agroecosystems. *Agricultural and Forest Meteorology* **150**: 510–18.

Makumba, W., Akinnifesi, F., Janssen, B., & Oenema, O. (2007) Long-term impact of a gliricidia-maize intercropping system on carbon sequestration in southern Malawi. *Agriculture, Ecosystems and Environment* **118**: 237–43.

Martius, C., Höfer, H., Garcia, M.V.B., Römbke, J., Förster, B., & Hanagarth, W. (2004) Microclimate in agroforestry systems in central Amazonia: does canopy closure matter to soil organisms? *Agroforestry Systems* **60** (3): 291–304.

Millennium Ecosystem Assessment (MEA) (2005) *Ecosystem and Human Well-Being: Synthesis.* Island Press, Washington, DC.

Modis. [Online] http://modis.gsfc.nasa.gov/ (accessed 14 February 2011).

Moreira, F.M.S., Huising, E.J., & Bignell, D.E. (2008) *A Handbook of Tropical Soil Biology: Sampling and characterization of below-ground biodiversity.* Earthscan, London.

Oerke, E.C. & Dehne, H.W. (2004) Safeguarding production—losses in major crops and the role of crop protection. *Crop Protection* **23**: 275–85.

Palm, C.A. (1995) Contribution of agroforestry trees to nutrient requirements of intercropped plants. *Agroforestry Systems* **30**(1–2): 105–24.

Palm, C.A., Gachengo, C.N., Delve, R.J., Cadisch, G., & Giller, K.E. (2001) Organic inputs for soil fertility management in tropical agroecosystems: Application of an organic resource database. *Agriculture, Ecosystems and Environment* **83**: 27–42.

Pauli, N., Oberthur, T., Barrios, E., & Conacher, A.J. (2010) Fine-scale spatial and temporal variation in earthworm surface casting activity in agroforestry fields, western Honduras. *Pedobiologia* **53**: 127–39.

Pauli, N., Barrios, E., Conacher, A.J., & Oberthur, T. (2011) Soil macrofauna in agricultural landscapes dominated by the Quesungual Slash-and-Mulch Agroforestry System, western Honduras. *Applied Soil Ecology* **47**: 119–32.

Pearl, J. (1988) *Probabilistic Reasoning in Intelligent Systems.* Morgan Kaufmann Publishers, San Mateo, CA.

Phiri, S., Barrios, E., Rao, I.M., & Singh, B.R. (2001) Changes in soil organic matter and phosphorus fractions under planted fallows and a crop rotation system on a Colombian volcanic-ash soil. *Plant and Soil* **231**: 211–23.

Rillig, M.C. (2004) Arbuscular mycorrhizae and terrestrial ecosystem processes. *Ecology Letters* **7**: 740–54.

Rowe, E.C., Hairiah, K., Giller, K.E., van Noordwijk, M., & Cadisch, G. (1999) Testing the safety-net role of hedgerow tree roots by ^{15}N placement at different soil depths. *Agroforestry Systems* **43**: 81–93.

Sanchez, P.A., Ahamed, S., Carré, F., *et al.* (2009) Digital soil map of the world. *Science* **325**: 680–1.

Sanginga, N., Bowen, G.D., & Danso, S.K.A. (1990) Assessement of the genetic variability for N_2 fixation between and within provenances of *Leucaena leucocephala* and *Acacia albida* estimated by ^{15}N labeling techniques. *Plant and Soil* **127**: 169–78.

Sanginga, N., Danso, S.K.A., Zapata, F., & Bowen, G.D. (1994) Field validation of intraspecific variation in phosphorus use efficiency and nitrogen fixation by provenances of *Gliricidia sepium* grown in low P soils. *Applied Soil Ecology* **1**: 127–38.

Sharrock, R.A., Sinclair, F.L., Gliddon, C., et al. (2004) A global assessment using PCR techniques of mycorrhizal fungal populations colonising *Tithonia diversifolia*. *Mycorrhiza* **14**: 103–9.

Shepherd, K.D., & Walsh, M.G. (2007) Infrared Spectroscopy—enabling an evidence-based diagnostic surveillance approach to agricultural and environmental management in developing countries. *Journal of Near Infrared Spectroscopy* **15**: 1–19.

Shepherd, K.D, Vagen, T.G, Gumbricht, T., & Walsh, M.G. (2008) Land Degradation Surveillance: Quantifying and Monitoring Land Degradation. In: *Sustainable Land Management Sourcebook*, pp. 141–7. The World Bank, Washington DC.

Sileshi, G., Kenis, M., Ogol, C.K.P.O., Sithanantham, S. (2001) Predators of *Mesoplatys ochroptera* Stål in sesbania-planted fallows in eastern Zambia. *BioControl* **46**: 289–310.

Sileshi, G., Mafongoya, P.L., Kwesiga, F., Nkunika, P. (2005) Termite damage to maize grown in agroforestry systems, traditional fallows and monoculture on Nitrogen-limited soils in eastern Zambia. *Agricultural and Forest Entomology* **7**: 61–9.

Sileshi, G., & Mafongoya, P.L. (2006a) Variation in macrofaunal communities under contrasting land-use systems in eastern Zambia. *Applied Soil Ecology* **33**: 49–60.

Sileshi, G., & Mafongoya, P.L. (2006b) Long-term effect of legume-improved fallows on soil invertebrates and maize yield in eastern Zambia. *Agriculture, Ecosystem and Environment*, **115**: 69–78.

Sileshi, G., & Mafongoya, P.L. (2007) Quantity and quality of organic inputs from coppicing leguminous trees influence abundance of soil macrofauna in maize crops in eastern Zambia. *Biology and Fertility of Soils* **43**: 333–40.

Sileshi, G., Schroth, G., Rao, M.R., & Girma, H. (2008) Weeds, diseases, insect pests and tri-trophic interactions in tropical agroforestry. In: D.R. Batish, R.K. Kohli, S. Jose, & H.P. Singh (eds.) *Ecological Basis of Agroforestry*, pp. 73–94. CRC Press, Boca Raton, FL.

Sinclair, F.L. (1999) A general classification of agroforestry practice. *Agroforestry Systems* **46**: 161–80.

Sinclair, F.L. (2004) Agroforestry. In: J. Burley, J. Evans & J.A. Youngquist (eds.) *Encyclopedia of Forest Sciences*, pp. 27–32. Elsevier, Amsterdam.

Six, J., Elliot, E.T., Paustian, K., & Doran, J.W. (1998) Aggregation and soil organic matter accumulation in cultivated and native grassland soils. *Soil Science Society of America Journal* **62**: 1367–77.

Six, J., Feller, C., Denef, K., Ogle, S.M., Moraes Sa, J.C., & Albrecht, A. (2002) Soil organic matter, biota and aggregation in temperate and tropical soils—Effects of no-tillage. *Agronomie* **22** (7/8): 755–75.

Sprent, J.I. (1999) Not all nitrogen-fixing legumes have high requirement for phosphorus. *Agroforestry Forum* **9**(4): 7–10.

Susilo, F.X., Neutel, A.M., van Noordwijk, M., Hairiah, K., Brown, G., & Swift, M.J. (2004) Soil biodiversity and food webs. In: M. van Noordwijk, G. Cadisch, & C.K. Ong. (eds.) *Below-ground Interactions in Tropical Agroecosystems*, pp. 285–302. CAB International, Wallingford.

Swift, M.J., Izac, A.M.N., & van Noordwijk, M. (2004) Biodiversity and ecosystem services in agricultural landscapes—are we asking the right questions? *Agriculture, Ecosystem and Environment* **104**: 113–34.

Tian, G., Kang, B.T., & Brussaard, L. (1997) Effect of mulch quality on earthworm activity and nutrient supply in the humid tropics. *Soil Biology and Biochemistry* **29**: 369–73.

Tian, G., Olimah, J.A., Adjeoye, G.O., & Kang, B.T. (2000) Regeneration of earthworm populations in a degraded soil by natural and planted fallows under humid tropical conditions. *Soil Science Society of America Journal* **64**: 222–8.

Vanlauwe, B., Swift, M.J., & Merck, R. (1996) Soil litter dynamics and N use in a leucaena (*Leucaena leucocephala* Lam. (De Witt)) alley cropping system in southwestern Nigeria. *Soil Biology and Biochemistry* **28**: 739–49.

Wall, D.H. (2004) *Sustaining biodiversity and ecosystem services in soils and sediments*. SCOPE 64. Island Press, Washington DC.

Wardle, D.A., Bardgett, R.D., Klironomos, J.N., Setala, H., Van der Putten, W.H., & Wall, D.H. (2004) Ecological linkages between aboveground and belowground biota. *Science* **304**: 1629–33.

Wu, T., Ayres, E., Li, G., Bardgett, R.D., Wall, D.H., & Garey, J.R. (2009) Molecular profiling of soil animal diversity in natural ecosystems: incongruence of molecular and morphological results. *Soil Biology and Biochemistry* **41**: 849–57.

Young, A. (1997) *Agroforestry for soil management* (2nd Edn). CAB International, Wallingford.

Zomer, R.J., Trabucco, A., Coe, R., & Place, F. (2009) *Trees on Farm: Analysis of Global Extent and Geographical Patterns of Agroforestry*. ICRAF Working Paper no. 89. Nairobi, Kenya: World Agroforestry Centre.

CHAPTER 5.3

Soil Health: The Concept, Its Role, and Strategies for Monitoring

Douglas L. Karlen

5.3.1 The concept of soil health

Soil health is not a new concept. Greek and Roman philosophers were aware of the importance of soil health to agricultural prosperity and demonstrated this in their treatises on farm management more than 2,000 years ago. An example from Hillel (1991) is an account whereby Plato has Critias proclaim:

> What now remains of the formerly rich land is like the skeleton of a sick man, with all the fat and soft earth having wasted away and only the bare framework remaining. Formerly, many of the mountains were arable. The plains that were full of rich soil are now marshes. Hills that were once covered with forests and produced abundant pasture now produce only food for bees. Once the land was enriched by yearly rains, which were not lost, as they are now, by flowing from the bare land into the sea. The soil was deep, it absorbed and kept the water in the loamy soil, and the water that soaked into the hills fed springs and running streams everywhere. Now the abandoned shrines at spots where formerly there were springs attest that our description of the land is true.

During Biblical times, many references and comparisons were also expressed in relation to the health of soil or land resources. Several examples of these with regard to the relationship between civilization and the soil can also be found in Hillel (1991).

During the early 20th century, the concept of soil tilth emerged (Karlen *et al.* 1990) with an emphasis on soil structure (Fig. 5.3.1) and the processes it influences (e.g. aeration, root growth, infiltration, and water retention). Several authors began to describe how climatic conditions and various soil, crop, and animal management practices affected soil tilth. Keen (1931) provided a summary of previous observations regarding the importance of tilth for the planting of crops into a proper seedbed, the impact of frost on soil, how hard rains after plowing would create a crust, and how grazing animals could compact soils. Ultimately, the term tilth was recognized by the soil science community for its ability to describe a fundamental understanding of soil structure and how it changes in response to tillage, organic matter additions, crop rotations, and environmental factors such as rainfall or temperature.

Warkentin and Fletcher (1977) introduced the term soil quality (Fig. 5.3.2) and argued that multiple land uses must be considered, even when the primary use for a specific soil resource is for intensive production agriculture. They also stressed that (1) soil resources are constantly being evaluated for an ever increasing range of uses (e.g. food, feed, or fiber production, recreation, forestry, urban development, etc.), (2) several different stakeholder groups are concerned about soil resources, (3) the priorities and demands of society are changing, and (4) soil-resource and land-use decisions are made in a human or institutional context.

Soil Ecology and Ecosystem Services. First Edition. Edited by Diana H. Wall *et al.*
© 2012 Oxford University Press. Published 2012 by Oxford University Press.

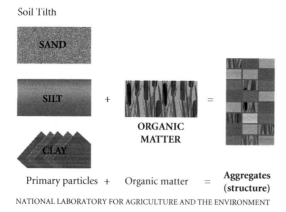

Figure 5.3.1 Soil tilth emphasizes the physical condition of a soil as related to ease of tillage, fitness as a seedbed, and its impedance to seedling emergence and root penetration (i.e. aggregation and structure). (From the USDA National Laboratory for Agriculture and the Environment, Ames, Iowa.)

Changes in soil quality and its impact on people are probably most quickly recognized by gardeners and small farmers because they generally work their soil either by hand or with small tools. Effects of improving soil quality are readily apparent to them because hard (i.e. dense or compacted) soil requires more energy to prepare a seedbed and generally produces less vigorous plants.

The intimate relationship between farmers and their soil resources became evident to Romig et al. (1996) when they identified a set of soil health (Fig. 5.3.3) indicators through their conversations with farmers in Wisconsin. The indicators that were identified included both the capacity of a soil to perform certain functions (infiltrate, decompose, cycle nutrients, etc.) and attributes or vital statistics about the soil resource (e.g. soil color, structure, root morphology, presence of earthworms, animal health, and others). Romig *et al.* (1996) subsequently used this information to develop one of the first soil health scorecards that could be used to classify each indicator as being:

1. Healthy—performance of function is optimal and structure is normal;
2. Impaired—an abnormality in function and/or structure;
3. Unhealthy—severe restriction or inability to perform normal function, severe deformity or loss of structure, disabled.

Unfortunately, all three concepts (soil tilth, soil quality, and soil health) remain controversial among soil scientists because of the difficulty in providing an exact definition or agreed-upon protocol for measuring any one of them. A major reason for this difficulty is that the soil resource characteristics that most accurately characterize all three terms are qualitative and quantitative as well as static and dynamic.

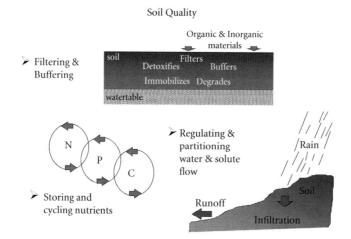

Figure 5.3.2 Soil quality emphasizes physical, chemical, and biological attributes with a focus on the "capacity of soil to function" with regard to maintaining productivity, storing and cycling nutrients, regulating and partitioning water flow and filtering, buffering, and detoxifying applied organic and inorganic materials.

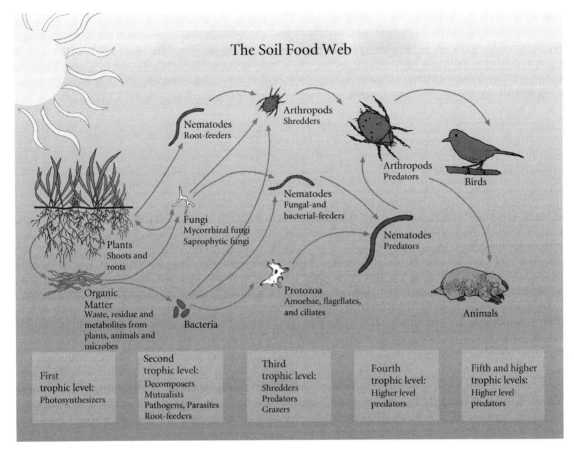

Figure 5.3.3 Soil health focuses on the "ecological attributes of soil" with an emphasis on biodiversity, food web structure, biotic activity and transformation of solar energy to biochemical energy. (Redrawn from Ingham 2007.)

As evidenced by Harris *et al.* (1996), Karlen *et al.* (2001, 2003) and others, the conceptual development of soil tilth, soil quality, and soil health have a history of intriguing people because of their connection between the land and human health. Unfortunately, due to the qualitative factors associated with all three terms, some have argued that defining any of the terms is akin to defining beauty. Many of those persons prefer to use the "science of agriculture" with a focus on plant nutrients as being the essential tools to sustain biological productivity, soil fertility, plant nutrition, soil management, and productivity as the indicators for assessing the quality of a specific soil resource (Patzel *et al.* 2000). Others prefer to focus on soil management and the production practices (e.g. tillage, crop rotation, applying animal or green manures, etc.) that farmers use to sustain not only their soil resources but the economic health of their operation (e.g. Letey *et al.* 2003; Sojka *et al.* 2003). This is especially true for systems that have a high reliance on purchased fertilizers, irrigation, and intensive tillage.

5.3.2 The evolution of soil health

The post-World War II development of synthetic fertilizers, an improved understanding of plant nutrition, and an infrastructure for delivering fertilizers to farmers improved tillage, planting, and harvest equipment, made subsurface drainage more cost-effective, increased efficiencies for both animal and crop production, and contributed to the development of global markets. Together these all contributed to rapid changes in agriculture during the latter half of the 20th century (Karlen *et al.* 2008). Unfortunately, these rapid changes also resulted in several unintended consequences, especially with regard to agriculture's impact on the global environment.

Increased awareness of the unintended consequences of agriculture and other soil management practices rekindled an interest in and an awareness of soil health that led to the initial efforts to define the concept based on the multiple functions associated with soil quality (Karlen *et al.* 2008). The soil health concept was discussed in relation to alternative and/or sustainable agriculture (Parr *et al.* 1992); used synonymously with soil quality by many authors (e.g. Larson & Pierce 1991; Karlen *et al.* 1997) as they identified and selected various chemical, physical, and biological indicators and developed management strategies (Doran & Parkin 1994, 1996) for evaluating and sustaining soil resources. Gradually the concept of soil health was differentiated from soil quality by emphasizing the ecological connections beyond those associated with either soil tilth or soil quality. Farmers, conservationists, and land managers still use all three terms, as they become more aware of the interconnected functions that soils provide and recognize both the long- and short-term effects of their soil and crop management decisions.

5.3.3 Monitoring soil health

Having defined the role of soil health as an ecological basis for assessing or monitoring long-term effects of agricultural and other soil management practices, a protocol was needed to integrate the multiple chemical, physical, and biological indicators that could be affected. Several strategies including soil health scorecards (Romig *et al.* 1996) and soil quality test kits (Sarrantonio *et al.* 1996) were developed, tested, and used by growers. Building upon the principles of ecology and systems engineering, a soil quality assessment framework known as the Soil Management Assessment Framework (SMAF) was also developed (Andrews *et al.* 2004), tested (e.g. Karlen *et al.* 2006; Wienhold *et al.* 2006; Zobeck *et al.* 2008), expanded to accommodate additional indicators (Wienhold *et al.* 2009; Stott *et al.* 2010), and referred to synonymously by some as a tool that could be adapted for the ecological assessment of soil health.

A SMAF or soil health assessment consists of three steps: indicator selection, indicator interpretation, and integration into a soil quality index (Andrews *et al.* 2004). The indicator selection step uses an expert system of decision rules to recommend indicators for inclusion in the assessment based on the user's stated management goals, location, and current management practice. The indicator interpretation step measures indicator data which are transformed to unitless scores ranging from zero to one based on clearly defined, site-specific relationships to soil function. The soil functions of interest include crop productivity, nutrient cycling, physical stability, water and solute flow, contaminant filtering and buffering, and biodiversity. The indicator interpretation step uses various factors (i.e. organic matter, texture, climate, slope, region, mineralogy, weathering class, crop, sampling time, and analytical method) to adjust threshold values in the scoring curves that are then used to assign the relatives zero to one value for each type of data being collected. The integration step allows for the individual indicator scores to be combined into a single index value. This can be done with equal or differential weighting for the various indicators depending upon the relative importance of the soil functions for which they are being measured (Karlen *et al.* 2008).

Most recently, a SMAF analysis was used to monitor effects of harvesting corn stover as a bioenergy feedstock. Using data for 1998 and 2007 from near Mead, Nebraska, USA, the analysis showed that although use of no-tillage had increased soil organic matter slightly, decreases in soil pH and increases in soil bulk density caused the overall Soil Quality Index (SQI or "soil health") to decline (Karlen *et al.* 2010). The concept of soil health (quality) assessment has thus evolved substantially since it was first defined. It will soon be possible for the scientific community to quantitatively measure and understand soil health through assessments such as those made using the SMAF. Multifactor assessments are crucial to understanding and measuring soil health effects on human and animal health.

5.3.4 Summary and conclusions

The concept of soil health continues to evolve with its primary purpose serving as an ecological assessment tool or protocol for evaluating short and long-term effects of soil management on the health or quality and sustainability of soil resources. Soil health is an integrative term which links soil biological, chemical, and physical properties, and ecological processes together in a manner that allows the user to determine how well a specific soil resource is performing the wanted and needed critical functions. The journey to understand and monitor soil health (quality) has just begun, but undoubtedly through the efforts of those reading this book, tremendous advancements will continue to be made.

References

Andrews, S.S., Karlen, D.L., & Cambardella, C.A. (2004) The soil management assessment framework: a quantitative soil quality evaluation method. *Soil Science Society of America Journal* **68**: 1945–62.

Doran, J.W., & Parkin, T.B. (1994) Defining and assessing soil quality. In: J.W. Doran, D.C. Coleman, D.F. Bezdicek and B.A, Stewart (ed.) *Defining soil quality for a sustainable environment* (SSSA Special Publication No. 35), pp. 3–21. Soil Science Society of America, Inc., Madison, WI.

Doran, J.W., & Parkin, T.B. (1996) Quantitative indicators of soil quality: a minimum data set. In: J.W. Doran and A.J. Jones (eds.) *Methods for assessing soil quality* (SSSA Special Publication No. 49), pp. 25–39. Soil Science Society of America, Inc., Madison, WI.

Harris, R.F., Karlen, D.L., & Mulla, D.J. (1996) A conceptual framework for assessment and management of soil quality and health. In: J.W. Doran and A.J. Jones (eds.) *Methods for Assessing Soil Quality* (SSSA Special Publication No. 49), pp. 61–82. Soil Science Society of America, Inc., Madison, WI.

Hillel, D. (1991) *Out of the Earth*. University of California Press, Berkeley, CA.

Ingham E.R. (2007) *Soil Food Web* [Online] http://soils.usda.gov/sqi/concepts/soil_biology/soil_food_web.html.

Karlen, D.L., Andrews, S.S., & Doran, J.W. (2001) Soil quality: Current concepts and applications. *Advances in Agronomy* **74**: 1–40.

Karlen, D.L., Andrews, S.S., Wienhold, B.J., & Doran, J.W. (2003) Soil quality: Humankind's foundation for survival—a research editorial by conservation professionals. *Journal of Soil and Water Conservation* **58**: 171–9.

Karlen, D.L., Andrews, S.S., Wienhold, B.J., & Zobeck T.M. (2008) Soil quality assessment: past, present and future. *Electronic Journal of Integrated Bioscience* **6**: 3–14.

Karlen, D.L., Erbach, D.C., Kaspar, T.C., Colvin, T.S., Berry, E.C., & Timmons, D.R. (1990) Soil tilth: A review of past perceptions and future needs. *Soil Science Society of America Journal* **54**: 153–61.

Karlen, D.L., Hurley, E.G., Andrews, S.S., et al. (2006) Crop rotation effects on soil quality at three northern corn/soybean belt locations. *Agronomy Journal* **98**: 484–95.

Karlen, D.L., Mausbach, M.J., Doran, J.W., Cline, R.G., Harris, R.F., & Shuman, G.E. (1997) Soil quality: A concept, definition, and framework for evaluation (A guest editorial). *Soil Science Society of America Journal* **61**: 4–10.

Karlen, D.L., Varvel, G.E., Johnson, J.M.F., et al. (2010) Monitoring soil quality to assess the sustainability of harvesting corn stover. *Agronomy Journal* **103**: 288–95.

Keen, B.A. (1931) *The physical properties of the soil*. Rothamsted Monograph on Agricultural Science. Ser. Longmans, Green and Coi., London.

Larson, W.E., & Pierce, F.J. (1991) Conservation and enhancement of soil quality. In: *Evaluation for sustainable land management in the developing world*. Vol. 2. IBSRAM Proc. 12(2). International Board for Soil Research and Management, Bangkok, Thailand.

Letey, J., Sojka, R.E., Upchurch, D.R., et al. (2003) Deficiencies in the soil quality concept and its application. *Journal of Soil and Water Conservation* **58**: 180–7.

Parr, J.F., Papendick, R.I., Hornick, S.B., & Meyer, R.E. (1992) Soil quality: attributes and relationship to alternative and sustainable agriculture. *American Journal of Alternative Agriculture* **7**: 5–11.

Patzel, N., Sticher, H., & Karlen, D.L. (2000) Soil fertility—phenomenon and concept. *Journal of Plant Nutrition and Soil Science* **163**: 129–42.

Romig, D.E., Garlynd, M.J., & Harris, R.F. (1996) Farmer-based assessment of soil quality: A soil health scorecard. In: J.W. Doran & A.J. Jones (eds.) *Methods for Assessing Soil Quality*. (SSSA Special Publication No. 49), pp. 39–60. Soil Science Society of America, Inc., Madison, WI.

Sarrantonio, M., Doran, J.W., Liebig, M.A., & Halvorson, J.J. (1996) On-farm assessment of soil quality and health. In: J.W. Doran & A.J. Jones (eds.) *Methods for Assessing Soil Quality* (SSSA Special Publication No. 49), pp. 83–105. Soil Science Society of America, Inc., Madison, WI.

Sojka, R.E., Upchurch, D.R., & Borlaug, N.E. (2003) Quality soil management or soil quality management: Performance versus semantics. In: D.L. Sparks (ed.)

Advances in Agronomy. Vol. 79, pp. 1–68. Academic Press, New York.

Stott, D.E., Andrews, S.S., Liebig, M.A., Wienhold, B.J., & Karlen, D.L. (2010) Evaluation of β-glucosidase activity as a soil quality indicator for the Soil Management Assessment Framework (SMAF). *Soil Science Society of America Journal* **74**: 107–19.

Warkentin, B.P., & Fletcher, H.R. (1977) Soil quality for intensive agriculture. In: *Proceedings of the International Seminar on Soil Environment and Fertility Management in Intensive Agriculture*, pp. 594–8. Society of Science of Soil and Manure, Japan.

Wienhold, B.J., Karlen, D.L., Andrews, S.S., & Stott, D.E. (2009) Protocol for indicator scoring in the soil management assessment framework (SMAF). *Renewable Agricultural Food Systems* **24**: 260–6.

Wienhold, B.J., Pikul Jr., J.L., Liebig, M.A., *et al.* (2006) Cropping system effects on soil quality in the Great Plains: Synthesis from a regional project. *Renewable Agricultural Food Systems* **21**: 49–59.

Zobeck, T.M., Halvorson, A.D., Wienhold, B.J., Acosta Martinez, V., & Karlen, D.L. (2008) Comparison of two soil quality indexes to evaluate cropping systems in northern Colorado. *Journal of Soil and Water Conservation* **63**: 329–38.

CHAPTER 5.4

Managing Soil Biodiversity and Ecosystem Services

Michel A. Cavigelli, Jude E. Maul, and Katalin Szlavecz

5.4.1 Introduction

Managed ecosystems (cropland, managed grasslands, permanent crops) represent about 40% of terrestrial ecosystems on a global basis (FAOSTAT 2011a). Managing lands for economic production generally involves reducing plant diversity considerably compared to native systems and substituting off-site inputs for ecosystem services provided in native systems, such as the provision of plant nutrients, regulation of pest populations, and introduction of new genetic diversity in crop species (Swift et al. 2004). Due to concerns about the sustainability of many existing managed systems (e.g. Vitousek et al. 1997; Galloway & Cowling 2002; Tilman et al. 2002; Green et al. 2005b; Butler et al. 2007), there is increasing interest among land managers, scientists, and others in developing managed systems that more fully take advantage of inherent ecosystem processes, augment ecosystem services, and reduce ecosystem disservices compared to most existing systems (National Research Council 2010). A fundamental question in developing such systems is whether biodiversity matters to the provisioning of ecosystem services. This question is of particular importance in light of the on-going and predicted massive losses of biodiversity at multiple scales and in diverse ecosystems around the world in response to human activity (Hooper et al. 2005).

If soil biodiversity impacts the provisioning of ecosystem services, then it is important that we understand how management impacts soil biodiversity. Since current management decisions rarely consider soil biodiversity beyond selection of plant species and perhaps some microorganisms, such as symbiotic nitrogen (N) fixing bacteria, considering soil biodiversity in management decisions represents a significant shift in thinking for land managers. Notable exceptions, however, include traditional systems where social customs support the maintenance of soil biodiversity (Swift et al. 2004), and organic farming, whose proponents explicitly suggest that managing soil biodiversity to create resistant and resilient systems will provide multiple ecosystem services such as efficient nutrient cycling, healthy and productive crops, and pest control (Soil Association 2002; National Research Council 2010). The goal of this chapter is to critically explore links among management practices, soil biodiversity, and ecosystem services.

While we recognize that changes in soil biodiversity and ecosystem services upon converting native ecosystems to managed lands are usually substantial (Tilman et al. 2002; Green et al. 2005b), we do not address these issues; they are addressed in a separate chapter on land use changes. Our goal here is to explore the impact of existing management options on ecosystem services and soil biodiversity.

A challenge in defining relationships between soil biodiversity and ecosystem function is that while such links are increasingly being recognized as resulting from functional rather than taxonomic diversity (Hooper et al. 2005; Reiss et al. 2009), our knowledge of the functional traits of the vast majority of soil organisms is severely limited (Swift et al. 2004). Therefore, we also discuss measures of taxonomic biodiversity such as species richness and evenness.

Soil Ecology and Ecosystem Services. First Edition. Edited by Diana H. Wall *et al.*
© 2012 Oxford University Press. Published 2012 by Oxford University Press.

We also recognize that landscape level diversity can contribute substantially to the provision of ecosystem services such as biocontrol, pollination, and nutrient retention, (Dinnes *et al.* 2002; Tscharntke *et al.* 2005). However, herein we focus on "plot/field" scale effects. We also focus on long-term rather than short-term impacts of management since short-term responses, albeit often significant, may be ephemeral and not reflect long-term responses. Alternatively, ecosystem services may respond episodically and on time scales not captured in short-term field trials. Our review also tends to be biased toward agricultural management of arable lands in temperate regions, in part reflecting a bias in the literature and in part due to author expertise. However, we think our conclusions are broadly applicable.

We use the Millennium Ecosystem Assessment (2005) definitions of ecosystem services and focus our discussion on terrestrially-based provisioning services of managed ecosystems (food and fiber), and four regulating and supporting services (soil carbon (C) sequestration and climate regulation, nutrient and water cycling, purification of water and air, and pest control), which comprise a subset of the broader category, ecosystem functions (Hooper *et al.* 2005).

Relationships addressed in this chapter are illustrated in Fig. 5.4.1. We first address impacts of management (plant selection, tillage, chemical applications, organic amendment application, and cropping systems) on soil biodiversity (A in Fig. 5.4.1) and the provision of ecosystem services (B in Fig. 5.4.1), keeping in mind that the impact of management is often strongly site dependent since soil biodiversity and ecosystem services are both constrained by soil properties (including those impacted by prior management) and climate (C and D in Fig. 5.4.1). We then address the link between biodiversity and ecosystem services (E in Fig. 5.4.1) where applicable, making specific comments in a concluding section regarding our current level of understanding of this link in managed systems.

5.4.2 Edible crop diversity

Crop plant biodiversity in the majority of the world's agroecosystems has decreased since the 1940s as the frequency and diversity of crop rotations have decreased. Of the approximately 7000 edible species of plants on earth (Fern 2000), 309 were grown in 2010 at measureable scales (FAOSTAT 2011b). Of these 309 species, 13 crops accounted for over 60% of the daily crop caloric intake by humans worldwide (Fig. 5.4.2). Since the values in Fig. 5.4.2 do not include calories from meat, fish, eggs, or milk, these data underestimate the proportional reliance on such crops as maize, soybean, wheat, and sorghum, which humans also consume indirectly after conversion to animal protein.

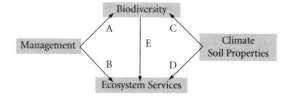

Figure 5.4.1 Impacts of management on biodiversity (A) and provision of ecosystem services (B) are constrained by climate and soil properties (C, D). Each of these impacts, along with the impact of biodiversity on ecosystem services (E) varies considerably among taxa, ecosystem type, and ecosystem service as discussed in the text.

Fig. 5.4.2 illustrates how dependent we are on a very small number of crops. This global crop diversity pattern is reflected in on-farm diversity, particularly in industrialized agricultural systems. Low diversity is due in part to the need for uniformity in crop phenology, including flowering time, seed size and time of ripening, to allow for the harvest of huge swaths of the landscape with highly automated equipment. There are many logistical reasons farmers choose to reduce on-farm biodiversity, including 1) timing of planting and harvesting, 2) water conservation, 3) meeting contract requirements, 4) land limitations, and 5) weed control (Aref & Wander 1997; Tilman *et al.* 2002). The efficiency of scale associated with mechanized agriculture has contributed to the consolidation of agricultural enterprises into larger and less diverse operations.

5.4.3 Plant selection impacts on ecosystem services

Plants provide provisioning services directly while providing regulating and supporting services directly and indirectly such that plant selection can

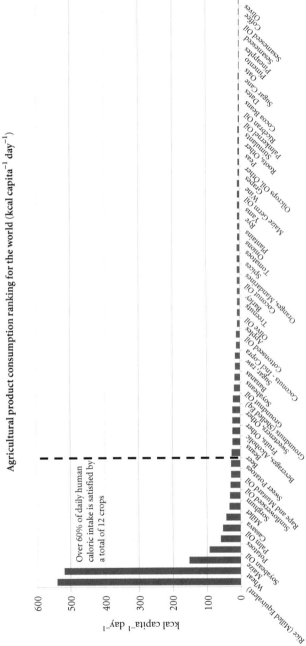

Figure 5.4.2 Agricultural product consumption ranking for the world (kcal capita^{-1} day^{-1}). Graph constructed using data found in FAOSTAT (2011a). Food and Agriculture Organization of the United Nations, Food supply, Crops primary equivalent webpage: http://faostat.fao.org/site/609/DesktopDefault.aspx?PageID=609

have a large impact on ecosystem services. Plant diversity in managed systems usually reflects the ecosystem services the land manager desires. Provisioning services tend to dominate these decisions but regulatory and supportive services provided by plants are also considered, such as biological N fixation by legumes and associated rhizobia, stabilization of soil structure by the fibrous root systems of grasses, or provision of habitat that supports organisms in the broader agroecosystem food web.

Plants also provide a number of regulating and supporting services that are often unplanned. Perhaps most significantly, plants convert solar energy into chemical energy (organic matter). Some of this organic matter is transferred to the soil while plants are alive through root death, sloughed tissue, and exudates and additional C is added to soil following plant senescence and death (Oades 1984; Janzen et al. 1997). Plants impact the distribution and dynamics of soil organic C (SOC) and nutrients by concentrating them into fertility islands, and by priming mineralization of SOC and nutrients. Plants also exert a strong direct control over soil moisture via their impacts on transpiration, evaporation, and interception of rainfall. Since plants are the ultimate source of almost all C inputs to soil, plants can be used to increase SOC, which provides a climate-regulating service by reducing atmospheric concentrations of CO_2. Increased SOC also characteristically results in increases in soil quality: increased soil water holding capacity, macroporosity, infiltration capacity, hydraulic conductivity, and aggregation, and decreased bulk density and surface crusting (Haynes & Naidu 1998). Many plant impacts on soils are long-lasting, such that plants have a legacy effect on soil properties.

The selection of plant species in managed systems impacts provision of ecosystem services because plants vary substantially in functional traits. For example, the amount of biomass and the lignin:N and C:N ratios of crop plants vary such that crop plant selection impacts the quantity and quality of residue returned to soil, which, in turn, impacts SOC levels (Paustian et al. 1997). The C:N ratio of plant residues controls residue decomposition rate during early stages of decomposition while the lignin:N ratio controls decomposition rate during later stages in residues with relatively high lignin content (Taylor et al. 1989). However, there is no inherent positive relationship between plant diversity and SOC. For instance, SOC is usually greater under a continuous corn rotation than a corn-soybean rotation since soybean produces less residue that has lower C:N and lignin:N ratios than corn (Paustian et al. 1997).

5.4.4 Plant selection impacts on soil biodiversity

Plant selection has a direct effect on soil biodiversity in that plant roots compete with other soil organisms for nutrients, niche space, and water, but it is not clear whether plants or other soil organisms are the more efficient competitors (Zak et al. 1990; Schimel & Bennett 2004). Changes in soil nutrient concentrations in the soil induced by rhizodeposition or nutrient depletion can influence soil microbial process rates and microbial community structure (Dunn et al. 2006; Hrynkiewicz et al. 2009; Bradford et al. 2008). Roots of different plant species also impact microfauna differentially. As an example, Collembola abundance is usually much greater under wheat and barley than tuber crops, reflecting the greater abundance of mycorrhiza and exudates under the grasses than tubers (Larink 1997).

Plants also have important indirect effects on soil biodiversity, including providing C and energy for the soil food web (Minoshima et al. 2007; Treonis et al. 2010), and regulating soil moisture dynamics, including infiltration and evapotranspiration (Bardgett et al. 2001). Soil biodiversity is also impacted by plant nutrient acquisition, which involves complex coordination between the plant, fungi, and bacteria. While plants generally rely on mineralization of soil organic matter by microbes to supply essential nutrients such as N and P, plants can indirectly affect mineralization rates by supplying the microbial community with labile C energy sources, thus "priming" soil metabolism and increasing the rate of soil organic matter mineralization (Dijkstra et al. 2006). Subsequent soil trophic cascades can result in further increases in plant available N and P (Paterson 2003).

The impact of plant species on biogeochemical cycling and soil biodiversity is illustrated by an example from California grasslands where exotic grasses resulted in a doubling of gross nitrification rates compared to native plant species (Hawkes et al. 2005). These changes were associated with increased abundance and changes in composition of ammonia-oxidizing bacterial communities in soils. Even within species, different plant cultivars may select for structurally and functionally distinct soil microbial communities. Briones et. al. (2002) found that gross nitrification rate in the rhizosphere was greater for a modern rice variety than two traditional varieties. In addition, ammonia-oxidizer communities were different among rice cultivars; both impacts may have been due to microscale differences in oxygen availability in the rhizosphere. These examples indicate that specific interactions between plants and microbes at the species level are not uncommon (and most likely the norm) and these interactions can have ecosystem scale impacts on biogeochemical cycling.

Leguminous plants impact soil biodiversity in unique ways. They can regulate symbiotic N fixation, thereby controlling the energetic cost of N fixation by limiting C flow to non-mutualistic symbionts (Johnson 1993). There is some evidence, however, that the plant-rhizobial symbiotic relationship can transition from a purely mutualistic to a parasitic one depending on soil nutrient status and the ability of the host plant to sanction non-mutualistic symbionts (Wall & Moore 1999; Kiers et al. 2002). Including leguminous plants in managed systems can also result in the buildup of populations of rhizobia and other symbiotic bacteria over time (Lange & Parker 1961).

Plants also impact soil biodiversity through their interactions with arbuscular mycorrhizal fungi (AMF). It is estimated that over 70% of terrestrial plants, including many domesticated species, maintain symbiotic biochemical relationships with AMF, exchanging photosynthetically fixed C for fungally-acquired P (Asimi et al. 1980; Harrison et al. 2002). It is unclear whether plants can regulate mycorrhizal associations to the same degree as they can control interactions with rhizobia (Klironomos 2003; Harrison 2005).

Direct and indirect mechanisms of plant nutrient acquisition occur in concert in the soil and there appear to be feedback mechanisms affecting the dominance of direct versus indirect processes (Fitter 2005). Increasing our understanding of plant functional traits and their regulation, especially in association with soil microorganisms and the rest of the soil food web, will allow farmers and other land managers to select plants based on a more complete understanding of their functional characteristics such that managing plant diversity can better provide regulating and supporting services in addition to provisioning services (Jackson et al. 2008).

5.4.5 Managing plant diversity

Despite the various constraints limiting plant diversity in managed systems, feasible strategies exist for increasing plant diversity. One of the most tractable and proven approaches is using cover crops (i.e. crops grown to prevent erosion and improve soil quality) in rotation with cash crops (i.e. crops grown for profit; Sarrantonio 1994; Abdul-Baki & Teasdale 2007; Clark 2007). Cover crops can be chosen to provide specific ecosystem services that can benefit farming operations such as biological N fixation, weed control, reduction of soil erosion and runoff, and retention of soil nutrients and C (Brandsæter & Netland 1999; Eviner & Chapin 2001). There are a number of examples of cover crops influencing soil microbial diversity, food web structure, and community composition on both short and long time scales, especially following incorporation of the cover crops (Schutter et al. 2001; Buyer et al. 2010). Maul and Drinkwater (2010) showed that some cover crop species (e.g. *Fagopyrum esculentum* and *xTriticosecale* spp.) increased total soil microbial community richness and diversity while other species (e.g. *Vicia villosa* and *Lolium multiflorum*) reduced community diversity. Soil microbial community diversity was, in part, positively related to root surface area.

Including perennial crops in crop rotations is another way to increase plant functional diversity. For example, a crop rotation that includes a perennial forage crop along with annual crops provides more ecosystem services than a crop rotation composed of just annual crops, including increased SOC, retention of soil N, decreased soil erosion and nutrient runoff, and even increased economic stability (Paustian et al. 1997; Cavigelli et al. 2009b). Recent advances in plant breeding suggest that economically viable perennial grain crops may be developed within the near future, which could revolutionize ecosystem services provided by grain cropping systems (National Research Council 2010; van Tassel et al. 2010).

Non-crop plant diversity in managed systems has declined concomitantly with crop plant diversity due to substantially improved weed control. Weeds, however, may provide some of the same ecosystem services as cover crops including reduced soil erosion, especially after crop harvest, and provide additional sources of SOC that complement those provided by crop plants. One potential strategy to improve plant diversity in managed systems is to allow for weed populations below a yield penalty threshold (Ryan et al. 2009). However, there is little information on the effects, and ecosystem services, of "tolerable weeds" in agroecological systems (Teasdale et al. 2003).

Polycultures have a long history in agriculture (Altieri & Trujillo 1987), especially in tropical regions (Moorhead et al. 2010) and in pasture systems (e.g. Murray et al. 2006). There is increasing interest in using mixtures and intercropping strategies due to benefits that range from additional crop yield, habitat for beneficial insects, and improved N fixation (Broadbent et al. 1982; Butler et al. 2007; Scherr & McNeely 2008). In particular, intercropping cover crops into cash crops may be a viable means of increasing plant diversity and providing indirect ecosystem services such as weed suppression, N fixation, and habitat for beneficial insects (Abdul-Baki et al. 1997; Fiedler et al. 2008; Bergkvist et al. 2010).

Increasing the evenness of the crop diversity presented in Fig. 5.4.2 may be one strategy to improve overall system biodiversity (Maneepitak 2007). When increasing plant diversity in managed systems it is important to consider that biodiversity and ecosystem stability are not linearly related due to an increase in system level variance as the biodiversity of a system increases (Tilman 1982). In light of recent advances in our level of understanding of the functional significance of plant biodiversity (Schwartz et al. 2000; Loreau et al. 2001) we may be at a point where ecological concepts such as "diversity imparts ecosystem stability" could be tested in managed systems.

5.4.6 Tillage impacts on ecosystem services

Tillage, which is used to kill weeds, prepare seedbeds, and reduce disease incidence is one of the most disruptive management practices (Fig. 5.4.3). Tillage disperses soil macroaggregates that are built over long time periods (Six et al. 2000), redistributes crop residues, fertilizers, soil amendments, and organisms within the soil matrix, and alters soil water (infiltration, holding capacity, and leaching), temperature, and gas dynamics (Kladivko 2001).

Repeated tillage has resulted in decreases in SOC levels due to accelerated decomposition and soil erosion such that in North America, for example, SOC levels in agricultural lands are lower than in uncultivated lands by 22–36% (Franzluebbers & Follett 2005). Loss of SOC on a global basis due to land use changes, including tillage and other management practices, has contributed significantly to the increase in the atmospheric concentration of CO_2 (Intergovernmental Panel on Climate Change 2007). Tillage also alters the physical location and chemical characteristics of SOC such that C in tilled systems tends to be more recalcitrant and contained in smaller soil aggregates than that in untilled systems (Six et al. 2000). Since tillage increases the soils, susceptibility to water and wind erosion, tillage plays an important role in increasing loss of soil and nutrients to surrounding bodies of water, thus decreasing water quality (Ghidey & Alberts 1998; Rhoton et al. 2002).

Concerns about the environmental impacts of tillage led to the development of reduced or no-tillage (NT) systems (Triplett & Dick 2008). While impacts of conventional tillage (CT) vs. NT management on crop yields varies with climate and soil type (DeFelice et al. 2006), reduced input costs often result in NT production being more profitable than CT methods. Widespread adoption of NT management in some areas has led to substantial reductions in soil erosion and surface nutrient losses, providing an important ecosystem service by improving surface water quality (Ghidey & Alberts 1998; Rhoton et al. 2002). However, NT systems substitute herbicide applications for tillage to kill weeds, and can sometimes contribute to increased herbicide leaching to groundwater or loss to surface bodies of water (Alletto et al. 2009).

Figure 5.4.3 Photographs illustrating a) tillage using a moldboard plow to incorporate agricultural amendments and control weeds, b) soybean plants emerging through the residues of a previous crop in a no-tillage system, c) herbicide application, and d) manure application in modern agricultural systems. Photos courtesy of a) Michel Cavigelli and b)–d) USDA NRCS.

NT cropping systems can be an effective means of helping to regulate climate by sequestering C in soil (Robertson *et al.* 2000; Franzluebbers & Follet 2005), although this is not always the case (Venterea *et al.* 2006). The extent of C sequestration provided by eliminating tillage depends on climate, soil texture, soil drainage, and time in NT (Six *et al.* 2004; Gregorich *et al.* 2005). Managed systems also impact climate regulation through emissions of nitrous oxide (N_2O), a greenhouse gas that is 298 times as potent as CO_2 and is involved in stratospheric ozone degradation (Intergovernmental Panel on Climate Change 2007). Emissions of N_2O can be higher in NT than CT systems due to creation of less aerobic soil conditions in NT that favor N_2O production by microbial nitrification and denitrification (Robertson & Grace 2004). Eliminating tillage can also increase soil N retention, thus reducing off-site impacts of nutrient losses and increasing plant N availability (Beare *et al.* 1997; Spargo *et al.* 2008) and gross N mineralization, nitrification, and immobilization (Muruganandam *et al.* 2010).

5.4.7 Tillage impacts on soil biodiversity

Recurring tillage, a common feature of annual cropping systems, maintains systems at an early stage of succession, which favors early successional weed and perhaps other species (Swift *et al.* 2004). Soil disturbance resulting from tillage impacts organisms indirectly by altering their habitats and directly by redistributing them in the profile and/or killing them such as when plant roots and fungal hyphal networks are broken (Miller & Jastrow 1990).

Wardle (1995) shows that the impact of tillage on soil organisms tends to be proportional to their body size. In CT soils, especially where the crop rotation does not include perennial crops—which provide earthworm habitat—and where there are no additions of manure—an earthworm food source—earthworm abundance and biomass are usually less than in NT soils (Kladivko 2001). This seems to be the case in large areas of the US Midwest. The impact of this loss is so noticeable that farmers in these areas are interested in reintroducing the European earthworm species, *Lumbricus terrestris*, to improve soil water infiltration, one of many ecosystem services provided by earthworms (Kladivko 2001). This is a clear, albeit rare, case where management has resulted in the elimination of a functional group that subsequently diminished an ecosystem service.

While tillage generally favors organisms with short generation times and high metabolic rates (Andrén & Lagerlof 1983), smaller organisms show varied patterns of response to tillage (Wardle 1995). Biomass of bacteria and fungi and fungal:bacterial ratios tend to be greater in NT than CT soils, especially near the soil surface, where fungi are particularly important in the decomposition of surface litter. Greater fungal biomass in NT than CT soils increases soil aggregate formation (Beare *et al.* 1997), likely contributes to soil C sequestration (Six *et al.* 2006), and is associated with greater N mineralization enzyme activities (Muruganandam *et al.* 2009). On the other hand, tillage tends to favor nematodes and some mites, especially near the soil surface, where tillage tends to have a negative impact on the biomass of most soil organisms (Wardle 1995). Since microflora serve as the base of much of the soil food web, differences in fungal:bacterial ratios due to tillage might be thought to be reflected in the rest of the food web. However, there is no consistent evidence of this, reflecting the complexity and our limited understanding of soil food web interactions and dynamics (Wardle 1995; Kladivko 2001).

The impact of tillage on soil community structure and function is less well studied. There is evidence that tillage can alter soil microbial and microfaunal community structure both at the surface and at depth but impacts on function are not clear (Treonis *et al.* 2010). Pest complexes (plant pathogens, nematodes, and insects) are often different in NT and CT systems, indicating a shift in pest community composition with changes in tillage (Kladivko 2001). For example, tillage can reduce populations of *Longidorus* and *Trichodorus* nematodes, many of which are vectors for plant viruses. Tillage also reduces root-feeding nematode abundance but increases microbivore populations (Gupta & Yeates 1997). In one recent study, despite greater gross N cycling rates (mineralization, nitrification, immobilization) in NT than CT systems, only gross immobilization was related to differences in soil microbial community structure while gross mineralization was related only to soil microbial biomass (Muruganandam *et al.* 2010).

5.4.8 Chemical application impacts on ecosystem services

A broad range of chemicals are used in managed systems to augment provisioning services, such as when fertilizers provide limiting nutrients to increase yields directly. Other chemicals augment provisioning services indirectly by increasing soil nutrient availability (lime), or reducing competition (herbicides) and herbivory (insecticides, fungicides, or nematicides; Fig. 5.4.3).

While lime and fertilizers have some direct impacts on soil physical properties under specific conditions, indirect impacts resulting from increased production of non-harvested plant parts and the consequent increase in C inputs into the soil and the attendant ecosystem services previously described are likely to be significantly more important than direct effects (Haynes & Naidu 1998). Lime contributes directly to increasing atmospheric concentrations of CO_2 since CO_2 is a product of acid neutralization reactions when lime is added to soil (Robertson *et al.* 2000).

Fertilizers, especially those containing N and P, and pesticides have contributed substantially to the degradation of surface and ground water quality via losses in runoff, erosion, or leachate in many parts of the world. Losses of N fertilizer are also the largest contributor to a doubling in the rate of N input into the global terrestrial N cycle by humans, which has resulted in a cascade of environmental impacts around the world: eutrophication of streams, lakes, and estuaries; acidification of soils, streams, and lakes; loss of soil nutrients; increased atmospheric concentration of N_2O; losses in terrestrial and aquatic biodiversity; and long-term declines in coastal marine fisheries (Vitousek *et al.* 1997; Galloway & Cowling 2002; Millennium Ecosystem Assessment 2005).

A recent study highlights a potential effect of relatively toxic and persistent pesticides—organochlorine insecticides—on an important ecosystem service, symbiotic biological N fixation (Fox *et al.*

2007). This study indicates that methyl parathion and pentachlorophenol disrupt the chemical signaling between alfalfa and its symbiotic N-fixing bacterium, *Sinorhizobium meliloti*, such that N-fixation and plant growth are retarded compared to a control not exposed to these pesticides. If the results of this research, conducted in a soil-less medium in the laboratory, are applicable to the field when pesticides are applied at recommended rates, the impact on ecosystem function could be substantial.

5.4.9 Chemical application impacts on soil biodiversity

The pH effect of liming results in increases in microbial biomass, net mineralization of soil N and S, and earthworm number and activity (Haynes & Naidu 1998), and a decrease in the fungal:bacterial biomass ratio; the latter being due to increased competition by bacteria as pH increases (Rousk *et al.* 2010). Increased biological activity tends to increase aggregate stability and provide associated ecosystem services. However, it is difficult to attribute many of these benefits to direct effects of lime on specific functional groups since liming also increases plant growth and associated return of non-harvested biomass to the soil. This indirect effect of lime on soil organisms is likely to be more important than direct effects (Haynes & Naidu 1998).

The primary impact of fertilizers on soil organisms also appears to be through an indirect effect on plant productivity and associated C inputs to soil, which usually results in increased microbial numbers and biomass, activity of some enzymes, and earthworm and nematode numbers and biomass, sometimes in proportion to the amount of N applied (Gupta & Yeates 1997; Haynes & Naidu 1998). In addition, increased plant residue production resulting from fertilization can increase soil organic matter and soil water holding capacity, which can also benefit soil organisms. However, by increasing plant productivity, fertilizers can also decrease soil moisture, which can decrease soil microbial activity (Murray *et al.* 2006). Fertilizers can also reduce plant diversity in managed polycultures such as pastures, thereby indirectly impacting soil organisms (Murray *et al.* 2006). One recent study shows what seems to be a direct impact of fertilizers on soil communities. While there were no differences in alfalfa yield, bacterial and fungal diversity indexes, and biomass and species richness among three P fertilizer application rates, fungal and bacterial community composition differed among treatments (Beauregard *et al.* 2010).

Pesticides are a very broad category of chemicals designed to disrupt metabolism according to various modes of action. Since pesticides are designed to be toxic to particular groups of organisms, it is perhaps surprising that they may not adversely impact gross measures of soil microbial diversity in the long term. Most changes in microbial diversity following one-time applications of pesticides and fumigants are short-lived such that microbial diversity returns to pre-application levels within a few weeks or months following application (Zelles *et al.* 1997; Engelen *et al.* 1998; Ibekwe *et al.* 2001). Repeated applications of pesticides, however, may lead to more durable changes in microbial populations but results are inconsistent (Stromberger *et al.* 2005). Pesticides usually have only a short-term effect on soil microarthropod abundance as a whole but there are clear species-specific responses leading to changes in community composition (Petersen & Krogh 1987). Repeated exposure to pesticides ultimately leads to a decrease in microarthropod species richness but an increase in abundance (Larink 1997). This increase is associated with a decline in predatory mites (e.g. *Gamasina*), which are more sensitive to pesticides than detritivorous mites. Herbicides affect soil fauna indirectly by reducing plant cover and thus available food resources (Prasse 1985).

The impacts of individual pesticides may be greater on particular soil functional groups than on the entire microbial community. Recent findings related to the use of glyphosate, the most commonly used pesticide in the world, illustrate this point. Numerous studies have shown that glyphosate, like other herbicides, has no long-term effect on soil microbial biomass, diversity and community composition, soil respiration, and soil enzyme activity (Wardle 1995; Kremer & Means 2009). However, recent studies indicate that glyphosate is released in the rhizosphere, where it can result in increased populations of *Fusarium*, *Phytophthora*, and *Pythium* plant pathogens, and increased incidence of the diseases caused by these organisms (Kremer & Means 2009). Glyphosate also seems to predispose plants to attack by soilborne fungal pathogens since the shikimate pathway that is disrupted by glyphosate also leads to the synthesis of plant defense compounds. These research findings are corroborated by farmer reports of elevated disease in glyphosate-resistant compared to other soybean varieties (Kremer & Means 2009). Glyphosate can also lead to an increase in detrimental manganese (Mn)-oxidizing bacteria, and decreases in beneficial microorganisms such as rhizobia, arbuscular mycorrhiza, and pseudomonads, though these effects are not consistent and impacts on ecosystems services are mixed (Kremer & Means 2009). Thus, glyphosate seems to impact specific organisms involved in pest control services but does not seem to impact broader measures of soil biodiversity and ecosystem function consistently, indicating that further research is warranted.

5.4.10 Organic material application impacts on ecosystem services

Since manure, biosolids, biochar, compost, and other forms of organic soil amendments contain plant available macro- and micronutrients they provide provisioning services and indirect benefits to soil quality via increased plant residue return to soils. Organic amendments differ from synthetic fertilizer inputs in that they also augment SOC levels and soil quality parameters directly. Thus, in concert with tillage, the positive effects of organic amendments may outweigh some of the negative effects of tillage on soil physical properties (Treonis *et al.* 2010). Since animal manures, many composts, and other organic amendments often contain more P than N relative to plant needs, careful management of these inputs in accordance with agronomic recommendations, as with fertilizers, is necessary to avoid negative environmental impacts.

5.4.11 Organic material application impacts on soil biodiversity

Applying organic amendments to soil (Fig. 5.4.3) leads directly to an increase in biomass of many groups of soil organisms and changes in the community composition of microflora, microfauna, and mesofauna (Edwards & Lofty 1969; Peacock *et al.* 2001; Treonis *et al.* 2010) since detritus and SOC are often limiting to soil organisms in managed systems. Organic amendments may impact the community structure of soil organism functional groups but responses are complex. For example, Enwall *et al.* (2005) showed that long-term (47-year) applications of sewage sludge and ammonium sulfate, but not cattle manure and calcium nitrate, resulted in differences in composition of denitrifying and total bacterial communities compared to those found in unfertilized plots. However, function, measured as potential denitrification and basal respiration rates, were not related to community composition but may have been related to differences in soil pH, C, and heavy metal content that were also impacted by the various treatments.

Applying organic amendments has been proposed as a means of increasing the abundance of predaceous nematodes and other organisms to regulate plant parasitic nematodes and reduce or eliminate nematicide use (Ferris & Matute 2004). The effectiveness of this strategy, however, is inconsistent (Oka 2009). For example, Treonis *et al.* (2010) showed that application of a combination of organic amendments decreased plant-parasitic nematodes but did not increase predaceous/omnivorous nematodes or any nematode functional indexes.

Organic amendments have also been shown to suppress soil-borne plant diseases by decreasing population densities of soilborne pathogens and increasing population densities of antagonistic microorganisms (Garbeva *et al.* 2004). Composts can suppress *Pythium*, *Phytophthora*, and *Ralstonia solanacearum* and can also be effective in conjunction with biocontrol agents such as *Trichoderma* or *Gliocladium* (Garbeva *et al.* 2004). While the mechanisms of disease suppression are not worked out, in at least some cases, changes in soil microbial community structure seem to be involved. Van Bruggen and Semenov (2000) suggest that the range in the ratio between K-strategists (oligotrophs which prefer low nutrient substrate concentrations) and r-strategists (copiotrophs which prefer high nutrient substrate concentrations) is associated with general disease suppression. Our understanding of mechanistic links between applying organic amendments, soil biodiversity, and disease suppressive soils, however, is limited (Garbeva et al. 2004).

5.4.12 Organic cropping system impacts on ecosystem services

Understanding the impacts of individual management practices on ecosystem services allows us to target particular management practices that could increase the delivery of ecosystem services. However, most management decisions involve altering more than one factor at a time such that studying cropping systems that reflect the mix of practices used by land managers can provide insights not available by conducting factorial studies alone (Drinkwater 2002).

We focus our discussion of cropping systems on comparing organic and conventional systems because these systems provide a contrast in the degree of dependence on ecosystem functions and the purported benefits to soil biodiversity (Soil Association 2002; Hole et al. 2005; National Research Council 2010). While we recognize that both organic and conventional management systems encompass a broad range of practices, distinct differences between the two systems include that synthetic fertilizers and pesticides are generally not allowed in organic farming and organic practices are usually explicitly required to include practices that promote the activity of soil organisms such as reliance on organic sources of nutrients and diverse crop rotations. It is important to note that conventional cropping systems may be managed using NT or CT but the vast majority of comparisons between organic and conventional systems in the literature include only CT systems.

Provisioning services provided by organic systems tend to be greater than those provided by conventional systems in developing countries but lesser in developed countries due to differences in production intensity in conventional systems in the two regions (Badgley et al. 2007). Soil type also seems to impact the relative productivity of organic and conventional systems. In the high fertility Mollisols of the central USA, for example, crop yields in organic systems tend to equal those in conventional systems (Delate & Cambardella 2004) while in the less fertile Ultisols of the southeast USA, crop yields tend to be lower in organic than in conventional systems (Cavigelli et al. 2008; C. Crozier, personal communication).

Organic systems often have greater SOC than conventional CT systems (Drinkwater et al. 1998; Marriott & Wander 2006). Soil C in organic systems has been shown to be greater than in NT systems when manure is used in organic systems (Cavigelli et al. 2009a) but not when no manure is used (Robertson et al. 2000). The few studies comparing N_2O emissions between organic and conventional systems show few, if any, consistent differences between systems (Robertson et al. 2000; Cavigelli et al. 2009a). However, organic systems sometimes (Kramer et al. 2006; Cavigelli et al. 2009a), but not always (Phillips 2007), show lower N_2O flux as a function of N inputs than conventional systems. A lower N_2O flux:N input ratio in organic than conventional systems might be due to greater soil $C:NO_3$ ratios in organically-managed soils, or possibly to differences in soil denitrifier community composition (Kramer et al. 2006). Likewise, nitrate leaching in organic systems is sometimes lower than in conventional systems, reflecting generally lower soil nitrate levels (Drinkwater et al. 1998; Kramer et al. 2006), but many studies show few differences (Stolze et al. 2000).

Soil erosion is sometimes lower in organic than conventional CT systems (Reganold et al. 1987). One of the few long-term studies that include a comparison of organic and conventional NT systems is the USDA-ARS Beltsville Farming Systems Project in Maryland, USA. A 100 year model simulation using the Water Erosion Prediction Project model showed that predicted soil and associated nutrient losses were highest in a conventional CT treatment (64 Mg soil, 43 Mg P, 94 Mg N ha^{-1} year^{-1}), lowest in a conventional NT treatment (9 Mg soil, 6 Mg P, 15 Mg N ha^{-1} year^{-1}) and intermediate in an organic treatment (43 Mg soil, 29 Mg P, 62 Mg N ha^{-1} year^{-1}) (Green et al. 2005a; Green & Cavigelli, unpublished data). Thus, while organic farming might reduce sediment and nutrient contamination of surface waters compared to conventional CT but not NT, pesticide contamination of surface and ground waters is essentially eliminated with organic farming (Stolze et al. 2000).

5.4.13 Organic cropping system impacts on soil biodiversity

Two review papers indicate that organic farming usually increases soil organism species richness and abundance compared with conventional systems, but results vary by taxonomic group (Bengtsson et al. 2005; Hole et al. 2005). Both reports found that, for the most part, abundance and species richness of carabids and abundance of earthworms, soil fauna, and fungi were greater in organic than conventional systems. Greater abundance of carabids in organic than conventional systems was attributed to greater plant (cover crop and/or weeds) cover, while greater numbers of earthworms, other soil fauna, and fungi were attributed to greater residue and/or manure inputs in organic systems. Community composition of carabid beetles was also different between organic and conventional systems, with some species being favored by organic management while others were favored by conventional management. Staphylinid beetle abundance was lower in organic than conventional systems, possibly due to competition with carabids (Hole et al. 2005). Bengtsson et al. (2005), who conducted a meta-analysis of 42 comparative studies, found no differences in microbial biomass and activity between organic and conventional systems while Hole et al. (2005), who conducted a more qualitative analysis of 76 studies, reported greater bacterial abundance and activity and greater abundance of bacterial feeding nematodes in organic than conventional systems. Hole et al. (2005) reported no differences in collembolan abundance or in fungal-feeding nematode abundance between systems. Bengtsson et al. (2005) noted that a lack of difference between systems in plant-feeding nematodes suggests that pest control in organic systems does not necessarily suffer from lack of pesticides. Both sets of authors mention that the benefits of organic farming for biodiversity might not be the result of organic farming per se, but rather of the limited use of agricultural chemicals and greater dependence on cover crops and animal manures and emphasis on greater plant diversity, all practices that, while intrinsic, are not exclusive, to organic farming.

Studies conducted at the USDA-ARS Beltsville Farming Systems Project provide some of the few data comparing soil biodiversity in organic and conventional NT systems. Results show more similarities in soil invertebrate communities between conventional NT and CT systems than between these systems and the organic system for most taxa studied. Total abundance and diversity (Shannon and Simpson indexes) of carabids were generally greater in the organic than the CT and NT systems and carabid assemblages were similar in the CT and NT systems but distinct in the organic system. Two of three species found in greater abundance in organic than conventional systems are weed seed consumers, suggesting that differences in community patterns might have some functional role (Clark et al. 2006). For ants, Shannon diversity and evenness indexes were also greater, while total abundance was lower in the organic than in the two conventional systems (Kjar et al. 2010). Unique patterns for carabid and ant communities in the organic systems probably had more to do with the use and management of cover crops in the organic system rather than other management factors (Clark et al. 2006; Kjar et al. 2010). Isopod and diplopod assemblages (all non-native species) were similar in CT and NT systems and most distinct in organic systems for one of two years of the study (data not shown). Isopod and diplopod abundance followed a different pattern in that numbers were greater in the NT and the organic systems than in the CT system (Fig. 5.4.4). Earthworms showed a different pattern, consistent with reports cited earlier in which biomass was greater in NT than in tilled systems and greater in the organic than the conventional CT system (Csuzdi et al. 2003). The high biomass in the NT system was due to the dominance of *Lumbricus friendi*, an anecic species of European origin (Csuzdi & Szlavecz 2003). The other species (common peregrine earthworms) had higher densities in the NT system. Despite differences in earthworm communities, water infiltration was more strongly impacted by timing of tillage than by difference in earthworm abundance among systems (Pitz et al. 2004).

Table 5.4.1 shows that species richness was similar for invertebrates in the three management systems and was relatively high for ants and ground beetles and low for isopods, millipedes, and earthworms. Species with low richness showed differences in abundance among management systems

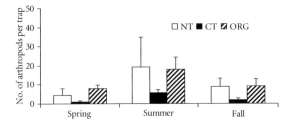

Figure 5.4.4 Abundance and activity of Isopoda and Diplopoda in conventional no-tillage (NT), conventional tillage (CT), and organic (ORG) cropping systems at the long-term Beltsville Farming Systems Project, MD, USA. Data were collected in 2002 in soybean plots.

but no or few differences in composition. Species with greater richness showed differences in both abundance and community composition. Additional work is needed to determine if differences in biodiversity at this site impact ecosystem function.

Research conducted after Hole et al. (2005) and Bengtsson et al. (2005) compiled their reviews provides some additional insight on soil biodiversity and ecosystem functions on organic and conventional farms. For example, recently published studies conducted on eight paired organic and conventional commercial strawberry farms in California, USA, show that the genes for 11 microbial functional and taxonomic groups were more abundant (had greater DNA microarray signal intensities) and there was greater genetic diversity for ten of these groups on organic than conventional farms (Reeve et al. 2010; Reganold et al. 2010). In addition,

233 of 1,171 individual genes were found only in the organic soils while only two genes were endemic to the conventional soil. While these studies provide good evidence of correlation between genetic diversity and functional capacity, the increased functional capacity on the organic farm might simply reflect the greater soil C and N resource levels reported for the organic farms, rather than being a function of greater diversity.

Pest control is commonly cited as a benefit of biodiversity in organic systems. Letourneau and Bothwell (2008) point out that at least six steps need to be demonstrated to determine if organic management supports biodiversity that is functionally significant with respect to pest control. While organic farming often results in greater soil biodiversity (step 1), that biodiversity must reflect an increase in the number of beneficial organisms (step 2), which must cause higher mortality among pests (step 3) found in similar abundance in organic and conventional systems (step 4), and reduce damage by pests (step 5) leading to yield increases compared to conventional methods such as pesticide application (step 6). Letourneau and Bothwell (2008) note that among the papers cited by Hole et al. (2005) and Bengtsson et al. (2005), fewer than three papers address at least one of steps 3, 4, 5, and 6.

One recently published study, however, seems to address at least five of these steps. Crowder et al. (2010) showed that evenness of predators and soil pathogens (though the latter was not statistically

Table 5.4.1 Species richness of soil macrofauna in conventional no-tillage (NT), conventional tillage (CT) and organic (ORG) plots at the USDA-ARS Beltsville Farming System Project, MD, USA

	NT	CT	ORG	Total	Percent non-native species
Ants[a] (Hymenoptera: Formicidae)	10	12	13	20	15
Ground beetles[b] (Coleoptera: Carabidae)	28	26	25	31	3
Isopods (Isopoda: Oniscidea)	3	2	2	3	100
Millipedes (Diplopoda)	3	2	3	4	100
Earthworms (Oligochaeta)	4	4	4	4	100

[a] Source: Kjar et al. (2010);
[b] Source: Clark et al. (2006).

significant) of the potato beetle, *Leptinotarsa decemlineata*, was greater on organic than conventional potato farms (steps 1 and 2). Abundance of the potato beetle was similar in the two systems (step 3). In a separate controlled experiment reported in the same manuscript, potato plant growth rate and potato beetle mortality increased with increasing evenness of both predators and soil pathogens (steps 4 and 6; plant size is related to yield in potato). These impacts translated into an 18% decrease in pest densities and a 35% increase in plant size for plants grown in environments with equivalent evenness values as found on organic and conventional farms. While the controlled study was conducted in enclosures that created a semi-artificial environment, this is a rare example of a study that attempts to provide a very complete approach to linking management, biodiversity, ecosystem function and an important ecosystem service, biocontrol.

5.4.14 Conclusions

Many management practices impact soil organisms by altering their environment and therefore favoring some species over others. Management only rarely results in the elimination of an entire functional group that then impacts an ecosystem service, as in the example of repeated tillage eliminating earthworms and reducing water infiltration. Instead, management impacts on ecosystem services and soil biodiversity are often complex such that responses may be subtle and vary with soil type, climate, ecosystem, taxonomic and/or functional group, and ecosystem service.

There is a strong need for additional studies carefully designed to test mechanistic links between various management practices, biodiversity of functional groups, and ecosystem functions. While a growing number of studies show management impacts on both soil biodiversity and ecosystem function, there is little solid evidence that biodiversity per se impacts ecosystem functions in managed systems. Concurrent changes in soil biodiversity and ecosystem function may simply reflect independent impacts of management (A and B in Fig. 5.4.1). Demonstrating a link between soil community structure and function (E in Fig. 5.4.1) probably requires a manipulative experimental approach. For example, Crowder *et al.* (2010), after demonstrating differences in biodiversity of potato beetle predators and pathogens in response to management, conducted separate biodiversity manipulation experiments to show that potato beetle mortality and potato plant growth rate increased with increasing evenness of predators and pathogens. Alternatively, environmental regulators (C and D in Fig. 5.4.1) of a given function (or functions; Reiss *et al.* 2009) can be controlled in mesocosm studies to isolate biodiversity as an independent variable impacting function (Cavigelli & Robertson, 2000) in systems where functional variability among the taxa has been demonstrated (Cavigelli & Robertson, 2001).

While there are a number of well-designed studies addressing biodiversity using various manipulative approaches (e.g. Bell *et al.* 2005; Brussaard *et al.* 2007; Strickland *et al.* 2009), the ranges of biodiversity or environmental disturbance selected often reflect a greater range than is typically seen in managed systems. This gap in the literature suggests that making greater inroads into understanding the link between biodiversity and ecosystem function in managed systems would benefit from greater collaborations between ecologists interested in fundamental ecological principles and agronomists, soil scientists and others interested in understanding the functional significance of the range of biodiversity commonly found in managed systems.

In addition, if links between soil biodiversity and ecosystem function are to be addressed in a manner that is useful to land managers, there is a great need to increase our understanding of the basic biology and ecology of diverse groups of soil organisms. We need greater knowledge of 1) the functional traits of large numbers of soil organisms (Allison & Martiny 2008; Reiss *et al.* 2009), 2) how complex interactions among soil organisms impact ecosystem functions (Reiss *et al.* 2009), and 3) the spatial and temporal distributions of soil organisms, including how to sample this variability representatively (Giller *et al.* 1997; Ettema & Wardle 2002; Cavigelli *et al.*, 2005).

Long-term research sites provide unique locations to conduct research on soil biodiversity and ecosystem function since they provide 1) ancillary data and metadata to help interpret factors controlling biodiversity and ecosystem services (Stres & Tiedje 2006), 2) an environment where temporal stability can be tested (Hooper *et al.* 2005), and 3) long-term management, which often impacts soil processes and communities differently than short-term management (Mallory & Griffin 2007; Strickland *et al.* 2009).

Despite our current incomplete understanding of the role of biodiversity on ecosystem functioning, management practices recommended to increase provisioning of ecosystem services are likely also to increase soil biodiversity. These recommendations include: 1) minimizing bare soil by using cover crops and perennial crops in a rotation or between perennial woody species such as in orchards, 2) reducing tillage, 3) applying organic amendments, albeit wisely (i.e. in synchrony with crop and system needs), and 4) reducing chemical inputs by increasing nutrient use efficiency and using integrated pest management concepts. Additional research is needed to further develop these management recommendations in diverse managed systems, including 1) developing integrated pest management systems that rely less heavily on chemical pesticides (Letourneau & Bothwell 2008), 2) developing reduced tillage systems that decrease or eliminate pesticide applications (Mirsky *et al.* 2009), 3) improving nutrient use efficiency by managing diverse nutrient sources in concert (soil reserves, organic amendments, cover crops, fertilizers) (Giller *et al.* 1997; Spargo *et al.* 2011), and 4) developing perennial grain crops (van Tassel et al 2010). In addition, government policies could facilitate adoption of more sustainable management practices. For example, polices that favor the provisioning of ecosystem services are likely also to help foster soil biodiversity.

References

Abdul-Baki, A.A., Morse, R.D., Teasdale, J.R., & Devine, T.E. (1997) Nitrogen requirements of broccoli in cover crop mulches and clean cultivation. *Journal of Vegetable Crop Production [USA]* **3**: 85–100.

Abdul-Baki, A.A., & Teasdale, J.R. (2007) *Sustainable Production of Fresh-Market Tomatoes and Other Vegetables with Cover Crop Mulches*. Farmers' Bulletin FB-2280, U.S. Dept. of Agriculture, Agricultural Research Service, Washington, DC.

Alletto, L., Coquet, Y., Benoit, P., Heddadj, D., & Barriuso, E. (2009) Tillage management effects on pesticide fate in soils. A review. *Agronomy for Sustainable Development* **30**: 367–400.

Allison, S.D., & Martiny, J.B.H. (2008) Resitance, resilience, and redundancy in microbial communities. *Proceedings of the National Academy of Sciences of the United States of America* **105**: 11512–19.

Altieri, M.A., & Trujillo, J. (1987) The agroecology of corn production in Tlaxcala, Mexico. *Human Ecology* **15**: 189–221.

Andrén, O., & Lagerlof, J. (1983) Soil fauna (microarthropods, enchtraeids, nematodes) in Swedish agricultural cropping systems. *Acta Agriculturae Scandinavica* **33**: 33–52.

Aref, S., & Wander, M.M. (1997) Long-term trends of corn yield and soil organic matter in different crop sequences and soil fertility treatments on the Morrow Plots. In: D.L. Sparks (ed.) *Advances in Agronomy*, pp. 153–97. Academic Press, San Diego, CA.

Asimi, S., Gianinazzi-Pearson, V., & Gianinazzi, S. (1980) Influence of increasing soil phosphorus levels on interactions between vesicular-arbuscular mycorrhizae and *Rhizobium* in soybeans. *Canadian Journal of Botany* **58**: 2200–5.

Badgley, C., Moghtader, J., Quintero, E., et al. (2007) Organic agriculture and the global food supply. *Renewable Agriculture and Food Systems* **22**: 86–108.

Bardgett, R.D., Anderson, J.M., Behan Pelletier, V., et al. (2001) The influence of soil biodiversity on hydrological pathways and the transfer of materials between terrestrial and aquatic ecosystems. *Ecosystems* **4**: 421–9.

Beare, M.H., Hu, S., Coleman, D.C., & Hendrix, P.F. (1997) Influences of mycelia fungi on soil aggregation and organic matter storage in conventional and no-tillage soils. *Applied Soil Ecology* **5**: 211–19.

Beauregard, M.S., Hamel, C., Atul-Nayyar, & St-Arnaud, M. (2010) Long-term phosphorus fertilization impacts soil fungal and bacterial diversity but not AM fungal community in alfalfa. *Microbial Ecology* **59**: 379–89.

Bell, T., Newman, J.A., Silverman, B.W., Turner, S.L., & Lilley, A.K. (2005) The contribution of species richness and composition to bacterial services. *Nature* **436**: 1157–60.

Bengtsson, J., Ahnström, J., & Weibull, A.C. (2005) The effects of organic agriculture on biodiversity and abundance: a meta-analysis. *Journal of Applied Ecology* **42**: 261–9.

Bergkvist, G., Stenberg, M., Wetterlind, J., Bāth, B., & Elfstrand, S. (2010) Clover cover crops under-sown in winter wheat increase yield of subsequent spring barley. Effect of N dose and companion grass. *Field Crops Research* **120**: 292–8.

Bradford, M.A., Fierer, N., & Reynolds, J.F. (2008) Soil carbon stocks in experimental mesocosms are dependent on the rate of labile carbon, nitrogen and phosphorus inputs to soils. *Functional Ecology* **22**: 964–74.

Brandsæter, L.O., & Netland, J. (1999) Winter annual legumes for use as cover crops in row crops in northern regions: I. Field experiments. *Crop Science* **39**: 1369–79.

Briones, A.M., Okabe, S., Umemiya, Y., Ramsing, N.B., Reichardt, W., & Okuyama, H. (2002) Influence of different cultivars on populations of ammonia-oxidizing bacteria in the root environment of rice. *Applied and Environmental Microbiology* **68**: 3067–75.

Broadbent, F.E., Nakashima, T., & Chang, G.Y. (1982) Estimation of nitrogen fixation by isotope dilution in field and greenhouse experiments. *Agronomy Journal* **74**: 625–8.

Brussaard, L., de Ruiter, P.C., & Brown, G.G. (2007) Soil biodiversity for agricultural sustainability. *Agriculture, Ecosystems Environment* **121**: 233–44.

Butler, S.J., Vickery, J.A., & Norris, K. (2007) Farmland biodiversity and the footprint of agriculture. *Science* **315**: 381–4.

Buyer, J.S., Teasdale, J.R., Roberts, D.P., Zasada, I.A., & Maul, J.E. (2010) Factors affecting soil microbial community structure in tomato cropping systems. *Soil Biology & Biochemistry* **42**: 831–41.

Cavigelli, M.A., Djurickovic, M., Mirsky, S.B., Maul, J.E., & Spargo, J.T. (2009a) Global warming potential of organic and conventional grain cropping systems in the mid-Atlantic region of the U.S. 2009 Farming Systems Design Proceedings, 23–26 August, Monterey, California, pp. 51–52. http://www.iemss.org/farmsys09/uploads/2009_FSD_Proceedings.pdf

Cavigelli, M.A., Hima, B.L., Hanson, J.C., Teasdale, J.R., Conklin, A.E., & Lu, Y.C. (2009b) Long-term economic performance of organic and conventional field crops in the mid-Atlantic region. *Renewable Agriculture and Food Systems* **24**: 102–19.

Cavigelli, M.A., Lengnick, L.L., Buyer, J.S., et al. (2005) Landscape level variation in soil resources and microbial properties in a no-till corn field. *Applied Soil Ecology* **29**: 99–123.

Cavigelli, M.A., & Robertson, G.P. (2000) The functional significance of denitrifier community composition in a terrestrial ecosystem. *Ecology* **81**: 1402–14.

Cavigelli, M.A., & Robertson, G.P. (2001) Role of denitrifier diversity in rates of nitrous oxide consumption in a terrestrial ecosystem. *Soil Biology & Biochemistry* **33**: 297–310.

Cavigelli, M.A., Teasdale, J.R., & Conklin, A. (2008) Long-term agronomic performance of organic and conventional field crops in the mid-Atlantic region. *Agronomy Journal* **100**: 785–94.

Clark, A. (2007) *Managing Cover Crops Profitably*, 3rd Ed. Sustainable Agriculture Network, Beltsville, MD.

Clark, S., Szlavecz, K., Cavigelli, M.A., & Purrington, F. (2006) Ground beetle (Coleoptera: Carabidae) assemblages in organic, no-till, and chisel-till cropping systems in Maryland. *Environmental Entomology* **35**: 1304–12.

Crowder, D.W., Northfield, T.D., Strand, M.R., & Snyder, W.E. (2010) Organic agriculture promotes evenness and natural pest control. *Nature* **466**: 109–12.

Csuzdi, C., & Szlavecz, K. (2003) *Lumbricus friendi* Cognetti, 1904 a new exotic earthworm from North America. *Northeastern Naturalist* **10**: 77–82.

Csuzdi, C., Szlavecz, K., & Cavigelli, M. (2003) Effects of crop management systems on species composition and abundance of earthworm communities (Oligochaeta). 5th Ecological Congress of Hungary, Budapest, Hungary.

DeFelice, M.S., Carter, P.R., & Mitchell, S.B. (2006) Influence of tillage on corn and soybean yield in the United States and Canada. *Crop Management* doi:10.1094/CM-2006-0626-01-RS. [Online] http://www.plantmanagementnetwork.org/pub/cm/research/2006/tillage (accessed 19 February 2011).

Delate, K., & Cambardella, C.A. (2004) Agroecosystem performance during transition to certified organic grain production. *Agronomy Journal* **96**: 1288–98.

Dijkstra, F.A., Cheng, W., & Johnson, D.W. (2006) Plant biomass influences rhizosphere priming effects on soil organic matter decomposition in two differently managed soils. *Soil Biology & Biochemistry* **38**: 2519–26.

Dinnes, D.L., Karlen, D.L., Jaynes, D.B., et al. (2002) Nitrogen management strategies to reduce nitrate leaching in tile-drained Midwestern soils. *Agronomy Journal* **94**: 153–71.

Drinkwater, L.E. (2002) Cropping systems research: Reconsidering agricultural experimental approaches. *HortTechnology* **12**: 355–61.

Drinkwater, L.E., Wagoner, P., & Sarrantonio, M. (1998) Legume-based cropping systems have reduced carbon and nitrogen losses. *Nature* **396**: 262–5.

Dunn, R., Mikola, J., Bol, R., & Bardgett, R. (2006) Influence of microbial activity on plant–microbial competition for organic and inorganic nitrogen. *Plant and Soil* **289**: 321–34.

Edwards, C.A., & Lofty, J.R. (1969) The influence of agricultural practices on soil micro-arthropod populations. In: J.G. Sheals (ed.) *The soil ecosystem*, pp. 237–48. Publs. Syst. Assoc., London.

Engelen, B., Meinken, K., von Wintzingerode, F., et al. (1998) Monitoring impact of a pesticide treatment on bacterial soil communities by metabolic and genetic fingerprinting in addition to conventional testing procedures. *Applied and Environmental Microbiology* **64**: 2814–21.

Enwall, K., Philippot, L., & Hallin, S. (2005) Activity and composition of the denitrifying bacterial community respond differently to long-term fertilization. *Applied and Environmental Microbiology* **71**: 8335–43.

Ettema, C.H., & Wardle, D.A. (2002) Spatial soil ecology. *Trends in Ecology & Evolution* **17**: 177–83.

Eviner, V.T., & Chapin III, F.S. (2001) Plant species provide vital ecosystem functions for sustainable agriculture, rangeland management and restoration. *California Agriculture* **55**: 54–9.

FAOSTAT. (2011a). Food and Agriculture Organization of the United Nations, Resources, ResourceSTAT [Online] http://faostat.fao.org/site/377/DesktopDefault.aspx?PageID=377 (accessed 17 February 2011).

FAOSTAT. (2011b) Food and Agriculture Organization of the United Nations, Food supply, Crops primary equivalent [Online] http://faostat.fao.org/site/609/DesktopDefault.aspx?PageID=609 (accessed 17 February 2011).

Fern, K. (2000) *Plants for a future: edible & useful plants for a healthier world*. Permanent Publications, Hampshire.

Ferris, H., & Matute, M.M. (2004) Structural and functional succession in the nematode fauna of a soil food web. *Applied Soil Ecology* **23**: 93–110.

Fiedler, A.K., Landis, D.A., & Wratten, S.D. (2008) Maximizing ecosystem services from conservation biological control: The role of habitat management. *Biological Control* **45**: 254–71.

Fitter, A.H. (2005) Darkness visible: Reflections on underground ecology. *Journal of Ecology* **93**: 231–43.

Fox, J.E., Gulledge, J., Engelhaupt, E., Burow, M.E., & McLachlan, J.A. (2007) Pesticides reduce symbiotic efficiency of nitrogen-fixing rhizobia and host plants. *Proceedings of the National Academy of Sciences of the United States of America* **104**: 10282–7.

Franzluebbers, A.J., & Follett, R.F. (2005) Greenhouse gas contributions and mitigation potential in agricultural regions of North America: Introduction. *Soil & Tillage Research* **83**: 1–8.

Galloway, J.M., & Cowling, E.B. (2002) Reactive nitrogen and the world: 200 years of change. *Ambio* **31**: 64–71.

Garbeva, P., van Veen, J.A., & van Elsas, J.D. (2004) Microbial diversity in soil: Selection of microbial populations by plant and soil type and implications for disease suppressiveness. *Annual Review of Phytopathology* **42**: 243–70.

Ghidey, F., & Alberts, E.E. (1998) Runoff and soil losses as affected by corn and soybean tillage systems. *Journal of Soil and Water Conservation* **53**: 64–70.

Giller, K.E., Beare, M.H., Lavelle, P., Izac, A.M.N., & Swift, M.J. (1997) Agricultural intensification, soil biodiversity and agroecosystem function. *Applied Soil Ecology* **6**: 3–16.

Green, R.E., Cornell, S.J., Scharlemann, J.P.W., & Balmford, A. (2005b) Farming and the fate of wild nature. *Science* **307**: 550–5.

Green, V.S., Cavigelli, M.A., Dao, T.H., & Flanagan, D. (2005a) Soil physical properties and aggregate-associated C, N, and P distributions in organic and conventional cropping systems. *Soil Science* **170**: 822–31.

Gregorich, E.G., Rochette, P., VandenBygaart, A.J., & Angers, D.A. (2005) Greenhouse gas contributions of agricultural soils and potential mitigation practices in Eastern Canada. *Soil & Tillage Research* **83**: 53–72.

Gupta, V.V.S.R., & Yeates, G.W. (1997) Soil microfauna as bioindicators of soil health. In: C.E. Pankhurst, B.M. Doube, & V.V.S.R. Gupta (eds.) *Biological indicators of soil health*, pp. 201–34. CAB International, Wallingford.

Harrison, M.J. (2005) Signaling in the arbuscular mycorrhizal symbiosis. *Annual Review of Microbiology* **59**: 19–42.

Harrison, M.J., Dewbre, G.R., & Liu, J. (2002) A phosphate transporter from *Medicago truncatula* involved in the acquisition of phosphate released by arbuscular mycorrhizal fungi. *Plant Cell* **14**: 2413–30.

Hawkes, C.V., Wren, I.F., Herman, D.J., & Firestone, M.K. (2005) Plant invasion alters nitrogen cycling by modifying the soil nitrifying community. *Ecology Letters* **8**: 976–85.

Haynes, R.J., & Naidu, R. (1998) Influence of lime, fertilizer and manure applications on soil organic matter content and soil physical conditions: A review. *Nutrient Cycling in Agroecosystems* **51**: 123–37.

Hole, D.G., Perkins, A.J., Wilson, J.D., Alexander, I.H., Grice, P.V., & Evans, A.D. (2005) Does organic farming benefit biodiversity? *Biological Conservation* **122**: 113–30.

Hooper, D.U., Chapin III, F.S., Ewel, J.J., et al. (2005) Effects of biodiversity on ecosystem functioning: A consensus of current knowledge. *Ecological Monographs* **75**: 3–35.

Hrynkiewicz, K., Baum, C., & Leinweber, P. (2009) Mycorrhizal community structure, microbial biomass P and phosphatase activities under *Salix polaris* as influenced by nutrient availability. *European Journal of Soil Biology* **45**: 168–75.

Ibekwe, A.M., Papiernik, S.K., Gan, J., et al. (2001) Impact of fumigants on soil microbial communities. *Applied and Environmental Microbiology* **67**: 3245–57.

Intergovernmental Panel on Climate Change (2007) *Climate Change 2007: Synthesis Report. Contribution of Working Groups I, II and III to the Fourth Assessment Report of the Intergovernmental Panel on Climate Change* pp. 104.

[Core Writing Team, R.K. Pachauri, & A. Reisinger (eds.)]. IPCC, Geneva.

Jackson, L.E., Burger, M., & Cavagnaro, T.R. (2008) Roots, nitrogen transformations, and ecosystem services. *Annual Review of Plant Biology* **59**: 341–63.

Janzen, H.H., Campbell, B.H., & Bremer, E. (1997) Soil organic matter dynamics and their relationship to soil quality. *Developments in Soil Science* **25**: 277–91.

Johnson, N.C. (1993) Can fertilization of soil select less mutualistic mycorrhizae? *Ecological Applications* **3**: 749–57.

Kiers, E.T., West, S.A., & Denison, R.F. (2002) Mediating mutualisms: Farm management practices and evolutionary changes in symbiont co-operation. *Journal of Applied Ecology* **39**: 745–54.

Kjar, D., Szlavecz, K., Cavigelli, M., Phillips, J., & Scace, C. (2010) Ant (Hymenoptera: Formicidae) community differences associated with organic, no-till, and chisel-till cropping systems (maize and soybean). XVI International Congress of the International Union for the Study of Social Insects, Copenhagen, Denmark.

Kladivko, E.J. (2001) Tillage systems and soil ecology. *Soil & Tillage Research* **61**: 61–76.

Klironomos, J.N. (2003) Variation in plant response to native and exotic arbuscular mycorrhizal fungi. *Ecology* **84**: 2292–301.

Kramer, S.B., Reganold, J.P., Glover, J.D., Bohannan, J.M., & Mooney, H.A. (2006) Reduced nitrate leaching and enhanced denitrifier activity and efficiency in organically fertilized soils. *Proceedings of the National Academy of Sciences of the United States of America* **103**: 4522–7.

Kremer, R.J., & Means, N.E. (2009) Glyphosate and glyphosate-resistant crop interactions with rhizosphere microorganisms. *European Journal of Agronomy* **31**: 153–61.

Lange, R.T., & Parker, C.A. (1961) Effective nodulation of *Lupinus digitatus* by native rhizobia in South-Western Australia. *Plant and Soil* **15**: 193–8.

Larink, O. (1997) Springtails and mites: Important knots in the food web of soils. In: G. Benckiser (ed.) *Fauna in soil ecosystems: recycling processes, nutrient fluxes, and agricultural production*, pp. 225–64. Marcel Dekker, New York.

Letourneau, D.K., & Bothwell, S.G. (2008) Comparison of organic and conventional farms: Challenging ecologists to make biodiversity functional. *Frontiers in Ecology and the Environment* **6**: 430–8.

Loreau, M., Naeem, S., Inchausti, P., et al. (2001) Biodiversity and ecosystem functioning: Current knowledge and future challenges. *Science* **294**: 804–8.

Mallory, E.B., & Griffin, T.S. (2007) Impacts of soil amendment history on nitrogen availability from manure and fertilizer. *Soil Science Society of America Journal* **71**: 964–73.

Maneepitak, S. (2007) Relative value of agricultural biodiversity on diversified farms: A case study in Donjaedee district, Suphanburi province, central Thailand. *Journal of Developments in Sustainable Agriculture* **2**: 86–91.

Marriott, E.E., & Wander, M.M. (2006) Total and labile soil organic matter in organic and conventional farming systems. *Soil Science Society of America Journal* **70**: 950–9.

Maul, J.E., & Drinkwater, L.E. (2010) Short-term plant species impact on microbial community structure in soils with long-term agricultural history. *Plant and Soil* **330**: 369–82.

Millennium Ecosystem Assessment. (2005). *Ecosystems and Human Well-Being: Synthesis*. Island Press, Washington DC.

Miller, R.M., & Jastrow, J.D. (1990) Hierarchy of root and mycorrhizal fungal interactions with soil aggregation. *Soil Biology & Biochemistry* **22**: 579–84.

Minoshima, H., Jackson, L.E., Cavagnaro, T.R., et al. (2007) Soil food webs and carbon dynamics in response to conservation tillage in California. *Soil Science Society of America Journal* **71**: 952–63.

Mirsky, S.B., Curran, W.S., Mortensen, D.A., Ryan, M.R., & Shumway, D.L. (2009) Control of cereal rye with a roller/crimper as influenced by cover crop phenology. *Agronomy Journal* **101**: 1589–96.

Moorhead, L.C., Philpott, S.M., & Bichier, P. (2010) Epiphyte biodiversity in the coffee agricultural matrix: Canopy stratification and distance from forest fragments. *Conservation Biology* **24**: 737–46.

Murray, P.J., Cook, R., Currie, A.F., et al. (2006) Interactions between fertilizer addition, plants and the soil environment: Implication for soil faunal structure and diversity. *Applied Soil Ecology* **33**: 199–207.

Muruganandam, S., Israel, D.W., & Robarge, W.P. (2009) Activities of nitrogen-mineralization enzymes associated with soil aggregate size fractions of three tillage systems. *Soil Science Society of America Journal* **73**: 751–9.

Muruganandam, S., Israel, D.W., & Robarge, W.P. (2010) Nitrogen transformations and microbial communities in soil aggregates from three tillage systems. *Soil Science Society of America Journal* **74**: 1–10.

National Research Council (2010) *Toward Sustainable Agricultural Systems in the 21st Century*. National Academies Press, Washington, DC.

Oades, J.M. (1984) Soil organic matter and structural stability mechanisms and implications for management. *Plant and Soil* **76**: 319–38.

Oka, Y. (2009) Mechanisms of nematode suppression by organic amendments—a review. *Applied Soil Ecology* **44**: 101–15.

Paterson, E. (2003) Importance of rhizodeposition in the coupling of plant and microbial productivity. *European Journal of Soil Science* **54**: 741–50.

Paustian, K., Collins, H.P., & Paul, E.A. (1997) Management controls on soil carbon. In: E.A. Paul et al. (eds.) *Soil organic matter in temperate agroecosystems*, pp. 15–49. CRC Press, Boca Raton, FL.

Peacock, A.D., Mullen, M.D., Ringelberg, D.B., et al. (2001) Soil microbial community responses to dairy manure or ammonium nitrate applications. *Soil Biology & Biochemistry* **33**: 1011–19.

Petersen, H., & Krogh, P.H. (1987) Effects of perturbing microarthropod communities of a permanent pasture and a rye field by an insecticide and a fungicide. In: B.R. Striganova (ed.) *Soil fauna and soil fertility. Proceedings of the 9th International Colloquium on Soil Zoology*, pp. 217–29. Moscow.

Phillips, R.L. (2007) Organic agriculture and nitrous oxide emissions at sub-zero soil temperatures. *Journal of Environmental Quality* **36**: 23–30.

Pitz, S.L., Szlavecz, K., & Cavigelli, M.A. (2004) *Hydrology and earthworms in agroecosystems*. Ecological Society of America Mid-Atlantic Ecology Conference, Lancaster, PA.

Prasse, I. (1985) Indications of structural changes in the communities of microarthropods of the soil in an agroecosystem after applying herbicides. *Agriculture, Ecosystems and Environment* **13**: 205–15.

Reeve, J.R., Schadt, C.W., Carpenter-Boggs, L., Kang, S., Zhou, J., & Reganold, J.P. (2010) Effects of soil type and farm management on soil ecological functional genes and microbial activites. *The ISME Journal* **4**: 1099–107.

Reganold, J.P., Andrews, P.K., Reeve, J.R., et al. (2010) Fruit and soil quality of organic and conventional strawberry agroecosystems. *PLoS ONE* **5**: 1–14.

Reganold, J.P., Elliot, L.F., & Unger, Y.L. (1987) Long-term effects of organic and conventional farming on soil erosion. *Nature* **330**: 370–2.

Reiss, J., Bridle, J.R., Montoya, J.M., & Woodward, G. (2009) Emerging horizons in biodiversity and ecosystem functioning research. *Trends in Ecology & Evolution* **24**: 505–14.

Rhoton, F.E., Shipitalo, M.J., & Lindbo, D.L. (2002) Runoff and soil loss from midwestern and southeastern US silt loam soils as affected by tillage practice and soil organic matter content. *Soil & Tillage Research* **66**: 1–11.

Robertson, G.P., & Grace, P.R. (2004) Greenhouse gas fluxes in tropical and temperate agriculture: The need for a full-cost accounting of global warming potentials. *Environment, Development and Sustainability* **6**: 51–63.

Robertson, G.P., Paul, E.A., & Harwood, R.R. (2000) Greenhouse gases in intensive agriculture: Contributions of individual gases to the radiative forcing of the atmosphere. *Science* **289**: 1922–5.

Rousk, J., Brookes, P.C., & Baathe, E. (2010) Investigating the mechanisms for the opposing pH relationships of fungal and bacterial growth in soil. *Soil Biology & Biochemistry* **42**: 926–34.

Ryan, M.R., Smith, R.G., Mortensen, D.A., et al. (2009) Weed-crop competition relationships differ between organic and conventional cropping systems. *Weed Research* **49**: 572–80.

Sarrantonio, M. (1994) *Northeast Cover Crop Handbook*. The Rodale Institute, Emmaus, PA.

Scherr, S.J., & McNeely, J.A. (2008) Biodiversity conservation and agricultural sustainability: Towards a new paradigm of "ecoagriculture" landscapes. *Philosophical Transactions of the Royal Society B: Biological Sciences* **363**: 477–94.

Schimel, J.P., & Bennett, J. (2004) Nitrogen mineralization: Challenges of a changing paradigm. *Ecology* **85**: 591–602.

Schutter, M.E., Sandeno, J.M., & Dick, R.P. (2001) Seasonal, soil type, and alternative management influences on microbial communities of vegetable cropping systems. *Biology and Fertility of Soils* **34**: 397–410.

Schwartz, M.W., Brigham, C.A., Hoeksema, J.D., Lyons, K.G., Mills, M.H., & van Mantgem, P.J. (2000) Linking biodiversity to ecosystem function: Implications for conservation ecology. *Oecologia* **122**: 297–305.

Six, J., Frey, S.D., Thiet, R.K., & Batten, K.M. (2006) Bacterial and fungal contributions to carbon sequestration in agroecosystems. *Soil Science Society of America Journal* **70**: 555–69.

Six, J., Ogle, S.M., Breidt, F.J., Conant, R.T., Mosier, A.R., & Paustian, K. (2004) The potential to mitigate global warming with no-tillage management is only realized when practiced in the long term. *Global Change Biology* **10**: 155–60.

Six, J., Paustian, K., Elliott, E.T., & Combrink, C. (2000) Soil structure and soil organic matter: I. Distribution of aggregate size classes and aggregate associated carbon. *Soil Science Society of America Journal* **64**: 681–9.

Soil Association (2002) *Soil Association Organic Standards and Certification (Revision 14 2002/03)*. Soil Association, Bristol.

Spargo, J.T., Alley, M.M., Follett, R.G., Wallace, J.V. (2008) Soil nitrogen conservation with continuous no-till management. *Nutrient Cycling in Agroecosystems* **82**: 283–97.

Spargo, J.T., Cavigelli, M.A., Mirsky, S.B., Maul, J.E., & Meisinger, J.J. (2011) Mineralizable soil nitrogen and labile soil organic matter in diverse long-term cropping systems. *Nutrient Cycling in Agroecosystems* **90**: 253–66.

Stolze, M., Piorr, A., Haring, A., & Dabbert, S. (2000) *The Environmental Impacts of Organic Farming in Europe—Organic Farming in Europe: Economics and Policy*, vol. 6. University of Hohenheim, Stuttgart.

Stres, B., & Tiedje, J.M. (2006) New frontiers in soil microbiology: How to link structure and function in microbial communities? *Nucleic Acids and Proteins in Soil* **8**: 1–22.

Strickland, M.S., Lauber, C., Fierer, N., & Bradford, M.A. (2009) Testing the functional significance of microbial community composition. *Ecology* **90**: 441–51.

Stromberger, M.E., Klose, S., Ajwa, H., Trout, T., & Fennimore, S. (2005) Microbial populations and enzyme activities in soils fumigated with methyl bromide alternatives. *Soil Science Society of America Journal* **69**: 1987–99.

Swift, M.J., Izac, A.M.N., & van Noordwijk, M. (2004) Biodiversity and ecosystem services in agricultural landscapes—are we asking the right questions? *Agriculture, Ecosystems Environment* **104**: 113–34.

Taylor, B.R., Parkinson, D., & Parsons, W.F.J. (1989) Nitrogen and lignin content as predictors of litter decay rates: a microcosm test. *Ecology* **70**: 97–104.

Teasdale, J.R., Mangum, R.W., Radhakrishnan, J., & Cavigelli, M.A. (2003) Factors influencing annual fluctuations of the weed seed bank at the long-term Beltsville Farming Systems Project. *Aspects of Applied Biology* **69**: 93–9.

Tilman, D. (1982) *Resource Competition and Community Structure*. Princeton University Press, Princeton, NJ.

Tilman, D., Cassman, K.G., Matson, P.A., Naylor, R., & Polasky, S. (2002) Agricultural sustainability and intensive production practices. *Nature* **418**: 671–7.

Treonis, A.M., Austin, E.E., Buyer, J.S., Maul, J.E., Spicer, L., & Zasada, I.A. (2010) Effects of organic amendment and tillage on soil microorganisms and microfauna. *Applied Soil Ecology* **46**: 103–10.

Triplett, G.B., & Dick, W.A. (2008) No-tillage crop production: A revolution in agriculture! *Agronomy Journal* **100**: S153–65.

Tscharntke, T., Klein, A.M., Kruess, A., Steffan-Dewenter, I., & Thies, C. (2005) Landscape perspectives on agricultural intensification and biodiversity—ecosystem service management. *Ecology Letters* **8**: 857–74.

van Bruggen, A.H.C., & Semenov, A.M. (2000) In search of biological indicators for soil health and disease suppression. *Applied Soil Ecology* **15**: 13–24.

van Tassel, D.L., DeHaan, L.R., & Cox, T.S. (2010) Missing domesticated plant forms: can artificial selection fill the gap? *Evolutionary Applications* **3**: 434–52.

Venterea, R.T., Baker, J.M., Dolan, M.S., & Spokas, K.A. (2006) Carbon and nitrogen storage are greater under biennial tillage in a Minnesota corn-soybean rotation. *Soil Science Society of America Journal* **70**: 1752–62.

Vitousek, P.M., Aber, J.D., Howart, R.W., *et al.* (1997) Human alteration of the global nitrogen cycle: Sources and consequences. *Ecological Applications* **7**: 737–50.

Wall, D.H., & Moore, J.C. (1999) Interactions underground: Soil biodiversity, mutualism, and ecosystem processes. *Bioscience* **49**: 109–17.

Wardle, D.A. (1995) Impacts of disturbance on detritus food webs in agro-ecosystems of contrasting tillage and weed management practices. *Advances in Ecological Research* **26**: 105–85.

Zak, D.R., Groffman, P.M., Pregitzer, K.S., Christensen, S., & Tiedje, J.M. (1990) The vernal dam—plant microbe competition for nitrogen in northern hardwood forests. *Ecology* **71**: 651–6.

Zelles, L., Palojarvi, A., Kandeler, E., *et al.* (1997) Changes in soil microbial properties and phospholipid fatty acid fractions after chloroform fumigation. *Soil Biology & Biochemistry* **29**: 1325–36.

CHAPTER 5.5

Soil Ecosystem Resilience and Recovery

A. Stuart Grandy, Jennifer M. Fraterrigo, and Sharon A. Billings

5.5.1 Introduction

Soils are widely recognized for their important role in plant growth and productivity, are the foundation upon which sustainable agricultural systems have historically been built (Howard 1945), and are the key to sustaining a global population that is expected to soon reach 10 billion people (Sanchez 2002; UN Millennium Project 2005). Many other key ecosystem functions, along with food and fiber production, depend on soils, including trace gas mitigation, carbon (C) sequestration, and clean water (Haygarth & Ritz 2009). Soils contain more than twice as much C than either the atmosphere or vegetation, such that even small changes in soil C storage will have a substantial effect on atmospheric carbon dioxide (CO_2) concentrations, and play a key role in atmosphere-biosphere exchange of methane (CH_4) and nitrous oxide (N_2O) (CAST 2004). Clearly, soils are one of our most valuable natural resources given the numerous processes they support and their capacity to mitigate the harmful consequences of human activities.

In spite of the critical role that soils play in human civilization and, more broadly, ecosystem vitality, history reveals that most major efforts to understand and protect soils have been reactionary rather than anticipatory (Robertson et al. 2004). For example, soil degradation—a process defined by Lal (1997) as a decline in soil productivity and capacity to regulate environmental processes—was a problem for decades before the dust bowl lead to the development of the Soil Conservation Districts in 1935. The reactionary nature of humanity's relationship with soil is of growing concern given the host of services humanity requires from it and the challenges generated by current soil management practices. Consider that more than half of the Earth's 13 billion hectares of soil are being used for agricultural production or are otherwise disturbed (Richter & Markewitz 2001), and that the term "endangered," so often used in the past to describe plants and animals, is now being used to describe many soils around the world (Pimental & Sparks 2000; Amundson et al. 2003). Indeed, Amundson et al. (2003) found that in some agricultural regions as much as 80% of soils classified as rare have been reduced to less than half their original area. Fertilization, a key feature of agricultural practices in most regions, drives a significant portion of soil-derived N_2O (Robertson et al. 2000; McSwiney & Robertson 2005) and has led directly to declines in water quality around the globe (Howarth et al. 1996; David et al. 2010). Globally, agricultural practices have lead to reductions in soil organic matter (SOM) in topsoil by up to 50%, a phenomenon directly tied to increased erosion rates, changes in soil decomposer process rates, and increased dependence on inorganic fertilizers to sustain crop productivity (Lal 2004; Drinkwater et al. 2008). This suite of ecological problems induced by soil management practices underscores the need to prioritize the conservation of soils before they reach a state of extreme degradation.

Understanding how soils respond to and recover from disturbance can help guide the development of soil conservation strategies. Given the importance of soil to sustaining our civilization, prioritizing

Soil Ecology and Ecosystem Services. First Edition. Edited by Diana H. Wall et al.
© 2012 Oxford University Press. Published 2012 by Oxford University Press.

efforts to understand soils and their responses to disturbance is critical for preventing declines in their functioning. To accomplish this, we need to develop a more comprehensive theoretical and applied framework for understanding soil resilience and recovery in changing environments. In this chapter, we define disturbance as it relates to soil ecosystems, develop a framework in which we can assess a soil's capacity to recover from such disturbances, and describe the responses of soils to multiple types of disturbances. We do not intend to provide an exhaustive account of how soils respond to all the disturbances generated by humans, nor do we assess every soil function important for ecosystem processing or humanity. Instead, we develop ideas about soil recovery from disturbance as a means to guide future investigations of soil resilience as humanity's demands on soil increase over the coming decades.

5.5.2 Soil disturbance, resilience, and recovery

Human activities are altering the interplay between chemical, physical, and biological processes in ways that change the structure and function of the soil ecosystem. To develop a framework for better understanding soil change, and to enhance capacity to predict its direction and magnitude, we can look to ecological theories describing how ecosystems respond to external forces. We focus on areas in which ecological theory can provide insights into understanding soils in changing environments: disturbance, resilience, and recovery. When considered together, these concepts provide the foundation for developing a unifying framework to understand how external forces, both human and natural, are influencing soils and their functions.

5.5.2.1 Disturbance

Disturbance refers to an event or series of events that disrupts ecosystem, community, or population structure and changes resources or the physical environment (White & Pickett 1985). Although disturbance can originate from either natural or anthropogenic sources, a distinction has recently been made between disturbance events that are discrete in time and human-caused global-change drivers that chronically alter resource availability (Smith et al. 2009). Discrete disturbances are events that may occur only once or periodically, but that exhibit clearly defined periods of disturbance and disturbance cessation. Although the duration of a disturbance can vary, discrete disturbances typically occur over a limited time period. Gap-producing storm events in forests and anthropogenic forest clearing with subsequent regrowth represent examples of discrete disturbances. In contrast, chronic disturbances continue through time and cannot be depicted as isolated events. These disturbances are increasing in abundance as anthropogenic influences on Earth's systems become more prominent and include some types of erosion, elevated atmospheric CO_2, deposition of N and other pollutants, and climate change.

With regard to soil ecosystems, there are several key differences between discrete disturbances and chronic disturbances from global-change. Chronic resource alterations, for example, are the result of chronic disturbances and are not generated directly by changes to the biota, experience minimal feedback from the biota, and do not diminish over time but may increase or accumulate (Smith et al. 2009). Due to this lack of biotic regulation and consequent lack of biological feedback mechanisms that buffer ecosystems against rapid changes, chronic resource alterations have the potential to produce abrupt changes in ecosystem state as ecological thresholds are reached (Smith et al. 2009). In other words, a system may appear to be exhibiting resilience to chronic disturbances until a threshold is surpassed, which brings about a rapid and dramatic change in ecosystem conditions. However, alternative ecosystem states brought about by surpassing an ecosystem's capacity for resilience can also emerge from changes in state variables owing to discrete events (Beisner et al. 2003), and can be strongly reinforced by variable interactions and feedbacks across scales (Peters et al. 2004). Because of this, an ecosystem may remain permanently in an alternative state, even if some external forces are returned to their pre-disturbance condition.

Unifying these concepts to understand ecosystem recovery and resilience is an important challenge in ecology, particularly given the potential for interactions between discrete and chronic disturbance events. Similarities already exist between the conceptual frameworks used to appreciate ecological responses to discrete and chronic disturbances, as both incorporate a hierarchical perspective to integrate responses across scales (Fraterrigo & Rusak 2008; Smith *et al.* 2009). The potential for differences in ecosystem sensitivity to perturbation are also uniformly recognized. Whether resources are altered by discrete or chronic events, it is clear that pre-existing conditions, such as community composition or pre-disturbance soil nutrient status, can create heterogeneity in ecosystem response to disturbance or chronic stressors. With respect to ecosystem recovery and resilience, such factors may shorten or lengthen the time it takes ecosystems to return to their pre-disturbance state after the perturbation ceases, or may increase the likelihood that an ecosystem will reach a novel state. Additional research is needed to evaluate how well we can understand and predict the response of ecological systems to chronic alterations based on studies of discrete disturbances.

5.5.2.2 Resilience

Ecosystem resilience was defined by Holling (1973) as the extent to which an ecosystem can withstand disturbance before dramatic shifts in its structure and function occur. Building on this definition, Folke *et al.* (2004) and Walker *et al.* (2004) defined resilience as the capacity of an ecosystem to absorb a disturbance and reorganize with minimal changes in structure and function. The concept of resilience is not based on a steady state ideal but recognizes that ecosystems will vary in time and space, while still maintaining key biological interactions and functions. Even in natural ecosystems, conditions are never constant and stochasticity results from phenomena such as fire, insect or disease outbreaks, and extreme climate events. Forces such as these may affect the biota or ecosystem functions, but resilient ecosystems will eventually recover to function within their normal range. Resilient systems are thus capable of renewal after a disturbance, and do this through self-organization.

To illustrate resilience concepts, Scheffer *et al.* (2001) described the region in which stable ecosystems fluctuate in response to external factors as a "basin of attraction." Ecosystem characteristics such as community structure and functions (e.g. C storage) can be envisioned as a ball that moves about the basin center in response to external forcings, but will always return to the basin depression due to interactions and feedbacks within the system. For example, the C cycle in soils is tightly regulated by interactions between soil organisms, substrate availability, and the physical and chemical soil environment. In late successional communities, these interactions define SOM concentrations at equilibrium and, although there is seasonal and annual variation in C cycling rates, equilibrium conditions persist unless there is a substantial outside disturbance. Even if there is an outside force such as a fire or pest outbreak that influences C cycling, SOM concentrations will eventually return to within the range of variability unless there is a profound change in the processes regulating C cycling.

When an ecosystem's capacity for resilience is exceeded, it may not return to its original state, or may take a long time to do so, following a very different path to recovery than to degradation. Scheffer *et al.* (2001) demonstrated this as a ball representing some function being pushed from one basin of attraction into another; a push back in the other direction will not always return the ball to the original basin. If it does, the return path may be difficult to predict. Soil legacies of historically farmed forests are an example of functional stasis despite a return to pre-existing conditions. Nearly a century after agricultural abandonment, nitrification rates remain higher in forested sites previously used for agriculture than in forested sites that were previously in pasture or woodlot (Compton & Boone 2000). This phenomenon can similarly apply to the biota of an ecosystem. A pertinent example is the shift from grass- to shrub-dominated plant communities in the American southwest as a result of overgrazing by livestock and historical changes in climate (Grover & Musick 1990). Despite cessation of grazing, grasslands have been slow to recover (Valone *et al.* 2002). This path-dependent process, termed hysteresis, is a common feature of ecosystem degradation and recovery and highlights the importance of trying to maintain ecosystem resilience, and to not exceed an ecosystem's capacity to absorb disturbance.

5.5.2.3 Hysteresis

Hysteresis describes how a system's memory of past inputs and processes influences its responses to current conditions. In systems experiencing hysteresis, current behavior is a function of conditions both past and present. More specific to ecosystem science, ecosystems experiencing hysteresis exhibit a recovery trajectory different from that followed during ecosystem degradation. This is because the factors leading to ecosystem decline, and the new ecosystem conditions these factors create, leave behind a legacy that influences how an ecosystem responds to current stimulus. The processes of soil degradation and recovery often follow a pattern of hysteresis. Almost all soil properties and processes degrade more quickly following disturbance than they recover following the cessation of disturbance, and there remains debate over whether or not some soil properties are capable of ever fully recovering to their original condition.

The decline in soil organic matter following agricultural conversion and subsequent recovery after agricultural abandonment is one example of hysteresis in soils with important implications (Fig. 5.5.1). In temperate regions, soils lose 30–50% of their topsoil C within decades of agricultural conversion (Davidson & Ackerman 1993; West & Post 2002). In contrast to SOM loss, chronosequence studies as well as long-term experiments show that the recovery of this SOM after agricultural abandonment may take decades to centuries (Richter & Markewitz 2001; Grandy & Robertson 2007; Foote & Grogan 2010). In the tropics, hysteresis may be even more pronounced. Following agricultural conversion, SOM loss is very rapid, with losses of more than 50% sometimes occurring within years of agricultural conversion. Changes in the texture and structure of surface soils, as well as the soil biota, change the inherent capacity of formerly cultivated soils to recover C (Buckley & Schmidt 2001; Churchman et al. 2010)

Hysteresis may strongly influence the recovery of ecosystems from disturbance and some ecosystems

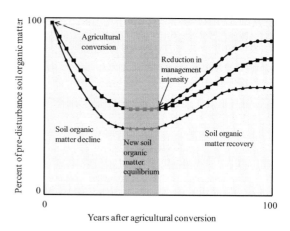

Figure 5.5.1 Hysteresis in soil organic matter loss and recovery following agricultural disturbance (adapted from Schlesinger 1997). In temperate ecosystems (squares) ~50% of topsoil organic matter is lost in a few decades. The recovery rate is typically slower than the loss, but can be faster (circles) depending on site conditions, particularly climate, vegetation, and soil characteristics, as well as management (e.g. the use of organic amendments such as compost). Generally, in tropical ecosystems (triangles) the rate of soil organic matter loss is faster and the recovery typically slower and less extensive.

may experience permanent state shifts if their capacity for resilience is exceeded. This concept, in conjunction with the critical nature of many soil functions for ecosystem processes and humanity, prompt the need for a more comprehensive understanding of soil ecosystem resilience and its fundamental controls. Surprisingly, however, few studies have examined soil resilience and hysteresis in the context of ecological theory (Seybold et al. 1999; Chaer et al. 2009; Royer-Tardif et al. 2010; Churchman et al. 2010). To a large extent, this is because we have not had the necessary long-term soil monitoring programs in place to understand how soils function within their normal range of variability. It is also much harder to observe changes in soil ecosystems than it is in other ecosystems and theoreticians struggle with the complexity of soil ecosystems. There is no limit to how the interactions between different soil properties and local climatic conditions will play out, which makes every soil in some ways unique. Because of this complexity, soil scientists often take a "black box" approach for which inputs and outputs of matter and energy are predicted or measured but the internal controls remain essentially unknown. Black boxes and their resulting model functions are typically not mechanistic, and are thus limited in their predictive capacity. For example, we may model soil respiration as a function of moisture and temperature (Davidson & Janssens 2006) but very few efforts have attempted to incorporate into models the enzymes that generate the substrates that drive respiration (Schimel & Weintraub 2003; Moorhead & Sinsabaugh 2006; Lawrence et al. 2009). In the following sections, we provide examples of soil resilience and recovery, discuss the need for studying soil biota, and describe the role that long-term soil experiments play in understanding soil ecosystem resilience in a world that is rapidly changing.

5.5.3 Resilience and recovery: soil organic matter dynamics

Soil organic matter is linked to a wide array of key soil functions relevant to soil ecosystem resilience and recovery. Soil organic matter processes help define a soil's capacity to retain and provide nutrients and moisture for plant uptake and govern soil-atmosphere fluxes of key greenhouse gases (Schlesinger 1997). Soil organic matter dynamics—SOM inputs, transformations, and outputs—also dictate net SOM pool sizes that, in turn, can govern the resilience of entire ecosystems following disturbance (Adl 2008; Lal 2010). Here we address both discrete and chronic disturbances as a way to explore issues of SOM dynamics and their resilience across multiple timescales. These examples are by no means exhaustive, but they demonstrate the varying concepts of resilience that are relevant to SOM dynamics over different time scales.

5.5.3.1 Discrete disturbance

Forest clearing, a discrete disturbance whether via storm-related disturbance or human activities, can impart significant changes in SOM dynamics. The sudden change in temperature and moisture conditions associated with clearing can result in dramatic changes in SOM decomposition rates (Ritter 2005; Scharenbroch & Bockheim 2008). Warmer temperatures can result in greater rates of SOM decomposition, which can be further promoted with soil moisture increases as transpirational demand declines (Abd Latif & Blackburn 2010). The resilience of the SOM system—its ability to return to its previous process rates—depends on the abiotic conditions that regrowth imposes on the recovering soil profile, the rapidity with which regrowth occurs, and the vegetation that develops.

Studies suggest that some aspects of SOM dynamics can quickly return to pre-disturbance levels following forest clearing. If environmental conditions promote the rapid development of dense vegetative cover, abiotic conditions can return to pre-disturbance levels on a timescale of weeks to months (Frolking et al. 2009), and soil respiration, a general measure of belowground activity, can return to values similar to those prior to the disturbance (Litton et al. 2003) regardless of the vegetation community that colonizes the site. However, as a site recovers from a disturbance, shifts across plant successional stages in the relative abundances of plant populations can drive changes in some SOM transformations. Typically, in forested regions, shade-intolerant species colonize clearings soon after disturbances (Perry 1994), sometimes representing a significant change from the previous vegetation. Such changes in dominant vegetation can result in altered decomposer community dynamics via changing quality of organic matter inputs. Pine litter, for example, which can dominate organic matter inputs to soil profiles in early successional forests of eastern North America, is typically of lower quality (i.e. higher C:N ratio) than litterfall from hardwoods (Aber & Melillo 1991). Further, some early successional species such as alder (Alnus spp.) take part in symbiotic relationships with N_2-fixing bacteria (Mitchell & Ruess 2009) that can result in significant inputs of plant-available N following disturbance. Thus turnover rates of organic matter can change as an indirect result of disturbance recovery, driven by changes in substrate quality and nutrient availability as well as by abiotic conditions (DeGryze et al. 2004).

The plowing that accompanies agricultural conversion can also impart significant changes in SOM dynamics. Increased aeration of surface horizons, changes in soil temperature, and the breakdown of aggregates can lead to enhanced SOM oxidation rates (Elliott 1986; Cambardella & Elliott 1994; Calderon & Jackson 2002). This, in conjunction with enhanced erosion rates and the repeated removal of nutrients during harvest, can result in significant depletion of SOM. The loss of SOM following plowing is associated with the almost immediate destruction of soil aggregates (Grandy & Robertson 2006) and following agricultural abandonment, or the adoption of less intensive management practices including no-till, the recovery of SOM is related to the reformation of soil aggregates, which protect SOM from microbial attack (Jastrow 1996; Jastrow et al. 1996; Six et al. 2002; Puget et al. 2005). Within a couple years after agricultural abandonment, changes in aggregate structure can be detected, and in some early successional systems aggregation may return to pre-cultivation levels within about a decade (DeGryze et al. 2004; Grandy & Robertson 2007). In general, soils with a combination of texture and mineralogy that confers to them a high cation exchange capacity and aggregation potential will undergo more rapid recovery than weekly structured, sandy soils (Baer et al. 2010). However, there is typically a significant lag time between the recovery of aggregation and the recovery of aggregate-associated SOM pools. The slow recovery of aggregate-associated SOM has a long-term effect on soil ecosystem functioning, as well as on an ecosystem's capacity to exhibit resilience following any future disturbance.

5.5.3.2 Chronic disturbance

In the earlier examples we explored SOM responses following a discrete disturbance; however, many external forcings on soil systems represent chronic disturbances. In the case of chronic disturbances, cessation does not seem likely in the near future, challenging our understanding of soil recovery from them. Both naturally occurring and anthropogenically enhanced erosion, altered SOM quantity and quality entering soil profiles due to rising atmospheric CO_2 concentrations, deposition of nitrogen (N) and other pollutants, and climate change all can serve as examples of chronic disturbances to soil systems, in addition to the perhaps most obvious chronic disturbance affecting SOM dynamics—agricultural activities. Assessing the impacts of chronic disturbances such as these can help us assess the capacity of SOM to maintain vital ecosystem functions. Further, given the inherent challenges of temporal scale when studying the dynamics of slow-turnover SOM pools, chronic disturbances provide opportunities to explore the impact of altered external forcings on long-term processes otherwise only rarely available, and to index changes in SOM pools with relatively long mean residence times.

Erosion is an example of an often-chronic disturbance that influences SOM dynamics. Following the removal of topsoil, the SOM dynamics of the remaining material reflect the net result of simultaneous phenomena. Typically, newly exposed surfaces contain lower SOM concentrations and a greater proportion of relatively slow-turnover SOM compounds (Lal 2005). These effects may be challenging to detect after only one year of low to moderate intensity erosion, but chronic erosion clearly exposes these trends (Montgomery 2007). Thus, lower reactivity material becomes increasingly closer to the surface as erosion proceeds, influencing the rate and type of SOM dynamics that take place. The ability of SOM processes to recover from this disturbance depends upon the ability of the site to produce SOM inputs. This dynamic replacement of SOM through ecosystem productivity (Stallard 1998) dictates the resulting carbon dynamics at the eroding site (van Oost et al. 2007; Billings et al. 2010).

Rising levels of atmospheric CO_2 is an example of a disturbance to which all soils are subjected and that is gradually increasing in intensity. Though this disturbance will likely intensify in the foreseeable future, prohibiting investigators from assessing an ecosystem capacity to bounce back to pre-industrial patterns, we can observe resilience to continuous elevated CO_2 concentrations. Elevated CO_2-induced changes in the quantity and composition of organic substrates and enhancements in plant uptake of nutrients can alter microbial processes and long-term SOM dynamics (Bernhardt et al. 2006; Reich et al. 2006; Schlesinger et al. 2006; Feng et al. 2010). For example, the apparent stimulation of organisms adept at accessing C contained within relatively recalcitrant OM compounds with elevated CO_2

(Carney et al. 2007; Billings & Ziegler 2008) could, if sustained, result in enhanced degradation of large, otherwise stable SOM pools. However, it is unknown if the net effect of altered ecosystem processes, including enhanced belowground inputs and priming of extant SOM pools, will result in increases or decreases in SOM stocks. Because of the slow turnover time of much of Earth's SOM pools, decades of consistent experimental manipulations and associated observations will be needed to develop the capacity to predict elevated CO_2 effects on soil ecosystems.

The deposition of N and other pollutants is another chronic disturbance to soil ecosystems in many regions of the globe. Much N is deposited on ecosystems downwind of industrial and feedlot sources (Galloway & Cowling 2002; Matson et al. 2002; Vitousek et al. 2002). The ecosystem consequences of N deposition have been experimentally mimicked in myriad environments via the addition of inorganic N fertilizers. These experiments have revealed a host of responses that vary by ecosystem type. Forest soils exposed to N enrichment exhibit reductions in potential oxidative enzyme activities, which result in the preservation of slow-turnover SOM and increases in SOM, particularly when litter inputs contain high lignin concentrations (Dijkstra et al. 2004; Zak et al. 2008). In some grassland systems, chronic N fertilization can result in enhanced exo-enzyme activity and SOM breakdown (Stursova et al. 2006; Tiemann & Billings 2011). These responses to chronic N deposition imply that soil profiles, regardless of the ecosystem type they support, maintain basic functioning with N additions but specific effects on long-term SOM dynamics vary with the quality of litter inputs. Anthropogenic climate change is perhaps the most complex chronic disturbance that imparts an influence on SOM dynamics. General circulation models suggest that we are shifting in the coming century and beyond to a warmer world in which precipitation events will be generally more intense (Solomon et al. 2007). As with elevated CO_2, we cannot truly assess the capacity of soil profiles to exhibit resilience of SOM dynamics after cessation of this disturbance. However, the response of SOM dynamics to changing climate is a key feature of ecosystem stability.

The enzyme-substrate interactions that drive the mineralization of SOM compounds proceed according to the Arrhenius relationship

$$V_{max} = v * e^{\frac{Ea}{R*T}}$$

where V_{max} is the maximum decomposition rate, v is the temperature-independent maximum decomposition rate, E_a is the activation energy of a particular enzyme-substrate pairing, R is the gas constant, and T is the temperature. This relationship indicates that with increasing E_a, the temperature sensitivity of a reaction increases and is a critical feature of our understanding of SOM dynamics (Davidson & Janssens 2006). It becomes particularly important when trying to predict decomposition rates in a warmer world because of Earth's vast stores of relatively slow-turnover SOM (Trumbore 2000) that likely possess high E_a. To what extent will enhanced release of CO_2 from slow-pool SOM provide a positive feedback to global warming? This unanswered question remains central to understanding the long-term implications of global temperature change. Two features, one abiotic and one biotic, may mitigate the responses predicted by the Arrhenius relationship. First, the limitation of many soil microbial populations by C or nutrients (Schimel & Weintraub 2003) may mute microbial responses to temperature, and such an influence would be exacerbated by moisture limitations (Davidson & Janssens 2006). Second, microbial populations may experience acclimation or adaptation to new temperature regimes that reduce biomass-specific respiration rates (Bradford et al. 2008; Bradford et al. 2010) and thus mitigate the temperature response predicted by enzyme kinetics. These features—declines in substrate availability driven by limits on physical diffusion, and microbial acclimation or adaptation—will ultimately modify the kinetically-driven response of SOM dynamics to temperature.

The combination of temperature-driven enzyme kinetics, substrate availability, and microbial acclimation and adaptation will ultimately govern the transformations of SOM compounds within soil profiles. In concert with any climate-induced change in biomass input rates to soil profiles, these transformations will determine future SOM pool sizes.

Organic matter stocks, particularly those exhibiting relatively slow turnover rates, can serve as a buffer against ecosystem-scale disturbances by providing a source of mineralizable nutrients. As a result, phenomena that portend enhanced rates of SOM degradation such as climate change may represent a negative influence on ecosystem resilience to disturbance.

Agricultural activities represent an obvious disturbance to many soil profiles, and at sites where agriculture has ceased and more natural vegetation has developed, we have an opportunity to assess patterns of profile recovery. Repeated plowing and agricultural extraction can drive a soil profile to such a degraded state that, at the very least, require decades of forest regrowth to re-develop soil A horizons (Richter & Markewitz 2001). Therefore, site conditions prior to the beginning of successional processes, which are a function of intensity and duration of disturbance and edaphic soil properties, are a critical feature to consider when assessing the ability of SOM dynamics to recover from disturbance and exhibit characteristics of an undisturbed site. Knowledge of such conditions, in conjunction with long-term soil sampling efforts during forest regrowth, have afforded investigators remarkable opportunities to understand the resilience of soil profiles following centuries of disturbance. For example, as forests recover after centuries of agricultural disturbance, their demand for nutrients can drive net reductions in SOM content of their soil profiles (Richter et al. 1999). After a period of decline, SOM content can subsequently begin to increase again, reflecting continued inputs from the aggrading biomass (Richter et al. 1999). After four decades of forest regrowth, soil profiles in the Calhoun Experimental Forest in South Carolina, USA, still have not attained the SOM content of a nearby old-growth hardwood stand similar to what likely grew on Calhoun's pine stands (Richter & Markewitz 2001). In spite of the relatively slow recovery period of SOM concentrations and content in these profiles, the stable isotopic profiles ($\delta^{13}C$ and $\delta^{15}N$) exhibit remarkable resilience; the distribution of these stable isotopic signatures with depth after four decades of regrowth is indistinguishable from that observed in mature forests, and reflects both the great degree of SOM turnover fueling the aggrading forest's growth and the apparent limits on SOM $\delta^{15}N$ and $\delta^{13}C$ values in relatively old forests (Billings & Richter 2006). Rapid ^{15}N and to a lesser extent ^{13}C enrichment in SOM with forest development indicates the importance of SOM mineralization as a source of nutrients, and suggests that the relatively rapid accrual of ^{15}N-enriched microbial necromass is a driver of SOM ^{15}N enrichment as mineralization proceeds. These divergent patterns—rapid recovery of stable isotopic signatures of SOM vs. relatively slow recovery of SOM content after disturbance—emphasize how our perceptions of resilience are governed by the parameters we choose to measure.

5.5.4 Resilience and recovery: soil nutrient cycling

Soil nutrient cycling and SOM dynamics often parallel one another. Like SOM dynamics, there is substantial variability in the resilience and recovery of nutrient dynamics that are linked to the intensity and duration of disturbance and differences in state variables. However, while nutrient cycling and SOM dynamics following disturbance are often linked, they operate at different time scales and may exhibit different degrees of response to changes in state factors. For example, both SOM dynamics and nutrient cycling are influenced by plant communities but these effects play out at very different time scales. Plant community dynamics have an immediate effect on nutrient cycling that, at least in the short to intermediate term, can be stronger than the influence of soil texture, mineralogy or other factors. In contrast, plant community effects on SOM dynamics play out over a longer period of time.

Vegetation composition can have a large effect on nutrient dynamics through its influence on microbial community activity, and thus may influence the resilience of nutrient cycling. Microbial communities mediate many nutrient transformations, which can be influenced by processes linked to plant species or functional characteristics. For example, plant tissue nutrient concentrations and plant-derived extracellular enzymes feedback to nutrient cycling rates through decomposition and mineralization (Mellilo et al. 1982; Wedin & Tilman 1990; Sinsabaugh

1994). Feedback strength can differ among plant species, however, due to intrinsic differences in nutrient cycling characteristics. Lovett et al. (2004) showed that some tree species, such as hemlock, show characteristics of slow N cycling, resulting in a lack of responsiveness to exogenous increases in N. Other plant species show characteristics associated with rapid N cycling, and are highly responsive to exogenous changes in N availability (Lovett et al. 2004). Consequently, the potential for plant species or functional traits to modulate disturbance effects on nutrient cycling via feedbacks may depend on the dominant plant species. Plant species associated with slow nutrient cycling may promote ecosystem resilience, whereas those associated with fast cycling may contribute to rapid ecosystem change. To illustrate, Lovett and Rueth (1999) found a significant correlation between atmospheric N deposition and N cycling on plots dominated by maple trees but not on plots dominated by beech trees, presumably due to beech's N-conserving dynamics and limited feedback to increased N availability. Under chronic resource addition, nutrients may accumulate to the extent that the ability of slow nutrient cycling plant species to modulate impacts is surpassed. In a study conducted in a moist-meadow alpine tundra, for example, Suding et al. (2008) found that although the slow N cycling perennial *Geum* initially conferred resilience on N dynamics, accumulation of N contributed to the breakdown of positive plant–soil feedbacks that facilitated the growth of *Geum*. This breakdown led to the decline of *Geum* and an abrupt shift in N cycling (Suding et al. 2008). Negative plant–soil feedbacks, which stabilize plant communities (Bever et al. 2010; Mangan et al. 2010), may also contribute to the resilience of nutrient dynamics by preventing local extinctions of plant species with slow cycling characteristics, but we are unaware of any studies that address this topic.

Direct effects of disturbance on microbial community structure can also affect soil nutrient cycling. For example, there is evidence that two very different types of disturbance, atmospheric N deposition and agricultural conversion, can reduce arbuscular mycorrhizal fungi (AMF), fungal to bacterial biomass ratio, and total microbial biomass (Bedini et al. 2007; Treseder 2008; van Diepen et al. 2010). Previous work has shown that the loss of AMF and saprotrophic fungi can slow organic matter decomposition and reduce N mineralization rates (Beare 1997; Smith & Read 1997), although ecosystem N status is likely to govern transformation rates and pool sizes (Schimel & Bennett 2004). Loss of microbial biomass may also reduce P cycling because the mineralization of organic P occurs as a result of demand for P and is not necessarily released as a consequence of organic matter decomposition (McGill & Cole 1981; Tiessen et al. 1982; Smeck 1985).

The importance of vegetation for the recovery of soil nutrient cycling following disturbance has long been recognized, but new insights continue to be generated about the mechanisms underlying this relationship. In their Hubbard Brook study, Bormann and Likens (1979) demonstrated the critical role that vegetation plays in conserving soil nutrients following disturbance. Following experimental removal and suppression of forest cover, they observed elevated NO_3^- concentrations in the stream draining the experimental watershed and then a decline in stream water NO_3^- concentrations when vegetation was allowed to regrow (Bormann & Likens 1979). Because of the scale and manipulative nature of their work, the results were quickly integrated into conceptual models of ecosystem recovery. In the three decades since the publication of this classic work, however, cross-site research on soil N retention has shown that ecosystems vary substantially in their response to vegetation removal due to differences in plant community composition and the influence of other state factors. For instance, because some N-fixing species readily invade newly disturbed and exposed mineral soils (e.g. red alder in the western USA), vegetation removal can lead to overall increased soil N availability in some systems (Van Miegroet & Johnson 2009). Conversely, removal of N-fixing species can diminish soil nutrient loss in some cases by reducing system inputs (Van Miegroet & Cole 1984). Studies such as these highlight the need to consider species identity and functional traits in addition to biomass to understand vegetation effects on soil nutrient cycling in the context of disturbance.

Closer scrutiny of specific nutrient pools and their rates of change have yielded a more mechanistic understanding of how other state factors alone and in combination with plant communities affect ecosystem response to disturbance. Soil physical characteristics, particularly those associated with parent material, may have important consequences for soil recovery. Clay content and mineralogy can mitigate the loss of NO_3^- and soil organic carbon following disturbance via fixation, adsorption, and physical protection (Vitousek & Melillo 1979; McLauchlan 2006). More generally, there is evidence that differences in soil particle-size distributions can alter soil N recovery rates (Burke et al. 1995). Parent material also determines the original supply of inorganic nutrients that are released by weathering, and can therefore influence nutrient availability following disturbance (Anderson 1988). Finally, disturbance-driven changes in vegetation may interact with parent material via chemical and mechanical weathering processes to generate differences in nutrient cycling rates, although we currently lack empirical support for this hypothesis.

The effects of climate on the recovery and resilience of soil nutrient cycling are poorly understood. Research from arid ecosystems subjected to grazing suggests slow recovery of several nutrients as a result of changes in soil organic matter content and loss of rock-derived nutrients due to wind erosion (Neff et al. 2005). There is also a perception that such systems have little resilience, possibly because of strong feedbacks across scales (Schlesinger et al. 1990; Peters et al. 2004). For example, disturbance can damage microbiotic crusts, which are important for both N fixation and stabilizing soils; as a result, ecosystem N can be severely diminished and soils can become more vulnerable to erosion, leading to further nutrient loss (Evans & Belnap 1999; Chaudhary et al. 2009). On the other hand, warm and humid climates are associated with high rates of SOM decomposition, which suggests that nutrients will be rapidly mineralized and available for incorporation into aboveground biomass following disturbance. Vegetation recovery may lead to a substantial fraction of nutrients being translocated from the mineral soil to the aboveground biomass. If the resultant litter is nutrient poor and becomes a nutrient sink, nutrient cycling rates will decline and the system may experience persistent nutrient deficiencies. Such dynamics may explain why some aggrading forests in the southeastern USA that receive large N inputs due to atmospheric deposition remain persistently deficient in N (Richter et al. 2000).

We have shown that in the broadest sense, nutrient cycling and SOM dynamics exhibit distinctive responses to disturbance. The state factors provide a framework for understanding and predicting these responses, and are thus central to characterizing soil ecosystem resilience and recovery. However, we have only just begun to understand how soils change over time in response to disturbance, in large part because the underlying interactions between edaphic soil properties, soil ecosystem processes, and disturbance remain general. Below we discuss what we believe are the essential elements to developing a finer-scale understanding of these interactions and thus soil ecosystem resilience and recovery.

5.5.5 Future directions

Predicting soil ecosystem resilience, rates of change, and recovery following or during disturbance requires additional insights into soil change over time and its underlying process level controls across a range of ecosystem types. We know, in general, that the state factors—parent material, climate, organisms, topography, and time—are important, but a finer scale understanding is needed to develop a unifying theory of ecosystem resilience. Though multiple soil forming factors have an obvious influence over soil characteristics (Jenny 1980) and associated responses to disturbance, we suggest that a sustained effort in two critical research arenas is needed to move forward. First, we need to continue to unravel the complexity of the soil biota—a central feature of the main soil forming factors (Bardgett 2005). Recent advances in molecular techniques and their geochemical applications reveal both the astonishing degree of soil biotic diversity, and the importance of those organisms for driving soil responses to disturbance. Second, long-term studies are needed to explore the relationships between state variables and soil ecosystem processes. An international commitment to long-term experiments combined with a sustained effort to understand soil biota would help us achieve a greater understanding of soil ecosystem resilience and recovery in changing environments.

5.5.5.1 Soil biota

With the rapid development of new methods, we are gaining insights into the structure and function of soil communities as well as the key role that they play in regulating ecosystem processes. Soils contain a vast array of microbes and animals that together make up the soil food web. The soil microbial population, consisting of the archaea, bacteria, and fungi, is the most diverse and abundant group of soil organisms. Along with the microbes, soils contain a myriad of fauna such as microarthropods, nematodes, and earthworms. For example, the oribatid mites are a common micro-arthropod that feed on micro-organisms and plant litter and may be found in populations as large as 25,000–500,000 individuals m^{-2} soil (Coleman et al. 2004). Overall, these secondary consumers and other higher taxa possess a tremendous diversity of feeding preferences and life history strategies and interact with the microbes to control soil ecosystem functions.

Human activities can substantially change the composition of the soil food web (Jangid et al. 2008; Maul & Drinkwater 2010). Buckley and Schmidt (2003) found that microbial community composition was similar in conventionally managed agricultural fields and in early successional grass-dominated communities abandoned from agricultural production for nine years. Microbial communities in both of these systems differed significantly from those in adjacent, never cultivated grassland soils. Another successional ecosystem >45 years old had microbial communities similar to those in the never-cultivated grassland, suggesting that decades were needed for the recovery of microbial communities. Cleveland et al. (2003) found that conversion of tropical forest to pasture reduced microbial biomass by ~40–60%, and that declines in soil C alone did not explain these changes. Fraterrigo et al. (2006), studying the effects of past land use on microbial communities in southern Appalachian forests abandoned from logging or agriculture >50 years before reforestation, found that soil microbial communities and potential net N mineralization rates varied significantly with past land use. Many other studies have also shown that land use intensification typically reduces microbial biomass and alters microbial community structure (e.g. Nüsslein & Tiedje 1999; Stark et al. 2008; Buyer et al. 2010; Enwall et al. 2010). The soil fauna, including nematodes, earthworms, micro-arthropods, termites, and many other types of organisms, also play a key role in nutrient cycling, soil structural formation, and organic matter dynamics and are sensitive to a range of chronic and discrete soil disturbances.

The specific links between soil communities and biogeochemical processes remain poorly understood, but there is growing evidence that variation in soil microbial community size, activity, and composition is in part responsible for observed variation in nutrient cycling and soil organic matter dynamics among ecosystems (Fierer et al. 2007; Reeve et al. 2010; Miki et al. 2010). In the recent past it was widely thought that ubiquitous and general soil microbial processes including decomposition were not influenced by microbial community structure, but recent studies have questioned this tenant. In one such study, Strickland et al. (2009) used a "common-garden" approach to test whether microbial community composition influences the decomposition of litter. They were specifically interested in testing whether microbial communities perceive litter quality differently based on their historical exposure to different litter types. They found that microbial communities did influence decomposition rates, and that the variation in rates was, in part, explained by whether or not the communities had prior exposure to a particular litter type. These results suggest that microbial communities perceive litter quality differently based on their physiological characteristics, which are determined in part by historical site conditions. The preference decomposer communities show for the plant litter produced above them, termed "home-field advantage" has now been shown in a number of field and laboratory experiments (Cragg & Bardgett 2001; Ayres et al. 2006; Hobbie et al. 2006; Ayres et al. 2009) and, in general, microbial community structure may account for a significant amount of the variation in C mineralization rates (Strickland et al. 2010).

Other decomposition processes, including changes in litter chemistry over time, are also influenced by variation in soil communities. Wickings et al. (2010) observed that different types of land use result in functionally distinct decomposer communities that vary in their ability to metabolize different substrates (Adair et al. 2008; Grandy et al. 2008; Preston et al. 2009). This variation in substrate utilization resulted in differences in the chemistry of decomposing organic matter, such that the same litter decomposing in different ecosystems exhibited variation in chemical changes over time. Other key ecosystem processes, including nitrification, denitrification, and methane oxidation are also influenced by microbial community structure (Cavigelli & Robertson 2000; Bengtson et al. 2009; Jia & Conrad 2009). Further, changes in soil animals can also have an overriding influence on ecosystem functions (Hättenschwiler et al. 2005; Ekschmitt et al. 2005). For example, microbial grazing by collembolans and other microarthropods can influence microbial community structure as well as nutrient cycling (Newell 1984; Wolters 2000). Clearly, new insights into the soil biota will improve our understanding of soil ecosystem resilience, and long-term experiments provide the ideal environment for studying the soil biota and its role in ecosystem change.

5.5.5.2 Long-term studies

Long-term soil research is essential to understanding soil ecosystem resilience and recovery but, unfortunately, remains rare (Poulton 1996; Robertson et al. 2004; Richter et al. 2007). This is due to a number of obstacles to long-term research including the expense of maintaining long-term sites, the short duration of most funding cycles, and the length of typical graduate degree programs. Further, in the agricultural sciences, the overwhelming preponderance of short-term studies also reflects the need to generate applied information that can be used by land managers. Nonetheless, the need for long-term soil studies is dictated by the nature of the disturbances humans impose on soils, as well as by key characteristics of soil itself.

Some of the greatest challenges faced by soils are from chronic disturbances such as changes in temperature and atmospheric CO_2 concentrations, as well as ecosystem N enrichment. The entire range of possible effects of these kinds of chronic, external forces can only be understood by making observations through time. For example, studies of nitrogen saturation in forest ecosystems reveal that although forests in the northeastern USA tend to be limited or co-limited by N, chronic N enrichment can induce long-term declines in productivity and increases in NO_3^- leaching (Magill et al. 2000; Aber et al. 2002). Such results cannot be discerned without long-term soil ecosystem monitoring. The rise in atmospheric CO_2 concentrations provides another example of how the nature of chronic disturbances requires long-term studies (Lichter et al. 2008). Unless dramatic cuts in fossil fuel emissions are achieved, this disturbance will increase in intensity in the coming decades. The long-term nature of this disturbance, the changes in SOM input quantities and quality that it can impose, and the altered climate with which it is associated combine to prompt recognition of the need for long-term, observational studies of soil responses.

Temperature effects on soil processes are another chronic disturbance that can only be determined long-term. For example, recent studies around the globe point to long-term declines in soil organic matter that cannot be attributed to changes in land use or patterns of soil disturbance but appear to be associated with climate changes (e.g. VandenBygaart et al. 2002; Bellamy et al. 2005; Varvel 2006; Stevens & van Wesemael 2008;). Recent research at W.K. Kellogg Biological Station LTER site in Michigan suggests that between the mid 1980s and 2006, soils of long-term, conventionally managed row-crop ecosystems that were considered at equilibrium have lost SOC. Further, SOC losses were even observed in no-till systems over this same period, although they were not as great as those in conventional till (Senthilkumar et al. 2009). Warmer temperatures observed during the past 20 years at the KBS LTER site and across Michigan in general may explain these declines in SOM (Senthilkumar et al. 2009; Andresen & Winkler 2009).

In addition to the nature of some chronic disturbances governing the need for long-term soil studies, characteristics of soils themselves dictate this need. Many soil properties only show changes on the timescale of years to decades following changes in patterns of soil disturbance. The slow-turnover rates of multiple SOM pools (Trumbore 2000), and their spatial variability, generate immense challenges for addressing many soil ecosystem questions (Robertson *et al.* 1997; Ettema & Wardle 2002). All of these factors show that studying SOM resilience requires an appreciation of the immense variation in time scales at which SOM dynamics take place (Fig. 5.5.2), and point to the need for more data sets generated from long-term soil studies. Because direct observation of soils across millennia is impossible, and is only rare across centuries or decades (Fig. 5.5.3), the archiving of soil samples can serve a critical means of assessing how soils change across temporal scales meaningful in "ecosystem time." Long-term soil experiments (Richter *et al.* 2007) are a critical means of achieving this objective.

The NSF Long-term Ecological Research Program (LTER) has provided many examples of the advantages of long-term soil research. At the W.K. Kellogg Biological Station row-crop LTER site a decade of monitoring showed the potential for certain land use practices to reduce agricultural trace gas emissions (Robertson *et al.* 2000) and thus, help mitigate atmospheric increases in trace gases. This long-term

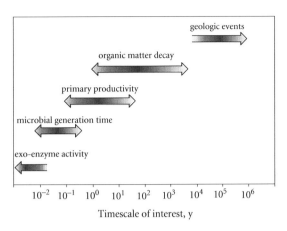

Figure 5.5.2 An adaptation of Janzen's (2004) conception of response time variation for key features related to soil organic matter dynamics. Similar variation in response times exist for soil organic matter responses to ecosystem disturbance.

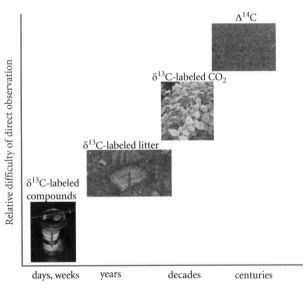

Figure 5.5.3 Conceptual relationship between time scales of interest for questions about SOM dynamics and the relative difficulty of directly observing those dynamics, displaying examples of radio- and stable isotopic experiments frequently employed for quantifying rates of C flow through SOM pools in laboratory and field settings. Data points associated with $\delta^{13}C$ experiments represent ^{13}C-labeling of compounds or of photosynthate; data point associated with $\Delta^{14}C$ represents archived soil samples from the 19th century in Rothamsted, England. $\Delta^{14}C$ assessments of these soils have generated estimates of turnover times of both labile and slow-turnover SOC pools.

monitoring program also helped identify N_2O as a major component of the greenhouse gas budget of managed soils. In a related study, Grandy *et al.* (2006) analyzed soil N_2O emissions from the KBS LTER site to determine whether responses to no-till management varied over 14 years. The effects of no-till on N_2O emissions remain contentious, and some studies have shown that no-till management increases N_2O emissions (Baggs *et al.* 2003), while other studies have found that no-till decreases N_2O emissions or has no effect (Elmi *et al.* 2003). The study by Grandy *et al.* (2006) found no difference between till and no-till, and also found no variation over time in the difference between till and no-till. This work contributed to our current understanding that the variability in N_2O responses to no-till depends on soil type, as well as the duration of no-till (Six *et al.* 2004). Additional studies at the KBS LTER site highlight long-term changes in SOM associated with different land use practices, and pinpoint the specific SOM pools that are particularly sensitive to management (Grandy & Robertson 2007; Syswerda *et al.* 2011).

The Rothamsted Experimental Station, U.K. has had long-term agricultural experiments in place for more than 150 years (Jenkinson 1991). These experiments have provided many key advances in understanding soil properties and processes in long-term agricultural sites. For example, the Rothamsted soil carbon model set the groundwork for many subsequent models based on SOM pools with different turnover times (Jenkinson & Rayner 1977). This model has been continually refined since its original publication, including recent efforts to link the conceptual model pools with measurable SOM pools (e.g. Skjemstad *et al.* 2004; Scharnagl *et al.* 2010). Other studies have shown changes in the organic chemistry of Rothamsted soils over a century and a half of management (e.g. Bull *et al.* 2000a,b), the turnover dynamics of subsoil C (Jenkinson *et al.* 2008), as well as changes in surface soil mineralogy due to different management practices (Tye *et al.* 2009). Perhaps Rothamsted's greatest achievement, however, has been in showing that soil fertility and agricultural productivity can be sustained for over 150 years with adequate use of either farmyard manure or inorganic nutrients (Poulton 1996), while also highlighting some of the environmental consequences of arable cropping (Goulding *et al.* 2000).

By focusing on soil biota and long-term experiments, we do not intend to imply that other fields of soil research are unimportant for developing key concepts of soil resilience following disturbance. Rather, we suggest that by pursuing these research foci, investigators necessarily will explore a broad diversity of questions, each of which could potentially add to emerging theories of soil ecosystem resilience. The remarkable array of studies that assess various facets of soil response to disturbances, both discrete and chronic, attests to the importance of the topic. Indeed, the argument can be made that successful progress within this discipline—soil resilience to and recovery from disturbance—will dictate humanity's ability to function well in the coming centuries.

References

Abd Latif, Z., & Blackburn, G.A. (2010) The effects of gap size on some microclimate variables during late summer and autumn in a temperate broadleaved deciduous forest. *International Journal of Biometeorology* **54**: 119–29.

Aber, J.D., & Melillo, J.M. (1991) *Terrestrial Ecosystems*. Saunders College Publishing, Philadelphia, PA.

Aber, J.D., Ollinger, S.V., Driscoll, C.T., *et al.* (2002) Inorganic nitrogen losses from a forested ecosystem in response to physical, chemical, biotic, and climatic perturbations. *Ecosystems* **5**: 648–58.

Adair, E.C., Parton, W.J., Del Grosso, S.J., *et al.* (2008) Simple three-pool model accurately describes patterns of long-term litter decomposition in diverse climates. *Global Change Biology* **14**: 2636–60.

Adl, S.M. (2008) Setting the tempo in land remediation: Short-term and long-term patterns in biodiversity recovery. *Microbes and Environments* **23**: 13–19.

Amundson, R., Guo, Y., & Gong, P. (2003) Soil diversity and land use in the United States. *Ecosystems* **6**: 470–82.

Anderson, D. (1988) The effect of parent material and soil development on nutrient cycling in temperate ecosystems. *Biogeochemistry* **5**: 71–97.

Andresen, J.A., & Winkler, J.A. (2009) Weather and Climate. In: R.J. Schatzl, T. Darden, & D. Brandt (eds.) *Michigan Geography and Geology*, pp. 288–314. Pearson Custom Publishing, Boston, MA.

Ayres, E., Dromph, K.M., & Bardgett, R.D. (2006) Do plant species encourage soil biota that specialise in the rapid decomposition of their litter? *Soil Biology & Biochemistry* **38**: 183–6.

Ayres, E., Steltzer, H., Berg, S., & Wall, D.H. (2009) Soil biota accelerate decomposition in high-elevation forests by specializing in the breakdown of litter produced by the plant species above them. *Journal of Ecology* **97**: 901–12.

Baer, S.G., Meyer, C.K., Bach, E.M., Klopf, R.P., & Six, J. (2010) Contrasting ecosystem recovery on two soil textures: implications for carbon mitigation and grassland conservation. *Ecosphere* **1**(1): art. 5.

Baggs, E.M., Stevenson, M., Pihlatie, M., Regar, A., Cook, H., & Cadisch, G. (2003) Nitrous oxide emissions following application of residues and fertiliser under zero and conventional tillage. *Plant and Soil* **254**: 361–70.

Bardgett, R. (2005) *The biology of soil: a community and ecosystem approach*. Oxford University Press, New York.

Beare, M.H. (1997) Fungal and bacterial pathways of organic matter decomposition and nitrogen mineralization in arable soils. In: Brussaard, L., & Ferrera-Cerrato, R. (eds.) *Soil Ecology in Sustainable Agricultural Systems*, pp. 37–70. Lewis Publishers, Boca Raton, FL.

Bedini, S., Avio, L., Argese, E., & Giovannetti, M. (2007) Effects of long-term land use on arbuscular mycorrhizal fungi and glomalin-related soil protein. *Agriculture Ecosystems & Environment* **120**: 463–6.

Beisner, B.E., Haydon, D.T., & Cuddington, K. (2003) Alternative stable states in ecology. *Frontiers in Ecology and the Environment* **1**: 376–82.

Bellamy, P.H., Loveland, P.J., Bradley, R.I., Lark, R.M., & Kirk, G.J.D. (2005) Carbon losses from all soils across England and Wales. *Nature* **437**: 245–8.

Bengtson, P., Basiliko, N., Dumont, M.G., *et al.* (2009) Links between methanotroph community composition and CH_4 oxidation in a pine forest soil. *Fems Microbiology Ecology* **70**: 356–66.

Bernhardt, E., Barber, J., & Pippen, J. (2006) Long-term effects of free air CO_2 enrichment (FACE) on soil respiration. *Biogeochemistry* **77**: 91–116.

Bever, J.D., Dickie, I.A., Facelli, E., *et al.* (2010) Rooting theories of plant community ecology in microbial interactions. *Trends in Ecology & Evolution* **25**: 468–78.

Billings, S.A., Buddemeier, R.W., Richter, D.D., Van Oost, K., & Bohling, G. (2010) A simple method for estimating the influence of eroding soil profiles on atmospheric CO_2. *Global Biogeochemical Cycles* **24**: GB2001.

Billings, S.A., & Richter, D.D. (2006) Changes in stable isotopic signatures of soil nitrogen and carbon during 40 years of forest development. *Oecologia* **148**: 325–33.

Billings, S.A., & Ziegler, S.E. (2008) Altered patterns of soil carbon substrate usage and heterotrophic respiration in a pine forest with elevated CO_2 and N fertilization. *Global Change Biology* **14**: 1025–36.

Bormann, F.H., & Likens, G.E. (1979) *Pattern and Process in a Forested Ecosystem: Disturbance, Development and the Steady State based on the Hubbard Brook Ecosystem Study*. Springer-Verlag, New York.

Bradford, M.A., Davies, C.A., Frey, S.D., *et al.* (2008) Thermal adaptation of soil microbial respiration to elevated temperature. *Ecology Letters* **11**: 1316–27.

Bradford, M.A., Watts, B.W., & Davies, C.A. (2010) Thermal adaptation of heterotrophic soil respiration in laboratory microcosms. *Global Change Biology* **16**: 1576–88.

Buckley, D.H., & Schmidt, T.M. (2001) The structure of microbial communities in soil and the lasting impact of cultivation. *Microbial Ecology* **42**: 11–21.

Buckley, D.H., & Schmidt, T.M. (2003) Diversity and dynamics of microbial communities in soils from agroecosystems. *Environmental Microbiology* **5**: 441–52.

Bull, I.D., Nott, C.J., van Bergen, P.F., Poulton, P.R., & Evershed, R.P. (2000a) Organic geochemical studies of soils from the Rothamsted Classical Experiments—VI. The occurrence and source of organic acids in an experimental grassland soil. *Soil Biology & Biochemistry* **32**: 1367–76.

Bull, I.D., van Bergen, P.F., Nott, C.J., Poulton, P.R., & Evershed, R.P. (2000b) Organic geochemical studies of soils from the Rothamsted classical experiments—V. The fate of lipids in different long-term experiments. *Organic Geochemistry* **31**: 389–408.

Burke, I.C., Lauenroth, W.K., & Coffin, D.P. (1995) Soil organic matter recovery in semiarid grasslands: implications for the conservation reserve program. *Ecological Applications* **5**: 793–801.

Buyer, J.S., Teasdale, J.R., Roberts, D.P., Zasada, I.A., & Maul, J.E. (2010) Factors affecting soil microbial community structure in tomato cropping systems. *Soil Biology & Biochemistry* **42**: 831–41.

Calderon, F.J., & Jackson, L.E. (2002) Rototillage, disking, and subsequent irrigation: Effects on soil nitrogen dynamics, microbial biomass, and carbon dioxide efflux. *Journal of Environmental Quality* **31**: 752–8.

Cambardella, C.A., & Elliott, E.T. (1994) Carbon and nitrogen dynamics of soil organic matter fractions from cultivated grassland soils. *Soil Science Society of America Journal* **58**: 123–30.

Carney, K.M., Hungate, B.A., Drake, B.G., & Megonigal, J.P. (2007) Altered soil microbial community at elevated CO_2 leads to loss of soil carbon. *Proceedings of the National Academy of Sciences of the United States of America* **104**: 4990–5.

CAST. (2004) *Climate Change and Greenhouse Gas Mitigation: Challenges and Opportunities for Agriculture Council for Agricultural Science and Technology*. Ames, Iowa, USA.

Cavigelli, M.A., & Robertson, G.P. (2000) The functional significance of denitrifier community composition in a terrestrial ecosystem. *Ecology* **81**: 1402–14.

Chaer, G., Fernandes, M., Myrold, D., & Bottomley, P. (2009) Comparative resistance and resilience of soil microbial communities and enzyme activities in adjacent native forest and agricultural soils. *Microbial Ecology* **58**: 414–24.

Chaudhary, V.B., Bowker, M.A., O'Dell, T.E., Grace, J.B., Redman, A.E., Rillig, M.C., & Johnson, N.C. (2009) Untangling the biological contributions to soil stability in semiarid shrublands. *Ecological Applications* **19**: 110–22.

Churchman, G.J., Foster, R.C., D'Acqui, L.P., *et al.* (2010) Effect of land-use history on the potential for carbon sequestration in an Alfisol. *Soil & Tillage Research* **109**: 23–35.

Cleveland, C.C., Townsend, A.R., Schmidt, S.K., & Constance, B.C. (2003) Soil microbial dynamics and biogeochemistry in tropical forests and pastures, southwestern Costa Rica. *Ecological Applications* **13**: 314–26.

Coleman, D.C., Crossley, D.A.J., & Hendrix, P.F. (2004) *Fundamentals of Soil Ecology*. Elsevier, New York.

Compton, J.E., & Boone, R.D. (2000) Long-term impacts of agriculture on soil carbon and nitrogen in New England forests. *Ecology* **81**: 2314–30.

Cragg, R.G., & Bardgett, R.D. (2001) How changes in soil faunal diversity and composition within a trophic group influence decomposition processes. *Soil Biology & Biochemistry* **33**: 2073–81.

David, M.B., Drinkwater, L.E., & McIsaac, G.F. (2010) Sources of nitrate yields in the Mississippi River Basin. *Journal of Environmental Quality* **39**: 1657–67.

Davidson, E.A., & Ackerman, I.L. (1993) Changes in soil carbon inventories following cultivation of previously untilled soil. *Biogeochemistry* **20**: 161–93.

Davidson, E.A., & Janssens, I.A. (2006) Temperature sensitivity of soil carbon decomposition and feedbacks to climate change. *Nature* **440**: 165–73.

DeGryze, S., Six, J., Paustian, K., Morris, S.J., Paul, E.A., & Merckx, R. (2004) Soil organic carbon pool changes following land-use conversions. *Global Change Biology* **10**: 1120–32.

Dijkstra, F.A., Hobbie, S.E., Knops, J.M.H., & Reich, P.B. (2004) Nitrogen deposition and plant species interact to influence soil carbon stabilization. *Ecology Letters* **7**: 1192–8.

Drinkwater, L.E., Schipanski, M., Snapp, S.S., & Jackson, L.E. (2008) Ecologically based nutrient management. In: Snapp, S. & Pound, B. (eds.) *Agricultural Systems: Agroecology and Rural Innovation for Development*, pp. 161–210. Academic Press, Elsevier, Burlington, MA.

Ekschmitt, K., Liu, M.Q., Vetter, S., Fox, O., & Wolters, V. (2005) Strategies used by soil biota to overcome soil organic matter stability—why is dead organic matter left over in the soil? *Geoderma* **128**: 167–76.

Elliott, E.T. (1986) Aggregate structure and carbon, nitrogen, and phosphorous in native and cultivated soils. *Soil Science Society of America Journal* **50**: 627–33.

Elmi, A.A., Madramootoo, C., Hamel, C., & Liu, A. (2003) Denitrification and nitrous oxide to nitrous oxide plus dinitrogen ratios in the soil profile under three tillage systems. *Biology and Fertility of Soils* **38**: 340–8.

Enwall, K., Throback, I.N., Stenberg, M., Soderstrom, M., & Hallin, S. (2010) Soil resources influence spatial patterns of denitrifying communities at scales compatible with land management. *Applied and Environmental Microbiology* **76**: 2243–50.

Ettema, C.H., & Wardle, D.A. (2002) Spatial soil ecology. *Trends in Ecology & Evolution* **17**: 177–83.

Evans, R.D., & Belnap, J. (1999) Long-term consequences of disturbance on nitrogen dynamics in an arid ecosystem. *Ecology* **80**: 150–60.

Feng, X.J., Simpson, A.J., Schlesinger, W.H., & Simpson, M.J. (2010) Altered microbial community structure and organic matter composition under elevated CO_2 and N fertilization in the Duke forest. *Global Change Biology* **16**: 2104–16.

Fierer, N., Bradford, M.A., & Jackson, R.B. (2007) Toward an ecological classification of soil bacteria. *Ecology* **88**: 1354–64.

Folke, C., Carpenter, S., Walker, B., *et al.* (2004) Regime shifts, resilience, and biodiversity in ecosystem management. *Annual Review of Ecology Evolution and Systematics* **35**: 557–81.

Foote, R.L., & Grogan, P. (2010) Soil carbon accumulation during temperate forest succession on abandoned low productivity agricultural lands. *Ecosystems* **13**: 795–812.

Fraterrigo, J.M., Balser, T.C., & Turner, M.G. (2006) Microbial community variation and its relationship with nitrogen mineralization in historically altered forests. *Ecology* **87**: 570–9.

Fraterrigo, J.M., & Rusak, J.A. (2008) Disturbance-driven changes in the variability of ecological patterns and processes. *Ecology Letters* **11**: 756–70.

Frolking, S., Palace, M.W., Clark, D.B., Chambers, J.Q., Shugart, H.H., & Hurtt, G.C. (2009) Forest disturbance and recovery: A general review in the context of spaceborne remote sensing of impacts on aboveground biomass and canopy structure. *Journal of Geophysical Research-Biogeosciences* **114**: G00E02.

Galloway, J.N., & Cowling, E.B. (2002) Reactive nitrogen and the world: 200 years of change. *Ambio* **31**: 64–71.

Goulding, K.W.T., Poulton, P.R., Webster, C.P., & Howe, M.T. (2000) Nitrate leaching from the Broadbalk Wheat Experiment, Rothamsted, UK, as influenced by fertilizer and manure inputs and the weather. *Soil Use and Management* **16**: 244–50.

Grandy, A.S., Loecke, T.D., Parr, S., & Robertson, G.P. (2006) Long-term trends in nitrous oxide emissions, soil nitrogen, and crop yields of till and no-till cropping systems. *Journal of Environmental Quality* **35**: 1487–95.

Grandy, A.S., & Robertson, G.P. (2006) Initial cultivation of a temperate-region soil immediately accelerates aggregate turnover and CO_2 and N_2O fluxes. *Global Change Biology* **12**: 1507–20.

Grandy, A.S., & Robertson, G.P. (2007) Land-use intensity effects on soil organic carbon accumulation rates and mechanisms. *Ecosystems* **10**: 58–73.

Grandy, A.S., Sinsabaugh, R.L., Neff, J.C., Stursova, M., & Zak, D.R. (2008) Nitrogen deposition effects on soil organic matter chemistry are linked to variation in enzymes, ecosystems and size fractions. *Biogeochemistry* **91**: 37–49.

Grover, H.D., & Musick, H.B. (1990) Shrubland encroachment in southern New Mexico, USA-an analysis of desertification processes in the American Southwest. *Climatic Change* **17**: 305–30.

Hättenschwiler, S., Tiunov, A.V., & Scheu, S. (2005) Biodiversity and litter decomposition in terrestrial ecosystems. *Annual Review of Ecology Evolution and Systematics* **36**: 191–218.

Haygarth, P.M., & Ritz, K. (2009) The future of soils and land use in the UK: Soil systems for the provision of land-based ecosystem services. *Land Use Policy* **26**: S187–97.

Hobbie, S.E., Reich, P.B., Oleksyn, J., Ogdahl, M., Zytkowiak, R., Hale, C., & Karolewski, P. (2006) Tree species effects on decomposition and forest floor dynamics in a common garden. *Ecology* **87**: 2288–97.

Holling, C.S. (1973) Resilience and stability of ecological systems. *Annual Review of Ecology and Systematics* **4**: 1–23.

Howard, A. (1945) *An Agricultural Testament*. Oxford University Press, New York.

Howarth, R.W., Billen, G., Swaney, D., et al. (1996) Regional nitrogen budgets and riverine N&P fluxes for the drainages to the North Atlantic Ocean: Natural and human influences. *Biogeochemistry* **35**: 75–139.

Jangid, K., Williams, M.A., Franzluebbers, A.J., et al. (2008) Relative impacts of land-use, management intensity and fertilization upon soil microbial community structure in agricultural systems. *Soil Biology & Biochemistry* **40**: 2843–53.

Janzen, H.H. (2004) Carbon cycling in earth systems—a soil science perspective. *Agriculture Ecosystems & Environment* **104**: 399–417.

Jastrow, J.D. (1996) Soil aggregate formation and the accrual of particulate and mineral-associated organic matter. *Soil Biology & Biochemistry* **28**: 665–76.

Jastrow, J.D., Boutton, T.W., & Miller, R.M. (1996) Carbon dynamics of aggregate-associated organic matter estimated by carbon-13 natural abundance. *Soil Science Society of America Journal* **60**: 801–7.

Jenkinson, D.S. (1991) The Rothamsted long-term experiments—are they still of use? *Agronomy Journal* **83**: 2–10.

Jenkinson, D.S., Poulton, P.R., & Bryant, C. (2008) The turnover of organic carbon in subsoils. Part 1. Natural and bomb radiocarbon in soil profiles from the Rothamsted long-term field experiments. *European Journal of Soil Science* **59**: 391–9.

Jenkinson, D.S., & Rayner, J.H. (1977) Turnover of soil organic matter in some of the Rothamsted classical experiments. *Soil Science* **123**: 298–305.

Jenny, H. (1980) *The Soil Resource: Origin and Behavior*. Ecological Studies 37. Springer-Verlag, New York.

Jia, Z.J., & Conrad, R. (2009) Bacteria rather than Archaea dominate microbial ammonia oxidation in an agricultural soil. *Environmental Microbiology* **11**: 1658–71.

Lal, R. (1997) Degradation and resilience of soils. *Philosophical Transactions of the Royal Society B: Biological Sciences* **352**: 997–1008.

Lal, R. (2004) Soil carbon sequestration impacts on global climate change and food security. *Science* **304**: 1623–7.

Lal, R. (2005) Soil erosion and carbon dynamics. *Soil & Tillage Research* **81**: 137–42.

Lal, R. (2010) Enhancing eco-efficiency in agro-ecosystems through soil carbon sequestration. *Crop Science* **50**: S120-31.

Lawrence, C.L., Neff, J.C., & Schimel, J.S. (2009) Does adding microbial mechanisms of decomposition improve soil organic matter models? A comparison of four models using data from a pulsed rewetting experiment. *Soil Biology & Biochemistry* **41**: 1923–34.

Lichter, J., Billings, S.A., Ziegler, S.E., et al. (2008) Soil carbon sequestration in a pine forest after nine years of atmospheric CO_2 enrichment. *Global Change Biology* **14**: 2910–22.

Litton, C.M., Ryan, M.G., Knight, D.H., & Stahl, P.D. (2003) Soil-surface carbon dioxide efflux and microbial biomass in relation to tree density 13 years after a stand replacing fire in a lodgepole pine ecosystem. *Global Change Biology* **9**: 680–96.

Lovett, G.M., & Rueth, H. (1999) Soil nitrogen transformations in beech and maple stands along a nitrogen deposition gradient. *Ecological Applications* **9**: 1330–44.

Lovett, G.M., Weathers, K.C., Arthur, M.A., & Schultz, J.C. (2004) Nitrogen cycling in a northern hardwood forest: Do species matter? *Biogeochemistry* **67**: 289–308

Magill, A.H., Aber, J.D., Berntson, G.M., et al. (2000) Long-term nitrogen additions and nitrogen saturation in two temperate forests. *Ecosystems* **3**: 238–53.

Mangan, S.A., Schnitzer, S.A., Herre, E.A., et al. (2010) Negative plant-soil feedback predicts tree-species relative abundance in a tropical forest. *Nature* **466**: 752–5.

Matson, P., Lohse, K.A., & Hall, S.J. (2002) The globalization of Nitrogen deposition: consequences for terrestrial ecosystems. *Ambio* **31**: 113–19.

Maul, J., & Drinkwater, L. (2010) Short-term plant species impact on microbial community structure in soils with long-term agricultural history. *Plant and Soil* **330**: 369–82.

McGill, W.B., & Cole, C.V. (1981) Comparative aspects of cycling of organic C, N, S and P through soil organic matter. *Geoderma* **26**: 267–86.

McLauchlan, K. (2006) The nature and longevity of agricultural impacts on soil carbon and nutrients: A review. *Ecosystems* **9**: 1364–82.

McSwiney, C.P., & Robertson, G.P. (2005) Nonlinear response of N_2O flux to incremental fertilizer addition in a continuous maize (*Zea mays* L.) cropping system. *Global Change Biology* **11**: 1712–19.

Melillo, J.M., Aber, J.D., & Muratore, J.F. (1982) Nitrogen and lignin control of hardwood leaf litter decomposition dynamics. *Ecology* **63**: 621–6

Miki, T., Ushio, M., Fukui, S., & Kondoh, M. (2010) Functional diversity of microbial decomposers facilitates plant coexistence in a plant-microbe-soil feedback model. *Proceedings of the National Academy of Sciences of the United States of America* **107**: 14251–6.

Mitchell, J.S., & Ruess, R.W. (2009) Seasonal patterns of climate controls over nitrogen fixation by *Alnus viridis* subsp *fruticosa* in a secondary successional chronosequence in interior Alaska. *Ecoscience* **16**: 341–51.

Montgomery, D.R. (2007) Soil erosion and agricultural sustainability. *Proceedings of the National Academy of Sciences of the United States of America* **104**: 13268–72.

Moorhead, D.L., & Sinsabaugh, R.L. (2006) A theoretical model of litter decay and microbial interaction. *Ecological Monographs* **76**: 151–74.

Neff, J.C., Reynolds, R.L., Belnap, J., & Lamothe, P. (2005) Multi-decadal impacts of grazing on soil physical and biogeochemical properties in southeast Utah. *Ecological Applications* **15**: 87–95.

Newell, K. (1984) Interaction between two decomposer Basidiomycetes and a collembolan under Sitka spruce: grazing and its potential effects on fungal distribution and litter decomposition. *Soil Biology & Biochemistry* **16**: 227–33.

Nusslein, K., & Tiedje, J.M. (1999) Soil bacterial community shift correlated with change from forest to pasture vegetation in a tropical soil. *Applied and Environmental Microbiology* **65**: 3622–6.

Perry, D.A. (1994) *Forest Ecosystems*. Johns Hopkins University Press, Baltimore, MD.

Peters, D.P.C., Pielke, R.A., Bestelmeyer, B.T., Allen, C.D., Munson-McGee, S., & Havstad, K.M. (2004) Cross-scale interactions, nonlinearities, and forecasting catastrophic events. *Proceedings of the National Academy of Sciences of the United States of America* **101**: 15130–5.

Pimental, D., & Sparks, D.L. (2000) Soil as an endangered ecosystem. *Bioscience* **50**: 947–947.

Poulton, P.R. (1996) The Rothamsted long-term experiments: Are they still relevant? *Canadian Journal of Plant Science* **76**: 559–71.

Preston, C.M., Nault, J.R., Trofymow, J.A., & Smyth, C. (2009) Chemical changes during 6 years of decomposition of 11 litters in some Canadian forest sites. Part 1. Elemental composition, tannins, phenolics, and proximate fractions. *Ecosystems* **12**: 1053–77.

Puget, P., Lal, R., Izaurralde, C., Post, M., & Owens, L. (2005) Stock and distribution of total and corn-derived soil organic carbon in aggregate and primary particle fractions for different land use and soil management practices. *Soil Science* **170**: 256–79.

Reeve, J.R., Schadt, C.W., Carpenter-Boggs, L., Kang, S., Zhou, J.Z., & Reganold, J.P. (2010) Effects of soil type and farm management on soil ecological functional genes and microbial activities. *Isme Journal* **4**: 1099–107.

Reich, P., Hungate, B., & Luo, Y. (2006) Carbon-nitrogen interactions in terrestrial ecosystems in response to rising atmospheric carbon dioxide. *Annual Review of Ecology Evolution and Systematics* **37**: 611–36.

Richter, D.D., Hofmockel, M., Callaham, M.A., Powlson, D.S., & Smith, P. (2007) Long-term soil experiments: Keys to managing Earth's rapidly changing ecosystems. *Soil Science Society of America Journal* **71**: 266–79.

Richter, D.D., Markewitz, D., Heine, P.R., et al. (2000) Legacies of agriculture and forest regrowth in the nitrogen of old-field soils. *Forest Ecology and Management* **138**: 233–48.

Richter, D.D., Markewitz, D., Trumbore, S.E., & Wells, C.G. (1999) Rapid accumulation and turnover of soil carbon in a re-establishing forest. *Nature* **400**: 56–8.

Richter, D.D., & Markewitz, V. (2001) *Understanding Soil Change*. Cambridge University Press, Cambridge.

Ritter, E. (2005) Litter decomposition and nitrogen mineralization in newly formed gaps in a Danish beech (*Fagus sylvatica*) forest. *Soil Biology & Biochemistry* **37**: 1237–47.

Robertson, G.P., Broome, J.C., Chornesky, E.A., et al. (2004) Rethinking the vision for environmental research in US agriculture. *Bioscience* **54**: 61–5.

Robertson, G.P., Klingensmith, K.M., Klug, M.J., Paul, E.A., Crum, J.R., & Ellis, B.G. (1997) Soil resources, microbial activity, and primary production across an agricultural ecosystem. *Ecological Applications* **7**: 158–70.

Robertson, G.P., Paul, E.A., & Harwood, R.R. (2000) Greenhouse gases in intensive agriculture: Contributions of individual gases to the radiative forcing of the atmosphere. *Science* **289**: 1922–5.

Royer-Tardif, S., Bradley, R.L., & Parsons, W.F.J. (2010) Evidence that plant diversity and site productivity confer stability to forest floor microbial biomass. *Soil Biology & Biochemistry* **42**: 813–21.

Sanchez, P.A. (2002) Ecology—Soil fertility and hunger in Africa. *Science* **295**: 2019–20.

Scharenbroch, B.C., & Bockheim, J.G. (2008) The effects of gap disturbance on nitrogen cycling and retention in late-successional northern hardwood-hemlock forests. *Biogeochemistry* **87**: 231–45.

Scharnagl, B., Vrugt, J.A., Vereecken, H., & Herbst, M. (2010) Information content of incubation experiments for inverse estimation of pools in the Rothamsted carbon model: a Bayesian perspective. *Biogeosciences* **7**: 763–76.

Scheffer, M., Carpenter, S., Foley, J.A., Folke, C., & Walker, B. (2001) Catastrophic shifts in ecosystems. *Nature* **413**: 591–6.

Schimel, J.P., & Bennett, J. (2004) Nitrogen mineralization: Challenges of a changing paradigm. *Ecology* **85**: 591–602.

Schimel, J.P., & Weintraub, M.N. (2003) The implications of exoenzyme activity on microbial carbon and nitrogen limitation in soil: a theoretical model. *Soil Biology & Biochemistry* **35**: 549–63.

Schlesinger, W.H. (1997) *Biogeochemistry: An Analysis of Global Change*. Academic Press, New York.

Schlesinger, W.H., Bernhardt, E.S., DeLucia, E.H., et al. (2006) The Duke Forest FACE experiment: CO_2 enrichment of a loblolly pine forest. In: J. Nosberger, S.P. Long, R.J. Norby, M. Stitt, G.R. Hendrey, & H. Blum (eds.) *Managed Ecosystems and CO_2: Case Studies, Processes, and Perspectives, Ecological Studies*, Vol. 187, pp. 197–212. Springer, Berlin.

Schlesinger, W.H., Reynolds, J.F., Cunningham, G.L., et al. (1990) Biological feedbacks in global desertification. *Science* **247**: 1043–8.

Senthilkumar, S., Basso, B., Kravchenko, A.N., & Robertson, G.P. (2009) Contemporary evidence of soil carbon loss in the US corn belt. *Soil Science Society of America Journal* **73**: 2078–86.

Seybold, C.A., Herrick, J.E., & Brejda, J.J. (1999) Soil resilience: A fundamental component of soil quality. *Soil Science* **164**: 224–34.

Sinsabaugh, R.L. (1994) Enzymatic analysis of microbial pattern and process. *Biology and Fertility of Soils* **17**: 69–74

Six, J., Conant, R.T., Paul, E.A., & Paustian, K. (2002) Stabilization mechanisms of soil organic matter: Implications for C-saturation of soils. *Plant and Soil* **241**: 155–76.

Six, J., Ogle, S.M., Breidt, F.J., Conant, R.T., Mosier, A.R., & Paustian, K. (2004) The potential to mitigate global warming with no-tillage management is only realized when practised in the long term. *Global Change Biology* **10**: 155–60.

Skjemstad, J.O., Spouncer, L.R., Cowie, B., & Swift, R.S. (2004) Calibration of the Rothamsted organic carbon turnover model (RothC ver. 26.3), using measurable soil organic carbon pools. *Australian Journal of Soil Research* **42**: 79–88.

Smeck, N.E. (1985) Phosphorus dynamics in soils and landscapes. *Geoderma* **36**: 185–99.

Smith, M.D., Knapp, A.K., & Collins, S.L. (2009) A framework for assessing ecosystem dynamics in response to chronic resource alterations induced by global change. *Ecology* **90**: 3279–89.

Smith, S.E., & Read, D.J. (1997) *Mycorrhizal Symbiosis*. Academic Press, San Diego, CA.

Solomon, S., Qin, D., Manning, M., et al. (eds.) (2007) *Contribution of Working Group I to the Fourth Assessment Report of the Intergovernmental Panel on Climate Change*. Cambridge University Press, New York.

Stallard, R.F. (1998) Terrestrial sedimentation and the carbon cycle: coupling weathering and erosion to carbon burial. *Global Biogeochemical Cycles* **12**:231–57.

Stark, C.H., Condron, L.M., O'Callaghan, M., Stewart, A., & Di, H.J. (2008) Differences in soil enzyme activities, microbial community structure and short-term nitrogen mineralisation resulting from farm management history and organic matter amendments. *Soil Biology & Biochemistry* **40**: 1352–63.

Stevens, A., & van Wesemael, B. (2008) Soil organic carbon dynamics at the regional scale as influenced by land use history: a case study in forest soils from southern Belgium. *Soil Use and Management* **24**: 69–79.

Strickland, M.S., Callaham, M.A., Davies, C.A., et al. (2010) Rates of *in situ* carbon mineralization in relation to land-use, microbial community and edaphic characteristics. *Soil Biology & Biochemistry* **42**: 260–9.

Strickland, M.S., Osburn, E., Lauber, C., Fierer, N., & Bradford, M.A. (2009) Litter quality is in the eye of the beholder: initial decomposition rates as a function of inoculum characteristics. *Functional Ecology* **23**: 627–36.

Stursova, M., Crenshaw, C.L., & Sinsabaugh, R.L. (2006) Microbial responses to long-term N deposition in a semiarid grassland. *Microbial Ecology* **51**: 90–8.

Suding, K.N., Ashton, I.W., Bechtold, H., Bowman, W.D., Mobley, M.L., & Winkleman, R. (2008) Plant and microbe contribution to community resilience in a directionally changing environment. *Ecological Monographs* **78**: 313–29.

Syswerda, S.P., Corbin, A.T., Mokma, D.L., Kravchenko, A.N., & Robertson, G.P. (2011) Agricultural management and soil carbon storage in surface vs. deep layers. *Soil Science Society of America Journal* **75**: 92–101.

Tiemann, L.K., & Billings, S.A. (2011) Indirect effects of nitrogen amendments on organic substrate quality increase enzymatic activity driving decomposition in a mesic grassland. *Ecosystems* **14**: 234–47.

Tiessen, H., Stewart, J.W., & Bettany, J.R. (1982) Cultivation effects on the amounts and concentrations of carbon, nitrogen, and phosphorus in grassland soils. *Agronomy Journal* **74**: 831–5.

Treseder, K.K. (2008) Nitrogen additions and microbial biomass: a meta-analysis of ecosystem studies. *Ecology Letters* **11**: 1111–20.

Trumbore, S. (2000) Age of soil organic matter and soil respiration: Radiocarbon constraints on belowground C dynamics. *Ecological Applications* **10**: 399–411.

Tye, A.M., Kemp, S.J., & Poulton, P.R. (2009) Responses of soil clay mineralogy in the Rothamsted Classical Experiments in relation to management practice and changing land use. *Geoderma* **153**: 136–46.

UN Millennium Project (2005) *Halving Hunger: It Can Be Done*. Summary version of the report of the Task Force on Hunger. The Earth Institute at Columbia University, New York.

Valone, T.J., Meyer, M., Brown, J.H., & Chew, R.M. (2002) Timescale of perennial grass recovery in desertified arid grasslands following livestock removal. *Conservation Biology* **16**: 995–1002.

van Diepen, L.T.A., Lilleskov, E.A., Pregitzer, K.S., & Miller, R.M. (2010) Simulated nitrogen deposition causes a decline of intra- and extraradical abundance of arbuscular mycorrhizal fungi and changes in microbial community structure in northern hardwood forests. *Ecosystems* **13**: 683–95.

Van Miegroet, H., & Cole, D.W. (1984) The impact of nitrification on soil acidification and cation leaching in a red alder ecosystem. *Journal of Environmental Quality* **13**: 586–90.

Van Miegroet, H., & Johnson, D.W. (2009) Feedbacks and synergism among biogeochemistry, basic ecology, and forest soil science. *Forest Ecology and Management* **258**: 2214–23.

Van Oost, K., Quine, T.A., Govers, G., et al. (2007) The impact of agricultural soil erosion on the global carbon cycle. *Science* **318**: 626–9.

VandenBygaart, A.J., Yang, X.M., Kay, B.D., & Aspinall, J.D. (2002) Variability in carbon sequestration potential in no-till soil landscapes of southern Ontario. *Soil & Tillage Research* **65**: 231–41.

Varvel, G.E. (2006) Soil organic carbon changes in diversified rotations of the western corn belt. *Soil Science Society of America Journal* **70**: 426–33.

Vitousek, P.M., Hättenschwiler, S., Olander, L., & Allison, S. (2002) Nitrogen and nature. *Ambio* **31**: 97–101.

Vitousek, P.M., & Melillo, J.M. (1979) Nitrate losses from disturbed forests: patterns and mechanisms. *Forest Science* **25**: 605–19.

Walker, B., Hollin, C.S., Carpenter, S.R., & Kinzig, A. (2004) Resilience, adaptability and transformability in social-ecological systems. *Ecology and Society* **9**: art. 5.

Wedin, D.A., & Tilman, D. (1990) Species effects on nitrogen cycling: a test with perennial grasses. *Oecologia* **84**: 433–41.

West, T.O., & Post, W.M. (2002) Soil organic carbon sequestration rates by tillage and crop rotation: a global data analysis. *Soil Science Society of America Journal* **66**: 1930–46.

White, A.S., & Pickett, S.T.A. (1985) Natural disturbance and patch dynamics: An introduction. In: S.T.A. Pickett, & P.S. White (eds.) *The Ecology of Natural Disturbance and Patch Dynamics*, pp. 3–13. Academic Press, New York.

Wickings, K., Grandy, A.S., Reed, S., & Cleveland, C. (2010) Management intensity alters decomposition via biological pathways. *Biogeochemistry* **104**: 365–79.

Wolters, V. (2000) Invertebrate control of soil organic matter stability. *Biology and Fertility of Soils* **31**: 1–19.

Zak, D.R., Holmes, W.E., Burton, A.J., Pregitzer, K.S., & Talhelm, A.F. (2008) Simulated atmospheric NO_3^- deposition increases soil organic matter by slowing decomposition *Ecological Applications* **18**: 2016–27.

CHAPTER 5.6

Applying Soil Ecological Knowledge to Restore Ecosystem Services

Sara G. Baer, Liam Heneghan, and Valerie T. Eviner

5.6.1 Introduction

Environmental degradation resulting from resource extraction, land-use change, and invasion by exotic species alters numerous functions and services provided by intact and unexploited ecosystems. Ecological restoration is the human-facilitated improvement of a degraded ecosystem and represents an important means to repair or reinstate many natural services in degraded ecosystems (Table 5.6.1). Restoration can be initiated from any point along a continuum of degradation and restoration goals can vary from focused improvements (e.g. amelioration of unsuitable pH, soil stabilization, and re-establishing rare species) to holistic recovery of biological diversity, efficient nutrient cycling, and complex energy flow pathways (Hobbs & Harris 2001). Restorations that aim to re-establish ecosystem structure and function "prior to degradation" may find that historical and extant targets can be difficult to define considering the variability of natural systems in space and time (White & Walker 1997) or unrealistic to attain if multiple abiotic and/or biotic factors have been highly modified through human activity (Hilderbrand et al. 2005). Furthermore, restorations are now conducted under novel conditions including invasive species pressure, greater inputs of nutrients through atmospheric deposition, and higher atmospheric CO_2 levels that may restrict the ability to restore systems to some state in the past (Hobbs & Harris 2001; Hobbs et al. 2009).

Degradation of terrestrial ecosystems is often strongly reflected as damage to the soil system. It can take tens to thousands of years for some soil properties to develop through the interaction of parent material, climate, topography, and organisms (Jenny 1941). Soil degradation can result in the loss of or alteration to many soil properties and functions. We define *soil legacy* as the physical, chemical, and biological attributes and interactions that remain following a significant change to an ecosystem. This definition of legacy is aligned with *ecological legacy* (White & Jentsch 2004), but differs from *disturbance legacy*, which has been used to indicate the residual effects of an abiotic or biotic disturbance on ecosystem properties (e.g. Reinhart & Callaway 2006). Disturbance legacy, Soil legacy is the degree to which soil properties (e.g. horizonation, porosity, texture, nutrient storage, organic matter content, aggregation, etc.) and functions (e.g. nutrient supply, infiltration, etc.) at the onset of restoration reflect characteristics before the degrading influence.

The range of variation in ecosystem degradation produces varying soil legacy at the onset of restoration. Heneghan et al. (2008) proposed that more soil ecological knowledge (defined as the integrated understanding of soil physical, chemical, and biological factors and processes in the context of plant-soil feedback) may be required to restore complex interactions following disturbance (Fig. 5.6.1). This chapter presents the utility of soil ecological knowledge to the practice of ecological restoration along a continuum of ecosystem degradation that results in varying legacy of the plant and/or soil system. Mineral resource extraction can result in severe ecosystem degradation that leaves *little to no soil legacy*, initially. Restoration of these highly degraded lands

Soil Ecology and Ecosystem Services. First Edition. Edited by Diana H. Wall et al.
© 2012 Oxford University Press. Published 2012 by Oxford University Press.

Table 5.6.1 Ecosystem goods and services provided through ecological restoration.

Ecosystem service[a]	How ecological restoration can provide ecosystem goods (functions) and services
Gas and climate regulation	Revegetation of degraded lands through reforestation, grassland establishment, and wetland restoration can aid in mitigating atmospheric CO_2; conversion of agricultural land to perennial vegetation can reduce N_2O emissions if nitrification rates are reduced.
Disturbance regulation	Restoration of degraded lands to perennial vegetation increases transpiration of water to the atmosphere and improved soil structure promotes infiltration to reduce runoff and provide flood control.
Water supply	Improved soil structure promotes infiltration to groundwater; hydrologic manipulations in wetland and floodplain restorations can promote groundwater recharge.
Erosion control and sediment retention	Conversion of highly erodible arable lands and riparian buffer zones to perennial vegetation reduces erosion and traps sediment.
Soil formation	Restoration of degraded lands to perennial vegetation promotes organic matter accrual; organic acids from root exudates and decomposition can facilitate weathering and soil formation.
Nutrient cycling	Establishing plants associated with N-fixing microorganisms on degraded lands can promote N and organic matter accrual; developing root systems, microbial biomass, and organic matter during restoration of degraded soil can promote nutrient conservation; conversion of agricultural lands to perennial vegetation reduces nutrient inputs and pollution.
Biological control	Perennial vegetation restored within agricultural landscapes can provide refugia for predators of crop pests.
Pollination	Floristically diverse restorations can provision pollinators for plant populations.
Refugia	Restoration can reduce landscape fragmentation and provide resources and/or reproduction habitat requirements for local and transient wildlife populations.
Food production	Restoration of degraded land and wetlands with perennial vegetation can increase game populations.
Raw materials	Reforestations can be managed for production of lumber; perennial grasslands can be used for biofuel and forage production.
Genetic resources	Restoration conducted with high fidelity to local gene pools represents a means to preserve genetic variation for medicinal purposes or crop improvement (e.g. resistance to plant pathogens).
Recreation	Forest, wetland, and grassland restoration increase habitat for wildlife and opportunity for hunting; restoration can improve water quality (increase clarity through less sedimentation, reduced eutrophication through nutrient abatement) and conditions for fish populations; restorations provide areas for hiking and nature appreciation.
Cultural	Restorations and re-creations, especially when conducted to achieve a historic community, can preserve cultural heritage.

[a] List modified from Costanza et al. (1997).

has revealed the importance of physical, chemical, and biological properties of soil to revegetation, as well as the role of soil heterogeneity in providing refugia for the recolonization of soil biota. Addition of topsoil to mined land can rapidly increase the amount of soil legacy at the onset of restoration with consequence for improved ecosystem functions. Restoration of agricultural systems often represents *moderate soil legacy* at the onset of restoration. Soil degradation resulting from agriculture varies with the type of production system. Most development and application of soil ecological knowledge to the restoration of agricultural systems has been gleaned from those that have been cultivated. Although recovery of many aspects of soil structure and function can proceed passively following revegetation of cultivated soils, there are increasing efforts to restore biological and physical complexity to better represent historic or extant systems. Restoring biodiversity and structural complexity may

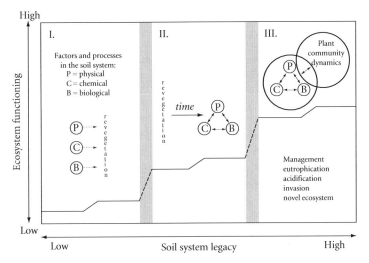

Figure 5.6.1 Relationship between the degree of soil legacy in degraded lands (including interaction among physical (P), chemical (C), and biological (B) properties and processes) and ecosystem functioning in the context of restoration. Restoration is human-facilitated improvement of a degraded ecosystem's functioning (indicated by stair-step line). Abiotic and biotic constraints may need to be alleviated to achieve proportionally greater improvement in ecosystem functioning (indicated by the dashed line within gray windows). Three general circumstances of varying soil legacy are depicted, although a continuum exists within the realm of each example (regions I-III). *Region I* represents restoration of highly disturbed lands (e.g. some mine land reclamations), where P, C, and/or B properties and processes may be disconnected, and reconnection of these components is required to restore vegetation. *Region II* depicts restoration from disturbance such as agriculture, where P, C, and B properties and processes may be present and interactive, but altered in status and/or composition. Revegetation drives the recovery of P, C, and B properties and processes towards a target state in region II, but recovery may be constrained by limited species, development of complex plant-soil feedback, or altered representation of higher trophic levels (e.g. aboveground herbivores). *Region III* represents restoration scenarios where soil legacy may be significant, but plant communities are undergoing dynamic change (invasion) or have been replaced by self-perpetuating assemblages of species that have never before coexisted (novel ecosystems). In Region III, aspects of soil biota, nutrient status, and/or biogeochemical processes may be highly altered, but P, C, and B properties and interactions are considered more complex and tightly coupled to aboveground community dynamics. (Modified from Heneghan *et al.* 2008, with permission from John Wiley & Sons.)

require more soil ecological knowledge and an explicit focus on plant–soil feedbacks (where a plant species or community alters soils in a way that impacts its own success as well as that of other species). These plant–soil feedbacks can be key mechanisms by which species become invasive. Thus, ecosystems under dynamic change through invasion may contain high soil legacy that has become altered in subtle ways. Restoration of these systems may require considerable soil ecological knowledge to establish rare, historically dominant, or a diversity of species that depend on specific or historic soil properties, processes, and/or biotic composition (Heneghan *et al.* 2008; Eviner *et al.* 2010), particularly if that system has attained a self-organizing alternative stable state (Suding *et al.* 2004). In no-analog environments, large shifts in the abiotic and biotic components of an ecosystem have occurred, resulting in new and self sustaining assemblages of plant species that have never before coexisted and perhaps a novel soil legacy. In these novel systems, it is critical to recognize that the new communities can provide ecosystem services (i.e. beneficial functions, as defined by the Millennium Ecosystem Assessment 2005), and attempts to restore these systems to a former state could result in failure and compromise ecosystem services provided.

Disturbance to the soil system has potential to alter ecosystem services (Daily *et al.* 1997) such as efficient and conservative nutrient cycling, soil formation, and/or water holding capacity (to name only a few) that are provided by structurally heterogeneous soils with a thriving biota. In the absence of restoration, soil structure and functioning, as well as ecosystem services, would remain in a degraded state, continue to decline, or recover slowly (Insam

Table 5.6.2 Knowledge or manipulation of soil used to promote revegetation, recovery of soil properties or processes, and/or biodiversity during restoration with consequences for improving ecosystem functions and services listed in Table 5.6.1. This list largely reflects documented changes and successful responses to manipulations; it should be recognized that failures commonly go unreported and inconsistent responses may exist.

Soil property	Knowledge or manipulation during ecological restoration.
Structure	Severe soil compaction limits root growth and water infiltration. Soil ripping has been applied in extreme circumstances to promote establishment of vegetation (Ashby 1997)
	In soils degraded through cultivation, development of soil macroaggregates corresponds to soil microbial community recovery, particularly arbuscular mycorrhizae fungi in restored grassland (Jastrow et al. 1998; Bach et al. 2010)
Texture	Soil texture modulates recovery of soil properties (C and N accrual), microbial biomass, aggregate structure, and nutrient transformations during grassland restoration, with faster recovery occurring in soil with high clay content (Baer et al. 2010)
	Tropical forest regrowth is influenced by soil texture according to coarse soil classifications and soils with higher clay content exhibit faster forest regrowth than those with higher sand content and nutrient leaching (Lu et al. 2002)
Organic matter	Addition of organic chelators can enhance phytoextraction of metals in contaminated soils (Huang & Cunningham 1996)
	Low levels of soil organic matter can result in inadequate nutrient supply to plants. Addition of organic waste, humus material, and topsoil are used to increase organic matter and provide slow release nutrients in mined land (Bradshaw & Chadwick 1980)
	Allelopathic compounds are often organic, and application activated C has been shown to effectively reduce extractable organic C and N in soil (Kulmatiski & Beard 2006)
Nutrient status	Highly degraded soils may require nutrient inputs (fertilizer application) to support vegetation growth. Leguminous tree species have been used to revegetate areas degraded from mining in the tropics without addition of organic soils, but supplements of rock phosphate, gypsum, micronutrients and K are required (Franco & DeFaria 1997)
	High nutrient availability often corresponds with low plant species richness and dominance by a few species that monopolize resources. Cutting followed by haying, cropping addition of Al or Fe (ferric) sulfate, deep tillage and incorporation of inert material has been used to reduce or dilute soil P (Walker et al. 2004). Carbon amendments have been used to reduce available N and non-target species (Blumenthal et al. 2003)
pH	Soil pH affects plant growth through its regulation of nutrient solubility and metal toxicity. In general, soil with pH around 6.5 contains the highest nutrient availability and lowest metal toxicity (Wong 2003)
	Elemental S and pyritic peat have been used to lower pH of arable land to promote establishment of acid requiring plant communities of high conservation value (Owen et al. 1999)
Heterogeneity	Heterogeneous soils provide refugia for recolonization of soil flora and fauna in contaminated sites (Eijsackers 2004); soil heterogeneity can decrease during restoration, particularly if it is imparted by transient biota, i.e. ants (Lane & BassiriRad 2005)
	Microtopography can dictate variation small-scale soil heterogeneity that may influence plant community composition (Flinn & Marks 2007) and differentially affect recovery rates of belowground ecosystem properties and processes during restoration (Meyer et al. 2008)
	Manipulating soil heterogeneity can promote plant community heterogeneity (Baer et al. 2004)
Soil biota	Microbial biomass increases in response to soil organic matter accrual (Insam & Domsch 1988). Total richness and biomass of coarsely defined microbial groups (i.e. based on phospholipid fatty acid biomarkers) increases in response to establishing perennial vegetation on former cropland, as related to time and/or re-development of macroaggregates (Matamala et al. 2008; Bach et al. 2010)
	Re-application of topsoil following mining aims to provide symbiotic microorganisms needed for restoration of shrubs on mined land (Paschke et al. 2003); plant-dependent soil organisms (e.g. mycorrhizae) can be compromised with consequence for restoration of some plant species if topsoil is stockpiled (Mummey et al. 2002)
	Inoculating soil dominated by invasive species with soil derived from native plant communities can reduce invasive species cover and promoted native perennial species (Rowe et al. 2009)
	Soil fauna have been used as indicators of recovery of other soil properties, e.g. recolonization of restored sites by ants and correlation with microbial biomass (Andersen & Sparling 1995)

& Haselwandter 1989). Restoration can facilitate recovery or reinstate numerous ecosystem services in degraded lands and landscapes (Table 5.6.1) largely through revegetation efforts, and establishing key species that promote a specific ecosystem service (Eviner & Chapin 2001). This chapter synthesizes how knowledge of soil processes, belowground heterogeneity, roles of soil biota, controls by state factors, and potential feedbacks with the aboveground community can be applied to promote restoration success (Table 5.6.2).

5.6.2 Low to high legacy: lessons from restoration of mined land

Soil ecological knowledge is an asset to restoring severely disturbed ecosystems, particularly with respect to understanding the role of soil properties and plant–soil relationships that promote revegetation. Mineral resources exist underground and their extraction requires the removal of vegetation and soils, which can result in complete environmental destruction; however, a continuum of ecosystem degradation from mining activity does exist.

Mine workings (materials produced by mining) generally contain factors that limit ecosystem development over the short term in the absence of reclamation practices (Bradshaw 1997). In the extreme case, deposition of waste rocks (i.e. host rock with insufficient mineral amounts for extraction) and mine tailings (i.e. remnant slurry post mineral extraction) at the surface become sources of pollution, eliminate biodiversity, and drastically reduce the production capacity of the environment. Restoration of mine workings to an improved ecosystem state has elucidated that physically and chemically altered soils are critical constraints to plant growth. Transport of materials by heavy machinery inevitably results in severe soil compaction, and mechanical soil treatment (e.g. soil ripping) is commonly applied prior to revegetation to loosen compacted soil and promote root penetration (Ashby 1997). Mine waste materials can include weathered subsoils and overburdens with deficiencies in mineral amounts or unweathered to the extent that nutrients cannot be released quickly enough to sustain plant growth. Thus, nutrient deficiencies must be determined. Fertilizers or amendments (topsoil, residues, sewage sludge, etc.) are commonly added to alleviate nutrients that are too limiting to sustain plant growth (Bradshaw 1997; Wong 2003).

Knowledge of physico-chemical properties of soil is required for restoration of extremely disturbed mined environments. Mining for coal and metal ores results in the oxidation of sulphide minerals, and consequently, severe acidification of soil (pH 2–2.5; Bradshaw & Chadwick 1980; Bradshaw 1997). Soil pH may be one of the most important properties to consider in mined land restoration because it influences the solubility of nutrients and metal ions, activity of soil microorganisms and fauna, and pH-sensitive processes such as nitrification and nitrogen fixation (van Breeman 2004). Remedying pH levels requires knowledge of soil acidity, the potential acidity that can arise from further oxidation of metal sulfides, and the acid neutralizing capacity of the soil to determine effective amounts of calcium carbonate (lime) to increase and maintain pH for successful restoration (Costigan et al. 1981). Waste materials produced from the mining for heavy metals can contain significant amounts of residual metals, even after processing, with little to no potential for natural reduction (Li 2006). The concentration and chemical forms of metals (as affected by pH) determine soil toxicity. Many heavy metal ions exhibit reduced solubility at low pH; however, the solubility (i.e. availability) of essential metals (e.g. molybdenum) decreases at high pH and results in metal deficiencies (van Breeman 2004). Organic acids derived from soil organic matter and root exudates represent important sources of chelators present at low levels in mined soils. Phytoextraction of metals using hyperaccumulating plant species, coupled with the addition of metal-chelating compounds can enhance metal uptake and removal to 1) render the soil more suitable for growth of microorganisms that are inhibited by heavy metals (e.g. rhizobia), 2) promote development of symbioses important for plant growth (e.g. legume-rhizobia), and 3) facilitate desired plant communities (Wong 2003).

Heavy metals and persistent organic pollutants can render soil toxic to belowground organisms and plants. Physicochemical properties of soil are important for decontamination of organic pollutants, which bind to soil constituents (e.g. clay, organic matter, organomineral complexes) over time. Once contamination levels begin to decline, microbial populations can recover. For example, Yin et al. (2000) found that the arrival of bacterial species along a restoration gradient that included a mine spoil, restored mined land, and undisturbed forest was dependent on time since disturbance. Refugia for soil organisms, provided by uncontaminated soil microsites in physically heterogeneous soil, can serve as important sources of soil flora and fauna for recolonization (Eijsackers 2004) that might be necessary for some plants to establish. Delayed arrival of microbial species could represent a constraint to restoring plant diversity (Harris 2003).

Recognition that soil health (vital whole soil) is imperative for successful revegetation has resulted in formulation of policy in many countries to conserve and replace surface soils in mined sites. Mined land amended with topsoil can, in some cases, represent a rapid transition from a low to moderate or even high soil legacy starting point for restoration. Topsoil amendments provide biological or biologically-produced soil components such as organic matter, stored nutrients, available nutrients, soil biota, and plant propagules (Brenner et al. 1984). Topsoil replacement, however, has resulted in varied restoration success and care must be taken in how it is managed. For example, if topsoil is added to a dissimilar or compacted underlying material, then hydraulic discontinuity and instability can result (Bradshaw 1997). Depth of topsoil addition can also influence the trajectory of plant community recovery. Paschke et al. (2003) reported that deep additions of topsoil encouraged grasses and forbs to outcompete mountain shrubs that are adapted to the historically shallow soils. "Live handled" topsoil is preferred over stockpiled topsoil to provision restorations with symbiotic soil organisms. Preservation of topsoil through long-term stockpiling has been shown to reduce soil organic matter (according to many of the same mechanisms as cultivation) and be detrimental to symbiotic arbuscular mycorrhizae fungi. These physical and biological changes to soil during long-term stockpiling, coupled with uniform re-application of the stored soil, has revealed important soil-related constraints to restoration. Mummey et al. (2002) documented that disturbance to soil structure (through removal and stockpiling) and absence of an associated plant community to sustain symbiotic microorganisms during stockpiling imposed long-lasting effects on vegetation and spatial organization of surface mined land. Failure of native shrubs to establish in stockpiled and re-applied topsoil impacted root distribution, water relations, fire cycles, and spatial resource heterogeneity in the restored mined lands. This study underscores the importance of soil physicochemical properties and the soil microbial community to ecosystem restoration. In some cases (e.g. shallow soil on steep terrain), topsoil conservation and replacement is difficult and mine spoils are substituted for topsoil, despite that their properties and growth of native species therein can be different than native soils (Showalter et al. 2010).

5.6.3 Moderate legacy: restoration of agricultural systems

Agriculture represents the most globally widespread anthropogenic influence on ecosystems (Ellis & Ramankutty 2008). The degree of soil perturbation from agriculture depends upon the type of production system (e.g. annual crops, pasture, vs. timber) and historic management (e.g. tillage and rotation regimes, fertilizer application, grazing intensity, and selective harvest vs. forest clearcutting). Restoration goals for agricultural systems vary considerably, partly due to differences in historical communities (i.e. grassland, forest, or wetland) that were converted to production systems. Soil ecological knowledge is increasingly considered to restore multiple ecosystem services in these systems, particularly those provided by biodiversity.

5.6.3.1 Restoration to grassland

Long-term conventionally cultivated systems generally contain no legacy of the historic plant community and limited potential for colonization of historic species from the regional species pool due to few native propagules in landscapes dominated by agriculture. It is well recognized that long-term cultivation degrades soil structure, promotes more extreme wet–dry cycles, alters soil microbial composition, increases decomposition, and lowers soil carbon (C) and other nutrients to new equilibrium levels (Dick 1992). Furthermore, long-term fertilized soils may exhibit high residual nutrient levels, high nitrification rates, and lower pH. In general, simple goals such as revegetation to reduce erosion can be easily achieved and concomitant improvements in soil structure and functions can proceed passively. The development of perennial grass root systems in long-term conventionally cultivated soil can promote soil macroaggregate formation, microbial and fungal biomass, and soil C accrual on a decadal time scale (Matamala et al. 2008), but recovery can vary drastically between highly contrasting soil textures (Bach et al. 2010; Baer et al. 2010). Paustian et al. (1998) reviewed the potential for agricultural systems to mitigate increasing atmospheric CO_2 and suggested that planting perennial vegetation such as grass represents one of the most favorable scenarios for increasing soil C stocks following land degradation. Erosion reduces land productivity and ability to maintain or improve C stocks in biomass and soil; therefore, measures to reduce soil erosion are very important. Despite improvement in numerous ecosystem services provided by simple erosion control measures (functional restorations), these practices generally do not restore biodiversity and associated ecosystem services (e.g. genetic resources, pollination, cultural heritage).

Restoration of some plant communities may require specific soil conditions. In Europe, liming has been used to raise pH of naturally acidic soils to promote crop production; however, high soil pH limits restoration of grassland and heathland communities that prefer acidic soil conditions. Applications of elemental S and pyritic peat have been used to effectively lower pH, promote establishment of species that require acidic soil conditions, and reduce ruderal species (Owen et al. 1999). High residual soil fertility from nutrient management in agricultural soils can constrain the restoration of diverse plant communities, particularly if soil fertility results in asymmetric competition for nutrients and dominance by a few species (Baer et al. 2003). High species coexistence is generally associated with low levels of soil fertility (Janssens et al. 1998), further supported by the consistent phenomenon of declines in plant diversity under nutrient enrichment. Manipulating soil fertility has been explored as a restoration tool to increase plant species diversity, reduce invasive or non-target species, and increase soil heterogeneity prior to, or during restoration. Walker et al. (2004) reviewed efforts to reduce soil fertility in arable lands to recreate species-rich grasslands in Europe. Measures used to reduce phosphorus (P) availability include haying or cropping to promote off-take of P, addition of aluminum (Al) or iron (Fe) (ferric) sulfate to increase P adsorption capacity, and deep cultivation or addition of inert or organic materials to dilute nutrient pools. In the UK, nutrient reduction through deep cultivation combined with seed addition has been shown to increase plant community similarity to grassland targets with conservation value on a decadal time scale (Walker et al. 2004). Carbon addition can be an effective tool to reduce inorganic nitrogen (N) availability by promoting the growth of the soil microbial biomass and immobilization of N, with significant consequence for reducing non-target plant cover during grassland restoration (Baer et al. 2003; Blumenthal et al. 2003). Reducing N availability at the onset of restoration may be a valuable tool to prevent invasive species establishment. However, the effectiveness of C addition is variable, and efficacy of such manipulations should be considered in the context of long-term temporal dynamics of N availability modulated by development of soil C stocks and increasing plant–soil feedback over time (Baer & Blair 2008).

5.6.3.2 Restoration to forest

Forest restoration on agricultural land has generally employed fewer soil manipulations relative to restoring grasslands, but there are studies that demonstrate soil properties affect forest regrowth and CO_2 mitigation. Soil fertility and land-use history (as it has affected soil) are critical factors influencing forest regrowth (Tucker et al. 1998). The tropics, in particular, are subject to a continuous cycle of forest clearing for agricultural purposes and socioeconomic constraints to maintaining the productive capacity of soil, which results in rapid soil degradation and further clearing for agriculture (Lavelle

1987). Soil C loss due to cultivation of tropical soils occurs at a much faster rate than in subhumid regions due to faster decomposition, which is further exacerbated in mountainous regions by erosion. Due to the amount of C lost from forest conversion to agriculture, Paustian *et al.* (1998) contended that the most significant opportunity for mitigating CO_2 emissions is to reduce the rate of tropical land conversion to agriculture. Change in soil C stocks in response to reforestation is not well documented for the tropics. Paul *et al.* (2010) found no change in soil C pools in response to reforestation of rainforest species, but improvement in many other soil properties (i.e., extractable inorganic N, plant available inorganic N, nitrification rate, pH, and bulk density) occurred on a decadal time scale.

Our understanding of soil-related factors that influence forest regrowth and/or composition on former agricultural lands comes mostly from studies of reforestation following land abandonment. For example, differential rates of forest regrowth following agricultural land-use in the tropics have been attributed to a suite of soil properties associated with different soil orders. Clay-rich Alfisols hold more nutrients and support faster forest regrowth relative to Ultisols and Oxisols with higher sand content and lower fertility (Lu *et al.* 2002). Over half of the forest cover throughout Europe and eastern North America occurs on former agricultural land (Vellend 2003). Residual impacts of cultivation on soil have been documented in secondary forest following 90–120 years of abandonment from agriculture. Compton and Boone (2000) demonstrated that formerly cultivated secondary forest sites contained less forest floor C, more mineral soil N and P, lower C:N and C:P ratios, and higher nitrification rates relative to sites that were selectively logged, with no cultivation disturbance to mineral soil. Furthermore, forest composition of abandoned agricultural lands can remain distinct from forests that have never been cleared for agriculture for centuries, particularly with respect to herbaceous richness (Flinn & Marks 2007). Although the cause of compositional variation among secondary forests on formerly cultivated soil and uncultivated soil has not been deduced, Flinn and Marks (2007) speculated that soil properties and processes play an important role, particularly in the reduction of microtopography, which influences small scale species distributions and the variable history of agricultural intensity and management among sites.

5.6.3.3 Restoration to wetland

Globally, over half of wetland area has been lost, most of which has been converted to agriculture. Costanza *et al.* (1997) estimated that shallow waters, which cover <2% of the Earth's surface, provide up to 40% of global renewable ecosystem services; thus, restoration of these systems has important consequences for ecosystem services. Restoration of wetland systems generally involves restoring or manipulating hydrology to promote recovery of ecosystem services such as disturbance regulation, water supply, sediment retention, and nutrient cycling (Table 5.6.1). However, the vast array of restoration approaches coupled with variation in landscape factors, disturbance regime, invasive species pressure, historic seed banks, and nutrient supply constrain predictability of wetland restorations to achieve specific targets (Zedler 2000).

Soil moisture, as affected by texture properties and hydrology, is probably the most frequently considered soil-related factor in wetland restoration because it regulates biogeochemical processes and aboveground community structure. Wetting and drying, of soils affect nutrient dynamics. Nitrate supplied by mineralization is stimulated by soil drying, and removal of nitrate through denitrification requires temporary inundation to induce dissimilatory reduction of N conducted by facultative anaerobic microorganisms (Venterink *et al.* 2002). Drained and cultivated agricultural soils are generally considered to contain limited denitrification potential due to aerobic conditions and depleted soil organic matter. Restoring wetlands has been promoted as a mechanism to reduce surface water nutrient loads through vegetation uptake and denitrification (Mitsch *et al.* 2001). However, Orr *et al.* (2007) documented no change in actual and potential denitrification rates in former agricultural soils following cessation of agriculture and hydrologic reconnection in a leveed floodplain. Thus, processes assumed to self-repair under certain conditions may not always develop from recreating the physical template (Hilderbrand *et al.* 2005).

Water depth is a key controller of wetland structure and function, and long-term drainage and tillage of former wetland soils has led to organic matter loss and subsidence of soil (Verhoeven & Setter 2010). In addition to hydrology, soil wetness and variation in soil properties (e.g. texture, organic matter storage, and nutrient availability) are influenced by microtopography. Meyer *et al.* (2008) demonstrated differential recovery rates of belowground structure and function in re-created sloughs containing slight topographic variation between the temporally inundated central channels of sloughs and slightly elevated slough margins. Microtopographic variation also imparts variation in vegetation composition (VivianSmith 1997). Thus, soil texture and moisture characteristics will need to be considered (potentially re-created) to successfully restore all species and assemblages associated with topographic transition.

Attaining biodiversity targets and associated services through wetland restoration is generally impeded under nutrient enrichment, as species richness is commonly low where nutrient supply is high (Green & Galatowitsch 2002). Managing nutrients will require active consideration of inputs from the landscape, as well as soil properties and processes that modulate nutrient availability through adsorption and microbial transformations, respectively. There can be a great disparity between recovery rates of vegetation and soil properties during wetland restoration (Craft *et al.* 1999), which may impose temporal constraints on realizing ecosystem services that in-tact wetlands perform.

5.6.4 High legacy under dynamic change: preventing invasion and restoring invaded systems

Invasion of natural and managed lands by new, undesirable species can displace species and impair ecosystem services provided by diverse communities (e.g. efficient nutrient cycling, pollination, refugia, genetic resources, recreation, and cultural heritage). Efforts to understand biological invasion into habitats of biodiversity conservation concern have largely focused on 1) traits of invading species, 2) habitat and soil-related factors linked to facilitating or resisting invasion, 3) post-colonization effects of invasive species on a variety of soil-mediated ecosystem processes, and 4) complex interactions of all of these factors. There has been a large effort to quantify how invasive plants modify soil properties over the past two decades. Recent studies have provided detailed information on the effects of invasive species (not only plants) on ecosystem productivity, decomposition, soil nutrient dynamics, and soil food webs. Consequently, the prospect of developing a new array of restoration tools to manage native communities or control invasive species is growing.

Efforts to identify physiological or life history traits that differentiate invasive from non-invasive species, as well as highly invasible from less invasible habitats have been largely inconclusive. For instance, an analysis of 79 independent native versus invasive plant comparisons revealed that invaders did not consistently have higher growth rates, competitive abilities, or fecundity (Daehler 2003). In fact, the success of non-native plants generally depended on growing conditions, illustrating the importance of habitat and often soil-based factors influencing the invasion process. One particularly influential hypothesis on habitat factors facilitating invasion has been the "fluctuating resource hypothesis" (Davis *et al.* 2000). This hypothesis suggests that plant invasion depends on soil resource supply rates augmented by the availability of propagules of the invasive organism. Although there is some support for this hypothesis (Foster & Dickson 2004) other studies have been either less emphatically supportive (Kercher *et al.* 2007) or concluded the contrary (Walker *et al.* 2005). The search for patterns of habitat susceptibility to invasion has also examined whether species-rich or low resource environments are more resistant to invasion, but this potential mechanism has not been supported consistently (Lonsdale 1999; Funk & Vitousek 2007). Thus, a variety of life history strategies can be associated with invasive species and many habitat types are susceptible to invasion, which limits the development of generalities about the mechanisms driving biological invasion.

Intimate knowledge of local systems and their potential invaders remain essential to preventing invasion or restoring invaded environments. Such knowledge forms the basis for developing and implementing management. For instance, the invasion and establishment of *Alliaria petiolata* (garlic mustard), an herbaceous biennial prevalent in Eastern and Midwestern woodlands, is sensitive to both woodland density and site fertility, but light availability appears to be the most important factor affecting the proliferation of this species (Meekins & McCarthy 2000). Other biennial weeds in these systems, including *Dipsacus sylvestris* (common teasel) and *Barbarea vulgaris* (garden yellowrocket), flourish in response to soil disturbance (Roberts 1986). Managers can use these specific understandings to protect systems from a suite of potential invaders (e.g. managing for high woodland density with low light levels, coupled with minimal soil disturbance). Management based on well-understood ecological traits of invasive species, potential novel invaders in the regional species pool, and environmental conditions are needed to predict and prevent invasion.

Conservation programs should prioritize the protection of sites with high legacy of native biota, particularly if restoring highly invaded areas is not cost effective and probability of success is low (Hobbs & Harris 2001; Hobbs *et al.* 2009). For example, Chicago Wilderness is a consortium of over 250 conservation-oriented organizations that prioritizes habitat conservation and management in exactly this manner. Traditional management of areas retaining some residual biodiversity has been to cut and remove invasive plants, but other trophic levels that may interact with invasion processes are typically not managed, such as non-native earthworms that impact soil processes in ways that can have consequence for plant composition (Heneghan *et al.* 2006). The success of invasive species removals is largely determined by the degree of invasion, as the "early detection, rapid response" method to controlling invasive species is generally most effective. Ecosystem changes resulting from invasion increase with time since invasion; therefore, restoration in the early stages of invasion is likely to be more successful because there is less modification to the historic system (or higher legacy) relative to long after an invader has established (Strayer *et al.* 2006). Although removal of invaders followed by the re-establishment of historical disturbance regimes (e.g. fire) may conserve and enhance native biodiversity (Hobbs & Huenneke 1992), there has been growing concern that methods exclusively targeted at physical removal of an invader can leave a habitat vulnerable to rapid reinvasion (Iannone & Galatowitsch 2008). Furthermore, invasive species that share similar physiological traits and response to management as native species can pose a real dilemma for managers (Reed *et al.* 2005).

5.6.4.1 Using soil ecological knowledge to control invasion

On a large scale, there has been less active management of soil properties and/or processes to prevent or reduce biological invasion relative to active management of vegetation (Callaham *et al.* 2008). Several studies have quantified the effects of invaders on soil properties and processes, and a few studies have manipulated soil to reduce invasive species. Because soil and plants interact, a modification to the plant community, such as invasion and dominance by a new species, is expected to change soil conditions (Wardle *et al.* 2004; van der Putten *et al.* 2009). Common impacts of invaders on soil include alteration to microbial communities and decomposition rates with consequence for nutrient supply (Corbin & D'Antonio 2004b; Belnap *et al.* 2005), and these changes may persist even after removal of invasive species (Ehrenfeld & Scott 2001). There is limited knowledge about the passive recovery of soil physical, chemical, and biological components following removal of invasive species. If feedback processes develop, there is potential for the system to persist in a self-organizing alternative stable state (Suding *et al.* 2004).

Development of restoration strategies that ameliorate soil conditions modified by invasive species may be needed to successfully increase desired species. Invasive plants may benefit themselves by altering soil biota, soil structure, amount and quality of organic matter, form of available nutrients (e.g. nitrate vs. ammonium), and/or by adding allelochemicals to the soil. Restoration techniques that mediate these changes to soil include: using plant species that reverse invader-cultured soil conditions, adding microbial-containing inoculum to soil, and adding charcoal to bind allelochemicals (Eviner et al. 2010). Kulmatiski and Beard (2006) added 1% activated C to soil dominated by two exotic species and found reduced extractable organic C and N coupled with consistent shifts in community composition (reduction in cover of two target exotic species and an increase in overall cover of native species). Despite these promising results, not all native species responded positively and other (non-target) exotic species increased in cover. This example does illustrate the potential utility of soil-based tools to reduce invasive species.

A common observation is that invasive plant species enhance the availability of limiting nutrients, N and P in particular (Ehrenfeld 2003). Numerous methods, similar to ones used in restoring arable land, have been used to reduce nutrient availability in invaded environments. Carbon amendments (e.g. mulch, sawdust, and sugar additions) can be effective at reducing invasive species and promoting plant diversity. However, some studies found that benefits are short lived and that strategies targeted at manipulating propagule availability were more effective (Morghan & Seastedt 1999; Corbin & D'Antonio 2004a). Rowe et al. (2009) compared the efficacy of nutrient reduction through sucrose application to inoculation of invaded communities with soil from uninvaded communities and found that both tools reduced the focal invasive species' cover and increased perennial native species cover; however, nutrient reduction also increased non-native annual/biennial cover.

Although the tool box of soil-related approaches to reducing invasion is growing, there is not an immediate prospect of applicability, particularly on a large spatial scale. Because invasive species can alter many aspects of soil, the most promising management option will vary with invader. Conflicting outcomes of soil-based restoration tools to control invasive species limit the ability to make general recommendations. To successfully manage ecosystems undergoing dynamic change through the invasion of undesirable species it will be prudent to: 1) study invasion on a local scale, 2) acquire comprehensive information on life history strategies and resource requirements (including interactions with belowground flora and fauna) of local invaders, 3) know the environmental conditions prior to invasion, 4) quantify the impacts of invasion on biological communities and soil, and 5) develop restoration tools that target reducing invasive species and promoting desired communities, coupled with evaluating the efficacy of those tools. Finally, reporting failures of soil-related manipulations used to manage invasive species can be just as valuable as the knowledge gained from restoration success.

5.6.5 Novel legacy: no-analog ecosystems and environmental conditions

There is increasing recognition that restoring ecosystems of the past may not be feasible under rapid and widespread anthropogenic-driven changes to the Earth's atmosphere, climate, land, disturbance regimes, and available nutrients, in combination with invasion of non-native organisms and corresponding loss of diversity (Vitousek et al. 1997). These environmental changes have led to the development of novel ecosystems, defined as self-perpetuating communities that contain no historic analog in terms of species composition and potential functions (Fig. 5.6.1, Scenario III) (Hobbs et al. 2009). Novel ecosystems represent a self-sustaining stable state under new biotic and abiotic conditions. They are often characterized by exotic (either actively or formerly invasive) assemblages of species that have never before coexisted. Because invasive plants commonly alter soils in ways that benefit themselves (Kulmatiski & Kardol 2008), they can facilitate the creation of a new stable state and partly explain the emergence of some novel ecosystems.

The development of novel ecosystems can be due to 1) shifts in abiotic conditions, which then drive biotic changes, 2) shifts in the biotic community, which then alter abiotic conditions, or 3) interactions of biotic and abiotic changes (Hobbs et al. 2009). Increased N deposition in coastal sage scrub habitats of southern California has driven major shifts in the biological community, with consequences for ecosystem structure and functioning. Increased N input through deposition has increased productivity, fuel loads, and fire frequency, which has resulted in the conversion of native shrubland to grasslands dominated by exotic species (reviewed in Fenn *et al.* 2003; Fenn *et al.* 2010). Even with efforts to control exotic grasses and planting native species, these grasslands persist over the long-term, and represent a new stable state (Cox & Allen 2008). Wolkovich *et al.* (2010) reported that this grass-dominated system was nine times more productive than the native system, had lower erosion rates, and a 1.4-fold greater C storage in soil and litter, which shifted the system from a variable C source to a C sink.

Shifts in biotic composition can strongly affect abiotic components of the ecosystem. In fact, the direct impacts of environmental changes on ecosystem processes are often small compared to the indirect ecosystem effects mediated by changes in the plant community (Chapin 2003). For example, in Western Australia, clearing of native vegetation for agriculture leads to a shift from vegetation with deep roots and high evapotranspiration rate, to agricultural species that have shallower roots and use less water. Low evapotranspiration rate in the system dominated by agricultural species leads to a rise in the water table and salinization of the soil surface (with negative consequences for crop production). These abiotic shifts, caused by conversion to agriculture also limit which plant species can thrive in the surrounding natural landscape (George *et al.* 1997). Furthermore, the saline soils can contain 2.5-fold lower soil organic C stocks, lower soil N, decreased soil aggregation, and increased bulk density (Wong *et al.* 2008). Ecosystem transformations such as these can also be mediated by species extinctions, invasion of exotic species, and even shifts in dominant native species (Hobbs *et al.* 2009).

State changes can also be driven by an interaction between shifts in biotic and abiotic changes. For example, 80–87% of Inner Mongolia's grasslands are undergoing desertification due to overgrazing, which causes shifts in vegetation composition and exacerbates drought (Li *et al.* 2000). Plant composition shifts resulting from overgrazing in Inner Mongolia's grasslands leads to decreased plant diversity and cover, increased erosion, decreased soil C and nutrients, and increased aridity at a larger scale (Li *et al.* 2000). Nitrogen additions to these degraded grasslands can enhance the prevalence of perennial drought-tolerant species. Although N addition, in this circumstance, does not necessarily restore historic community composition, it does enhance ecosystem function and resilience of the system to drought. Conversely, N addition to non-degraded perennial grasslands in this region can lead to dominance by annual, early successional species that may not establish in a drought, and leave the system susceptible to massive erosional losses and potentially desertification after a large windstorm (Bai et al. 2010).

In novel systems, manipulations to soil structure, chemistry, and/or biota could be key to retaining some presence of historic species, but restoration to historic communities in most cases would require extreme modification. Effective restoration of novel ecosystems will require knowledge of factors that caused state changes (Beisner *et al.* 2003). Threshold changes in system states are more likely to occur in ecosystems with strong interactions (Suding & Hobbs 2009). Thus, understanding abiotic and biotic interactions is critical to predict and manage state changes. Restoration of a novel system to some past condition may not be feasible (or reasonable) if the environmental change needed to reverse a state change exceeds the environmental change that initiated the state change, a phenomenon known as "hysteresis." For example, successful restoration of coastal wetlands may require large decreases in salinity, because salinity tolerance of establishing seedlings is much less than that of mature wetland plants (Zedler 2005).

Key factors that contribute to the resilience/recoverability of an ecosystem include regulation of abiotic conditions (including nutrient supply), propagule availability in the the regional species pool, and landscape connectivity. All of these factors must be considered to conserve and restore communities and ecosystem services. The role of regulating abiotic conditions has proven critical for restoration of southern California salt marshes. In these systems, small patches cleared for restoration have been successful, but larger-scale restoration efforts have been unsuccessful because large expanses of bare ground heat up and lead to high evaporation, high levels of salinity, and minimal plant establishment (Zedler 2005). Securing ecosystem functions will require a diversity of genotypes and species tolerant of harsh abiotic conditions in the regional or restoration pool of propagules (Eviner & Chapin 2001). Finally, proximity and/or connectivity to remnant patches across the landscape can be critical to maintaining abiotic conditions, providing resources and species through migration and propagule dispersal, and mediating disturbance regimes critical to the maintenance of a system (Bengtsson et al. 2003).

Rapidly changing environmental conditions may necessitate shifting some restoration goals to maintaining or reinstating ecosystem services. Restoration and management to some historic plant communities could, in fact, promote species and communities that can no longer persist without intense management (and eliminate plants that can), potentially leading to a collapse in the provision of multiple ecosystem services (Hobbs et al. 2009). Although novel ecosystems have received negative attention because they are often dominated by non-native species, in some scenarios, these communities have and will represent the most effective way to provide ecosystem services in greatly altered environmental conditions (Ewel & Putz 2004).

5.6.6 Conclusions

The importance of soil ecology in the science and practice of ecological restoration has been repeatedly recognized (Bradshaw 1999; Young et al. 2005; Heneghan et al. 2008). To synthesize the role of soil ecological knowledge in guiding restoration science and practice is challenging because failures are commonly unreported and most published studies on ecological restoration reflect only a brief excerpt (temporally and spatially discrete) from a continuous process that inherently involves change over time. Furthermore, it is difficult to identify generalities across varying types and degrees of disturbance among an array of ecosystems that are restored and managed in unique ways for different purposes. This summary of restoration from the soil legacy continuum perspective elucidates the following:

1. Slower recovery of soil properties and processes relative to plant components (cover and productivity) is consistent across many types of restored ecosystems. Many studies report improved soil structure, C stocks, microbial populations, and nutrient cycling in the trajectory of a target state within decades, but full recovery is anticipated to take much longer.
2. Inconsistent patterns in plant community recovery are commonly attributed to soil properties and/or processes altered through disturbance, which can persist in some cases for over a century.
3. There is more knowledge about singular soil-related constraints to restoration than complex soil or plant-soil interactions. For example, soils with high nutrient availability typically support species-poor plant assemblages. Direct physical and chemical (reduction in P availability through dilution or adsorption) and indirect biological (C addition to increase microbial demand for N) methods used to reduce nutrient availability can be effective. Similarly, pH can be manipulated to alter metal toxicity to promote revegetation or promote restoration of communities adapted to specific pH conditions.
4. The proverbial "black box" of species and their interactions with physical and chemical properties of soil truly demonstrates that a whole soil connected to the aboveground community is greater than the sum of all the parts. This is evidenced by attempts to preserve vital topsoil for restoration, but discovery that displacement, disconnection (from plants), and replacement alters properties of soil enough to constrain restoration of key species.
5. Where substantial elements of the native biota persist in ecosystems undergoing dynamic change through biological invasion, it may be prudent to integrate knowledge of the soil factors that may promote invasion (i.e. disturbance or nutrient supply) and invasive species impacts on soil into management.
6. In highly invaded sites, where the native biota is vestigial, managers may want to consider amelioration of soil properties and processes affected by the invasive species. Novel tools have been tested in attempt to

ameliorate allelopathic effects of invasive species on soil and supply microbial propagules through inoculum additions, but they are limited thus far in knowledge of their effectiveness across ecosystems and at a large spatial scale.

7. Under changing environmental conditions, traditional restoration targets may need to be reconsidered, particularly if restoration requires continued intensive management to sustain species that are no longer suitable for the environment or restoration will require modification to the extent that will cause ecosystem degradation and compromise services. Novel ecosystems may maintain ecosystem services such as productivity, nutrient provision and retention, erosion control, soil C storage and protection, as well as water infiltration, storage, and supply.

Despite many uncertainties, the increasing number of studies that have monitored or manipulated soil properties and processes (successfully or not) tangibly achieve the ultimate goal of restoration ecology, which is to use ecological knowledge to steer restoration practice and test our basic understanding of ecology.

References

Andersen, A.N., & Sparling, G.P. (1995) Ants as indicators of restoration success: relationship with soil microbial biomass in the Australian seasonal tropics. *Restoration Ecology* **5**: 109–14.

Ashby, W.C. (1997) Soil ripping and herbicides enhance shrub restoration on strip mines. *Restoration Ecology* **5**: 169–77.

Bach, E.M, Baer, S.G., Meyer, C.K., & Six, J. (2010) Soil texture affects microbial and structural recovery during grassland restoration. *Soil Biology & Biochemistry* **42**: 2182–91.

Baer, S.G., & Blair, J.M. (2008) Grassland establishment under varying resource availability: a test of positive and negative feedback. *Ecology* **89**: 1859–71.

Baer, S.G., Blair, J.M., Collins, S.L., & Knapp, A.K. (2003) Soil resources regulate productivity and diversity in newly established tallgrass prairie. *Ecology* **84**: 724–35.

Baer, S.G., Blair, J.M., Collins, S.L., & Knapp, A.K. (2004) Plant community responses to resource availability and heterogeneity during restoration. *Oecologia* **139**: 617–29.

Baer, S.G., Meyer, C.K., Bach, E.M., Klopf, R.P., & Six, J. (2010) Contrasting ecosystem recovery on two soil textures: implications for carbon mitigation and grassland conservation. *Ecosphere* **1**: art. 5.

Bai, Y., Wu, J., Clark, C.M., et al. (2010) Tradeoffs and thresholds in the effects of nitrogen addition on biodiversity and ecosystem functioning: evidence from inner Mongolia Grasslands. *Global Change Biology* **16**: 358–72.

Beisner, B.E., Haydon, D.T., & Cuddington, K. (2003) Alternative stable states in ecology. *Frontiers in Ecology and the Environment* **1**: 376–82.

Belnap, J., Phillips, S.L., Sherrod, S.K., & Moldenke, A. (2005) Soil biota can change after exotic plant invasion: does this affect ecosystem processes? *Ecology* **86**: 3007–17.

Bengtsson, J., Angelstam, P., Elmqvist, T., et al. (2003) Reserves, resilience and dynamic landscapes. *Ambio* **32**: 389–96.

Blumenthal, D.M., Jordan, N.R., & Russelle, M.P. (2003) Soil carbon addition controls weeds and facilitates prairie restoration. *Ecological Applications* **13**: 605–15.

Bradshaw, A.D. (1997) Reclamation of mined land—using natural processes. *Ecological Engineering* **8**: 255–69.

Bradshaw, A.D. (1999) The importance of soil ecology in restoration science. In: K.M. Urbanska, N.R. Webb, P.J. Edwards (eds.) *Restoration Ecology and Sustainable Development*, pp. 33–64. Cambridge University Press, Cambridge.

Bradshaw, A.D., & Chadwick, M.J. (1980) *The Restoration of Land: The Ecology and Reclamation of Derelict and Degraded Land*. Blackwell Scientific Publications, Berkeley, CA.

Brenner, F.J., Werner, M., & Pike, J. (1984) Ecosystem development and natural succession in surface coal mine reclamation. *Minerals and the Environment* **6**:10–22.

Callaham, M.A., Rhoades, C.C., & Heneghan, L. (2008) A striking profile: soil ecological knowledge in restoration management and science. *Restoration Ecology* **16**: 604–7.

Chapin, F.S., III. (2003) Effects of plant traits on ecosystem and regional processes: a conceptual framework for predicting the consequences of global change. *Annals of Botany* **91**: 455–63.

Compton, J.A., & Boone, R.D. (2000) Long-term impacts of agriculture on soil carbon and nitrogen in New England forests. *Ecology* **81**: 2314–30.

Corbin, J.D., & D'Antonio, C.M. (2004a) Can carbon addition increase competitiveness of native grasses? A case study from California. *Restoration Ecology* **12**: 36–43.

Corbin, J.D., & D'Antonio, C.M. (2004b) Effects of exotic species on soil nitrogen cycling: implications for restoration. *Weed Technology* **18**: 1464–7.

Costanza, R., dArge, R., Farber, S., et al. (1997) The value of the world's ecosystem services and natural capital. *Nature* **387**: 253–60.

Costigan, P., Bradshaw, A.D., & Gemmell, R.P. (1981) The reclamation of acidic colliery spoil. I. Acid production potential. *Journal of Applied Ecology* **18**: 865–78.

Cox, R.D., & Allen, E.B. (2008) Stability of exotic annual grasses following restoration efforts in southern California coastal sage scrub. *Journal of Applied Ecology* **45**: 495–504.

Craft, C., Reader, J., Sacco, J.N., & Broome, S.W. (1999) Twenty-five years of ecosystem development of constructed *Spartina aterniflora* (Loisel) marshes. *Ecological Applications* **9**: 1405–19.

Daehler, C.C. (2003) Performance comparisons of co-occurring native and alien invasive plants: Implications for conservation and restoration. *Annual Review of Ecology Evolution and Systematics* **34**: 183–211.

Daily, G.C., Matson, P.A., & Vitousek, P.M. (1997) Ecosystem services supplied by soil. In: G.C. Daily (ed.) *Nature's Services: Societal Dependence on Natural Ecosystems*, pp. 113–32. Island Press, Washington, DC.

Davis, M.A., Grime, J.P., & Thompson, K. (2000) Fluctuating resources in plant communities: a general theory of invasibility. *Journal of Ecology* **88**: 528–34.

Dick, R.P. (1992) A review: long-term effects of agricultural systems on soil biochemical and microbial parameters. *Agriculture, Ecosystems & Environment* **40**: 25–36.

Ehrenfeld, J.G. (2003) Effects of plant invasions on soil nutrient cycling processes. *Ecosystems* **6**: 503–23.

Ehrenfeld, J.G., & Scott, N. (2001) Invasive species and the soil: effects on organisms and ecosystem processes. *Ecological Applications* **11**: 1259–60.

Eijsackers, H. (2004) Leading concepts towards vital soil. In: P. Doelman & H.J.P. Eijsackers (eds.) *Vital soil*, pp. 1–20. Developments in Soil Science, Volume 29. Elsevier B. V., Amsterdam.

Ellis, E.C., & Ramankutty, N. (2008) Putting people in the map: Anthropogenic biomes of the world. *Frontiers in Ecology & the Environment* **6**: 439–47.

Eviner, V.T., & Chapin, F.S., III. (2001) Plant species provide vital ecosystem functions for sustainable agriculture, rangeland management, and restoration. *California Agriculture* **55**: 54–9.

Eviner, V.T., Hoskinson, S.A., & Hawkes, C.V. (2010) Ecosystem impacts of exotic plants can feed back to increase invasion in western US rangelands. *Rangelands* **31**: 21–31.

Ewel, J.J., & Putz, F.E. (2004) A place for alien species in ecosystem restoration. *Frontiers in Ecology and the Environment* **7**: 354–60.

Fenn, M.E., Allen, E.B., Weiss, S.B., *et al.* (2010) Nitrogen critical loads and management alternatives for N-impacted ecosystems in California. *Journal of Environmental Management* **91**: 2404–23.

Fenn, M.E., Baron, J.S., Allen, E.B., *et al.* (2003) Ecological effects of nitrogen deposition in the Western United States. *Bioscience* **53**: 404–20.

Flinn, K.M., & Marks, P.L. (2007) Agricultural legacies in forest environments: tree communities, soil properties, and light availability. *Ecological Applications* **17**: 452–63.

Foster, B.L., & Dickson, T.L. (2004) Grassland diversity and productivity: The interplay of resource availability and propagule pools. *Ecology* **85**: 1541–7.

Franco, A.A., & DeFaria, S.M. (1997) The contribution of N_2-fixing tree legumes to land reclamation and sustainability in the tropics. *Soil Biology & Biochemistry* **29**: 897–903.

Funk, J.L., & Vitousek, P.M. (2007) Resource-use efficiency and plant invasion in low-resource systems. *Nature* **446**: 1079–81.

George, R., McFarlane, D., & Nulsen, B. (1997) Salinity threatens the viability of agriculture and ecosystems in Western Australia. *Hydrogeology Journal* **5**: 6–21.

Green, E.K., & Galatowitsch, S.M. (2002) Effects of *Phalaris arundinacea* and nitrate-N addition on the establishment of wetland plant communities. *Journal of Applied Ecology* **39**: 134–44.

Harris, J.A. (2003) Measurements of the soil microbial community for estimating the success of restoration. *European Journal of Soil Science* **54**: 801–8.

Heneghan, L., Miller, S.P., Baer, S., *et al.* (2008) Integrating soil ecological knowledge into restoration management. *Restoration Ecology* **16**: 608–17.

Heneghan, L., Steffen, J., & Fagen, K. (2006) Interactions of an introduced shrub and introduced earthworms in an Illinois urban woodland: Impact on leaf litter decomposition. *Pedobiologia* **50**: 543–51.

Hilderbrand, R.H., Watts, A.C., & Randle, A.M. (2005) The myths of restoration ecology. *Ecology and Society*, **10**: article 19 [online] http://www.ecologyandsociety.org/vol10/iss1/art19.

Hobbs, R.J., & Harris, J.A. (2001) Restoration ecology: repairing the Earth's damaged ecosystems in the new millennium. *Restoration Ecology* **9**: 239–46.

Hobbs, R.J., Higgs, E., & Harris, J.A. (2009) Novel ecosystems: implications for conservation and restoration. *Trends in Ecology and Evolution* **24**: 599–609.

Hobbs, R.J., & Huenneke, L.F. (1992) Disturbance, diversity, and invasion—implications for conservations. *Conservation Biology* **6**: 324–37.

Huang, J.W., & Cunningham, S.D. (1996) Lead phytoextraction: species variation in lead uptake and translocation. *New Phytologist* **134**: 75–84.

Iannone, B.V., & Galatowitsch, S.M. (2008) Altering light and soil N to limit *Phalaris arundinacea* reinvasion in sedge meadow restorations. *Restoration Ecology* **16**: 689–701.

Insam, H., & Domsch, K.H. (1988) Relationship between soil organic carbon and microbial biomass on chronosequences of reclamation sites. *Microbial Ecology* **15**: 177–88.

Insam, H., & Haselwandter, K. (1989) Metabolic quotient of the soil microflora in relation to plant succession. *Oecologia* **79** 174–8.

Janssens, F., Peeters, A., Tallowin, J.R.B., *et al.* (1998) Relationships between soil chemical factors and grassland diversity. *Plant and Soil* **202**: 69–78.

Jastrow, J.D., Miller, R.M., & Lussenhop, J. (1998) Contributions of interacting biological mechanisms to soil aggregate stabilization in restored prairie. *Soil Biology & Biochemistry* **30**: 905–16.

Jenny, H. (1941) Factors of soil formation. *Soil Science* **52**: 415.

Kercher, S.M., Herr-Turoff, A., & Zedler, J.B. (2007) Understanding invasion as a process: the case of *Phalaris arundinacea* in wet prairies. *Biological Invasions* **9**: 657–65.

Kulmatiski, A., & Beard, K.H. (2006) Activated carbon as a restoration tool: Potential for control of invasive plant in abandoned agriculture. *Restoration Ecology* **14**: 251–7.

Kulmatiski, A., & Kardol, P. (2008) Getting plant-soil feedbacks out of the greenhouse: experimental and conceptual approaches. *Progress in Botany* **69**: 449–72.

Lane, D.R., & BassiriRad, H. (2005) Diminishing effects of ant mounds on soil heterogeneity across a chronosequence of prairie restoration sites. *Pedobiologia* **49**: 359–66.

Lavelle, P. (1987) Biological processes and productivity of soils in the humid tropics. In: E.D. Robert (eds.) *The Geophysiology of Amazonia, Vegetation and Climate Interactions*, pp. 175–222. Wiley, New York.

Li, S.G. (2006) Ecological restoration of mineland with particular reference to metalliferous mine wasteland in China: A review of research and practice. *Science of the Total Environment* **357**: 38–53.

Li, S.G., Harazono, Y., Oikawa, T., Zhao, H.L., He, Z.Y., & Chang, X.L. (2000) Grassland desertification by grazing and the resulting micrometeorological changes in Inner Mongolia. *Agricultural and Forest Meteorology* **102**: 125–37.

Lonsdale, W.M. (1999) Global patterns of plant invasions and the concept of invasibility. *Ecology* **80**: 1522–36.

Lu, D., Moran, E., & Mausel, P. (2002) Linking Amazonian secondary succession forest regrowth to soil properties. *Land Degradation and Development* **13**: 331–43.

Matamala, R., Jastrow, J.D., Miller, R.M., & Garten, C.T. (2008) Temporal changes in C and N stocks of restored prairie: implications for C sequestration strategies. *Ecological Applications* **18**: 1470–88.

Meekins, J.F., & McCarthy, B.C. (2000) Responses of the biennial forest herb *Alliaria petiolata* to variation in population density, nutrient addition and light availability. *Journal of Ecology* **88**: 447–63.

Meyer, C.K., Baer, S.G., & Whiles, M.R. (2008) Ecosystem recovery across a chronosequence of restored wetlands in the Platte River Valley. *Ecosystems* **11**: 193–208.

Millennium Ecosystem Assessment (MEA) (2005) Ecosystems and human well-being: Policy Responses: Findings of the Responses Working Group of the Millennium Ecosystem Assessment. Island Press, Washington DC.

Mitsch, W.J., Day, J.W., Jr., Gilliam, J.W., *et al.* (2001) Reducing nitrogen loading to the Gulf of Mexico from the Mississippi River Basin, Strategies to counter a persistent ecological problem. *BioScience* **51**: 373–88.

Morghan, K.J.R., & Seastedt, T.R. (1999) Effects of soil nitrogen reduction on nonnative plants in restored grasslands. *Restoration Ecology* **7**: 51–5.

Mummey, D.L., Stahl, P.D., & Buyer, J.S. (2002) Soil microbiological properties 20 years after surface mine reclamation, spatial analysis of reclaimed and undisturbed sites. *Soil Biology & Biochemistry* **34**: 1717–25.

Orr, C.H., Stanley, E.H., Wilson, K.A., & Finlay, J.C. (2007) Effects of restoration and reflooding on soil dentrification in a leveed Midwestern floodplain. *Ecological Applications* **17**: 2365–76.

Owen, K.M., Marrs, R.H., Snow, C.S.R., & Evans, C. (1999) Soil acidification—the use of sulphur and acidic plant materials to acidify arable soils for the re-creation of heathland and acidic grassland at Minsmere, UK. *Biological Conservation* **87**: 105–21.

Paschke, M.W., Redente, E.F., & Brown, S.L. (2003) Biology and establishment of the mountain shrubs on mining disturbances in the Rocky Mountains, USA. *Land Degradation & Development* **14**: 459–80.

Paul, M., Catterall, C.P., Pollard, P.C., & Kanowski, J. (2010) Recovery of soil properties and functions in different rainforest restoration pathways. *Forest Ecology and Management* **259**: 2083–92.

Paustian, K., Cole, C.V., Sauerbeck, D., & Sampson, N. (1998) CO_2 mitigation by agriculture: an overview. *Climate Change* **40**: 135–62.

Reed, H.E., Seastedt, T.R., & Blair, J.M. (2005) Ecological consequences of C_4 grass invasion of a C_4 grassland: a dilemma for management. *Ecological Applications* **15**: 1560–9.

Reinhart, K.O., & Callaway, R.M. (2006) Soil biota and invasive plants. *New Phytologist* **170**: 445–57.

Roberts, H.A. (1986) Seed persistence in soil and seasonal emergence in plant species from different habitats. *Journal of Applied Ecology* **23**: 639–56.

Rowe, H.I., Brown, C.S., & Paschke, M.W. (2009) The influence of soil inoculums and nitrogen availability on restoration of high-elevation steppe communities invaded by *Bromus tectorum*. *Restoration Ecology* **17**: 686–94.

Showalter, J.M., Burger, J.A., & Zipper, C.E. (2010) Hardwood seedling growth on different mine spoil types with and without topsoil amendment. *Journal of Environmental Quality* **39**: 483–91.

Strayer, D.L., Eviner, V.T., Jeschke, J.M., & Pace, M. (2006) Understanding the long-term effects of species invasions. *Trends in Ecology & Evolution* **21**: 645–51.

Suding, K.N., Gross, K.L., & Houseman, G.R. (2004) Alternative states and positive feedbacks in restoration ecology. *Trends in Ecology & Evolution* **19**: 46–53

Suding, K.N., & Hobbs, R.J. (2009) Threshold models in conservation and restoration: a developing framework. *Trends in Ecology & Evolution* **24**:271–9.

Tucker, J.M., Brondizio, E.S., & Morán, E.F. (1998) Rates of forest regrowth in eastern Amazonia: a comparison of Altamira and Bragantina regions, Para State, Brazil. *Interciencia* **23**: 64–73.

van Breeman, N. (2004) The formation of soils. In: P. Doelman, & H.J.P. Eijsackers (eds.) *Vital soil*, pp. 21–40. Developments in Soil Science, Volume 29. Elsevier B.V., Amsterdam, The Netherlands.

van der Putten, W.H., Bardgett, R.D., de Ruiter, P.C., *et al.* (2009) Empirical and theoretical challenges in aboveground-belowground ecology. *Oecologia* **161**: 1–14.

Vellend, M. (2003) Habitat loss inhibits recovery of plant diversity as forests grow. *Ecology* **84**: 1158–64.

Venterink, H.O., Davidsson, T.E. Kiehl, K., & Leonardson, L. (2002) Impact of drying and re-wetting on N, P and K dynamics in a wetland soil. *Plant and Soil* **243**: 119–30.

Verhoeven, J.T.A., & Setter, T.L. (2010) Agricultural use of wetlands: opportunities and limitations. *Annals of Botany* **105**: 155–63.

Vitousek, P.M., Mooney, H.A., Lubchenco, J., & Melillo, M. (1997) Human domination of Earth's ecosystems. *Science* **277**: 494–9.

VivianSmith, G. (1997) Microtopographic heterogeneity and floristic diversity in experimental wetland communities. *Journal of Ecology* **85**: 71–82

Walker, K.J., Stevens, P.A., Stevens, D.P., Mountford, J.O., Manchester, S.J., & Pywell, R.F. (2004) The restoration and re-creation of species rich lowland grassland on land formerly managed for intensive agriculture in the UK. *Biological Conservation* **119**: 1–18.

Walker, S., Wilson, J.B., & Lee, W.G. (2005) Does fluctuating resource availability increase invasibility? Evidence from field experiments in New Zealand short tussock grassland. *Biological Invasions* **7**: 195–211.

Wardle, D.A., Bardgett, R.D., Klironomos, J.N., Setälä, H., van der Putten, W.H., & Wall, D.H. (2004) Ecological linkages between aboveground and belowground biota. *Science* **304**: 1629–33.

White, P.S., & Jentsch, A. (2004) Disturbance, succession, and community assembly in terrestrial plant communities. In: V.M. Temperton, R.J. Hobbs, T. Nuttle, & S. Hale (eds.) *Assembly rules and restoration ecology*, pp. 342–66. Island Press, Washington, DC.

White, P.S., & Walker, J.L. (1997) Approximating nature's variation: selecting and using reference information in restoration ecology. *Restoration Ecology* **5**: 338–49.

Wolkovich, E.M., Lipson, D.A., Virginia, R.A., Cottingham, K.L., & Bolger, D.T. (2010) Grass invasion causes rapid increases in ecosystem carbon and nitrogen storage in a semiarid shrubland. *Global Change Biology* **16**: 1351–65.

Wong, M.H. (2003) Ecological restoration of mine degraded soils, with emphasis on metal contaminated soils. *Chemosphere* **50**: 775–80.

Wong, V.N.L., Murphy, B.W., Koen, T.B., Greene, R.S.B., & Dalal, R.C. (2008) Soil organic carbon stocks in saline and sodic landscapes. *Soil Research* **46**: 378–89.

Yin, B., Crowley, D., Sparovek, G., De Melo, W.J., & Borneman, J. (2000) Bacterial functional redundancy along a soil reclamation gradient. *Applied and Environmental Microbiology* **66**: 4361–5.

Young, T.P., Petersen, D.A., & Clary, J.J. (2005) The ecology of restoration: historical links, emerging issues, and unexplored realms. *Ecology Letters* **8**: 662–73.

Zedler, J.B. (2000) Progress in wetland restoration ecology. *Trends in Ecology & Evolution* **15**: 402–7.

Zedler, J.B. (2005) Ecological restoration: guidance from theory. *San Francisco Estuary & Watershed Science* **3**: 1–31

Synthesis

Jeffrey E. Herrick and Johan Six

The chapters of Section 5 support the fundamental role that soil biota play in sustaining soil structure and nutrient cycling as the foundation for nearly all ecosystem services while illustrating the challenges of documenting explicit, consistent relationships. Provisioning services, including crop production, were the focus of the majority of the studies cited in this section and, as Cavigelli *et al.* point out, a high proportion of the management system comparisons are limited to conventional tillage against some form of organic or conservation tillage-based system. Despite these inherent limitations of the existing literature, a number of lessons can be drawn. Five that are particularly significant were each cited in at least two of the chapters: 1) the importance of spatial and temporal context for interpreting soil biota-sustainability relationships, 2) the importance of systematic approaches, common methods, and protocols to study soil biota—sustainability relationships, 3) the value of long-term studies for identifying and validating functional relationships, particularly where chronic, cumulative, and acute disturbances are involved, 4) the potential value of local knowledge for identifying, documenting, and monitoring these relationships, and 5) the extent to which hysteresis must be considered to understand and predict the potential contribution of soil biota to resilience and restoration.

(1) *Context is critical*. Cavigelli *et al.* conclude that, "management impacts on ecosystem services and soil biodiversity are often complex such that responses may be subtle and vary with soil type, climate, ecosystem, taxonomic and/or functional group, and ecosystem service." This statement can be extended to include soil biodiversity impacts on ecosystem services. The chapter by Van Oost and Bakker adds temporal context, and specifically the stage of soil degradation, as an additional factor that must be considered when interpreting soil biota—sustainability relationships. Several studies cited by Grandy *et al.* emphasize the fundamental importance of spatial context, and in particular soil texture. A lack of basic soil characterization data has been, and continues to be, a significant factor limiting meta-analyses of multiple datasets.

(2) *A systematic approach is needed*. The diversity of methods to quantify ecosystem services and soil biota sometimes seems to rival the diversity of the organisms themselves. While there are often good reasons for the selection or development of a new or esoteric method, in many cases our decision not to apply an existing method is limited only by ego or simply an unwillingness to make the time to identify common methods, and to recognize which modifications make it impossible to compare datasets that purportedly include the same parameter. National to regional efforts such as the African Soil Information System (AfSIS) cited by Barrios *et al.* should increase the availability of standardized information in the future.

(3) *Long-term studies*. Nearly all of the chapters cited the value of long-term studies involving multiple treatments for teasing apart complex relationships between management, soil biota, and ecosystem services in the longer-term. As Grandy *et al.* point out, long-term studies are particularly important for understanding the critical soil properties and processes that control ecosystem resilience and recovery. Chronic disturbances, including those associated with climate change and nitrogen deposition, both affect and are mediated by soil biota. These complex interactions are nearly impos-

sible to understand without longitudinal data. "Novel ecosystems" represent another case where long-term studies are essential as it is, by definition, impossible to substitute space-for-time experimental designs for new long-term studies. Baer *et al.* define these novel ecosystems as "self-perpetuating communities that contain no historic analog in terms of species composition and historic function" and "represent a self-sustaining stable state under new biotic and abiotic conditions." While the permanence of "stable states" continues to be debated by ecologists, the challenge of understanding ecosystems in the context of novel conditions is now widely accepted and forms a research priority a sustainable future.

(4) *Local knowledge and local communities can contribute to the identification, documentation and monitoring these relationships*. Both Barrios *et al.* and Karlen cite the value of farmers' knowledge in documenting relationships between management practices and soil health. New technologies, including GPS-enabled camera phones with data-input and transmission capabilities now allow individuals to share site-specific knowledge and information. With a limited amount of training in soil characterization (e.g. depth, sandiness, and stickiness), the information contributed by farmers could be increased beyond what is already possible by linking their geolocated observations to digital elevation models (for slope, slope shape, landscape position, and aspect), greenness indices that provide information on phenology and production, and other geospatial data derived from remote sensing imagery.

(5) Finally, *understanding hysteresis is key to understanding and predicting the potential contribution of soil biota to resilience and restoration*. Rangeland ecologists, in particular, have now concluded that stable states represented by a single climax plant community are the exception rather than the rule, and that hysteretic dynamics are to be expected. The soil biotic communities associated with transitions among states, and plant communities within states, are similarly complex and dynamic. While general rules are elusive, a number of studies have documented significant shifts in soil biotic communities associated with aboveground dynamics.

There is a virtually infinite need for research, and it is clearly impossible to always apply the comprehensive set of six steps necessary to confirm mechanistic relationships (listed by Barrios *et al.*). We argue that research will need to become increasingly strategic, focusing on the short- and long-term dynamics of key processes rather than simply attempting to establish the importance of particular groups of organisms. Studies must also be designed to allow multiple sources of spatial and temporal variability to be considered, if not addressed. We believe that this approach could also increase our understanding of key feedbacks and interactions between soil biota, pests, diseases, and plant nutrient and water status and, ultimately, plant and animal production.

Index

Page numbers in *italics* refer to Figures and Tables.

Acacia spp. 318, 322
Acari *see* mites
accelerated soil erosion 304, *305*
Achaeta 220
acidification 259
Acrobeloides sp. 139, *140*
acyl homoserine lactones (AHLs) 73
adaptive strategies 12–13, *14*
Africa Soil Information Service (AfSIS) 324, *325*
aggregation
   scale of 138–40
   soil aggregates 323
   see also spatial distribution
agriculture 52
   accelerated soil erosion 304, *305*
   agricultural land restoration 382–5
      to forest 383–4
      to grassland 382–3
      to wetland 384–5
   conservation agriculture 53
   entry points for biological management 54, *54*
   erosion-induced productivity losses 309–12, *310*
      future impacts 311–12
      past impacts 310–11
   impact on termite communities 211
   nitrogen enrichment effects 266
   organic cropping system impacts 347–50
      ecosystem services 347
      soil biodiversity 348–50, *349*
   soil erosion effect on productivity 305–9
      estimation of impact 306–9, *307*
      mechanisms 305–6

soil organic matter disturbance 361–4
working with nature 52–3
see also crop nutrients; tillage
agroforestry 315–16, *316*, 327–8
   influence on soil properties 316–18
      soil nutrients 317–18
      soil structure maintenance 323
   response ratio (RR) 318
   soil biota abundance promotion 318–19, *319*
   see also trees
Ailanthus altissima 91, *91*
alkaloids 162
allicin 162
Allolobophoridella 217
1-aminocyclopropane-1-carboxylate (ACC) deaminase 75
ammonification 71
ammonium
   anaerobic oxidation (ANAMMOX) 71
   availability 89
amphibiotic conditions 13
Amynthas 218
anaerobic patches 147
Andropogon 90
antagonists of soil-borne diseases 48
anti-nutrients 162
antioxidants 161
ants 33, 211
   biogeography 211–13
      continental scale 211–12
      local scale 213
      regional scale 212–13
   bioturbation 13
   global change impacts 213–14
   molecular phylogeny *212*

Aporrectodea 216
   A. longa 285
arbuscular mycorrhizal fungi (AM fungi) 156
   disturbance effects 365
   nitrogen enrichment effects 263–4
   soil biodiversity and 341
   soil structure maintenance role 323
   tree association 322–3
   use to boost crop nutrients 157–63, *160*
      increased nutrient uptake 157–60
      indirect effects 161–2
      iron chelation 161
      negative effects 162–3
      nutrient solubilization and transport 160–1
   see also mycorrhizal fungi
archaea 28, 32
   see also microbes
ascorbic acid 162
autocorrelation 138, 140

bacteria 28–32
   detrital succession 119
   nitrogen enrichment effects 261–3
   spatial distribution 143
   see also microbes
Bactris gasipaes 318
beetles
   assemblages on animal dung 120–1
   organic cropping system impacts 348
Bertholetia excelsa 318
Bimastos 216, 217

biodiversity
    management 337–8, *338*, 350–1
        chemical application
            impacts 345–6
        edible crop diversity 338, *339*
        organic cropping system
            impacts 348–50, *349*
        organic material application
            impacts 346–7
        plant selection impacts 340–1
        tillage impacts 343–4
    nitrogen enrichment effects 266–7
    urban biodiversity 274
    see also diversitwy of soil
        communities
biofortification of crops 155
biogeography
    ants 211–13
    earthworms 214–18
    ecosystem services and 50
    Enchytraeidae 219–33
    functional equivalence and 194–5
    island biogeography 204
    soil invertebrate ecosystem
        engineers 201–6, 235
        altitudinal gradients 205
        area–diversity relationships
            203–4
        landscape modification
            gradients 205
        latitudinal gradients 204–5
    termites 206–10, *208*
        fungus-growing termites 174–5
biological control 323–4
bioremediation 69–71
biotic resistance 101–2
bioturbation 13, 222
birch 141–2
Bromus spp. 286

Calliandra calothyrsus 318, 322
carbon cycling 70
    carbon pools 89
    climate change effects 242, 243,
        245–8
    grasslands 283–4, 285–7
    nitrogen enrichment effects on C
        sequestration 265
    soil organic carbon (SOC)
        management
            organic cropping system impact
                347

plant selection impacts 340
tree–biota interactions and 321–2
carbon dioxide concentrations
    plant community composition
        and 246
    plant community productivity
        and 246
    soil organic matter disturbance
        362–3, 368
    temperature effects 243
    tillage impact 342, 343
    see also carbon cycling; climate
        change
carotenoids 162
carrion decomposition 120, 123
cellulases 12
CENTURY model 10–11
Cephalobus sp. 140
Chronogaster sp. 139, *140*
ciliates, spatial distribution 139
cities *see* urbanization
climate change 241–2, *241*, 294–5
    impacts on soil communities
        213
        ants 213
        earthworms 218
        Enchytraeidae 222
        termites 210–11
    nitrogen enrichment effects 265
    plant–soil linkages and 242–50,
        242, 249–50
        making predictions 248–9, *249*
        multifactor effects 248
        plant community composition
            246–7
        plant community productivity
            246
        seasonality and phenology
            247–8
        temperature 243–4
        water 244–5
    soil processes and 241–2
        feedbacks 242
    urbanization and 275
cloning 68
clover, white 286
coevolution of plants and insects
    117
Cognettia sphagnetorum 221
Coleogyne ramosissima 141
collembolans (springtails) 33
    detrital succession 119–20

communication, microbial 72–3
community genetics 64
community level physiological
    profiling (CLPP) 35
community structure *see* soil
    communities
compost application *see* organic
    amendment impacts
conservation agriculture 53
cooperation
cover crops 341
crop nutrients
    low nutrient crops 153
    rhizosphere microbe use to
        boost levels 157–63,
        158–9, *160*
        changes in plant/soil
            chemistry 160
        increased nutrient uptake
            157–60
        indirect effects 161–2
        iron chelation 161
        negative effects 162–3
        nutrient solubilization and
            transport 160–1
    soil microbe roles 155–7
        cyanobacteria 156
        mycorrhizal fungi 156
        nitrogen fixing bacteria 156
        non-mycorrhizal fungal
            endophytes 156–7
        plant growth promoting
            rhizobacteria (PGPR) 156
        rhizobia 155
    traditional ways of boosting
        nutrient levels 154–5
        biofortification 155
        diversified diet 154
        fertilizer use 154
        fortification 155
crop rotation 341–2
C transformations 48
cultivation induced biological
    nitrogen fixation (C-BNF)
    258
cyanobacteria 156

Dactilenia barteri 323
decision-loop 104, *104*
decomposer communities 117–18,
    367–8
    home-field advantage 83, 90

plant genetic variation influence
    83, *84*, 86
decomposition *see* organic matter
    decomposition
deforestation
  impact on termite communities
    211
  soil organic matter disturbance
    361
denaturing gradient gel
    electrophoresis (DGGE) 35, 68
Dendrobaena 216
Dendrodrilus 217
denitrification 71, 147
  denitrifying microbial
    communities 67, 71–2
depositional landforms 302–3
desiccation resistance 13
detrital succession 118–21
  detrital ontogeny model 126, *126*
  facilitation model 121–3, *122*
  future directions 130–2
  inhibition model 121–3, *122*
  mechanisms 121–6
  processing chain model 123–6,
    124, 125
  tolerance model 121–3, *122*
detritivores 117–18
detritus 118–19
2,4-diacetylphloroglucinol
    (2,4-DAPG) 75
diet diversification 154
  see also crop nutrients
digestion, cooperation and 12
Diplocardia 216
diplopods, organic cropping system
    impacts 348–9
Diplura 33
disease control, agroforestry systems
    323–4
dispersal 203
  termites 175
disturbance 358–9
diversity of soil communities 7,
    28–33, 59–60, 117–18
  altitudinal gradients 205
  area–diversity relationships 203–4
  biogeography 201
  diversity–function relationships
    37–9, *41*
    holistic view 39–41
  edible crop diversity 338, *339*
  enigma of soil species diversity
    117–18, 136–7, 150, 233
  functional diversity 40–1, *41*
  horizontal diversity 40
  landscape modification gradients
    205
  latitudinal gradients 128–30, *129*,
    204–5
  microbes 28–32
  organic cropping system impacts
    348–50, *349*
  plant selection impacts 340–1
  spatial patterns 136–47, *138*,
    233–4
    fine or microscale 143–7
    small or mesoscale 138–43, *139*
  successional specialization role
    126–8
  tillage impacts 343–4
  vertical diversity 40
  see also biodiversity; soil
    communities
DNA chips 75–6
drainage 9
drought 244–5
dung
  decomposition 120
    succession of assemblages
      120–1
  impact on distribution of soil
    organisms 143, 150
dynamic equilibrium 15
dynamic global vegetation models
    (DGVMs) 248–9

earthworms 33, 214
  agroforestry effects on abundance
    319
  biogeography 214–18
    continental scale 215
    local scale 217–18
    regional scale 215–17
  bioturbation 13
  digestion and cooperation 12
  functional groups 214, *215*
  global change impacts 218
  grasslands 284–5, 287
  isotope analysis 128
  mucus 12
  nephridial symbiotic bacteria 217
  organic cropping system impacts
    348

phylogeny *216*
plant interactions 19, 118
  tree distribution relationships
    317, *317*
resting stages *14*
spatial distribution 140
tillage impacts 343–4
urban soils 273, 276
ecological restoration 377–81, *378,
    379, 380*, 389–90
  agricultural land restoration
    382–5
    to forest 383–4
    to grassland 382–3
    to wetland 384–5
  mined land restoration 381–2
  novel ecosystem development
    387–9
ecosystem engineers 33, 48, 320
  ants 211–14
    biogeography 211–13
    global change impacts 213–14
  bioturbation 13
  earthworms 214–18
    biogeography 215–18
    global change impacts 218
  Enchytraeidae 218–22
    biogeography 219–22
    global change impacts 222
  functional domains 15, 17
  impact on distribution of soil
    organisms 142, 146
  soil invertebrate ecosystem
    engineers *202*, 235
    altitudinal gradients 205
    area–diversity
      relationships 203–4
    biogeography 201–6
    landscape modification
      gradients 205
    latitudinal gradients 204–5
    trait-based ecology 222–4,
      223
  termites 206–11
    biogeography 206–10
    emergent landscape
      effects 182–3, *184*
    fungus-growing termites
      179–83, *184*, 234–5
    global change impacts 210–11
    mound-builder impacts 181–2,
      209–10

ecosystem functioning 46–9, *46*, *47*, *49*
  environmental driver effects 51–2
  impacts of global change 210–11, 213–14, 218
  feedback mechanisms 99–100
  land management effects 52
  plant genetic variation role 87–9, *88*
  soil biogeochemistry perspective 46–7
  soil biology perspective 47–8
  spatial distribution of soil functions 147–9
  trait–function relationships 51–2
ecosystem resilience 359–60
  see also soil
ecosystem services 45, *46*, *47*, *63*, *98*
  biogeographic effects 50
  biological levels of production 101–2
    community level 101–2
    ecosystem level 102
    gene level 101
  in cities 273–5, *274*
    cultural ecosystem services 275
    management for 276–8
    provisioning ecosystem services 274–5
    regulating ecosystem services 275
    supporting ecosystem services 274
  delivery of ecosystem services 99–101, *99*, 105–6
  ecological restoration 377–81, *378*, *379*, *380*
  environmental driver effects 51–2
  grasslands 284–8
    herbage production 287–8
    nutrient cycling 285–7
    soil structure maintenance 284
    trade-offs between 288–90
    water regulation 285
  land management effects 52
  management 337–8, *338*, 350–1
    chemical application impacts 344–5
    edible crop diversity 338, *339*
    organic cropping system impacts 347
    organic material application impacts 346
    plant selection impacts 338–40
    tillage impacts 342–3
  policy context 103–5
  soil biota relationships 319–20, *320*
  spatial scales 102–3, *103*
  supporting services 98
  tree–biota interactions and 320–4
    carbon transformations 321–2
    nutrient cycling 321–3
    pest and disease control 323–4
    soil structure maintenance 323
ectomycorrhizal fungi 156
  see also mycorrhizal fungi
Eisenia 217
Eisenoides 216
Enchytraeidae (potworms) 33, 39, 218–19
  biogeography 219–22
    continental scale 219–21
    local scale 221–2
    regional scale 221
  functional roles 222
  global change impacts 222
  phylogeny *220*
Enchytraeus 219
  E. lacteus 221
Enchytronia parva 221
energy flows, fungus-growing termite influence 178, *179*
enigma of soil species diversity 117–18, 136–7, 150, 233
  see also diversity of soil communities
environmental change 294–5
  impacts on ecosystem functioning 210–11, 213–14
  impacts on invertebrate ecosystem engineers
    ants 213–14
    earthworms 218
    Enchytraeidae 222
    termites 210–11
  predicting soil responses 189–90
    empirical models 190–1, 235
    including ecology in models 197–8
    mechanistic models 190–1, 235
    need for biology in models 190–2
  see also climate change

environmental filtering 202, *202*
enzymatic latch mechanism 244
erosion *see* soil
Eucalyptus globulus 86–7
Eudrilidae 215
eutrophication 266
extended phenotypes 90, 92, 100, *100*

Faidherbia albida 322
fauna
  detrital succession 119–21
  diversity–function relationships 38–9
  enigma of soil animal diversity 117–18, 150, 233
  grasslands 286–7
  macrofauna *30*, 33, 287
  mesofauna *30*, 32–3
  microfauna *30*, 32, 286–7
  study methods 35
  successional specialization 127–8
feces cascade 126
feeding in soils 10–12
fertilization of soils 154, 256–8
  grasslands 288, 291
  impacts on ecological services 344
  impacts on soil biodiversity 345
  see also nitrogen enrichment
flavonoids 161
flower diagrams *99*, 101
fluorescence *in situ* hybridization (FISH) 36
food production *see* agriculture; crop nutrients
food webs
  isotope analysis 128, 131
  micro food webs 16–17
forest restoration 383–4
fortification of crops 155
foundation species 111
functional domains
  connection within and across scales 20–1, *20*
  diversity–function relationships 37–9, *41*
  holistic view 39–41
  ecosystem engineers 15, 17
  functional diversity 40–1, *41*
  mosaics of 17
  processes at different scales 19–20
functional equivalence 193–5, *193*, 198
  experimental tests of 195–7

functional groups 29, 45–51, 223, 320
   earthworms 214, *215*
   ecosystem engineers 33
   macrofauna 33
   mesofauna 32–3
   microbes 28–32
   microfauna 32
   study methods 35–7
   trait-based ecology 50, 51–2, *51*, 54–5
functional marker genes 67
functional redundancy 192–3, *193*
functional similarity 193, *193*
functional traits 203
   diversity 50–1, *50*
fungi 28
   detrital succession 119
   nitrogen enrichment effects 261–4
      mycorrhiza 263–4
      saprotrophic fungi 261–3
   successional specialization 127
   tillage impacts 344
   see also microbes
fungus-growing termites 174–7, 206, 207, 234–5
   as ecosystem engineers 179–83
      emergent landscape effects 182–3, *184*
      mound-builder impacts 181–2
      soils 179–81, *179*
   evolution and biogeography 174–5
   fungiculture 175–7, *175*, *176*
   influences on ecosystem processes 177–8
      decomposition 177–8
      nutrient and energy flows 178, *179*
   see also termites

Gaia hypothesis 98, 112
   ecosystems services delivery 99–101, 105–6
Gatesona 216
genetically modified organisms (GMOs) 155
genetic analysis, microbial genes 66–8
   cloning 68
   decomposition networks 69–70
   future prospects 75–6

genes for interacting in the plant environment 73–5
methodological approaches 68–9
microbial communication 72–3
nitrogen turnover cascades 71–2
pollutant biodegradation networks 69–71
sequencing studies 68, 69
Gliricidia sepium 318, 322, 323
global atmospheric circulation models (GCMs) 249
global warming *see* climate change
Glossoscolecidae 215
glucosinolates 162
glyphosate 345–6
grasslands 282–3
   ecosystem services provided by soil biota 284–8
      herbage production 287–8
      nutrient cycling 285–7
      soil structure maintenance 284–5
      water regulation 285
   management intensity impact 288, *289*, 290
   plant selection impacts on soil biodiversity 340–1
   plant–soil interactions 282, 283–4, *283*
   restoration of agricultural land to grassland 382–3
   trade-offs between ecosystem services 288–90
grazing 53
   impact on distribution of soil organisms 141
greenhouse gas emissions (GHG) 248
   see also carbon dioxide concentrations; climate change
Green Revolution 309
green roofs 277
Growing Power, Inc. project 276
Guaranidrilus 220

Haber–Bosch process 256
   see also nitrogen enrichment
habitat fragmentation, community responses 205
   termites 211
heavy metals 162
Hemienchytraeus 220
herbicides 157

herbivore impact on distribution of soil organisms 141, 142–3
Heterodera trifolii 287
Hevea brasiliensis 318
home-field advantage 64, 83, 90, 111
horizontal heterogeneity in distribution
   fine or microscale 143–4
      drivers of 144–5
   small or mesoscale 138–41, *140*
      drivers of 141–3
human health
   nitrogen enrichment effects 266
   soil linkages 153–4, 234, 241
      microbe potential for health improvement 163–4
   see also crop nutrients
Hyperiodrilus africanus 323
hyphosphere 162
hysteresis 360–1, *360*

Indigofera constricta 318, 322
insurance hypothesis 66
Integrated valuation of ecosystem services and trade-offs (InVEST) tool 105
intergenic spacer analysis 69
Intergovernmental Panel on Climate Change (IPCC) 190
International Soil Metagenome Sequencing Project 76
International Tundra Experiment (ITEX) 246
invasive species 385
   invasion prevention 385–6
   management 386–7
island biogeography 204
isopods
   bioturbation 13
   organic cropping system impact 348, *349*
isotope analysis 128, 131

Komarekiona 216

land health surveillance 324–6, *325*
land management 50
   effects on ecosystem functioning and services 51–2
   entry points for biological management 54, *54*
   working with nature 52–3

landscape level 17, 54–5
legumes
  crops 258
  grasslands 284, 286, 291
  soil biodiversity and 341
Leucaena spp. 318
  L. leucocephala 318, 321, 322
lignin 10
linamarin 163
litter 10
  decomposition 50, 119, 367–8
    home-field advantage 90
    plant genetic variation
      influence 87–9, *88*
    nitrogen enrichment effects 261
  traits 50
  vertical stratification and 146
litter-transformers 48, 119
  Enchytraeidae 222
loess soils 302
Lolium perenne 288
Longidorus 344
Long-term Ecological Research
  Program (LTER) 368, 369–70
Lovelock, James 99
  see also Gaia hypothesis
Lumbricillus 219
Lumbricus
  L. friendi 348
  L. terrestris 218, 285, 344

macrofauna *30*, 33
  diversity–function relationships 39
  grasslands 287
  study methods 35
macropores 8–9
Macrotermes 175, *175*, *176*, 181
Macrotermitinae 174, 206
  as ecosystem engineers 179–83
    emergent landscape effects
      182–3, *184*
    mound-builder impacts 181–2
    soils 179–81, *180*
  evolution and biogeography 174–5
  see also fungus-growing termites;
    termites
manure application *see* organic
  amendment impacts
Martinique vertisols 11
mesofauna *30*, 32–3
  diversity–function relationships
    38–9

study methods 35
mesopores 8
metabolic cooperation 70
methane emissions 245
microarray-based analyses 69, 70
  high-density DNA microarrays
    75–6
microbes 28–32, *30*, 367
  biofilms 16
  biomass 11
  detrital succession 119
  disturbance effects 365
  endophytic microbes 74
  functional networks 65–6, *66*
  genetic analysis 66–8
    decomposition networks
      69–70
    functional marker genes 67
    future prospects 75–6
    genes for interacting in the
      plant environment 73–5
    methodological approaches
      68–9
    microbial communication 72–3
    nitrogen turnover cascades
      71–2
    pollutant biodegradation
      networks 69–71
  grasslands 285–6
  human health improvement
    potential 163–4
  plant–microbe interactions 29–32,
    73–5, 86
  plant functional trait effects
    84–5
  plant genetic variation
    relationships 86
  roles in ecosystem functioning
    28–32
    decomposition 67
    denitrifying communities 67
    disturbance effects 67–8
    diversity–function
      relationships 37–8
    nitrifying communities
      29–32, 67
  spatial distribution 143–7
  study methods 34–5
  successional specialization 127
microfauna *30*, 32
  diversity–function
    relationships 38

grasslands 286–7
study methods 35
micro food webs 16–17
micropores 8
Microtermes 175
Millennium Ecosystem Assessment
  (MEA) 5, 64, 98, 105–6,
  112, 338
mined land 381
  restoration 381–2
mineralization 71, 89, 259
mites 32–3, 118
  detrital succession 120
  spatial distribution 146
  successional specialization 127–8
mmoX gene 67
modeling approaches 37
mucus 12
multiple substrate-induced
  respiration (MSIR) 35
multi-pool soil organic carbon (SOC)
  models 197
mutualistic associations 12
  digestion and 12
  root associations 29–32
mycorrhizal fungi 29, 73–4, 101–2
  grasslands 286
  influence on crop nutrient levels
    156–63, *160*
    changes in plant/soil chemistry
      160
    increased nutrient uptake
      157–60
    indirect effects 161–2
    iron chelation 161
    negative effects 162–3
    nutrient solubilization and
      transport 160–1
  nitrogen enrichment effects
    263–4
  spatial distribution 144, 145
  urban soils 276

Natural Capital Project 105
nematodes 32
  community structure 19
  diversity–function
    relationships 38
  grasslands 286–7
  spatial distribution 139–40,
    140, 145
  tillage impacts 344

nested spatial scales 149–50
nif (nitrogen fixation) genes 74
Nile river valley 302
nitrogen 256
    as a limiting resource 256
    see also nitrogen cycling; nitrogen enrichment
nitrogen cycling 70, 71–2
    agroforestry and 322–3
    chemical application impacts 344–5
    denitrification 71, 147
        denitrifying microbial communities 67, 71–2
    grasslands 285–7
    human disturbance 67–8
    microbial gene studies 71–2
    mineralization 71, 89, 259
    nitrate ammonification 71
    nitrification 71, 89, 147–8
        nitrifying microbial communities 67
    nitrogen fixation 71, 322
        cultivation induced biological nitrogen fixation (C-BNF) 258
        industrial 67
        nitrogen-fixing bacteria 29–32, 74, 156
    nitrogen pools 89
    organic cropping system impacts 347
    resilience and recovery 365–6
    spatial distribution of soil functions 147–9, *148*
nitrogen enrichment 256–8, *257*, 267
    effects on ecosystem services 265–7
        biodiversity conservation 265–6
        food, biomass and forage production 266
        greenhouse gases and carbon sequestration 265
        human health 266
        water supply 266
    effects on soil biota 259–65, *262*, *264*
        multitrophic impacts 264–5
        mycorrhiza 263–4
        plants 259–61, *260*
        saprotrophic fungi and bacteria 261–3
        grasslands 288, 291

    impacts on soil chemistry 258–9
    soil organic matter disturbance 363
    see also nitrogen cycling
nitrous oxide (N_2O) emissions 370
no-tillage (NT) cropping systems 342–4
novel ecosystem development 387–9
nutrient flows, fungus-growing termite influence 178, *179*
nutritional value of crops *see* crop nutrients

Ocnerodrilidae 215
Odontotermes 175, 181–2
organic amendment impacts 346–7
    ecosystem services 346
    soil biodiversity 346–7
organic cropping system impacts 347–50
    ecosystem services 347
    soil biodiversity 348–50, *349*
organic matter decomposition
    decomposer communities
        home-field advantage 83, 90
        plant genetic variation influence 83, *84*, 86
    fungus-growing termite influence 177–8
    litter 50, 87–9, *88*, 90, 119
    microbial functions 67
        genetic studies 69–70
    temperature sensitivity 243–4
    wood 50
    see also soil organic matter (SOM)

perennial crops 341–2
permafrost 245
pest control, agroforestry systems 323–4
pesticides 157
    impacts on ecosystem services 344–5
    impacts on soil biodiversity 345–6
phenazine-1-carboxylic acid (PCA) 75
phenols 161
phospholipid fatty acids (PLFA) 34
phylogenetic community structure 203, 223
    ants 213

    termites 209–10
phylogeny
    ants *212*
    earthworms *216*
    Enchytraeidae *220*
    termites *208*
phytic acid 162
phytonutrients 161
plant growth promoting rhizobacteria (PGPR) 156
plants
    climate change effects 246–8
        community composition 246–7
        community productivity 246
        seasonality and phenology 247–8
    distribution of soil organisms and 141–2
        root influences 144–5, *145*
    diversity management 341–2
        impacts on soil biodiversity 340–1
    ecosystem services management 338–40
    edible crop diversity 338, *339*
    genetic variation
        influence on soil communities 83, *84*, 85–7, *85*, *87*
        role in ecosystem processes 87–9, *88*
    nitrogen enrichment effects 259–61, *260*
    plant–earthworm interactions 19, 118
    tree distribution relationships 317, *317*
    plant–insect coevolution 117
    plant–microbe interactions 29–32, 73–5, 86
        grasslands 286
    see also trees
plant–soil linkages 82–3, *84*, 92–3, 294
    climate change impacts 242–50, *242*, 249–50
    making predictions 248–9, *249*
    multifactor effects 248
    plant community composition 246–7
    plant community productivity 246

plant–soil linkages (cont.)
  seasonality and phenology 247–8
  temperature 243–4
  water 244–5
 evolutionary implications 89–92, *91*
 feedbacks 83, 89–93, 294
 grasslands 282, 283–4, *283*
 home-field advantage 64, 83, 90
 plant influences on soil communities 82–3
  plant genetic variation relationship 83, *84*, 85–7, *85*, *87*, 92–3, *93*
  role of plant functional traits 84–5, 92
 see also plants
pmoA gene 67
pollutants
 biodegradation 69–71
 soil organic matter disturbance 363
 urban soils 273
polycultures 342
polymerase chain reaction (PCR) 68
polyphenols 10
Pontoscolex corethrurus 19, 218
Populus (aspen)
 P. angustifolia 86, 88–9, *88*, 90
 P. fremontii 88–9, 90
 P. tremuloides 86, *88*
Porcellio scaber (sowbug) 118
porosity 5, 8–9, 323
 pore networks 9
 pore sizes 8–9, *9*
portfolio effect 193
potworms see Enchytraeidae (potworms)
priming 246
processing chain model 123–6, *124*, *125*
pro-nutrients 162
Prosellodrilus 216
protozoa 32
 grasslands 286
 spatial distribution 139, 144, 145
Protura 33
pyrosequencing 69

quantitative polymerase chain reaction (qPCR) 35

Quercus laevis 86
quorum quenching 73
quorum sensing 72–3

radioactive contamination 163
rain gardens 275
removal studies 36
resilience 359–60
 see also soil
resource quality 123
respiration in soils 9–10
resting stages 14
restoration see ecological restoration
Reticulitermes 210
rhizobia 155
rhizodeposition 11
rhizosphere biota 48, 73–4
 microbe use to boost crop nutrients 157–63, *158–9*
 spatial distribution 144–5, *145*
root associations 29–32
root biota 48
root exudates 11
Rothamsted Experimental Station, U.K. 370
rotifers 32

scales in soil function 16–17
 connection of functional units within and across scales 20–1, *20*
 processes at different scales 19–20
Scherotheca 216, 217
Schizolobium amazonicum 318
Second Green Revolution 153
self-organization 15, *16*, 18, 59
 soil communities 18–19
Senna siamea 318
sequencing studies 68
 next generation sequencing 69, 76
Sesbania sesban 321
single-strand-conformation polymorphism (SSCP) 68
Sitona spp. 287
size distribution 13
snow cover 245
soil 7
 challenge of eco-efficient use 16–21
 conditions in 7–12
 disturbance 358–9
  long-term studies 368–70

 ecological restoration 377–81, *378*, *379*, 380
 erosion 301, 303–5, *303*, 312–13, 362
  accelerated erosion 304, *305*
  effect on agricultural productivity 305–9, *306*
  importance of erosion-induced productivity losses 309–12, *310*
  natural erosion 303
  organic cropping system and 347
 feeding in 10–12
 formation 17, 271, 301–3, *303*
  urban soils 271, *272*
 fungus-growing termite effects 179–81, *180*
 grasslands 284–5
 human health relationships 153–4, 234, 241
  see also crop nutrients
 hysteresis 360–1, *360*
 importance of 357–8
 management 50
 physical structures 19
 predicting environmental change responses 189–90
  "black box" models 191–2, 194, 197, 198, 235
  empirical models 189–90, 235
  including ecology in models 197–8
  mechanistic models 189–90, 235
  need for biology in models 190–2
 research approaches 21
 resilience and recovery 359–70
  long-term studies 368–70, *369*
  nutrient cycling 364–6
  prediction 366–7
  soil biota 367–8
  soil organic matter dynamics 361–4
 respiration in 9–10
 scales in soil function 16, 19–21
 self-organization 15, *16*, 18, 59
 soils as extended phenotypes 100, *100*
 structure
  influence on distribution of organisms 144

maintenance in agroforestry
systems 323
maintenance in grasslands
284–5
sustainability 299–300
tree influence on properties and
biota 316–18
soil structure maintenance 323
urbanization effects 270–3
indirect effects 271–3
soil compaction 271
soil sealing 271, 277
see also plant–soil linkages; soil
communities; soil health; soil
organic matter (SOM); soil water
soil communities 367–8
agroforestry effects on abundance
318–19, *319*
disturbance 367
global change impacts
ants 213–14
climate change 210–11
deforestation 211
earthworms 218
habitat fragmentation 211
termites 210–11
invertebrates 203–6
altitudinal gradients 205
area–diversity
relationships 203–4
landscape modification
gradients 205
latitudinal gradients 204–5
nitrogen enrichment effects
259–65
organic cropping system impacts
348–50, *349*
phylogenetic community
structure 203
plant influences 82–3
management implications
340–1
plant genetic variation role 83,
84, 85–7, *85*, *87*, 92–3, *93*
role of plant functional traits
84–5
self-organization 18–19
spatial distribution 137–47, *138*
fine or microscale 143–7
horizontal heterogeneity
138–45, *140*
nested spatial scales 149–50

small or mesoscale 138–43, *139*
vertical heterogeneity 145–7
study methods 34–7
experimental manipulation 36
functional approaches 35–6
modeling approaches 37
taxonomic approaches 34–5
tillage impacts 344
see also diversity of soil
communities
soil health 331–3, *333*, 335, 382
evolution of 333–4
monitoring 324–7, 334
land health surveillance
324–6, *325*
local knowledge integration
326–7, *326*
soil quality 331–2, *332*
soil tilth 331, *332*
Soil Management Assessment
Framework (SMAF) 334
soil organic matter (SOM) 10–11
agroforestry and 321–2
chemically-protected pool 11
decline 360
dynamics related to disturbance
360, *360*, 361–4
chronic disturbance 362–4
discrete disturbance 361–2
long-term studies 368–70, *369*
metabolic pool 10–11
physically-protected pool 11
soil organic carbon (SOC)
management 340
see also organic matter decomposition
soil water 9–10
amphibiotic conditions 13
climate change and 244–5
grassland water regulation 285
urbanization effects 272–3
solanin 163
Solidago altissima 86, 88
spatial distribution
soil functions 147–9, *148*
soil organisms 136–47, *138*, 233–4
fine or microscale 143–7
horizontal heterogeneity
138–45, *140*
nested spatial scales 149–50
small or mesoscale 138–43, *139*
species distributions 203, *203*
vertical heterogeneity 145–7

species distributions, *see also*
biogeography
species distributions 203, *203*
see also biogeography
species richness 59–60
area–species richness
relationships 203–4
function relationships 37–9, *41*
organic cropping system impacts
348–9, *349*
see also biodiversity; diversity of
soil communities
springtails *see* collembolans
spruce 141–2
stable isotope probing (SIP) 68
Stipa hymenoides 141
storm water management 275
Striga spp. 324
successional specialization 126–8
sulfate-reducing bacteria 70–1
sustainability 299–300
symbiogenesis 19

Tardigrada (waterbears) 33
temperature effects 243–4, 368
urban heat islands 272
see also climate change
terminal restriction fragment length
polymorphism (T-RFLP) 35,
68–9
termites 33, 50, 173–4, 206
alates 178, *179*
biogeography 206–10, *208*
continental scale 206–10
local scale 210
regional scale 210
bioturbation 13
dispersal 175
evolutionary history 207
functional classification
207–10, *209*
global change impacts 210–11
phylogenetic community
structure 209–10
phylogenetic relationships *208*
see also fungus-growing termites
Termitomyces 174, 176, 177, 206
Terragenome 76
The Economics of Ecosystems and
Biodiversity program (TEEB)
104–5
Theobroma grandiflorum 318

thermal gradient gel electrophoresis
   (TGGE) 68
tillage 157, 342, *343*
   impacts on ecosystem services
      342–3
   impacts on soil biodiversity 343–4
   no-tillage (NT) systems 242–4
Tithonia diversifolia 318, 322
tracing studies 36
trait-based ecology 50, 54–5, 60, 202–3
   assembly rules 51
   environmental filters 202, *202*
   functional traits 203
      diversity 50–1, *50*
   implications for working with
      nature 52–3
   plant functional trait effects on
      soil communities 84–5, 92
   soil invertebrate ecosystem
      engineers 222–4, *223*
   trait–function relationships 51–2
trees
   influence on soil biota 316–18
      abundance of soil biota 318–19,
         319
      earthworm distribution 317,
         317, 319
   influence on soil properties
      316–18
   soil nutrients 317–18
   provision of soil-based ecosystem
      services and 320–4
      carbon transformations 321–2
      nutrient cycling 321–3
      pest and disease control 323–4
      soil structure maintenance 323
   tree planting programs 277
   see also agroforestry
Trichodorus 344
Trifolium repens 288
Tupidrilus 220

urban gardening 274–5
urbanization 270, 311
   ecosystem services in cities
      273–5, *274*
      cultural ecosystem services 275
      provisioning ecosystem services
         274–5
      regulating ecosystem services
         275
      supporting ecosystem services
         274
   effects on soils 270–3

   management for ecosystem
      services 276–8
   assessment for ecosystem
      services 277–8
   local scale improvements
      276–7
   urban scale intervention 277
urban heat islands 272
Urobenus 216
USDA-ARS Beltsville Farming
   Systems Project, Maryland
   347, 348

vertical heterogeneity in microscale
   distribution 145–6
   drivers of 146–7
viruses 28, 32
   see also microbes
vitamin C 162

waste treatment 275
water *see* soil water
waterbears 33
watershed level 17
weathering 301–2
wetland restoration 384–5
wood degradation 50

Printed and bound by CPI Group (UK) Ltd, Croydon, CR0 4YY